Masterclass in Neuroendocrinology

Volume 9

Series Editors

John A. Russell
Centre for Discovery Brain Sciences, Edinburgh University Medical School, CMVM, Edinburgh, UK

William E. Armstrong
Department of Anatomy and Neurobiology, The University of Tennessee Health Science Center, Memphis, TN, USA

Masterclass in Neuroendocrinology is a co-publication of the INF (International Neuroendocrine Federation) that aims to illustrate the highest standards and promote the use of the latest technologies in basic and clinical research, while also providing inspiration for further exploration into the exciting field of neuroendocrinology. It is intended for established researchers, trainees and students alike.

Each book

- is edited by leading experts in the field
- is written by a team of internationally respected researchers
- includes assessments of different experimental approaches, both in vivo and in vitro, and of how the resulting data are interpreted.

Founding Series Co-Editors: William E. Armstrong and John A. Russell

More information about this series at http://www.springer.com/series/15770

Susan Wray • Seth Blackshaw
Editors

Developmental Neuroendocrinology

Springer

Editors
Susan Wray
Cellular and Developmental Neurobiology
NINDS, NIH
Bethesda, USA

Seth Blackshaw
Department of Neuroscience
School of Medicine, Johns Hopkins University
Baltimore, MD, USA

ISSN 2662-2068 ISSN 2662-2076 (electronic)
Masterclass in Neuroendocrinology
ISBN 978-3-030-40001-9 ISBN 978-3-030-40002-6 (eBook)
https://doi.org/10.1007/978-3-030-40002-6

This Springer imprint is published by the registered company Springer Nature Switzerland AG.
The registered company address is: Gewerbestrasse 11, 6330 Cham, Switzerland

Series Preface

This series began publication as a joint venture between the International Neuroendocrine Federation and Wiley-Blackwell and now is continuing with Springer-Nature as publisher for the federation. The broad aim of the series is to provide established researchers, trainees, and students with authoritative up-to-date accounts of the present state of knowledge and prospects for the future across a range of topics in the burgeoning field of neuroendocrinology. The series is aimed at a wide audience as neuroendocrinology integrates neuroscience and endocrinology. We define neuroendocrinology as the study of the control of endocrine function by the brain and the actions of hormones on the brain. It encompasses the study of normal and abnormal function and the developmental origins of disease. It includes the study of the neural networks in the brain that regulate and form neuroendocrine systems. It also includes the study of behaviors and mental states that are influenced or regulated by hormones. It necessarily includes the understanding and study of peripheral physiological systems that are regulated by neuroendocrine mechanisms. Clearly, neuroendocrinology embraces many current issues of concern to human health and well-being, but research on these issues necessitates reductionist animal models.

Contemporary research in neuroendocrinology involves the use of a wide range of techniques and technologies, from the subcellular to systems at the whole-organism level. A particular aim of the series is to provide expert advice and discussion about experimental or study protocols in research in neuroendocrinology and to further advance the field by giving information and advice about novel techniques, technologies, and interdisciplinary approaches.

To achieve our aims, each book is on a particular theme in neuroendocrinology, and for each book we have recruited a pair of editors, expert in the field, and they have engaged an international team of experts to contribute chapters in their individual areas of expertise. Their mission was to give an update of knowledge and recent discoveries and to discuss new approaches, "gold-standard" protocols, translational possibilities, and future prospects. Authors were asked to write for a wide audience, to minimize references, and to consider the use of video clips and explanatory text boxes; each chapter is peer reviewed and has a Glossary and a detailed Index. We have been guided by an Advisory Editorial Board.

The MasterClass Series is open-ended; books in the series published to date are *Neurophysiology of Neuroendocrine Neurons* (2014, ed. WE Armstrong & JG Tasker); *Neuroendocrinology of Stress* (2015, ed. JA Russell & MJ Shipston); *Molecular Neuroendocrinology: From Genome to Physiology* (2016, ed. D Murphy & H Gainer); *Computational Neuroendocrinology* (2016, ed. DJ Macgregor & G Leng); *Neuroendocrinology of Appetite* (2016; ed. SL Dickson & JG Mercer); *The GnRH Neuron and its Control* (2018; ed. AE Herbison & TM Plant); *Model Animals in Neuroendocrinology* (2019, ed. M Ludwig and G Levkowitz); and *Neurosecretion: Secretory Mechanisms* (In Press, ed. J Lemos and G Dayanithi), #8 in the Series, and will be the first to be published by Springer Nature; books in preparation include *Glial-Neuronal Signaling in Neuroendocrine Systems* and *Neuroendocrine Clocks and Calendars*.

Feedback and suggestions are welcome.

International Neuroendocrine Federation—http://neuroendonow.com/

University of Edinburgh, Edinburgh, UK — John A. Russell
University of Tennessee Health Science Center, Memphis, TN, USA — William E. Armstrong

Volume Preface

The hypothalamus is the regulatory hub controlling neuroendocrine activity. An evolutionarily ancient structure, with multiple peptidergic systems contained within its boundaries, it is here that many innate behaviors and functions essential for survival are controlled. These include reproduction, homeostasis, feeding, drinking, growth, circadian rhythms, and resilience to psychological stress. Within the hypothalamus there are multiple distinct neuronal subtypes, distinguished by their functions and molecular expression patterns. These neurons are clustered within anatomical nuclei defined by classic histological approaches and also dispersed across nuclei. These neuronal subtypes form both intra-hypothalamic circuits and circuits with brainstem and other limbic structures and may influence daily behavior through their hypothalamic–pituitary connections. This volume presents the current state of knowledge for three aspects of developmental neuroendocrinology: Part I—the molecular specification of hypothalamic cells, Part II—developmental modulators and epigenetic factors that influence hypothalamic development, and Part III—the development of neuroendocrine circuits. Chapters are presented as succinct summaries of current progress and future prospects for the respective areas. The research often includes state-of-the-art techniques for delineating hypothalamic cell lineage and cell fate specification, for identifying genes altered by epigenetic factors, and for tracing of functional neural circuits. This book puts together a current and exciting perspective on developmental neuroendocrinology—covering topics ranging from genes that specify and mold the embryonic hypothalamus to molecular mechanisms that sculpt experience-dependent circuit formation in the postnatal period. Each chapter is written by one or more experts in the field.

Part I focuses on the molecular mechanisms that control specification of hypothalamic and pituitary cells. Cellular and molecular mechanisms that direct the integrated assembly and differentiation of the neuroendocrine hypothalamus are addressed in Chap. 1, while Chap. 2 discusses the hypothalamic Shh-Gli code, a region-specific set of rules governing the relationship between the morphogen Sonic hedgehog and its downstream effectors, the Gli family of transcription factors. Developmental patterns of gene expression in the developing hypothalamus that are conserved across vertebrates, including humans, are discussed in Chap. 3. Chapters 4 and 5 highlight the gene regulatory program that controls differentiation of the arcuate nucleus and the transcriptional network guiding specification and differentiation of the suprachiasmatic nucleus and sleep-regulating neurons of the dorsolateral hypothalamus. Chapter 6 focuses on transcription factors necessary for pituitary organogenesis.

The organizing theme for Part II is the role of environmental and epigenetic factors in regulating development of hypothalamic nucleogenesis and cell specification and survival. Chapter 7 discusses the role of GABA-mediated signaling events in controlling these processes. Chapter 8 addresses the dependence of mammalian puberty on the interaction of gene sets organized into functionally connected networks, controlled by transcriptional and epigenetic switches. In Chap. 9, the focus is on the role of epigenetic factors in the terminal differentiation of GnRH neurons. Chapter 10 presents an overview of functions of imprinted genes in the hypothalamus. Chapter 11 completes this section by discussing short- and long-term oscillations in DNA methylation and histone acetylation in relation to the control of rhythmic physiology and behavior.

Part III of the book examines development of neuroendocrine neural circuitry. Research into the neural mechanisms underlying specific behaviors has a long history that has been repeatedly invigorated with new tools, exciting data, and ideas. Specific topics discussed in this section include advances in neural circuitry tracing, opto- and chemogenetics, and the contribution of immune cells to neuronal maturation. Chapter 12 discusses the development of stress response-regulatory circuitry in the limbic system. Chapter 13 discusses the formation of neural circuitry relaying visceral sensory signals from the brainstem to the hypothalamus. Chapter 14 discusses the role of astrocytes in the development of neuroendocrine circuits, with an emphasis on regulation of metabolism. Part III closes with an overview of sexual differentiation and reproductive functions, with Chap. 15 discussing mechanisms of steroid-mediated sexual differentiation, and Chap. 16 discussing the development and modulation of reproductive function by circadian signals.

Many different structural, developmental, and functional features are found in the neuroendocrine brain: these involve structurally defined nuclei, cells that are broadly distributed by tangential migration, cells that show synchronized physiological rhythms, and a diverse collection of neural circuits fundamental to survival. We hope the reader is intrigued by the developmental mechanisms explained here that create the neuroendocrine hypothalamus and that the findings and ideas discussed will propel readers to look further into the exciting field of developmental neuroendocrinology.

Bethesda, MD Susan Wray
Baltimore, MD Seth Blackshaw

Contents

Part I

Molecular Specification of Hypothalamic/ Pituitary Cells

Development of the Neuroendocrine Hypothalamus

1

Marysia Placzek, Travis Fu, and Matthew Towers

Abstract

The neuroendocrine hypothalamus consists of neurosecretory neurons and specialized glia of the median eminence and posterior pituitary. Neurons and glia act together to support and regulate reciprocal brain–body communications that enable the hypothalamus to anticipate and adapt to changing physiological conditions. Here we summarize recent studies in model organisms that show how an embryonic program of hypothalamic development unfolds in a manner that underlies adult function. Fate-mapping studies in chick show that much of the basal hypothalamus is built from a multipotent *Fgf10(+)* progenitor population whose mode of growth in four dimensions suggests a new model for hypothalamic development that we term the 'Anisotropic growth' model. Conditional genetic analyses and pharmacological interventions in mouse, zebrafish, and chick suggest that a conserved molecular mechanism may mediate anisotropic growth and establish the basal hypothalamus. A handful of conserved signalling factors and transcription factors (TFs) direct the progressive development of *Fgf10(+)* progenitors to hypothalamic progenitors, hypothalamic neurons, and then infundibular progenitors of the median eminence and posterior pituitary in a program where space and time are intrinsically linked. These studies reveal additionally that *Fgf10(+)* cell populations are maintained throughout life and are vital to the long-term building, maintenance, and function of the adult hypothalamo–pituitary axis. We speculate on whether the embryonic *Fgf10(+)* progenitor population harbours a hypothalamic stem cell and discuss the implications for the ability of the hypothalamus to adapt to changing physiological conditions on different time scales through life.

M. Placzek (✉) · T. Fu · M. Towers
Bateson Centre and Department of Biomedical Science, University of Sheffield, Sheffield, UK
e-mail: m.placzek@sheffield.ac.uk

S. Wray, S. Blackshaw (eds.), *Developmental Neuroendocrinology*, Masterclass in Neuroendocrinology 9, https://doi.org/10.1007/978-3-030-40002-6_1

Keywords
Fgf10 · Shh · Anisotropic growth · Infundibulum · Median eminence · Posterior pituitary

List of Abbreviations

3V	3rd ventricle
AP	Anterior Pituitary (aka Adenohypophysis)
ARC	Arcuate nucleus
bHLH	Basic helix-loop-helix
BMP	Bone morphogenetic protein
CRH	Corticotropin releasing hormone
DA	Dopamine
DMN	Dorsomedial nucleus
Fgf	Fibroblast growth factor
GHRH	Growth hormone releasing hormone
HD	Homeodomain
MB	Mammillary Body
ME	Median Eminence
OXT	Oxytocin
PeVN	Periventricular nucleus
PP	Posterior pituitary (aka neurohypophysis)
PVN	Paraventricular nucleus
RDVM	Rostral diencephalic ventral midline
Rx	Retina and anterior neural fold homeobox
SCN	Suprachiasmatic nucleus
Shh	Sonic hedgehog
SON	Supraoptic nucleus
SST	Somatostatin
TMN	Tubero-mammillary nucleus
TRH	Thyroid releasing hormone
VMN	Ventromedial nucleus
VZ	Ventricular zone

1.1 Introduction: Neuroendocrine Cells of the Adult Hypothalamus

The evolutionarily ancient hypothalamus is crucial to life, governing an individual's survival and species' propagation. Its cells enable the body to respond, anticipate, and adapt to changing physiological conditions. These include physiological fluctuations

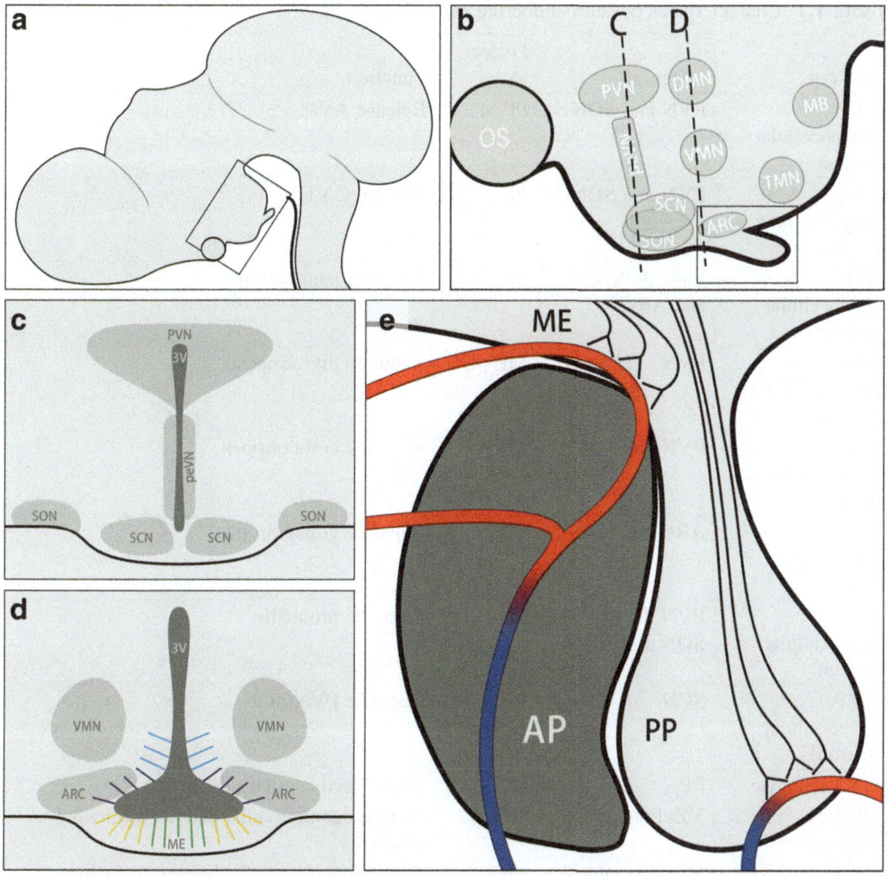

Fig. 1.1 The adult neuroendocrine hypothalamus. (**a**) Schematic side view of adult brain. (**b**) High-power view of boxed region in (**a**) to show hypothalamus and approximate position of hypothalamic nuclei relative to optic stalk (OS). (**c, d**) Transverse views through planes indicated in (**b**): at level of anterior hypothalamus (**c**) or ME (**d**) to show relative positions of neuroendocrine nuclei and glia. Different tanycyte subsets are indicated by colour: β2 green: β1 yellow; α2 purple, α1 blue. (**e**) High power view of boxed region in (**b**) to show approximate positions of ME, PP and AP. In the ME, parvocellular neuronal endfeet intimately contact tanycytes and portal capillaries. In the PP, magnocellular neuronal endfeet contact pituicytes and capillaries

over different time scales: for instance, short-term changes in stress/energy balance, or long-term seasonal or lifecourse changes in energy balance and reproduction.

Anatomically and functionally, the hypothalamus is complex. In much of the central nervous system, such as the cortex, cerebellum, hindbrain or spinal cord, functionally similar neurons are arranged in columns or layers. By contrast, the hypothalamus is composed of nuclei, or areas, arranged in a patchwork around the third ventricle. Each nucleus/area is located in a stereotyped position in the adult (Fig. 1.1a, b) and harbours functionally diverse neurons. Neuroendocrine neurons

Table 1.1 Characteristics of neuroendocrine neurons and glia

Cell type	Location	Project to	Function
AVP magnocellular neuron	PVN and SON	PP	Release AVP
OXT magnocellular neuron	PVN and SON	PP	Release OXT
SST parvocellular neuron	PVN, PeVN and ARC	ME	Regulate somatotropes
TRH parvocellular neuron	PVN	ME	Regulate thyrotropes
CRH parvocellular neuron	PVN and VMN	ME	Regulate corticotropes
GHRH parvocellular neuron	ARC and PVN	ME	Regulate gonadotropes
DA parvocellular neuron	PVN, PeVN, SON and ARC	ME	Regulate prolactin
AVP parvocellular neuron	SCN	PVN, SON	Regulate PVN/SON
Glial Pituicyte	PP	PP	Fenestrated capillary formation
Glial β1 tanycyte	VZ: Lateral ME	ARC	Stem/progenitor cells, barrier, trafficking functions
Glial β2 tanycyte	VZ: Central ME	ME	Stem/progenitor cells, barrier, trafficking functions, neurohormone release regulation
Glial α2 tanycyte	VZ: 3V	ARC, VMN	Stem/progenitor cells, circuit components, metabolite and hormone regulation

Summary of the position, projection and function of the major neuroendocrine neurons and glia

are a subset of hypothalamic neurons, located in multiple hypothalamic nuclei, and classically fall into two major categories. Magnocellular neurons are located in the paraventricular (PVN) and supraoptic (SON) nuclei and project to the posterior lobe of the pituitary (PP; aka posterior neurohypophysis) to release arginine vasopressin (AVP), and oxytocin (OXT). Parvocellular neurons are located in the PVN, SON, periventricular (PeVN), arcuate (ARC), and ventromedial (VMN) nuclei and project to the median eminence (ME) where they release corticotropin-releasing hormone (CRH), thyrotropin-releasing hormone (TRH), growth hormone-releasing hormone (GHRH), dopamine (DA), and somatostatin (SST) into the portal vasculature to control endocrine cells of the anterior pituitary (Swanson 1992; Fig. 1.1b–d; Table 1.1; List of Abbreviations). Recent studies show that additional neuronal subsets may be considered as part of the neuroendocrine system. Subsets of OXT(+) and AVP(+) neurons project internally within the hypothalamus, and mediate complex behaviours, including trust, aggression, and eating, or may indirectly regulate classic neuroendocrine behaviour through a network effect. For instance, subsets of neurons clustered within the suprachiasmatic nucleus (SCN) synthesise AVP and project to the SON and the PVN (Trudel and Bourque 2010; Gizowski et al. 2016).

The ME and PP contribute to the unique three-dimensional architecture of the hypothalamus (Fig. 1.1b, e). The ME and PP are composed of specialized glia cells, termed tanycytes and pituicytes (Table 1.1), microglia, the endfeet of neurosecretory neurons and a vast portal capillary network. Together these components provide blood–brain interfaces that enable communication between the body and the hypothalamus, supporting the release and transport of neurohormones to their endocrine targets, and providing the route through which the hypothalamus accesses circulating factors, including metabolites and hormones (Box 1.1). Despite its complex nature, the overall organization and 3-dimensional morphology of the hypothalamus, as well as the nature of its resident cell types have been highly conserved through evolution (Tessmar-Raible et al. 2007). However, understanding how the hypothalamus develops in embryogenesis has been challenging.

Box 1.1 Brain–Body Interfaces: Composition and Function of the Adult ME and PP

The ME and PP are composed of numerous cellular types, including glial cells (ME tanycytes and PP pituicytes), microglia, the endfeet of neurosecretory neurons, and a network of capillaries. Collectively these form the two-way circuits between brain and body that underlie hypothalamic functions. Glial pituicytes regulate the development of the permeable neurovascular interface in the PP (Anbalagan et al. 2018). Glial tanycytes regulate the proximity and access of neuronal endfeet to the portal capillaries, and hence the release and spread of neurotransmitters/neurohormones (reviewed in Clasadonte and Prevot 2018). At the same time, tanycytes are an active circuit component in acute homeostatic feedback systems from body to brain: through their dynamic barrier properties, hormone-regulating properties, transport functions, and potential to form parts of the neuroendocrine circuitry they support, modify, and respond to peripheral signals, including metabolites and hormones, to regulate body–brain communication and short-term/acute-fluctuating homeostatic mechanisms (reviewed in Rizotti and Lovell-Badge 2017; Lewis and Ebling 2017; Clasadonte and Prevot 2018). Different tanycyte subsets occupy different regions around the ME and 3V (Fig. 1.1d), project to particular hypothalamic regions (Table 1.1) and show particular properties (Table 1.1), including stem/progenitor behaviours (see text).

1.2 Models of Hypothalamic Development

The majority of hypothalamic neuroendocrine neurons and glial originate from progenitor cells located in ventral parts of the developing forebrain (the exception being GnRH neurons that migrate from the olfactory placode). Pinpointing the location of hypothalamic progenitor cells in the early embryo is important. It provides understanding of the tissues and signals that orchestrate the induction and specification of distinct hypothalamic cell subsets, potentially an understanding that is key for future regenerative therapies; it provides insight into how particular hypothalamic cells develop in concert, important for understanding hypothalamic circuits and networks; it suggests whether the hypothalamus harbours a stem cell that is capable of dividing and differentiating over life, able to restore or add cells in the adult brain; it provides insight into how particular cell subtypes of the hypothalamus might have evolved. However, understanding the location of progenitor cells, and how progenitor cells differentiate into hypothalamic neurons and glia has been challenging: the paucity of markers, the small size and complex morphology of the hypothalamus, together with the fact that neurons defined on the basis of neurotransmitter expression are located in multiple nuclei (Table 1.1) have all hampered its study. Nonetheless, previous studies have suggested two major models of hypothalamic development: the columnar model (Swanson 1992) and the (revised) prosomeric model (Puelles et al. 2012, 2013), summarized in Box 1.2. Each model has moved forward our understanding of this important brain region, but each has limitations. First, each assumes that adult hypothalamic neuroendocrine neurons are formed close to sites at which their progenitor cells are specified (as occurs in many other regions of the CNS: Placzek and Briscoe 2018). Second, each assumes that hypothalamic progenitor cell domains expand to an equivalent extent as development proceeds. Third, each has been developed through studies in rodents, where, for technical reasons, it is difficult to analyse the earliest stages of hypothalamic development.

Recent studies in the chick (Fu et al. 2017) suggest a new model for hypothalamic development. In the embryonic chick, a small *Fgf10(+)* progenitor cell population (which we refer to as bHyp (basal hypothalamic progenitor) cells) develops in the early-forming (10 somite stage) ventral diencephalon. Anteriorly, bHyp cells abut *Foxg1*(+) telencephalic progenitor cells, suggesting that they are nested within other diencephalic cell populations (Fig. 1.2a, b). Targeted DiI/DiO fate-mapping studies show that bHyp cells proliferate and give rise to other progenitor subtypes that downregulate *Fgf10* and extend throughout the basal hypothalamus, from the optic vesicle to the mammillary pouch (Fig. 1.2c; Fu et al. 2017). Growth from bHyp cells is anisotropic, and is greater than growth from adjacent diencephalic progenitor populations. Intriguingly, bHyp cells give rise to two major progenitor subsets, which grow tangentially and sequentially in opposite directions. First, bHyp cells generate 'anterior' progenitor cells that grow anteriorly (Fig. 1.3 orange), then bHyp cells generate mammillary progenitors that grow posteriorly (Fig. 1.3 blue). Throughout the generation of anterior and mammillary progenitors, population (s) of *Fgf10(+)* progenitor cells are centrally retained (Fig. 1.3 red). The extensive

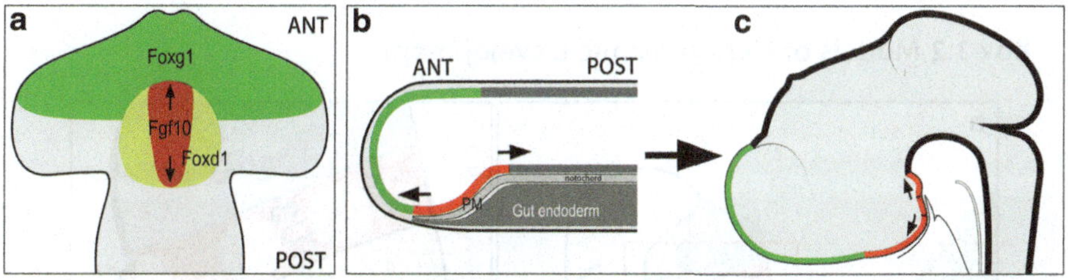

Fig. 1.2 Position and growth of chick bHyp cells. (**a**, **b**) Schematics of 10-somite stage chick embryos: ventral (**a**) and sagittal (**b**) views. *Fgf10(+)* bHyp progenitors (red) are nested within diencephalic progenitors (yellow: note in mouse these express *Foxd1*), abut *Foxg1(+)* telencephalic progenitor cells (green) anteriorly and lie above the prechordal mesendoderm (pm). Arrows indicate subsequent bidirectional growth (**c**) Side view, 25 somite embryo. bHyp progenitors give rise to descendants along the entire basal hypothalamus (red indicates all bHyp-derived cells: note *Fgf10* is maintained only centrally (between arrows). Arrows indicate bidirectional growth

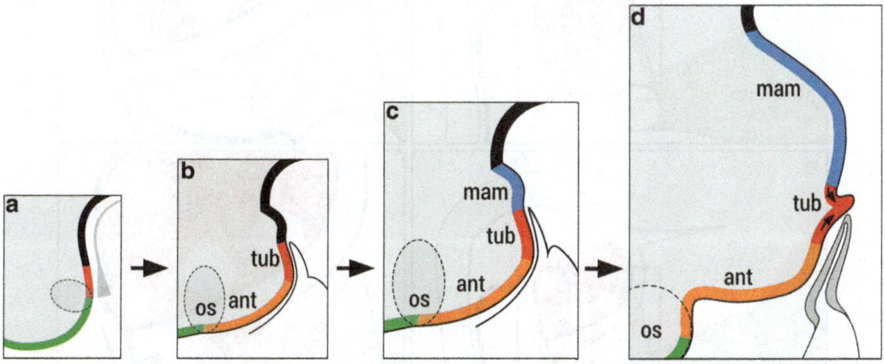

Fig. 1.3 Three-dimensional sequential anisotropic growth from bHyp cells. Schematic sagittal views of chick embryo (10–40 somites). (**a**) bHyp cells (red) abut the telencephalon (green) in the 10-somite embryo. (**b**) By 12 somites, bHyp cells begin to generate anterior progenitors (orange). (**c**) By 25 somites, mammillary progenitors are generated (blue): these extend posteriorly from bHyp cells that are now central (red). (**d**) Finally, infundibular glial cells are generated and grow ventrally (arrows). Dotted circle indicates optic stalk (os). Basal hypothalamus shown relative to underlying tissues: prechordal mesendoderm (**a**); developing (**b**) or definitive (**c**, **d**) Rathke's pouch

tangential growth of anterior and mammillary progenitor cells from bHyp cells means that the adult neurons to which they give rise, including neuroendocrine neurons, may ultimately lie distant from the sites of their progenitor cell specification.

Box 1.2 Models of Hypothalamic Development

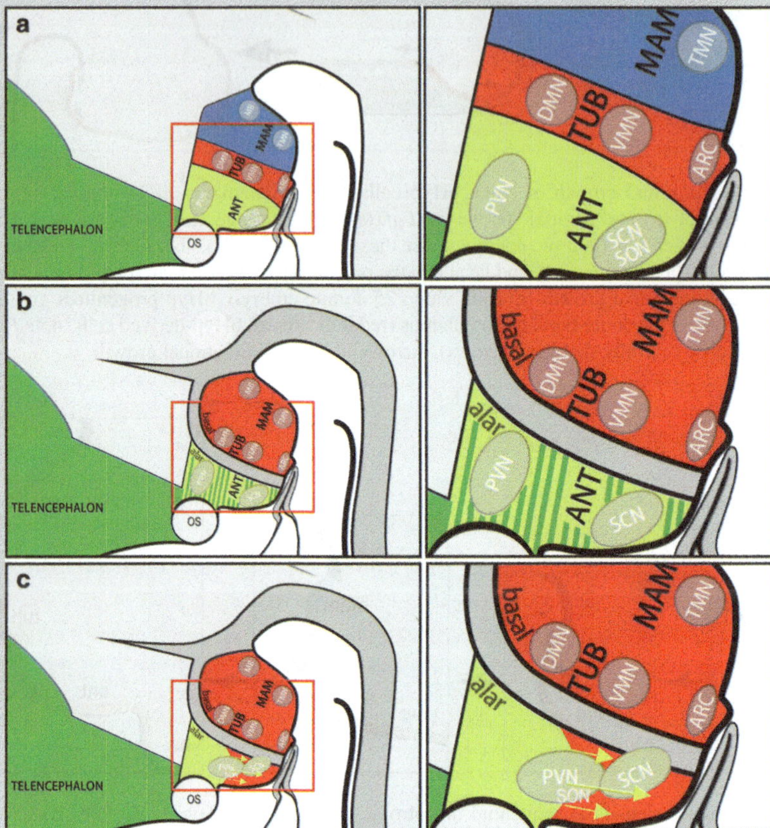

(A) The columnar model (Swanson 1992) derives from classical studies in the rat. It suggests that the hypothalamus is a diencephalic-derived structure whose neurons/nuclei fall into different areas along the rostro–caudal axis: preoptic/anterior, tuberal, and mammillary, reflecting distinct diencephalic progenitor subsets arranged in columns along the anterior–posterior (future rostro–caudal) axis. In this model, neuroendocrine neurons of the PVN, PeVN, SON and SCN derive from preoptic/anterior progenitors whereas those of the Arc derive from tuberal progenitors. Recent single cell RNA-seq studies of wild-type and mutant mice (Kim et al. 2019) support a version of the columnar model.

(B) The revised prosomeric model (Puelles et al. 2012, 2013), based on gene expression profiling and genetic inducible fate-mapping studies in mouse and chick (Sánchez-Arrones et al. 2009; Alvarez-Bolado et al. 2012)

(continued)

Box 1.2 (continued)

suggests that hypothalamic progenitors arise from a common diencephalic–telencephalic unit termed the secondary prosencephalon. This model suggests that neurons/nuclei differentiate from progenitors located anterior–dorsal, or within/posterior–ventral to a diagonal domain of Shh-expressing cells, and hence are either anterior–dorsal 'alar progenitors' or posterior–ventral 'basal progenitors'. In this model, neuroendocrine neurons of the PVN, PeVN, SON and SCN derived from alar progenitors that do not express Shh whereas neuroendocrine neurons of the Arc and VMN derive from basal progenitors that expressed Shh (Alvarez-Bolado et al. 2012). However, in most regions of the CNS, alar progenitors are dorsally located, so the idea that in the hypothalamus, anterior-most 'alar' progenitors occupy a basal position is controversial.

(C) The anisotropic growth model, described in the text, is based on gene expression profiling and fate-mapping studies in the early chick embryo (Fu et al. 2017) and suggests that the basal hypothalamus, at least, is a diencephalic-derived structure. This model suggests that the basal hypothalamus derives from progenitors that express both Fgf10 and Shh and abut telencephalic progenitors. In this model, adult neuroendocrine neurons/nuclei of the PVN/PeVN/SON/SCN may be of mixed progenitor origin, due to the extensive tangential migration of progenitor cells or differentiating cells. The figure shows one example, in which alar-derived neurons migrate ventrally after their specification (yellow arrows: see text).

Schematics show the location of different domains (ANT, anterior; TUB, tuberal; MAM, mammillary; alar; basal) and key hypothalamic nuclei. In B and C, these are shown relative to Shh-expressing cells (shown in grey).

1.3 Mechanisms of Anisotropic Growth

Molecular studies begin to reveal how chick bHyp cells develop and give rise to anterior and mammillary basal progenitors. These studies, and studies that suggest the conservation of anisotropic growth mechanisms in other species, have been described in detail in a recent review (Fu et al. 2019) and here we provide only a summary.

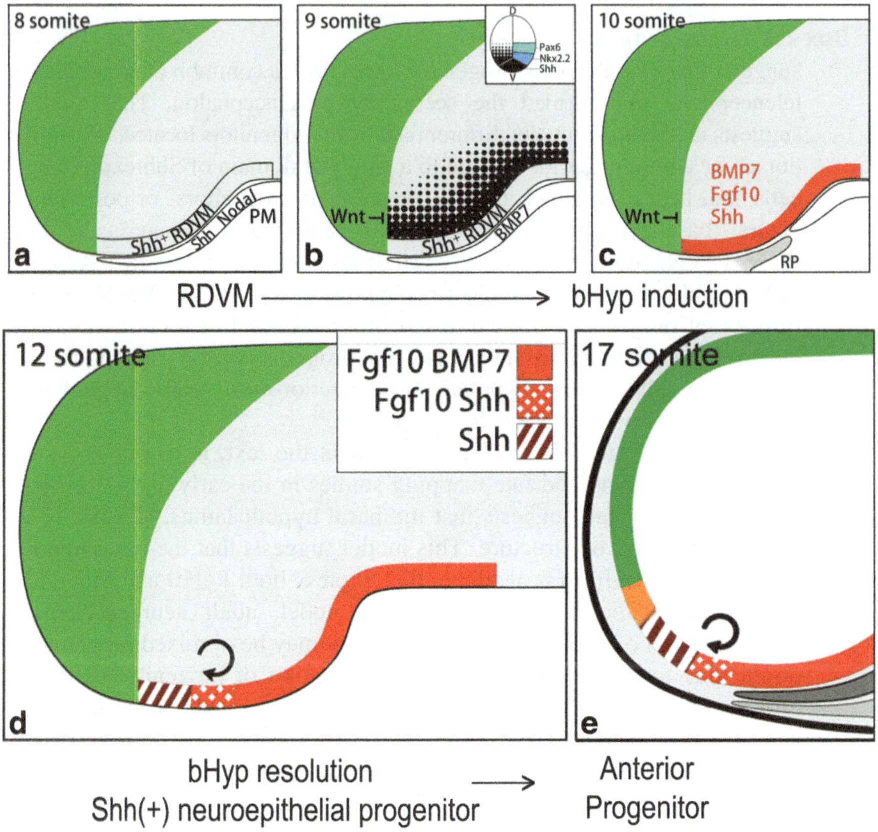

Fig. 1.4 Successive stages in bHyp cell induction and anterior progenitor growth and differentiation from bHyp cells. Schematic sagittal views of chick embryos (8–17 somites). (**a**) Induction of Shh(+) RDVM cells in the 8 somite embryo. (**b**) Establishment of dorsoventral pattern through a Shh morphogen gradient: inset shows patterned progenitor domains. (**c**) Differentiation of Shh(+) RDVM cells to bHyp cells that co-express *Shh*, *BMP7* and *Fgf10* (red area). (**d**) Resolution of bHyp into anterior *Fgf10/Shh(+)* cells (polka dot) and posterior *Fgf10/BMP7(+)* (red) domains and onset of growth (depicted by curved arrow). Shh(+) neuroepithelial progenitors (hatched) that downregulate Fgf10 grow anteriorly from *Fgf10/Shh(+)* cells. Note although shown in sagittal view, Shh(+) neuroepithelial progenitors form in an arc around *Fgf10(+)* cells. (**e**) Continued generation and differentiation of anterior progenitors (orange: from Fu et al. 2017)

1.3.1 bHyp Cell Induction

bHyp cells develop from a set of Shh(+) cells, termed rostral diencephalic ventral midline (RDVM) cells that are induced in the ventral midline of the developing forebrain (prior to the 10-somite stage in chick). Induction of RDVM cells is mediated by the signalling ligands Shh and Nodal (reviewed in Placzek and Briscoe 2005), deriving from the underlying prechordal mesendoderm (Fig. 1.4a). At their anterior limit, RDVM cells abut telencephalic progenitor cells (Fig. 1.4a). Tissue explant studies in chick suggest that Shh diffuses out of RDVM cells and forms a spatial morphogen gradient in neighbouring diencephalic neuroepithelial cells (Fig. 1.4b). This results in the establishment of a dorsoventral pattern. Progenitors that express the homeodomain transcription factor (TF) Nkx2.2 abut Shh(+) RDVM cells, and those expressing the homeodomain TF Pax6 are established adjacent to the Nkx2.2(+) domain (Fig. 1.4b inset). The induction of RDVM cells and their role in establishing an early dorsoventral pattern appear to be evolutionarily conserved. Genetic lineage-tracing studies in the mouse (Newman et al. 2018b) that build on ground-breaking large-scale transcriptomic analyses (Shimogori et al. 2010) indicate that the hypothalamus is of diencephalic origin. Knockout studies, or analysis of mutated genes, indicate that the prechordal mesendoderm induces Shh(+) RDVM cells in mouse, zebrafish, and humans, elegant work in mouse showing that Shh expression in RDVM cells is driven through a unique enhancer, SBE2 (Shh brain enhancer 2) (Jeong et al. 2006). Genetic studies in mouse and zebrafish show that Shh that derives from RDVM cells establishes Nkx2.2(+) and Pax6(+) progenitor domains. If *Shh* is selectively deleted in RDVM cells ($Shh^{\Delta hyp}$ mice; Zhao et al. 2012), the Shh-target gene Gli1 is absent, Nkx2.2 is reduced and Pax6 expands ventrally (Corman et al. 2018). Cross-repressive TF interactions may then sustain pattern: in zebrafish a ventral expansion of pax6 is detected after elimination/reduction of nkx2 genes (Manoli and Driever 2014).

As homeodomain progenitor domains become established adjacent to RDVM cells the prechordal mesendoderm changes its properties, downregulating *Shh* and upregulating *BMP7* (Fig. 1.4b). Tissue explant studies in chick suggest that BMP7 from the prechordal tissue has a major influence on overlying RDVM cells, causing them to upregulate *BMP7* and then *Fgf10*, i.e. to become bHyp cells (Fig. 1.4c; Manning et al. 2006). Upregulation of *BMP7/Fgf10* causes the cells to dramatically change their behaviour, switching from a quiescent to a proliferative state (Manning et al. 2006; Fu et al. 2017).

1.3.2 Basal Anterior Progenitor Growth

In the embryonic chick, proliferation is accompanied by a resolution of bHyp cells into two *Fgf10(+)* progenitor subtypes: a posterior population that expresses *Fgf10* and *BMP7* (Fig. 1.4d, red) and an anterior population that expresses *Fgf10* and *Shh* (Fig. 1.4d, polka-dots). A subset of *Fgf10+ Shh+* cells differentiate to progenitor cells that downregulate *Fgf10*, but retain/upregulate Shh and are displaced/migrate anteriorly (Fig. 1.4d, hatched). We term such cells, that derive from bHyp cells, 'neuroepithelial Shh(+) progenitors'. Neuroepithelial Shh(+) progenitors, in turn, give rise to wider sets of 'anterior' progenitors (Fig. 1.4d orange) an event that appears to be initiated as neuroepithelial Shh(+) progenitors become displaced/migrate further anteriorly and upregulate the cyclin-dependent kinase inhibitor, p57 (Fu et al. 2017). Upregulation of p57 is accompanied by upregulation of components of the Notch pathway (Ratie et al. 2013, 2014), likely to initiate a neurogenic differentiation program(s). As p57 and Notch components are upregulated, *Shh* is downregulated through an unknown mechanism, so that anterior progenitors express Shh only transiently. The constant generation and differentiation of neuroepithelial Shh(+) progenitors into anterior progenitors creates a temporal dimension that provides the opportunity to build complex arrays of basal hypothalamic neurons. This model predicts that basal hypothalamic anterior progenitor cell fate and position reflect the amount of time spent as an *Fgf10(+)* progenitor cell: cells that spend the shortest time within the *Fgf10(+)* domain give rise to cells that lie furthest from it.

Genetic and pharmacological studies suggest that the mechanisms that mediate the transition from quiescent RDVM to proliferating bHyp cells, then trigger the resolution of bHyp cells and anterior progenitor growth are evolutionarily conserved, and present in mouse as well as chick. In both species, dynamic inter-regulatory interactions between Shh, BMP and Fgf, and the precise balance of these three signalling ligands in time and space coordinates bHyp induction, resolution and the onset of proliferation. The transcriptional repressors, Tbx2 and/or Tbx3, play a critical role in these events (Manning et al. 2006; Trowe et al. 2013: details in Fu et al. 2019). Wnt signalling, deriving from telencephalic progenitor cells appears to restrict the bHyp domain (Fig. 1.4b, c) and studies in mouse, zebrafish and chick all indicate that Wnt signalling levels must be carefully regulated to establish the basal hypothalamic progenitor territory (Kapsimali et al. 2004; Brinkmeier et al. 2007; Potok et al. 2008; Manning et al. 2006; Newman et al. 2018b; Kim et al. 2019), although as yet, little is understood of how Wnt signalling integrates with Shh, BMP and Fgf signalling to mediate bHyp induction, resolution and proliferation.

Conditional genetic analyses and pharmacological interventions in mouse, zebrafish and chick indicate that the early anterior basal hypothalamus may be built from/have a contribution from neuroepithelial Shh(+) progenitors in all of these species. Feedforward–forward-back regulatory interactions between Shh and the paired-box transcription factor Rx (or its zebrafish homologue, rx3) appear to establish a growth loop that provides a dynamic stream of neuroepithelial Shh(+) progenitors, while maintaining population(s) of *Fgf10(+)* progenitors (Table 1.2; Fig. 1.5; Muthu et al. 2016; Orquera et al. 2016; Fu et al. 2017). Key to this growth loop is the ability of Shh to non-autonomously induce Rx, for Rx to autonomously induce Shh, and for Shh to autonomously downregulate Rx in order for cells to realize the anterior progenitor program. These studies begin to suggest that Shh, deriving from neuroepithelial Shh(+) progenitors, may drive the continued growth and differentiation of anterior basal hypothalamic progenitors, although this remains to be formally tested in the mouse. To date, caudal-most Shh(+) RDVM cells and neuroepithelial Shh(+) progenitors have been lineage traced (Szabó et al. 2009; Alvarez-Bolado et al. 2012). However, these studies are unlikely to report the fate of the first-generated neuroepithelial Shh(+) progenitors (i.e. those predicted to give rise to anterior basal progenitors): the time lag in recombination means that anterior progenitor cells are likely to have been generated before reporter expression is initiated.

In summary, Shh has three distinct roles in hypothalamic development, at different times in development and from different sources. First, it derives from prechordal

Table 1.2 Shh-Rx orchestrate growth and differentiation of the anterior basal hypothalamus

Species	Experiment	Result	References
Zebrafish	Rx3 (chokh) mutant or rx3 morpholino	Loss of Shh+pea3- progenitors; expansion of basal pea3+ domain; expansion of alar pax6+ domain; loss of anterior and tuberal neurons, including pomc, sf1/ff1b, avp, Otp and TH+ neurons	Tessmar-Raible et al. (2007); Dickmeis et al. (2007); Muthu et al. (2016)
Mouse	Rx2 conditional ablation	Loss/reduction of anterior territory (Lhx1, Nkx2.2); loss of anterior and tuberal neurons, including pomc, Otp, TH and Sst+ neurons	Orquera et al. (2016)
Chick	Timed cyclopamine (HH9–12)	Reduced Six3+p57+ anterior progenitor territory; reduced tract of post-optic commissure	Fu et al. (2017)
Zebrafish	Timed cyclopamine (30hpf)	Loss of anterior and tuberal neurons including pomc and sf1/ff1b	Muthu et al. (2016)
Mouse	Shh deletion ~E10.5 (Nkx2.1-Cre driver)	Loss of tuberal and anterior markers (Lhx1,Lhx6, Lhx9) and neurons (pomc, sf1)	Shimogori et al. (2010)
Mouse	Shh deletion ~9.0 (SBE2-Cre driver)	Loss/reduction of anterior markers (Six6, Nkx2.1, Nkx2.2); expansion of Pax6; loss of tuberal neurons (Hmx3, Sf1, Pomc, TH, Sst) and reduced mitotic progenitors	Zhao et al. (2012); Corman et al. (2018)
Mouse	Shh deletion ~E10 (Hesx1-Cre driver)	Loss of anterior territory (Nkx2,1, Wnt5a, Tcf4); loss of anterior and tuberal neurons (GHRH, Pomc, Sst, Oxt and Avp)	Carreno et al. (2017)
Mouse	Shh deletion ~E9.0 (Foxb1-Cre driver)	Reduced Foxd1, Nkx2.1; reduces anterior/tuberal neurons: Pomc, Hcrt, reduced mitotic progenitors	Szabó et al. (2009)

Summary of experiments that knockdown, or conditionally knockdown/reduce Rx/rx3 or Shh in hypothalamic neuroepithelial cells

mesendoderm and acts with Nodal to induce RDVM cells (Fig. 1.4a). Then it derives from RDVM cells and acts as a morphogen to establish an early dorsoventral pattern (Fig. 1.4b). Finally (at least in chick and zebrafish: Muthu et al. 2016; Fu et al. 2017), it derives from Shh(+) neuroepithelial cells to drive growth of the anterior basal hypothalamus, potentially by regulating cell cycle. Although not yet proved, it remains possible that Shh(+) neuroepithelial cells continue to act non-autonomously to pattern adjacent cells. This raises the possibility that progenitor cells in the basal hypothalamus are continually generated through two routes: from cells that are responsive to Shh signalling, but do not express Shh; or directly from Shh(+) neuroepithelial cells, a notion supported through recent work in mouse (Corman et al. 2018).

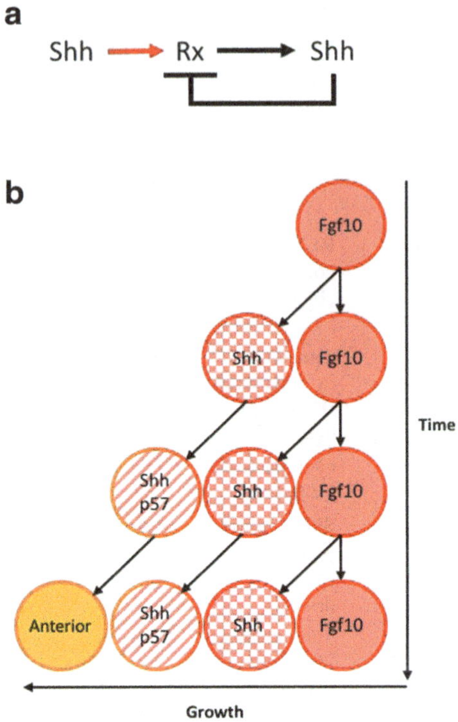

Fig. 1.5 Model for cellular homeostasis and generation of anterior progenitors from bHyp cells. (**a**) Shh induces Rx non-autonomously; Rx induces Shh autonomously; Shh represses Rx autonomously. (**b**) Over time, this loop builds a steady stream of anterior neuroepithelial Shh(+) progenitors (polka dots) while maintaining *Fgf10(+)* progenitors (red). Anterior neuroepithelial Shh(+) progenitors upregulate p57 (hatched), promoting anterior progenitor differentiation (orange), including the downregulation of Shh. A gradient of proliferation and differentiation is detected in the developing hypothalamus, emanating from bHyp progenitors (Fu et al. 2017)

1.3.3 Mammillary Progenitor Growth

Throughout the generation of anterior progenitors, a pool of *Fgf10(+)* progenitors is maintained (Fig. 1.3a, b). The mechanism that selects a steady supply of neuroepithelial Shh(+) progenitors simultaneously ensures a stable-size pool of *Fgf10(+)* progenitors (Fig. 1.5). Thus, genetic and pharmacological interventions suggest that Shh, deriving from neuroepithelial Shh(+) progenitors, feeds-back to regulate the size of the *Fgf10(+)* pool (Fu et al. 2017; Carreno et al. 2017), potentially acting in concert with Wnt signalling (Brinkmeier et al. 2007; Potok et al. 2008; Newman et al. 2018b). This concept is not new: in other parts of the brain, differentiating cells feedback to progenitor cells to maintain their appropriate numbers and behaviours.

The maintenance of pools of undifferentiated *Fgf10(+)* progenitor cells is important (Fig. 1.3 red area) because after generating anterior progenitors (Fig. 1.3 orange area), *Fgf10(+)* progenitors generate mammillary neurogenic progenitors that grow posteriorly (Fig. 1.3 blue area; Fu et al. 2017). At present, little is understood of mammillary progenitor growth, although there may be some commonalities with anterior progenitor growth (described in Fu et al. 2019). Further, it is not clear what promotes the switch from anterior to mammillary progenitor generation, nor why mammillary progenitors grow posteriorly. Regardless of the mechanism, the extensive growth of anterior then mammillary progenitors obscures earlier patterning. This means that the simple organization of the hypothalamus along the dorso-ventral axis, established through the early Shh morphogen gradient, is rapidly eroded through the subsequent extensive growth of anterior and mammillary progenitor populations. Progenitor cells that lie close to one another at early stages may ultimately give rise to neurons that lie far apart. This sheds new light on the different lineage relationships of hypothalamic neuronal and glial subsets (Sect. 1.4).

1.3.4 Infundibular Growth

After generating anterior and mammillary progenitors, *Fgf10(+)* progenitor cells generate the infundibulum, a small, but vital structure (Sect. 1.5) that is composed of glial progenitors that grow ventrally (Fig. 1.3d, arrows; Pearson et al. 2011; Goto et al. 2015). The sequential anisotropic growth in 3-dimensions from *Fgf10*(+) progenitor cells is peculiar and unprecedented within CNS development.

Loss-of-function studies in chick and mouse point to the vital role of Fgf10 in infundibular development. If the *Fgf10*(+) progenitor domain(s) is not maintained normally, or the intensity of *Fgf10* expression is altered, then the infundibulum does not develop (Fu et al. 2017; Carreno et al. 2017). Likewise, in *Fgf10*-null mice the infundibulum fails to form properly and infundibular cells undergo apoptosis (Ohuchi et al. 2000), and in mutant mice where *Fgf10* is downregulated, the infundibular-derived PP is hypoplastic (Mortensen et al. 2015). It is not known why *Fgf10*(+) progenitor cells switch from generating neurogenic anterior/mammillary to gliogenic infundibular progenitor cells, but potentially this neurogenic–gliogenic switch is triggered by the Notch pathway. Elsewhere in the hypothalamus Ascl1 represses oligodendrocyte differentiation (Marsters et al. 2016), so potentially, Ascl1 downregulation likewise triggers a switch from neurogenic to infundibular glial progenitor differentiation. In support of this idea, in Hes1(−/−); Hes5(+/−) mutant embryos, progenitor cells differentiate into neurons at the expense of pituicytes (derivatives of the infundibulum) (Goto et al. 2015).

Knockout studies in the mouse and analysis of human variants reveal a number of TFs that, like Hesx1, are required for infundibular formation, including *Nkx2.1*, *Tbx3*, *Sox2* and *Hes1/Hes5* (reviewed in Burbridge et al. 2016). Many of these are likely to affect early steps in the development of bHyp progenitors but conditional knockout studies reveal TFs that act directly downstream of Fgf signalling, including Rx, Lhx2 and Sox3 (Zhang et al. 2000; Orquera et al. 2016; Miranda-Angulo et al. 2014; Salvatierra et al. 2014; Zhao et al. 2010; Fig. 1.7 red flow-lines).

Growth of an infundibular-like structure can occur in isolated RDVM/bHyp cell explants (Pearson et al. 2011), suggesting that the instructions for infundibular growth are intrinsic to the bHyp population. This idea challenges previous models, which suggest that the infundibulum forms when an underlying tissue, Rathke's pouch (see Sect. 5.1) induces Hesx1 (a TF needed for infundibular development: see Davis et al 2013; Rizzoti 2015) in the overlying ventral diencephalon. One possible resolution of these findings is that Rathke's pouch maintains rather than induces

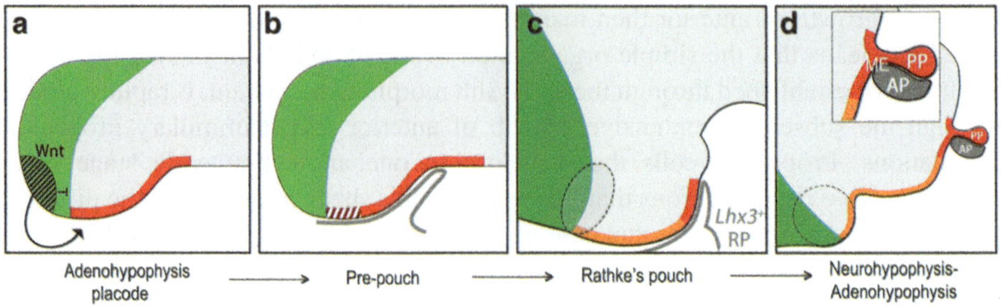

Fig. 1.6 (**a**) Development of the hypothalamo–pituitary axis. The adenohypophyseal (anterior pituitary) placode (black hatched) moves ventrally (curved arrow) beneath bHyp cells (red). (**b**) The placode invaginates to form a pouch as anterior neuroepithelial Shh(+) progenitors (red hatched) form. (**c**) Lhx3 is upregulated, forming the definitive RP as anterior progenitors (orange) form. (**d**) A subset of bHyp cells give rise to progenitors that grow ventrally and form the infundibulum, the precursor of the median eminence (ME) and posterior pituitary (PP). RP gives rise to the anterior pituitary (AP). Inset shows high power view of these regions. Dotted circle in (**c**, **d**) denotes optic stalk

Hesx1 expression. A second intriguing possibility is that the adenohypophyseal placode (the precursor to Rathke's pouch) lies so close to prospective hypothalamic cells that they interact (Fig. 1.6a hatched; Couly and Le Douarin 1985). Real-time analyses show cytoneme-like structures that extend between anterior-most midline cells and prospective hypothalamic cells (Patten et al. 2003), potentially regulating Wnt signalling between these two populations in the early neural plate (Mattes et al. 2018).

In summary, the bHyp population gives rise, sequentially, to anterior, mammillary and infundibular progenitor cells that grow in different dimensions: anterior, posterior and ventral, respectively (Fig. 1.3d). Future studies are needed to determine if bHyp cells are multipotent, contributing to all progenitor subtypes, or are more limited in their potential, and to establish whether potential correlates with a particular transcriptional signature. The future interrogation of bHyp cells through large-scale transcriptomic analyses is likely to reveal intriguing cellular mechanisms through which the hypothalamus is assembled in four dimensions (space and time) from this unique population.

1.4 Neuroendocrine Neurogenesis

The transcriptional programs that underpin neuroendocrine neurogenesis have been described in detail recently (in-depth reviews by Bedont et al. 2015; Burbridge et al. 2016; Alvarez-Bolado 2018), and here we provide only a summary. Gene knockout studies in the mouse suggest that downstream of Foxd1, the TFs, Six3/Six6 are required for hypothalamic specification, and are required either transiently, or in a sustained manner to support two major transcriptional programs for neuroendocrine neurogenesis: neuroendocrine neurons of the PVN, PeVN and SON differentiate via

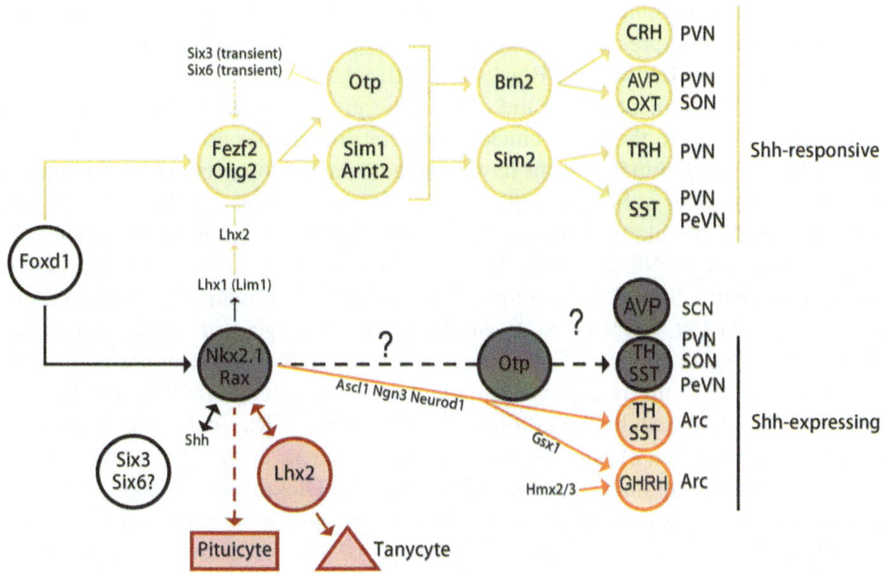

Fig. 1.7 Transcriptional programs underlying neuroendocrine neurogenesis and gliogenesis. Foxd1, Six3 and Six6 are expressed in many hypothalamic neuroendocrine progenitors. Upper yellow pathway: Many neurons of the PVN, PeVN and SON differentiate via the evolutionarily conserved zinc finger protein, Fezf2 and the bHLH protein, Olig2, which induce Otp and Sim1. Parallel activities of Sim1/Arnt2 (Aryl hydrocarbon receptor nuclear translocator 2, dimerization partner of Sim1) and Otp govern expression of (1) the pou-domain class 3 transcription factor Brain-2 (Brn2: required for expression of CRH, avp and oxt and (2) Sim2 and hence TRH and Sst expression. Lower orange pathway: Neurons of the Arc differentiate via an Nkx2.1-dependent ventral progenitor program that shows inter-regulatory interactions with the TFs Rx, Six3, Six6 and the signalling ligand Shh (see text). The Arc neurogenic program (orange) is dependent on the proneural gene Achaete-scute-like 1 (Ascl1: previously known as Mash1). Parallel to and/or downstream of Ascl1, Neurogenin 3 (Ngn3) and Neurod1 regulate Arc neurogenesis. Middle grey pathway: The finding that the PVN/PeVN/SON/SCN are reduced in size in Nkx2.1-null mice, and that ventral otp+ territory and associated TH+/Sst+ neurons, fail to form in rx3 mutant zebrafish raises the possibility that the PVN/PeVN/SON/SCN are of mixed origin, some neurons deriving from the ventral progenitor program. The two programs could exert cross-repressive interactions. *Six3* and *Six6* are expressed transiently in the PVN and SON, their downregulation concomitant with *Otp*, *Sim1* and Arnt2 upregulation. Additionally, the transcription factor Lhx2 may act downstream of the Nkx2.1 program as a repressor of *Otp* and *Sim1* and indeed of general anterior identities. After generating neurons, ventral progenitors give rise to glial pituicytes and tanycytes (red) the latter, via Rx/Lhx2 interactions

a Shh-responsive Fezf2/Olig2/Otp/Sim1 program (Fig. 1.7 yellow flow lines), whereas those of the ARC differentiate via a Shh-Nkx2.1/Rx program (Fig. 1.7 orange flow-lines). Initial reports suggested that the PVN/PeVN/SON but not the Arc can be detected in mice that lack functional Nkx2.1 (Kimura et al. 1996), while conversely, the PVN/PeVN/SON cannot be detected in mice that lack functional Sim1. Like neuroendocrine neurons of the PVN, PeVN and SON, Avp(+) neurons of the SCN are generally accepted to originate from Shh-responsive (rather than

Shh-expressing) progenitors, but via a Six3/Six6/Lhx1/Lhx2 program (Fig. 1.7 grey flow-lines) that suppresses the Fezf2/Olig2/Otp/Sim1 program [Fig. 1.7; see in-depth review by Bedont and Blackshaw (2015)].

What are the implications of the anisotropic growth model for neuroendocrine neurogenesis? As yet, the bHyp chick progenitor domain has not been fate-mapped to a stage when individual neurons have been generated. However, a prosaic (but not proved) view is that although many neuroendocrine neurons of the PVN, PeVN, SON and SCN are likely to derive from Shh-responsive progenitors (potentially those that expressed Nkx2.2 or Pax6: Fig. 1.4b), some may arise from bHyp cells, via neuroepithelial Shh(+) progenitors. In this model, some neuroendocrine neurons of the PVN, PeVN, SON or SCN would share a lineage with those of the Arc (Figs. 1.4e, 1.5). In support of this idea, the PVN/PeVN/SON/SCN are reduced in size in the Nkx2.1-null mouse (not remarked on in the original publication: Kimura et al. 1996). Additionally, conditional inactivation of Rx in mouse leads to a depletion of both Sst(+) and Avp(+) neurons (Orquera et al. 2016), and loss of function of rx3 in zebrafish depletes Avp(+) neurons (Dickmeis et al. 2007; Tessmar-Raible et al. 2007). While these studies do not distinguish between autonomous or nonautonomous mechanisms, loss of function of Rx/rx3 dramatically prevents/reduces the growth of basal anterior territories (Muthu et al. 2016; Orquera et al. 2016; Sect. 3.2), including domains that express the homeodomain TF, Otp (Table 1.2), required for the late development of multiple neuroendocrine neuronal subsets, including the PVN/PeVN/SON and dopaminergic Arc neurons (reviewed in Burbridge et al. 2016; Alvarez-Bolado 2018).

Together, these studies raise the possibility that bHyp-derived cells could migrate extensively into multiple nuclei and contribute to multiple neuroendocrine subsets, in addition to those of the Arc. The notion that hypothalamic nuclei may be of mixed progenitor contribution is further supported by lineage-tracing studies that show that alar progenitors, and/or their differentiating descendants can undergo extensive tangential migration (Manning et al. 2006; Zhao et al. 2008). Lineage-tracing studies (a tau-LacZ knockin allele at the Sim1 locus) of mice mutant for Sim1 show that while Sim1 mutant mice lack a structural PVN and SON, this is not due to the death of PVN/SON progenitor cells but instead to the fact that they do not migrate normally (Xu and Fan 2007). Therefore, the adult SON occupies the anterior basal hypothalamus because differentiating neurons migrate to this location (Box 1.2, panel C, yellow arrows), and not because its progenitor cells are specified there. In summary, a prediction is that adult hypothalamic nuclei or regions are populated by neurons of very distinct origin. Our poor understanding of hypothalamic neurogenesis highlights the need for state-of-the-art studies that can better resolve distinct progenitor and neuronal subtypes and their progression over time (Box 1.3).

Box 1.3 Defining Neuroendocrine Neurogenic and Gliogenic Programs
Researchers face many challenges in dissecting the programs that build neuroendocrine neurons. First little is known about the heterogeneity of either progenitor populations or neuronal subsets. Neurohormones such as Oxt, Sst, TH, and Avp, traditionally used as a proxy for cell fate, are widely expressed in multiple adult neuronal subtypes/nuclei, and so provide limited insight into the overall identity and generation of each subtype. This is likely to change quickly as large-scale omics/single cell RNA-seq studies shed light on the transcriptional profiles of different subtypes (e.g. Chen et al. 2017; Romanov et al. 2016; Campbell et al. 2017; Moffitt et al. 2018; Kim et al. 2019), and to do so in both wild-type and mutant mice (Kim et al. 2019). Second, while increasing numbers of studies emphasize the importance of progenitor and neuronal migrations/movements, many more conditional lineage-tracing studies from defined progenitor populations, linked to multiplex analysis of gene expression profiles, and real-time analyses of relative cell flows are needed to understand the full extent of early progenitor cell migrations, or mixing of progenitors of different origin, and the extent to which boundaries between hypothalamic regions are fluid or hard-set. Third, it is increasingly clear that TFs expressed early in development can exert multiple effects at different times (e.g. Newman et al. 2018a), so that loss of function can reflect a failure of early progenitor development or a failure of terminal differentiation (e.g. neurohormone maintenance). Conditional knockout studies and replacements of coding sequences with a reporter gene are needed to distinguish early versus late TF roles and to understand their cross-regulatory interactions that may maintain distinct domains.

1.5 Maintenance of Fgf10+ Populations for Longer-Term Building and Lifelong Neurogenesis

The chain of events that unfolds in a sequential manner to build hypothalamic neurons and the infundibulum and secure a supply of *Fgf10(+)* progenitor pools underlies longer-term building steps and lifelong hypothalamic neurogenesis.

1.5.1 Assembly of the Hypothalamo–Pituitary Axis

For many days, *Fgf10*$^{+ive}$ progenitor cells, and their infundibular daughters, are in intimate contact with the prospective anterior pituitary (AP, aka adenohypophyseal) cells. The AP derives from a placode that invaginates to form a rudimentary pouch that differentiates into Rathke's pouch, defined through the expression of Lhx3 (a master transcriptional regulator of the endocrine cells of the AP). Sequential signalling from the ventral-most hypothalamus induces Rathke's pouch: BMP4 induces the pouch rudiment, then Fgfs activate Lhx3 (Fig. 1.6). Explant recombination work shows that intimate contact between the infundibulum and the forming pouch is essential for differentiation of the rudimentary pouch to Rathke's pouch (reviewed in Davis et al. 2013; Rizzoti 2015).

New work indicates that Shh plays a role in this sequence. Transient blockade of Shh signalling, or conditional deletion of Shh from Hesx1(+) cells prevents both infundibular formation (Sect. 3.3) and the differentiation of Rathke's pouch from the rudimentary pouch. Further, instead of forming in intimate contact with the hypothalamus, the rudimentary pouch is separated from the hypothalamus by a loose meshwork of mesenchymal cells (Fu et al. 2017; Carreno et al. 2017). A prosaic view is that, in governing the normal movement and growth of anterior hypothalamic progenitor cells (see above), Shh directs the close apposition of the invaginating rudimentary pouch and the forming infundibulum (Fig. 1.6d). This close apposition may be important to induce Lhx3. This highlights the importance of early development to a longer-term building step that is crucial to function: assembly of the hypothalamo–pituitary neuraxis. The iterative deployment of BMPs, Fgfs and Shh for different steps in the assembly of the hypothalamo–pituitary neuraxis is likely to provide a platform through which to interpret human GWAS studies, and the wide array of mutations associated with pituitary defects.

1.5.2 Assembly of Neurovascular Interfaces in the ME/PP

Cells within the infundibulum, and at its neck, continue to express *Fgfs*, including *Fgf10* and *Fgf3*, long after the major neurogenic periods of hypothalamic development and play a role in guiding neurosecretory axon terminals and portal capillary vessels to set up the glial–neuronal–capillary interfaces that underlie the future functioning of the hypothalamo–pituitary axis and homeostatic balance (Fig. 1.1). These studies have been reviewed recently (Burbridge et al. 2016) and here we provide only a summary.

Genetic studies in mouse first indicated that Fgf signalling promotes the guidance of parvocellular neurosecretory axons (Tsai et al. 2011). Analyses of fgf3−/− zebrafish then revealed that avp(+) and oxt(+) magnocellular cells, and dopaminergic (DA) parvocellular axons fail to project to the ME-like neurohaemal region or PP, despite neurons initially developing normally (Gutnick et al. 2011; Liu et al. 2013). Expression of a dominant-negative Fgf receptor, in a temporally- or spatially controlled manner, revealed that Fgf3 directly affects growing axons. Parallel studies in chick demonstrated that Fgf3/Fgf10 deriving from the infundibulum act as long-range chemoattractants: low levels of Fgf3/Fgf10 stimulate and orient the growth of both AVP(+) and DA neurons. Intriguingly, high concentrations of Fgfs stall or repel growing axons, a mechanism that might help to ensure that neuroendocrine axons project to, but do not cross the ventral midline: uniquely in this region of the CNS, axons are non-commissural. Currently, little is known about subsequent mechanisms that ensure the correct projection of parvocellular axons to the median eminence, and of magnocellular axons to the posterior lobe. Future studies are needed to examine this, and to understand how Fgf ligands operate in conjunction with other factors, including Notch, Shh, BMP and Netrin (reviewed in Burbridge et al. 2016).

Concomitant with guiding axons, FGFs act with other factors to promote local endothelial vasculogenesis (Liu et al. 2013). These studies point to a role for FGF signalling during the early modelling of endothelial cells, and the initial formation of the capillary plexi that characterize the ME and PP (Gutnick and Levkowitz 2012, Fig.1.1) and suggest that, by attracting axons and stimulating endothelial cells, Fgfs ensure the coordinated growth of neuroendocrine axons and capillaries to the same general target. This is then another step in the self-organizing hypothalamic program, one that builds the neurovascular interface over time. Elegant studies in zebrafish show that as oxt(+) magnocellular neurons innervate the PP, they release oxt from their endfeet. Oxt acts as an angiogenic cue for nearby vascular sprouts, drawing them towards axon terminals, where they go on to form the hypophyseal arteries and veins (Gutnick et al. 2011). Thus, parvocellular axons actively participate in building the vascular networks required for their activity, another proof-of-principle for the idea that the hypothalamus self-organizes over time.

1.5.3 Pituicyte and Tanycyte Development

The long-term maintenance of *Fgf10*(+) progenitor populations through embryogenesis and into early postnatal life underlies the differentiation of pituicytes and tanycytes, specialized glial cells that arise in late prenatal/early postnatal life and that derive from infundibular glia (Fig. 1.7 red flow lines). Pituicytes constitute the PP: their position in the adult suggests that they derive from *Fgf10*(+) progenitors at the distal tip of the infundibulum. As yet, the transcriptional pathways that operate downstream of Fgf signalling and direct their differentiation remain enigmatic, but candidates are likely to be obtained through large-scale RNA analyses (Anbalagan et al. 2018).

Fgf10(+) cells in the proximal infundibulum differentiate first to radial glia and then to tanycytes—glial cells that are critical to body–brain communication and that regulate energy homeostasis and reproduction (Fig. 1.1; Table 1.1). Fgf induces and maintains the TF, Lhx2 (Fig. 1.7, red flow lines; Seth et al. 2006), which first directs infundibular development and then specifies tanycytes (Salvatierra et al. 2014; Samms et al. 2015). It is possible that sustained FGF signalling promotes the transition from infundibular precursors to radial glial cells by maintaining low levels of Wnt (Wang et al. 2012). An intimate bidirectional loop between Lhx2 and Rx2 operates into postnatal life to induce tanycyte-specific genes such as *Gpr50* and *DiO2* (deiodinase 2) in multiciliated tanycytes, and suppress ependymal fates (Salvatierra et al. 2014). By regulating Lhx2 and DiO2, Fgf signalling is therefore required throughout life for tanycyte identity and function: *DiO2* directly regulates thyroid hormone (TH) availability and hence appetite and energy expenditure (Samms et al. 2015, reviewed in Lewis and Ebling 2017).

1.5.4 Tanycytes: Fgf-Responsive Proliferating Progenitors

In addition to their pivotal roles in acute homeostasis (outside of the scope of this review: see Table 1.1), *Fgf10(+)* tanycyte subpopulations have been implicated in longer-term allostatic change through their ability to act as neural stem/progenitor (NSP) cells.

EdU-labelling studies in postnatal animals first showed that regions around the ME are hotspots of neurogenesis and gliogenesis, and subsequent studies revealed that β1, β2 and α2 tanycyte subsets can act as neurogenic and gliogenic NSP cells (summarized in Rizotti and Lovell-Badge 2017; Lewis and Ebling 2017; Ebling and Lewis 2018; Clasadonte and Prevot 2018). Lineage-tracing studies showed that *Fgf10(+)* tanycytes, although largely quiescent in adulthood, can respond and adapt to their changing environment by producing new neurons. Thus, *Fgf10(+)* tanycytes respond to a high-fat diet by generating neuropeptide Y (NPY)-neurons that regulate feeding behaviour (Haan et al. 2013). Other lineage-tracing studies showed that an FGF-responsive α-2 tanycyte subset can self-renew or give rise to other tanycyte populations, even in unstimulated mice, indicating that localized proliferation and differentiation to tanycytes maintains the integrity of the ventricular wall (Robins et al. 2013). Fgf10 and Fgf signal components are maintained in subsets of tanycytes (including β- and α-2 subsets) throughout life (Haan et al. 2013; Robins et al. 2013) and Fgf signalling can regulate tanycyte proliferation: infusion of FGF2 into the third ventricle stimulates proliferation of both β-1 and α-2 subtypes (Robins et al. 2013). Tanycyte neurogenesis and gliogenesis can be stimulated by a range of signals, hormones and metabolites, and ongoing studies are now seeking to determine whether/how Fgf signalling is involved in supporting adult neurogenic programs. One study suggests that tanycyte stimulation by FGF1 correlates with the sustained remission of diabetic hyperglycaemia in rodent models (Scarlett et al. 2016), suggesting that appropriate regulation of Fgf signalling in tanycytes may be critical to health and disease. Important questions for the future are: (1) whether adult programs of tanycyte homeostasis and neurogenesis will share features of embryonic *Fgf10(+)* progenitor homeostasis and neurogenesis; (2) whether physiological or epigenetic factors that are known to govern neurogenesis from adult tanycytes do so via interactions with developmental signal-TF pathways; and (3) whether neuroendocrine neurons can be generated throughout life.

1.6 Plasticity Through the Life Course: Is a Hypothalamic Stem Cell Established in the Embryo?

The presence of neurogenic/gliogenic tanycytes, in close proximity to the extensive capillary network of the ME, suggests that tanycytes are components of a hypothalamic NSP cell niche. Given their multipotent nature, a likely possibility is that *Fgf10(+)* populations that are established in the early embryo form the essential cell type, and organize the development of an emerging hypothalamic NSP cell niche.

An outstanding question is whether lineage-related progenitors within the niche share an equivalent state and so represent multipotent stem-like cells, or show inequivalence and represent adaptive progenitor populations. Currently, we lack in vivo evidence to distinguish between these possibilities, although ex vivo neurospherogenic studies demonstrate that some subsets of tanycytes exhibit a stem-like capacity to self-renew and differentiate (Robins et al. 2013).

The question of whether a multipotent stem cell is established in the early stages of embryogenesis has implications for how the neuroendocrine hypothalamus might adapt to changing physiological conditions on different time scales, and for the robustness of the hypothalamus through life. If a stem cell exists, then by definition, it has equivalent potential to differentiate throughout life. This implies that a single cell can respond to multiple stimuli to differentiate into multiple basal neurons over life. If, on the other hand, cells are not equivalent, i.e. have limited differentiation potential, this suggests that they become remodelled to anticipate and respond to the changing environment. How could such a cell become remodelled over time? We speculate that this could be achieved through a self-organization program that is driven by hormones that are themselves developmentally regulated by tanycytes. There is ample evidence that tanycytes can detect and respond to hormones. IGF, for instance, can stimulate tanycytes and provoke changes in neurogenic programs (reviewed in Rizotti and Lovell-Badge 2017; Lewis and Ebling 2017), and tanycytes both regulate (see Sect. 5.3) and respond to TH, through a mechanism that is intrinsically linked to developmental time (Mohácsik et al. 2016). Together, this raises the possibility that tanycytes may be key components of an intrinsic hypothalamic clock, responding to developmentally timed steroid signals and supporting long-term allostatic change.

An important area of future work will be to better understand whether, as suggested through emerging work, such tanycyte-regulated activity is sensitive to epigenetic change. This would form a platform to understand how environmental factors, in particular those that affect steroid signalling, could lead to maladaptive tanycyte behaviour and homeostatic dysfunction over life. It seems likely that a continued focus of research on bHyp cells and their derivatives will lead to a far better understanding of how the hypothalamus can effect both short-term physiological regulation of factors such as stress and energy balance, and long-term changes such as puberty and pregnancy, and to understand sensitive developmental time points for cells that regulate particular physiological processes.

Acknowledgments Work supported by the Wellcome Trust.

Key References

Alvarez-Bolado et al. (2012)—Lineage-tracing of a subset of Shh(+) neuroepithelial progenitors shows that they contribute to the tuberal and mammillary hypothalamus.

Carreno et al. (2017)—Genetic studies in the mouse reveal the integrated development of the ventral hypothalamus (including the infundibulum and Shh(+) neuroepithelial progenitors) and the adenohypophysis.

Chen et al. (2017)—A large scale single cell RNA seq analysis of the adult mouse hypothalamus confirms that neuroendocrine neurohormones/neurotransmitters are expressed in heterogeneous cell types.

Corman et al. (2018)—Sophisticated mouse studies, including analyses of conditional knock-out and reporter lines, show that tuberal neurons derive from both Shh-expressing and Shh-responsive progenitors.

Fu et al. (2017)—Fate-mapping and pharmacological studies in the early chick provides evidence for the anisotropic growth model of hypothalamic development.

Haan et al. (2013)—Evidence for de novo adult neurogenesis from *Fgf10(+)* tanycytes; together with Fu et al. (2017) and Robins et al. (2013), raises possibility that an Fgf10(+) population is important for building hypothalamic neurons throughout life.

Kim et al. (2019)—A new study that sets a benchmark for molecular analysis of the developing hypothalamus. Single cell RNA seq analyses of the hypothalamus at different time points in embryogenesis and postnatal life provide evidence for the early specification of key hypothalamic neuronal subsets. Comparison of wild type and mutant mice provides evidence that the hypothalamus is a diencephalic-derived structure.

Manning et al. (2006)—A study in the chick embryo that provided an understanding of the co-ordination of pattern and proliferation of ventral hypothalamic progenitor cells. Re-interpretation of this paper, in light of the anisotropic growth shown in Fu et al. (2019), suggests that the cells that express Tbx2/BMP are bHyp cells, and the Emx2+ cells are mammillary progenitors.

Muthu et al. (2016)—Anisotropic growth of anterior hypothalamus in zebrafish; absence of growth leads to failure of differentiation of both 'anterior' and 'tuberal' neuroendocrine neurons.

Newman et al. (2018a)—Sophisticated conditional knock-out and reporter lines show hypothalamus is of diencephalic origin.

Orquera et al. (2016)—Evidence for conserved mechanism of anterior progenitor growth and differentiation, via a Shh-Rax pathway.

Robins et al. (2013)—Analysis of adult mouse shows that Fgf10(+) stem/progenitor cells exist as tanycytes that line the ventricle; with Fu et al. (2019) and Haan et al. (2013), suggests that Fgf10(+) stem/progenitor cells are retained through life.

Shimogori et al. (2010)—A large-scale screen and in situ analysis identifies many hypothalamic progenitor domains; conditional knock-out of subsets of Shh(+) neuroepithelial progenitors shows importance in anterior/tuberal growth and neurogenesis.

Szabó et al. (2009)—Lineage-tracing of a subset of Shh(+) RDVM cells shows contribution to tuberal and mammillary hypothalamus.

Tessmar-Raible et al. (2007)—Conservation of neuroendocrine cells and conservation of key molecules.

Trowe et al. (2013)—Analysis of mouse, analysing control of Shh expression in RSVM cells. With Manning et al. (2006), suggests a mechanism for resolution of bHyp cells into Shh(+) and Shh(−) domains.

Xu et al. (2007)—Proof-of-principle that differentiating hypothalamic neurons undergo tangential migration.

Zhao et al. (2008)—Proof-of-principle that progenitors other than bHyp cells also undergo tangential migration.

References

Alvarez-Bolado G (2018) Development of neuroendocrine neurons in the mammalian hypothalamus. Cell Tissue Res. https://doi.org/10.1007/s00441-018-2859-1

Alvarez-Bolado G, Paul FA, Blaess S (2012) Sonic hedgehog lineage in the mouse hypothalamus: from progenitor domains to hypothalamic regions. Neural Dev 7:4. https://doi.org/10.1186/1749-8104-7-4

Anbalagan S, Gordon L, Blechman J, Matsuoka RL, Rajamannar P, Wircer E, Biran J, Reuveny A, Leshkowitz D, Stainier DYR, Levkowitz G (2018) Pituicyte cues regulate the development of permeable neuro-vascular interfaces. Dev Cell. https://doi.org/10.1016/j.devcel.2018.10.017

Bedont JL, Blackshaw S (2015) Constructing the suprachiasmatic nucleus: a watchmaker's perspective on the central clockworks. Front Syst Neurosci 9:74

Bedont JL, Newman EA, Blackshaw S (2015) Patterning, specification, and differentiation in the developing hypothalamus. Wiley Interdiscip Rev Dev Biol. https://doi.org/10.1002/wdev.187

Brinkmeier ML, Potok MA, Davis SW, Camper SA (2007) TCF4 deficiency expands ventral diencephalon signaling and increases induction of pituitary progenitors. Dev Biol 311:396–407

Burbridge S, Stewart I, Placzek M (2016) Development of the neuroendocrine hypothalamus. Compr Physiol 6(2):623–643

Campbell JN, Macosko EZ, Fenselau H et al (2017) A molecular census of arcuate hypothalamus and median eminence cell types. Nat Neurosci 20:484–496

Carreno G, Apps JR, Lodge EJ, Panousopoulos L, Haston S, Gonzalez-Meljem JM, Hahn H, Andoniadou CL, Martinez-Barbera JP (2017) Hypothalamic sonic hedgehog is required for cell specification and proliferation of LHX3/LHX4 pituitary embryonic precursors. Development 144(18):3289–3302

Chen R, Wu X, Jiang L, Zhang Y (2017) Single-cell RNA-Seq reveals hypothalamic cell diversity. Cell Rep 18(13):3227–3241

Clasadonte J, Prevot V (2018) The special relationship: glia-neuron interactions in the neuroendocrine hypothalamus. Nat Rev Endocrinol 14(1):25–44. https://doi.org/10.1038/nrendo.2017.124

Corman TS, Bergendahl SE, Epstein DJ (2018) Distinct temporal requirements for Sonic hedgehog signaling in development of the tuberal hypothalamus. Development 145(21). https://doi.org/10.1242/dev.167379

Couly GF, Le Douarin NM (1985) Mapping of the early neural primordium in quali-chick chimeras. Dev Bioll 110:422–439

Davis SW, Ellsworth BS, Peréz Millan MI, Gergics P, Schade V, Foyouzi N, Brinkmeier ML, Mortensen AH, Camper SA (2013) Pituitary gland development and disease: from stem cell to hormone production. Curr Top Dev Biol 106:1–47

Dickmeis T, Lahiri K, Nica G, Vallone D, Santoriello C, Neumann CJ, Hammerschmidt M, Foulkes NS (2007) Glucocorticoids play a key role in circadian cell cycle rhythms. PLoS Biol 5:e78

Ebling FJP, Lewis JE (2018) Tanycytes and hypothalamic control of energy balance. Glia 66(6):1176–1184

Fu T, Towers M, Placzek MA (2017) Fgf10+ progenitors give rise to the chick hypothalamus by rostral and caudal growth and differentiation. Development 144(18):3278–3288

Fu T, Pearson C, Towers M, Placzek, MA (2019) Development of the basal hypothalamus through anisotropic growth. J Neuroendocrinol 31(5):e12727
Gizowski C, Zaelzer C, Bourque CW (2016) Clock-driven vasopressin neurotransmission mediates anticipatory thirst prior to sleep. Nature 537:685–688
Goto M, Hojo M, Ando M, Kita A, Kitagawa M, Ohtsuka T, Kageyama R, Miyamoto S (2015) Hes1 and Hes5 are required for differentiation of pituicytes and formation of the neurohypophysis in pituitary development. Brain Res 1625:206–217
Gutnick A, Levkowitz G (2012) The neurohypophysis: fishing for new insights. J Neuroendocrinol 24(6):973–974
Gutnick A, Blechman J, Kaslin J, Herwig L, Belting HG, Affolter M, Bonkowsky JL, Levkowitz G (2011) The hypothalamic neuropeptide oxytocin is required for formation of the neurovascular interface of the pituitary. Dev Cell 21:642–654
Haan N, Goodman T, Najdi-Samiei A, Stratford CM, Rice R, El Agha E, Bellusci S, Hajihosseini MK (2013) Fgf10(+) tanycytes add new neurons to the appetite/energy-balance regulating centers of the postnatal and adult hypothalamus. J Neurosci 33(14):6170–6180
Jeong Y, El-Jaick K, Roessler E, Muenke M, Epstein DJ (2006) A functional screen for sonic hedgehog regulatory elements across a 1 Mb interval identifies long-range ventral forebrain enhancers. Development 133(4):761–772
Kapsimali M, Caneparo L, Houart C, Wilson SW (2004) Inhibition of Wnt/Axin/beta-catenin pathway activity promotes ventral CNS midline tissue to adopt hypothalamic rather than floorplate identity. Development 131:5923–5933
Kim DW, Washington PW, Wan ZQ, Lin S, Sun C, Jiang L, Blackshaw S (2019) Single cell RNA-Seq analysis identifies molecular mechanisms controlling hypothalamic patterning and differentiation. bioRxiv. https://doi.org/10.1101/657148
Kimura S, Hara Y, Pineau T, Fernandez-Salguero P, Fox CH, Ward JM, Gonzalez FJ (1996) The T/ebp null mouse: thyroid-specific enhancer-binding protein is essential for the organogenesis of the thyroid, lung, ventral forebrain, and pituitary. Genes Dev 10:60–69
Lewis JE, Ebling FJ (2017) Tanycytes as regulators of seasonal cycles in neuroendocrine function. Front Neurol 10:8–79
Liu F, Pogoda HM, Pearson CA, Ohyama K, Lohr H, Hammerschmidt M, Placzek M (2013) Direct and indirect roles of Fgf3 and Fgf10 in innervation and vascularisation of the vertebrate hypothalamic neurohypophysis. Development 140:1111–1122
Manning L, Ohyama K, Saeger B, Hatano O, Wilson SA, Logan M, Placzek M (2006) Regional morphogenesis in the hypothalamus: a BMP-Tbx2 pathway coordinates fate and proliferation through Shh downregulation. Dev Cell 11:873–885
Manoli M, Driever W (2014) nkx2.1 and nkx2.4 genes function partially redundant during development of the zebrafish hypothalamus, preoptic region, and pallidum. Front Neuroanat 8:145
Marsters CM, Rosin JM, Thornton HF, Aslanpour S, Klenin N, Wilkinson G, Schuurmans C, Pittman QJ, Kurrasch DM (2016) Oligodendrocyte development in the embryonic tuberal hypothalamus and the influence of Ascl1. Neural Dev 11(1):20
Mattes B, Dang Y, Greicius G, Kaufmann LT, Prunsche B, Rosenbauer J, Stegmaier J, Mikut R, Özbek S, Nienhaus GU, Schug A, Virshup DM, Scholpp S (2018) Wnt/PCP controls spreading of Wnt/β-catenin signals by cytonemes in vertebrates. Elife 7. https://doi.org/10.7554/eLife.36953
Miranda-Angulo AL, Byerly MS, Mesa J, Wang H, Blackshaw S (2014) Rax regulates hypothalamic tanycyte differentiation and barrier function in mice. J Comp Neurol 522:876–899
Moffitt JR, Bambah-Mukku D, Eichhorn SW, Vaughn E, Shekhar K, Perez JD, Rubinstein ND, Hao J, Regev A, Dulac C, Zhuang X (2018) Molecular, spatial, and functional single-cell profiling of the hypothalamic preoptic region. Science 362(6416). https://doi.org/10.1126/science.aau5324
Mohácsik P, Füzesi T, Doleschall M, Szilvásy-Szabó A, Vancamp P, Hadadi É, Darras VM, Fekete C, Gereben B (2016) Increased thyroid hormone activation accompanies the formation

of thyroid hormone-dependent negative feedback in developing chicken hypothalamus. Endocrinology 157(3):1211–1221

Mortensen AH, Schade V, Lamonerie T, Camper SA (2015) Deletion of OTX2 in neural ectoderm delays anterior pituitary development. Hum Mol Genet 24:939–953

Muthu V, Eachus H, Ellis P, Brown S, Placzek M (2016) Rx3 and Shh direct anisotropic growth and specification in the zebrafish tuberal/anterior hypothalamus. Development (14):2651–2663. https://doi.org/10.1242/dev.138305

Newman EA, Kim DW, Wan J, Wang J, Qian J, Blackshaw S (2018a) Foxd1 is required for terminal differentiaiton of anterior hypothalamic neuronal subtypes. Dev Biol 439(2):102–111. https://doi.org/10.1016/j.ydbio.2018.04.012

Newman EA, Wu D, Taketo MM, Zhang J, Blackshaw S (2018b) Canonical Wnt signaling regulates patterning, differentiation and nucleogenesis in mouse hypothalamus and prethalamus. Dev Biol 442(2):236–248

Ohuchi H, Hori Y, Yamasaki M, Harada H, Sekine K, Kato S, Itoh N (2000) FGF10 acts as a major ligand for FGF receptor 2 IIIb in mouse multi-organ development. Biochem Biophys Res Commun 277:643–649

Orquera DP, Nasif S, Low MJ, Rubinstein M, de Souza FSJ (2016) Essential function of the transcription factor Rax in the early patterning of the mammalian hypothalamus. Dev Biol 416 (1):212–224. https://doi.org/10.1016/j.ydbio.2016.05.021

Patten I, Kulesa P, Shen MM, Fraser S, Placzek M (2003) Distinct modes of floor plate induction in the chick embryo. Development 130:4809–4821

Pearson CA, Ohyama K, Manning L, Aghamohammadzadeh S, Sang H, Placzek M (2011) FGF-dependent midline-derived progenitor cells in hypothalamic infundibular development. Development 138:2613–2624

Placzek M, Briscoe J (2005) The floor plate: multiple cells, multiple signals. Nat Rev Neurosci 6:230–240

Placzek M, Briscoe J (2018) Sonic hedgehog in vertebrate neural tube development. Int J Dev Biol 62(1–2–3):225–234. https://doi.org/10.1387/ijdb.170293jb

Potok MA, Cha KB, Hunt A, Brinkmeier ML, Leitges M, Kispert A, Camper SA (2008) WNT signaling affects gene expression in the ventral diencephalon and pituitary gland growth. Dev Dyn 237:1006–1020

Puelles L, Martinez-de-la-Torre M, Bardet S, Rubenstein JLR (2012) Hypothalamus. In: Watson C, Paxinos G, Puelles L (eds) Mouse nervous system. Academic Press, San Diego, CA, pp 221–312

Puelles L, Harrison M, Paxinos G, Watson C (2013) A developmental ontology for the mammalian brain based on the prosomeric model. Trend Neurosci 36:570–578

Ratie L, Ware M, Barloy-Hubler F, Rome H, Gicquel I, Dubourg C, David V, Dupe V (2013) Novel genes upregulated when NOTCH signalling is disrupted during hypothalamic development. Neural Dev 8:25

Ratié L, Ware M, Jagline H, David V, Dupé V (2014) Dynamic expression of Notch-dependent neurogenic markers in the chick embryonic nervous system. Front Neuroanat 8:158

Rizotti K, Lovell-Badge R (2017) Pivotal role of median eminence tanycytes for hypothalamic function and neurogenesis. Mol Cell Endocrinol 445:7–13

Rizzoti K (2015) Genetic regulation of murine pituitary development. J Mol Endocrinol 54(2):R55–R73

Robins SC, Stewart I, McNay DE, Taylor V, Giachino C, Goetz M, Ninkovic J, Briancon N, Maratos-Flier E, Flier JS, Kokoeva MV, Placzek M (2013) α-Tanycytes of the adult hypothalamic third ventricle include distinct populations of FGF-responsive neural progenitors. Nat Commun 4:2049. https://doi.org/10.1038/ncomms3049

Romanov RA, Zeisel A, Bakker J et al (2016) Molecular interrogation of hypothalamic organization reveals distinct dopamine neuronal subtypes. Nat Neurosci 20:176–188

Salvatierra J, Lee DA, Zibetti C, Duran-Moreno M, Yoo S, Newman EA, Wang H, Bedont JL, de Melo J, Miranda-Angulo AL, Gil-Perotin S, Garcia-Verdugo JM, Blackshaw S (2014) The LIM

homeodomain factor Lhx2 is required for hypothalamic tanycyte specification and differentiation. J Neurosci 34:16809–16820
Samms RJ, Lewis JE, Lory A, Fowler MJ, Cooper S, Warner A, Emmerson P, Adams AC, Luckett JC, Perkins AC, Wilson D, Barrett P, Tsintzas K, Ebling FJ (2015) Antibody-mediated inhibition of the FGFR1c isoform induces a catabolic lean state in siberian hamsters. Curr Biol 25 (22):2997–3003
Sánchez-Arrones L, Ferrán JL, Rodríguez-Gallardo L, Puelles L (2009) Incipient forebrain boundaries traced by differential gene expression and fate mapping in the chick neural plate. Dev Biol 335:43–65
Scarlett JM, Rojas JM, Matsen ME, Kaiyala KJ, Stefanovski D, Bergman RN, Nguyen HT, Dorfman MD, Lantier L, Wasserman DH, Mirzadeh Z, Unterman TG, Morton GJ, Schwartz MW (2016) Central injection of fibroblast growth factor 1 induces sustained remission of diabetic hyperglycemia in rodents. Nat Med 22(7):800–806
Seth A, Culverwell J, Walkowicz M, Toro S, Rick JM, Neuhauss SC, Varga ZM, Karlstrom RO (2006) belladonna/(Ihx2) is required for neural patterning and midline axon guidance in the zebrafish forebrain. Development 133(4):725–735
Shimogori T, Lee DA, Miranda-Angulo A, Yang Y, Wang H, Jiang L, Yoshida AC, Kataoka A, Mashiko H, Avetisyan M, Qi L, Qian J, Blackshaw S (2010) A genomic atlas of mouse hypothalamic development. Nat Neurosci 13:767–775
Swanson LW (1992) Brain maps: structure of the rat brain. Elsevier, New York
Szabó NE, Zhao T, Cankaya M, Theil T, Zhou X, Alvarez-Bolado G (2009) Role of neuroepithelial Sonic hedgehog in hypothalamic patterning. J Neurosci 29(21):6989–7002. https://doi.org/10.1523/JNEUROSCI.1089-09.2009
Tessmar-Raible K, Raible F, Christodoulou F, Guy K, Rembold M, Hausen H, Arendt D (2007) Conserved sensory-neurosecretory cell types in annelid and fish forebrain: insights into hypothalamus evolution. Cell 129:1389–1400
Trowe MO, Zhao L, Weiss AC, Christoffels V, Epstein DJ, Kispert A (2013) Inhibition of Sox2-dependent activation of Shh in the ventral diencephalon by Tbx3 is required for formation of the neurohypophysis. Development 140:2299–2309
Trudel E, Bourque CW (2010) Central clock excites vasopressin neurons by waking osmosensory afferents during late sleep. Nat Neurosci 13:467–474
Tsai PS, Brooks LR, Rochester JR, Kavanaugh SI, Chung WC (2011) Fibroblast growth factor signaling in the developing neuroendocrine hypothalamus. Front Neuroendocrinol 32:95–107
Wang X, Kopinke D, Lin J, McPherson AD, Duncan RN, Otsuna H, Moro E, Hoshijima K, Grunwald DJ, Argenton F, Chien CB, Murtaugh LC, Dorsky RI (2012) Wnt signaling regulates postembryonic hypothalamic progenitor differentiation. Dev Cell 23:624–636
Xu C, Fan CM (2007) Allocation of paraventricular and supraoptic neurons requires Sim1 function: a role for a Sim1 downstream gene PlexinC1. Mol Endocrinol 21(5):1234–1245
Zhang L, Mathers PH, Jamrich M (2000) Function of Rx, but not Pax6, is essential for the formation of retinal progenitor cells in mice. Genesis 28:135–142
Zhao T, Szabo N, Ma J, Luo L, Zhou X, Alvarez-Bolado G (2008) Genetic mapping of Foxb1-cell lineage shows migration from caudal diencephalon to telencephalon and lateral hypothalamus. Eur J Neurosci 28:1941–1955
Zhao Y, Mailloux CM, Hermesz E, Palkóvits M, Westphal H (2010) A role of the LIM-homeobox gene Lhx2 in the regulation of pituitary development. Dev Biol 337(2):313–323. https://doi.org/10.1016/j.ydbio.2009.11.002
Zhao L, Zevallos SE, Rizzoti K, Jeong Y, Lovell-Badge R, Epstein DJ (2012) Disruption of SoxB1-dependent Sonic hedgehog expression in the hypothalamus causes septo-optic dysplasia. Dev Cell 22:585–596

Sonic hedgehog in Hypothalamus Development

2

Gonzalo Alvarez-Bolado

Abstract

Sonic hedgehog (SHH) is a secreted signaling protein able to confer ventral cell fate to the spinal cord on the basis of a concentration gradient that translates into activation of specific genes through Gli transcription factors. The intricate and dynamic expression pattern of *Shh* in the forebrain raises the question of its function in the developing hypothalamus. Experiments in chick and mouse show that SHH from the prechordal plate cooperates with Bone Morphogenetic Proteins (BMPs) to induce a special "hypothalamic floor plate" where *Shh* expression will be activated by three transcription factors, *Nkx2-1*, *Six3*, and *Sox2/3*. Conditional mouse mutants show that this new source of SHH is responsible for hypothalamic growth as well as full differentiation of most hypothalamic cell types and the formation of the lateral hypothalamus. Experiments on chick embryos indicate that for the specification of hypothalamic regions *Shh* cooperates with *Fgf10*. For the final differentiation of the basal hypothalamic regions and hypophysis, repression of *Shh* by T-box proteins is necessary. Precisely how SHH unlocks hypothalamic cell fates is not entirely clear. But in mouse and zebrafish, *Shh* bestows neuroendocrine cell fates through a complex relation with *Rax*/*rx3*.

Keywords

Floor plate, Gli, Hypothalamus, Lineage, Prechordal plate, Progenitor domain, Sonic hedgehog

G. Alvarez-Bolado (✉)
Department of Neuroanatomy, University of Heidelberg School of Medicine, Heidelberg, Germany
e-mail: Alvarez@uni-heidelberg.de

S. Wray, S. Blackshaw (eds.), *Developmental Neuroendocrinology*, Masterclass in Neuroendocrinology 9, https://doi.org/10.1007/978-3-030-40002-6_2

2.1 Introduction

Sonic hedgehog (*Shh*) encodes a secreted protein essential for the development of the central nervous system (CNS). SHH has fascinated developmental biologists since its discovery because of its ability to confer a variety of cell fates on the basis of a concentration gradient. The most important developmental role of SHH in the CNS is to confer specific character to the ventral half of the neural tube (Ericson et al. 1995). Progenitor cells in the early neural tube (the stem cells of the CNS) receive their "ventral character" from SHH secreted by a specific source tissue. In the spinal cord, the first source of SHH is the axial mesoderm underlying the presumptive spinal cord, i.e., the notochord. Neuroepithelial cells in the ventral midline of the presumptive spinal cord are induced by SHH to become floor plate cells and start expressing *Shh* and secreting SHH themselves into the neural tube. SHH then forms a concentration gradient effective over a relatively long distance (20 cell diameters from the source at least) and imparts ventral character to the progenitor cells that form the early neural tube according to this gradient. The highest concentration induces the most ventral structures and progressively lower concentrations of SHH specify progressively less ventral (more dorsal) cell fates. The complex spatial and temporal properties of the SHH gradient and the ways in which cells respond to it are well delineated. The resulting body of knowledge is an achievement of modern developmental biology (Dessaud et al. 2008; Sagner and Briscoe 2017). Most of this work was done using the spinal cord as a model. In the forebrain, the dynamic expression pattern of *Shh* suggested even a more intricate involvement of this signaling pathway in the development of the ventral forebrain and of the hypothalamus.

2.1.1 The Morphogen Protein SHH and Its Signaling Pathway

After translation, SHH cleaves itself into an active N-terminal fragment and is then covalently attached to two lipid moieties, palmitoyl and cholesterol (which probably confer SHH the peculiar capacity to signal through relatively long distances). SHH is then released by source cells through one or maybe several mechanisms. Essential for SHH to exert its influence on the target cell is a cellular compartment, the primary cilium (Huangfu et al. 2003), as well as several proteins acting as receptors, second messengers and effectors of the SHH signal. In vertebrates, the SHH transduction pathway ends on three effector proteins, the GLI transcription factors (GLI1, 2, and 3). Cells competent to respond to SHH express two transmembrane proteins, *Patched* (*Ptch1*) and *Smoothened* (*Smo*), as well as *Gli2* and *Gli3* (and other specific genes that regulate the pathway).

While GLI1 is a transcriptional activator, GLI2 and GLI3 can act as activators (GliA) or repressors (GliR). The GLI DNA-binding domain consists of five zinc fingers, and the three GLI proteins have similar DNA-binding specificity. The activator domain is the C-terminal, and the repressor domain of GLI2 and GLI3 is the N-terminal. In SHH responsive cells, GLI2 and GLI3 are translated and processed into their repressor forms and repress *Shh* pathway targets (the primary repressor is GLI3). Upon binding of SHH to the receptor complex, PTCH1 allows SMO to change into its active conformation. SMO then activates a signaling cascade that inhibits processing of GLI proteins into their repressor forms and instead they are processed into their activator forms (the activator form is mostly GLI2). In contrast, *Gli1* is the first gene expressed upon activation of the pathway and therefore its expression is a readout of *Shh* pathway activation. (Note: This is somewhat divergent in fishes). The components of the SHH transduction pathway are localized to the cilia, possibly allowing for greater sensitivity of recipient cells to SHH molecules. The SHH signaling cascade is still being unraveled, and the subject has been often and expertly reviewed (see for instance Briscoe and Thérond 2013).

2.1.2 Hypothalamic Nomenclatures (Fig. 2.1)

The adult hypothalamus has been classically subdivided into four transverse regions, from rostral to caudal: preoptic, anterior, tuberal, and mammillary (Fig. 2.1a). However, developmental gene expression patterns do not entirely conform to the neuroanatomy of the adult brain as it is traditionally understood (Puelles et al. 2012) (Fig. 2.1b; from Haddad-Tóvolli et al. 2015). If we take *Shh* expression (pink in Fig. 2.1b) as the ultimate ventral marker, the tuberal and mammillary regions are ventral (i.e., they express *Shh* during development) and the preoptic and anterior are dorsal (Fig. 2.1b). To avoid the resulting "clash of nomenclatures" (i.e., adult versus developmental neuroanatomy), the words "alar" and "basal" (Fig. 2.1b, yellow versus red, separated by a red line) are sometimes used instead of "dorsal" and "ventral" when speaking of the developing hypothalamus. The well-known term "ventral forebrain" becomes "anterior forebrain" and includes alar (dorsal) and basal (ventral) portions. The most anterior region has received the name of "acroterminal region" (green in Fig. 2.1b), the basal portion of which (green striped area in Fig. 2.1b) corresponds to the Rostral Diencephalic Ventral Midline cells (RDVM; also called hypothalamic floor plate and Medial Progenitor Domain (used throughout this chapter)).

For this reason, it is preferable to substitute "presumptive hypothalamus" for "ventral forebrain." Even more, developmental studies show that the preoptic region originates in the telencephalon and is not part of the diencephalon (see Sect. 2.7.3). These questions reflect the current discussion about the developmental and evolutionary origins of the four hypothalamic regions and are not yet settled.

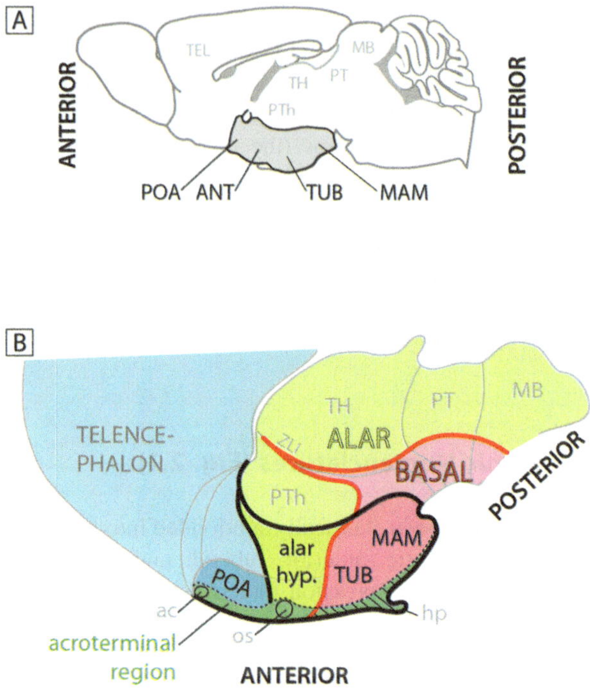

Fig. 2.1 Hypothalamic regional nomenclature and Shh expression. Conventional representation of the hypothalamus (gray) as ventral region with four regions aligned from rostral to caudal: POA, preoptic; ANT, anterior; TUB, tuberal; MAM, mammillary. Model of the hypothalamus considering *Shh* expression (pink) as basal (ventral) marker. The tuberal and mammillary regions would then be not "caudal" but rather basal ("ventral"), and the anterior region would form the alar ("dorsal") hypothalamus (yellow). In this model, the most anterior portion of the hypothalamus is called acroterminal region (green); its basal portion (striped green) corresponds to the Medial Progenitor Domain or hypothalamic floor plate. Additionally, the preoptic region is considered part of the telencephalon. The red line represents the boundary of Shh expression. ac, anterior commissure; hp, hypophysis; MB, midbrain; os, optic stalk; POA, preoptic region; PT, pretectum; PTh, prethalamus; TH, thalamus; ZLI, zona limitans (modified from Haddad-Tóvolli et al. 2015).

Box 2.1: Animal Models and Their Key Contributions

Our current knowledge on the role of *Shh* in the developing hypothalamus is based on work on chicken, mouse, and zebrafish. Precise surgical manipulations can be performed on chick embryos, so that very specific developmental issues can be asked. Three key contributions made on this model are those of Manning et al. (2006) on the importance of Shh repression by T-box2, Dale et al. (1997) on the cooperation between BMP and SHH to induce the hypothalamic floor plate, and the explanation of how hypothalamic regions appear through an interplay between *Fgf10* and *Shh* (Fu et al. 2017)

(continued)

Box 2.1 (continued)
(Sects. 2.6.4.4 and 2.7.3). The mouse model offers a tried-and-true way to learn about gene function: mutating genes and analyzing the resulting phenotypes on the mammalian brain. The analysis of mutant brains lacking Shh in the neuroepithelium (Szabó et al. 2009; Shimogori et al. 2010), the lineage of *Shh* in the hypothalamus (Alvarez-Bolado et al. 2012) and the function of *Gli* genes in hypothalamic development (Haddad-Tóvolli et al. 2015) are among the contributions made on this model. The zebrafish combines, in principle, the genetic accessibility of the mouse with the practicality and efficiency possible on the chick embryo. The complex relation between *Shh* and *rx3*, essential for the specification and differentiation of neuroendocrine cell fates has been unraveled in the zebrafish (Muthu et al. 2016). Needless to say, each model has its drawbacks. The chick embryo is not accessible genetically and the results of experiments cannot be evaluated on the adult brain. In mice, the questions that can be asked often lack refinement and obtaining results is a slow process. In fish, development and neuroanatomy of the hypothalamus are not as well known as they are in mammals.

2.2 *Shh* Expression in the Presumptive Hypothalamus of the Mouse (Fig. 2.2)

At E7.5 (neural fold), *Shh* is expressed in the prechordal plate, immediately underlying the midline of the neural plate (prechordal *Shh*; (Aoto et al. 2009; Echelard et al. 1993)) (Fig. 2.2a). This is followed immediately by the onset of *Gli1* expression in the midline of the overlying neural ectoderm, indicating activation of Shh signaling. *Gli1* expression appears in the midline; *Gli2* and *Gli3*, on the contrary, are expressed widely in the ectoderm, indicating competence to respond to secreted SHH (Hui et al. 1994).

At E8.5 *Shh* is expressed in a medial territory overlying the prechordal plate and extending through the presumptive mammillary and tuberal regions (the two regions forming the "basal hypothalamus," Sect. 2.1.2). This territory is the "Medial Progenitor Domain" (Sect. 2.3). This region is characterized not only by *Shh* expression like the floor plate in the spinal cord, but for the concomitant expression of *Nkx2-1* and *Bmp7* genes (while characteristic floor plate marker *Foxa2* is expressed here only transiently) (Dale et al. 1997) (Sect. 2.6.1). Expression of *Gli1* and *Gli2* overlap forming a paired lateral band (Fig. 2.2b), while *Gli3* is expressed in a peripheral, non-hypothalamic domain.

At E9.5 *Shh* expression in the ventral midline extends in the rostral direction but starts to be downregulated at the tuberal and mammillary level. In this way, a medial

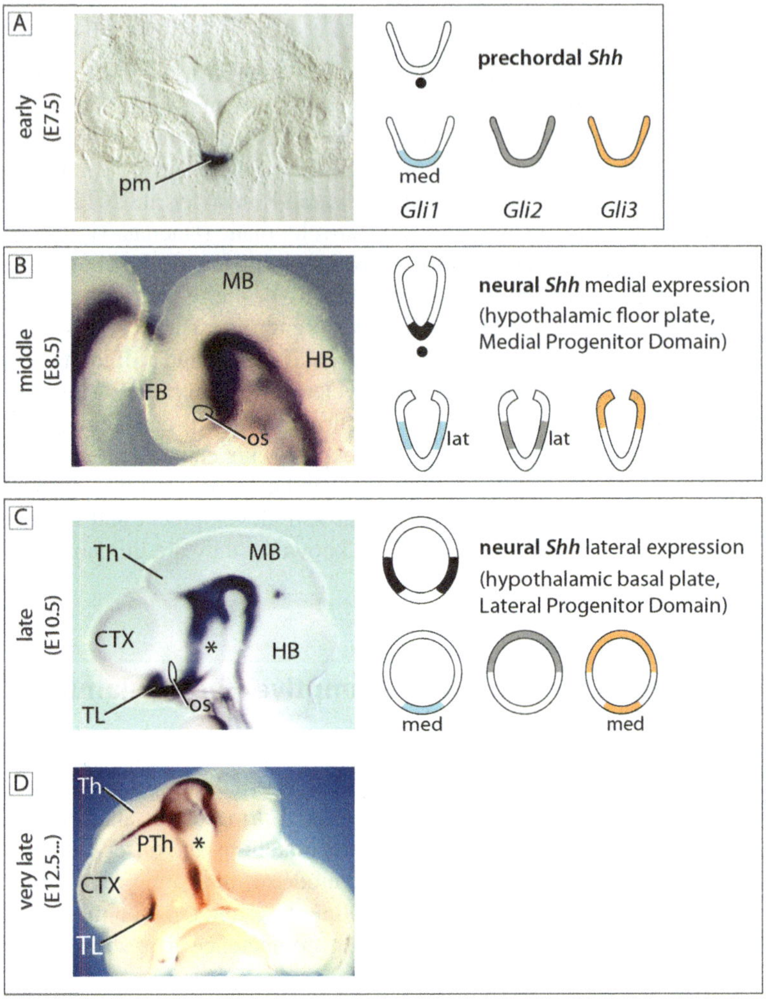

Fig. 2.2 *Shh* and *Gli* expression in the developing forebrain. In situ hybridization detection of *Shh* (left) and diagrams showing expression domains of the Gli factors and *Shh* in the presumptive hypothalamus at the early, middle, and late phases. "lat" and "med," lateral and medial domains, respectively. Diagrams (right side) of the corresponding Gli expression patterns with data from Haddad-Tóvolli et al. (2015) and Hui et al. (1994). The ages indicated (mouse embryonic days, E) are approximate (since they are based on an estimation of the time of fertilization). Note: Very similar expression patterns have been found in the mouse, chick, and zebrafish embryos. (**a**) *Shh* expression in the prechordal plate seen on a transverse section of the head folds of an E8.0 embryo. (**b–d**) *In situ* hybridization detection of Shh on wild-type mouse brains. In (**c**) and (**d**), the heads have been sagittally halved to show the neuroepithelium–ventricular zone on the inner side of the brain; rostral to the left. (**b**) *Shh* expression domain in the ventral forebrain at E8.5. (**c**) *Shh* expression is downregulated in the basal hypothalamus at E10.5. (**d**) *Shh* expression at E12.5. Asterisk (∗), basal hypothalamus; CTX, presumptive cortex; FB, forebrain; HB, hindbrain; MB, midbrain; os, optic stalk; pm, prechordal mesoderm; PTh, prethalamus; Th, Thalamus; TL, telencephalic *Shh* expression domain (modified from Blaess et al. 2014 and Haddad-Tóvolli et al. 2015)

ventral domain without *Shh* expression originates, while a lateral band of *Shh* expression (the Lateral Progenitor Domain; Sect. 2.3) appears (Fig. 2.2c). This is the "intriguing" and "dynamic" change in pattern that first attracted attention toward the role of *Shh* in the presumptive hypothalamus. The mechanistic explanation for it was obtained in the chick embryo and revealed one fundamental role of *Shh* (Manning et al. 2006) (Sects. 2.6.4.1 and 2.7.2). The lateral band of expression has been called the "hypothalamic basal plate" (as opposed to the "hypothalamic floor plate" where *Shh* was originally expressed). In chick embryos, these cells migrate "forward" from the diencephalon–mesencephalon boundary (Manning et al. 2006). A telencephalic domain of *Shh* expression appears at this stage. This is the Telencephalic Progenitor Domain and generates neurons for the cortex and basal ganglia as well as astrocytes for the preoptic region (Sect. 2.3).

At E10.5 (Fig. 2.2c), *Gli2* expression is now absent in the presumptive hypothalamus. *Gli3* and Shh-activation diagnostic marker *Gli1* overlap in the medial domain, indicating an activator function of *Gli3* (Gli3A) in the midline at this stage.

At E12.5 the general pattern of *Shh* in the forebrain does not change; however, the telencephalic domain becomes a clear separate domain (Fig. 2.2d). *Shh* expression in the forebrain peaks at this stage and starts fading, so that after E13.5 it has almost disappeared. The Lateral Progenitor Domain generates cells also at this stage (Sect. 2.3).

2.3 *Shh*-Expressing Progenitor Domains in the Hypothalamus and Their Derivatives

2.3.1 Shh Lineage and Fate-Mapping Strategy

The neuroepithelial cells expressing *Shh* are in principle stem cells, the cells that divide to originate neurons and glia for at least part of the hypothalamus. Which brain cells are born from *Shh*-expressing neuroepithelium in the ventral forebrain, i.e., which hypothalamic cells belong to the *Shh* lineage? (Alvarez-Bolado et al. 2012). In order to trace this cell lineage in mice, a mouse line carrying two different alleles was generated (Legué and Joyner 2010). The first allele encoded the recombinase *Cre* fused to the estrogen receptor (CreER) under the control of regulatory sequences of *Shh* (Shh^{CreER}); the corresponding fusion protein remains in the cytoplasm, not being able to enter the nucleus if not bound to a specific ligand, the synthetic drug tamoxifen. The second allele, inserted in a permanently active locus, encodes an easily identifiable and specific gene (e.g., β-galactosidase a.k.a. LacZ, or enhanced yellow fluorescent protein, EYFP) protected from activation by a floxed stop cassette. In mouse embryos carrying these two alleles, every cell producing SHH also produces the protein CreER. Approximately within 6 h of tamoxifen administration by injection, CreER translocates into the cell nucleus, where the Cre recombinase remains active for 24 h, enough time to eliminate the floxed stop cassette so that the cell (and its progeny) will permanently express β-galactosidase (or EYFP). To label developmental events in the mouse, tamoxifen is administered as an intraperitoneal injection to pregnant females at the desired embryonic day; injection at E8.5 (TM8.5) guarantees labeling of any cell expressing Shh between E9.0 and E10.0 and its progeny (lineage labeling).

2.3.2 The Medial Progenitor Domain (Fig. 2.3a, b)

Cells derived from *Shh*-expressing progenitors ("Shh-derived cells") marked between E8.0 and E9.0 (TM7.5) are found in the notochord and the prechordal plate and also, in very small numbers, in the midline of the mammillary and caudal tuberal region at E12.5. In the adult, these cells are found mostly in the mammillary body, with some in the posterior hypothalamus. Shh-derived cells from progenitors marked 1 day later (between E9.0 and E10.0, TM8.5; Fig. 2.3a, b) are very numerous. They occupy the entire mammillary region, the most ventral and medial portions of the tuberal region including most of the arcuate nucleus and the ventral and medial parts of the ventromedial nucleus (Fig. 2.3e, g) as well as the infundibulum and many median eminence tanycytes. This territory extends rostrally to the Retrochiasmatic Area, a small ventral and medial region between optic chiasm and arcuate nucleus representing the rostral tip of the *Shh* domain of expression in the hypothalamic ventral midline (Puelles et al. 2004; Shimamura et al. 1995). In summary, during the early and middle expression stages, Shh-derived hypothalamic cells form a caudal domain, which originates in the mammillary region and expands rostrally and ventrally into the ventral tuberal region. This distribution indicates an expansion of the *Shh*-expressing progenitor domain from the mammillary recess to a large, ventral, and caudal domain including all ventral midline neuroepithelium of the mammillary and tuberal regions (Fig. 2.3a, b, e, g). These progenitors form the Medial Progenitor Domain and co-expresses *Nkx2-1* (Dale et al. 1997). This domain is dependent on GLI2 acting as an activator (Gli2A) (Sects. 2.4 and 2.5). The Medial Progenitor Domain is largely restricted to the *Nkx2-1/Six3*-positive domain in the tuberal and mammillary regions at E12.5.

2.3.3 The Lateral Progenitor Domain

Beyond E9.5, the Medial Progenitor Domain does not express *Shh* anymore, while two novel expression domains appear (Fig. 2.2c). One of them, the Lateral Progenitor Domain, forms a paired lateral band spanning the tuberal and mammillary regions; these bands meet rostrally and ventrally immediately behind the optic fissure (Fig. 2.2c). This *Shh* expression domain has been called the hypothalamic basal plate. Analysis of mouse mutants shows that the formation of this domain is dependent on *Shh* expressed by the neuroepithelium (Sect. 2.5; Fig. 2.5g–i). From TM9.5 to TM12.5, the fate-mapped cells are found in the *Dbx1* expression domain in the tuberal hypothalamus, with the dorsal-most fate-mapped cells in the hypothalamus

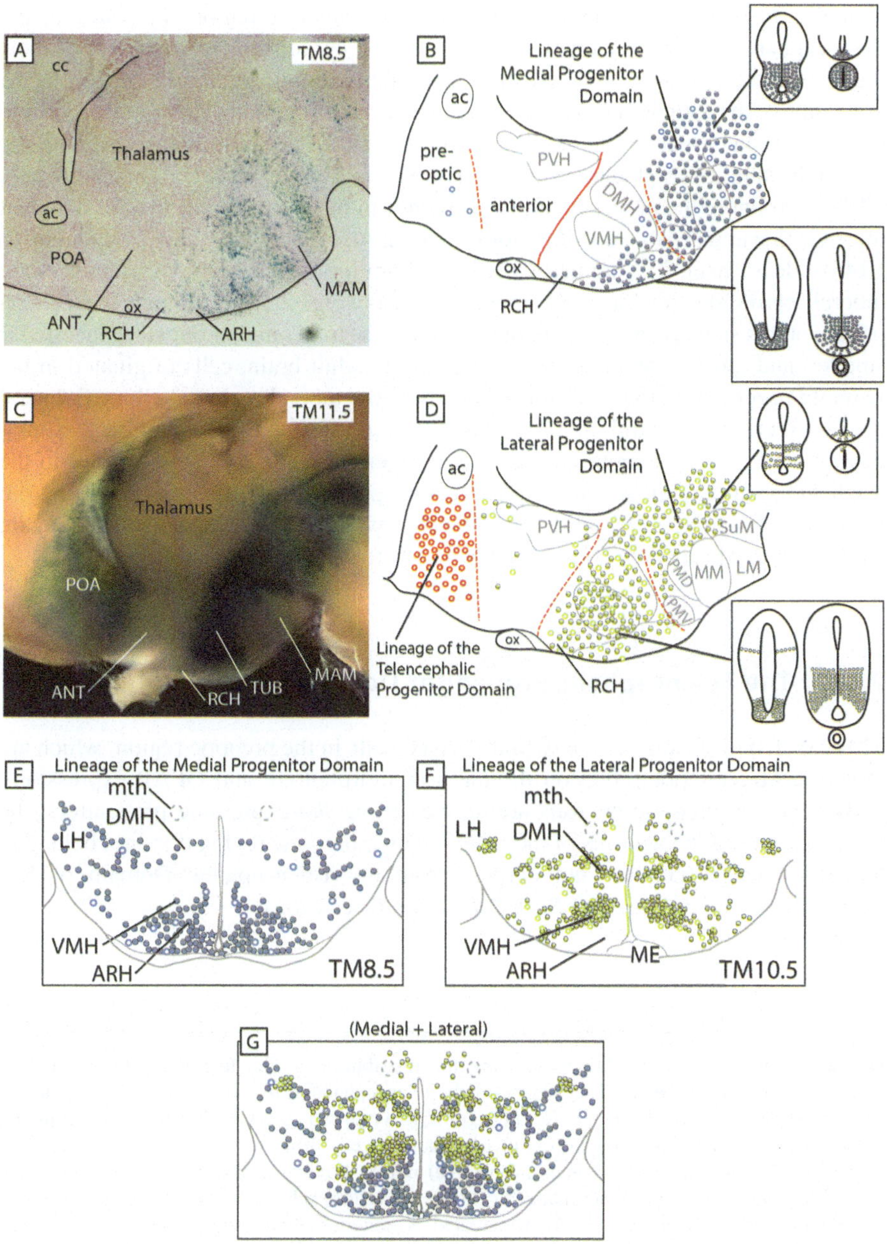

Fig. 2.3 Lineage of *Shh* in the mouse hypothalamus. (**a**) Sagittal section through the E18.5 mouse hypothalamus labeled for *Shh* lineage (X-gal reaction, blue) after tamoxifen injection at E8.5 to identify cells derived from the Medial Progenitor Domain. (**b**) Summary of A (Medial Progenitor Domain) in the adult mouse hypothalamus. Insets on the right side: Transverse sections of E12.5 mouse brains showing the distribution of the fate mapped cells at four different rostrocaudal levels. (**c**) Right side of hemi-dissected adult brain labeled for Shh-lineage (X-gal reaction, blue), after tamoxifen injection at E11.5 (Lateral Progenitor Domain). (**d**) Summary of C (Lateral Progenitor Domain) in the adult mouse hypothalamus. Insets on the right side: transverse sections of E12.5

abut the *Sim1*-positive domain localized to the anterior region. In addition, the Telencephalic Progenitor Domain starts to become obvious rostral to the eye primordium (Sects. 2.3.4 and 2.7.4). At this stage, the predominant source of Shh in the presumptive hypothalamus is the forebrain neuroepithelium (as opposed to the prechordal plate, where *Shh* expression dwindles after E10.0); this agrees with the loss of the lateral hypothalamus in neural *Shh* mutants (Szabó et al. 2009) (Sect. 2.5). GLI2 is probably the predominant Gli activator in this domain, but in *Gli2*-deficient mice GLI3 can partially substitute for it (Sect. 2.4). Finally, beyond E12.0 and up to E14.0 at least (Sect. 2.2), numerous cells are born for the lateral hypothalamus of the tuberal region and for the ventromedial nucleus (Fig. 2.3f). Although the general pattern at this stage is similar as in the previous, the total number of cells generated is smaller, and most of them are astrocytes. In the adult brain, cells originated in the Lateral Progenitor Domain are found mostly in the tuberal region at all mediolateral levels except the ventral midline (Fig. 2.3c, d, f, g). The mammillary, supramammillary, and premammillary dorsal and ventral nuclei originate mostly in the Medial Progenitor Domain (although part of the posterior hypothalamus derives from the Lateral Progenitor Domain). As is the case with the Medial Progenitor Domain, the Lateral produces only a few, scattered cells for the anterior region.

2.3.4 The Telencephalic Progenitor Domain

TM12.0–TM14.0 embryos show Shh-derived cells in the preoptic region, which are mostly astrocytes (identified on the basis of morphology and GFAP expression). Only very few preoptic neurons are derived from *Shh*-expressing progenitors; the large majority of the neurons of the hypothalamic preoptic region does not belong to the Shh lineage (i.e., are not derived from the Telencephalic Domain of *Shh* expression in the mouse) (see Sect. 2.7.4 for a discussion).

Fig. 2.3 (continued) mouse brains showing the distribution of the fate-mapped cells at four different rostro-caudal levels. The lineage of the telencephalic *Shh* domain in the hypothalamus (mostly astrocytes) is labeled in red. (**e, f**) Diagrams of a transverse section through the adult mouse tuberal region showing the distribution of fate-mapped cells from the Medial (**e**) and Lateral (**f**) Progenitor Domains. (**g**) Merged diagrams from (**e**) and (**f**) showing clear differential distribution of early (blue) and late (yellow) fate-mapped cells on the ventral side; at dorsal levels, the respective patterns of distribution are less clearly demarcated. ac, anterior commissure; ANT, anterior region; ARH, arcuate nucleus; cc, corpus callosum; DMH, dorsomedial nucleus; LH, lateral hypothalamus; LM, lateral mammillary nucleus; MAM, mammillary region; ME, median eminence; MM, medial mammillary nucleus; mth, mammillothalamic tract; ox, optic chiasm; PMD, dorsal premammillary nucleus; PMV, ventral premammillary nucleus; POA, preoptic region; PVH, paraventricular nucleus; RCH, retrochiasmatic area; SuM, supramammillary nucleus; VMH, ventromedial nucleus. In **b**, **d–g**, closed circles indicate neurons, open circles indicate astrocytes, and stars indicate tanycytes (modified from Alvarez-Bolado et al. 2012)

2.3.5 Conclusions

2.3.5.1 Differential Patterns of Neurogenesis Build Different Hypothalamic Nuclei

Since the area expressing *Shh* is very extensive, *Shh*-expressing hypothalamic progenitors do not show specific timing and rate of neurogenesis. Rather, different neurogenesis strategies are used which are typical for a given hypothalamic nucleus. This indicates that there are different *Shh*-expressing progenitor subpopulations generating neurons and glia according to specific programs. For instance, the dorsomedial nucleus and the lateral hypothalamus receive from the progenitors a slow and steady contribution of scattered cells from TM8.0 through TM12.5. The ventromedial nucleus, on the contrary, receives in a shorter time large numbers of neurons, which soon aggregate forming a recognizable entity. Additionally, neurons derived from progenitors expressing *Shh* after E10.5 or even after E11.5 (late-born cells), settle at all mediolateral levels of the tuberal hypothalamus. This indicates that, as a general rule, hypothalamic neurogenesis proceeds in a more complex way than three consecutive mediolateral waves generating the periventricular, medial, and lateral hypothalamic zones (the "three wave" hypothesis (Altman and Bayer 1986)).

2.3.5.2 The Preoptic and Anterior Regions

In the mouse preoptic region, the great majority of Shh-derived cells are astrocytes, and in the anterior region there are almost no Shh-derived cells at all. This means that neither the Medial nor the Lateral Progenitor Domains produce cells for the preoptic and anterior regions. Also, neither the preoptic nor the anterior regions receive Shh-derived neurons from caudal regions, that is, Shh-derived cells do not cross the alar–basal boundary. In this sense, the two "rostral" (alar) regions of the adult hypothalamus derive from the alar portion of the neural tube (which includes the telencephalon). This notwithstanding, both the preoptic and the anterior regions depend on Shh signaling (see below Sects. 2.5, 2.7.3, and 2.7.4).

Looking at the early *Shh* expression pattern (Fig. 2.2) it seems obvious that any part of the hypothalamus not derived from the Medial or the Lateral Progenitor Domains has to originate in the telencephalon. Expression of transcription factor *Foxg1* characterizes telencephalic progenitors (Xuan et al. 1995). The scarce data that we have on *Foxg1* lineage in the hypothalamus shows intriguing images of labeled cells in large parts of the anterior region, including the paraventricular nucleus (Kawaguchi et al. 2016). *Foxg1*-null heterozygous mutants show abnormal expression of *Oxt* and *Avp*, two neuropeptide genes specific of the anterior region (Frullanti et al. 2016). Detailed analyses on chick embryos have uncovered a detailed mechanism for the anterior region in that model (Fu et al. 2017) (see below, Sect. 2.7).

2.4 GLI Factors in the Shh Progenitor Domains of the Hypothalamus

The secreted protein SHH exerts its influence through transcription factors GLI1, 2, and 3, of which GLI2 and GLI3 can act as activators (GliA) or repressors (GliR). By comparing the hypothalamic phenotypes of mutant mice deficient in *Gli2* (*Gli2*$^{zfd/+}$, *Gli2* Zinc Finger-Deleted; (Mo et al. 1997)), *Gli3* (*Gli3*$^{Xt\text{-}J/+}$, Extra-Toes; (Hui and Joyner 1993)), or neural *Shh* and *Gli3* (*Foxb1-Cre;Shh*$^{loxP/+}$; *Gli3*$^{Xt\text{-}J/+}$; see Sect. 2.5), it is possible to deduce which factors act as GliA or GliR in the Medial and Lateral Progenitor Domains and if there are differences between them (Haddad-Tóvolli et al. 2015). We have tried to summarize a complex set of mutant phenotypes in the diagrams in Fig. 2.4.

2.4.1 The Only Role of *Shh* in the Alar Regions Is to Suppress Gli3R

The preoptic and anterior regions (alar hypothalamus) show strong expression of *Gli3*; despite this fact, the overall specification of these regions is not altered in the *Gli3* mouse mutant brain. It could be that, in the alar portions of the hypothalamus, Gli2A and Gli3A can substitute for each other. Alternatively, in these regions, the only role of *Shh* would be to suppress Gli3R; this would agree with observations that in double *Shh/Gli3* mutant mice the dorsoventral patterning of the telencephalon is appropriate (Rallu et al. 2002), and supports the hypothesis that both the preoptic and the anterior regions are telencephalic in origin (Sect. 2.7.4.2). Neither *Gli2* nor *Gli3* is required for the general patterning of the alar (anterior and preoptic) regions of the hypothalamus.

2.4.2 Differential Gli Requirements Between Medial and Lateral Progenitor Domains

In the tuberal and mammillary regions (the basal hypothalamus), *Gli2* is required for the development of the Medial Progenitor Domain (Fig. 2.4a) and its derivatives: median eminence, pituitary, most of the arcuate nucleus and ventromedial nucleus, and the mammillary and premammillary nuclei. The results show that the Medial Progenitor Domain requires nonneural *Shh* acting through Gli2A (this is in agreement with the phenotype of mice deficient in neural Shh (Shimogori et al. 2010; Szabó et al. 2009)) (Sect. 2.5). In contrast, the Lateral Progenitor Domain does not require *Gli2* (Fig. 2.4a): the lateral part of the ventromedial and the arcuate nuclei as well as the dorsomedial nucleus and the lateral hypothalamus. This could mean that GliA is not required by the Lateral Progenitor Domain. More likely, however, is that Gli3A is able to compensate in the *Gli2*$^{zfd/zfd}$ mutant. *Gli1* expression (diagnostic of Shh pathway activation) and *Gli3* expression overlap in the Lateral Progenitor Domain of the *Gli2*$^{zfd/zfd}$ mutants. In mutants deficient in neurally expressed *Shh*, Lateral Progenitor Domain derivatives disappear (Fig. 2.4a) indicating that *Shh* is

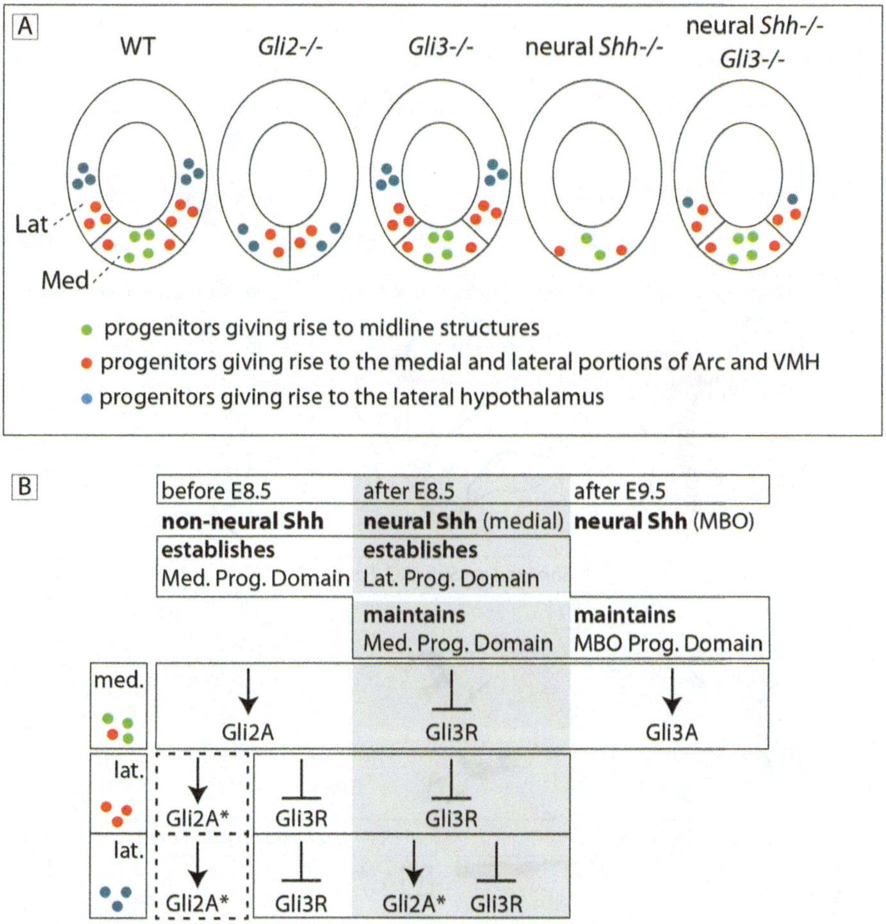

Fig. 2.4 Differential contribution of Gli factors to the Lateral and Medial Progenitor Domains. (**a**) Summary diagrams of progenitor domains (neuroepithelium) of the basal hypothalamus in wild type and mutants as deduced from the phenotypic analysis in the present study. (**b**) Contribution of Gli proteins to the specification of the medial and lateral progenitor domains in three successive stages of development. Colored dots in (**a**) and (**b**) represent different kinds of *Shh*-expressing progenitors, as indicated on the figure; dotted squares in (**b**) indicate that, although nonneural *Shh* determines the lateral domain through repression of Gli3, a direct influence of *Shh* through Gli2A on lateral progenitors before E8.5 is in principle possible (since there are some lateral progenitors in the double mutant deficient in neural *Shh* and *Gli3*). The asterisk (∗) means that loss of GliA2 could be compensated by Gli3A. MBO, mammillary body (modified from Haddad-Tóvolli et al. 2015)

needed to counter Gli3R. Therefore, the Lateral Progenitor Domain is patterned by neural *Shh* blocking Gli3R formation and either promoting Gli2A (in wild-type animals) or Gli3A (in *Gli2*-deficient mutants). In agreement, mouse mutants deficient in *Shh* expressed by the neuroepithelium do not develop a Lateral Expression Domain (Fig. 2.5g–i) and lack a lateral hypothalamus (Szabó et al. 2009) (Sect. 2.5).

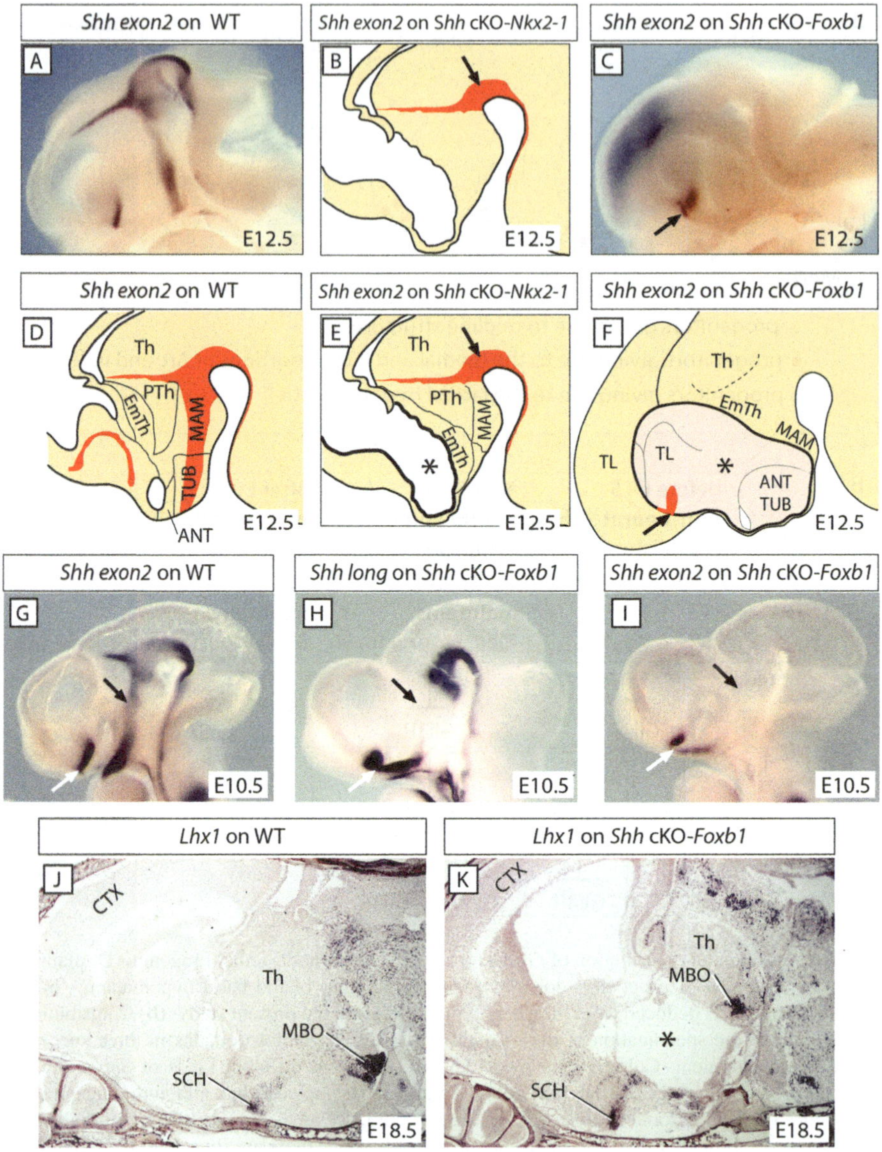

Fig. 2.5 Phenotype of mouse mutants deficient in neural *Shh*. (**a**) In situ detection of *Shh* expression with a probe detecting exon2, essential for SHH activity; E12.5 wild-type mouse embryo. (**b**) Diagram of *Shh* expression (red); E12.5 *Shh* cKO-*Nkx2-1* mouse embryo. (**c**) In situ detection of *Shh*-exon2; E12.5 *Shh* cKO-*Foxb1* embryo. (**d**) Diagram of *Shh* expression (red); sagittal section of an E12.5 wild type mouse embryo, showing forebrain regions of interest. (**e**) Diagram of *Shh* expression (red); sagittal section of an E12.5 *Shh* cKO-*Nkx2-1* mouse embryo, showing forebrain regions of interest. (**f**) Diagram of *Shh*-exon2 expression (red) on the right half of the brain of an E12.5 *Shh* cKO-*Foxb1* mouse embryo. (**g**) In situ detection of *Shh*-exon2 expression; E10.5 wild-type mouse embryo. (**h**) In situ detection of *Shh*, either intact or truncated (long probe; see text for details) expression; E10.5 5 *Shh* cKO-*Foxb1* mouse embryo. (**i**) In situ detection of *Shh*-exon2 expression; E10.5 *Shh* cKO-*Foxb1* mouse embryo. (**j**, **k**) In situ detection of *Lhx1*

2.4.3 Suppression of Gli3R by *Shh* Is Also Essential in the Medial and Lateral Progenitor Domains

Gli3 is not required for the overall specification of the hypothalamus in wild-type mice (although Gli3A is specifically required for mammillary proliferation). However, the ventral forebrain phenotypes of *Shh* mutants, be it the full *Shh* null mutant (Chiang et al. 1996) or the conditional *Shh* mutants lacking Shh expression in the presumptive hypothalamus ("neural *Shh* mutants") (Shimogori et al. 2010; Szabó et al. 2009) (Sect. 2.5), are much more severe than the $Gli2^{zfd/zfd}$ mutant phenotype. This indicates that suppression of Gli3R function by *Shh* (of neural or prechordal origin) is required for normal hypothalamic development. In support of this conclusion, the defects in the Medial and Lateral Progenitor Domains found in neural *Shh* mutants are largely rescued in double neural *Shh –/–; Gli3 –/–* mice (Fig. 2.4a).

2.4.4 *Orexin* (*Hcrt*)-Expressing Neurons of the Lateral Hypothalamus Are Particularly Dependent on GliA

The number of *Hcrt*-expressing neurons in the lateral hypothalamus is reduced in $Gli2^{zfd/zfd}$ mutants and completely lost in neural *Shh* mutants and in double neural *Shh –/–; Gli3 –/–* mutants. These results uncover a particular dependence of the progenitors of *Hcrt* neurons on GliA, maybe because they need GliA later than other progenitors in the Lateral Progenitor Domain, or for a longer time.

2.5 Role of Shh in Hypothalamic Development: Mouse Mutant Phenotypes

Box 2.2: From the Fruit Fly Larva to the Mouse Brain: The Pathway Leading to the Role of Shh in Hypothalamic Specification

The first hedgehog gene (hh) was described in the fruit fly as a gene involved in larval segmentation (Nüsslein-Volhard and Wieschaus 1980). Three homologs of hh were later found in chick, zebrafish, and mouse (Echelard

(continued)

Fig. 2.5 (continued) expression; sagittal sections of E18.5 wild type or *Shh* cKO-*Foxb1* mouse brains. Arrow in **b**, **c**, **e**, **f** shows remnant *Shh* expression domains in the conditional mutants. Arrows in **g**, **h**, and **i** show the lateral *Shh* expression domain or its theoretical location (black arrow) and the telencephalic *Shh* expression domain (white arrow). Asterisk in **e**, **f**, and **k** marks the abnormal size of the third ventricle in the mutants. ANT, anterior region; CTX, cortex; EmTh, thalamic eminence; MAM, mammillary region; MBO, mammillary body; PTh, prethalamus; SCH, suprachiasmatic nucleus; Th, thalamus; TUB, tuberal region (**b**, **d**, **e** with data from Shimogori et al. 2010)

Box 2.2 (continued)
et al. 1993; Krauss et al. 1993; Riddle et al. 1993). One of them, which received the name of *Sonic hedgehog* (*Shh*), was expressed in the floor plate of the neural tube and in the underlying axial mesoderm (notochord). Thanks to experiments on cultured embryo explants in the frog, rat, and chick model species, it was known that these two structures were essential for proper development of the ventral spinal cord (reviewed by Ruiz i Altaba and Jessell 1993). Ectopic expression experiments in transgenic mice suggested that Shh was responsible for this role (Echelard et al. 1993), and experiments on cultured explants of embryonic rat floor plate and chick spinal cord showed that Shh can specify ventral neural cell types (Roelink et al. 1994). Spectacular confirmation of the essential role of Shh in cell fate specification in the ventral neural tube at all rostro-caudal levels was provided by the phenotype of mouse embryos carrying null mutations of the *Shh* locus (Chiang et al. 1996). However, these mutants could not reveal the relative contributions to the hypothalamic development of Shh expressed by the neural tube versus Shh expressed by the prechordal plate. This question was addressed by means of conditional mutants more than a decade later (Szabó et al. 2009; Shimogori et al. 2010). Shortly after, the same approach was used to show that neural Shh is required for the appropriate development of hypophysis and eye (Zhao et al. 2012) and to specify glial and ependymal cells in the spinal cord (Yu et al. 2013).

The first *Shh* null mutant mouse (Chiang et al. 1996) confirmed the ventralizing role of *Shh* in the hypothalamus and extended it to the forebrain. However, the phenotype of this mutant was dominated by extreme holoprosencephaly, so that the entire hypothalamus together with most of the forebrain were missing. Obviously, *Shh* expression in the prechordal plate ("prechordal *Shh*") was indispensable for forebrain development. Conditional mutants eliminating *Shh* expression from the neural tube ("neural *Shh*") but not the prechordal plate have been much more informative. By "subtracting" the neural *Shh* phenotype we could also infer the role of prechordal *Shh*. The first such neural mutant removed *Shh* expression in the forebrain by using *Foxb1* regulatory sequences as the *Cre*-driver (*Foxb1-Cre* × $Shh^{fl/fl}$ or *Shh* cKO-*Foxb1* (Szabó et al. 2009). The method was used again by using a *Nkx2-1* as *Cre*-driver (*Nkx2-1-Cre* × $Shh^{loxP/loxP}$ or *Shh* cKO-*Nkx2-1* (Shimogori et al. 2010). Further observations on the role of Shh in the ventral forebrain were made using a *Cre* line driven by SBE2, an upstream regulatory element of *Shh* specific for the hypothalamus (Zhao et al. 2012). Finally, mutants of transcription factor *Six3* have thrown light on the complex role of neural *Shh* in the splitting of the eye field and telencephalic vesicle (Geng et al. 2008, 2016). These do not pertain directly to hypothalamic development (except inasmuch as the retina can be considered a hypothalamic derivative) but will be included here for the sake of completeness.

2.5.1 Summary of Mouse Mutant Phenotypes

According to the phenotype of mouse mutants deficient in *Shh* expressed in the neural tube, neural *Shh* is:

1. Not necessary to confer basic hypothalamic identity. Prechordal *Shh* must provide general hypothalamic specification, although is not enough for full growth and differentiation
2. Not necessary to specify the four hypothalamic regions and their positions relative to each other
3. Essential for the hypothalamic regions to differentiate and express their characteristic marker genes
4. Essential for the hypothalamus to reach appropriate size
5. Essential for the establishment of the Lateral Progenitor Domain and therefore for the development of the lateral hypothalamus (Sect. 2.3). Prechordal *Shh* can partially substitute for the Medial Progenitor Domain
6. Not essential for the specification of the neurohypophysis (but required for its correct development)
7. Essential for the development of the optic disk
8. Not essential for the splitting of the eye field and telencephalic vesicle, although it has a minor role in this process

2.5.2 The *Shh* cKO-*Foxb1 and Shh* cKO-*Nkx2-1* Conditional Mutants

2.5.2.1 Comparison of Mutants

The two conditional mutants removing Shh expression from the presumptive hypothalamus were generated using the same classical conditional strategy: crossing a mouse line carrying a Cre allele under the regulatory sequences of a “driver” gene expressed in an appropriate region, to a mouse line carrying an allele of *Shh* where exon 2, essential for *Shh* activity, has been floxed (Dassule et al. 2000). These mouse mutants express a truncated, inactive form of *Shh* in a specific portion of the neuroepithelium. The mutant based on *Foxb1* regulatory sequences as the *Cre*-driver (Szabó et al. 2009) abolishes SHH activity in the entire forebrain (including the presumptive hypothalamus and the zona limitans of the thalamus) as well as in the midbrain with one exception: a restricted region with weak full length *Shh* expression remains in the telencephalic *Shh* expression domain (Fig. 2.5a, c, f, i). The mutant hypothalamus was analyzed at E12.5 and at E18.5.The mutant based on *Nkx2-1* as driver (Shimogori et al. 2010) abolishes neural SHH activity in the presumptive hypothalamus and also in the telencephalic domain. Two SHH sources, the zona limitans (dorsal) and the midbrain (caudal) are not affected in this mutant (Fig. 2.5a, b, e). The mutant hypothalamus was analyzed at E10.5 and E12.5.

By using ISH probes against exon2, expression of active *Shh* can be detected, while a “long” probe based on other *Shh* sequences detects all *Shh* expression in the mutant (for instance, compare panels h and i from Fig. 2.5). The results found with both mutants agree, and some differences (but no contradictions) found are probably due to the *Cre*-driver genes used having differential expression and leaving different Shh expression domains intact (either the telencephalic domain or the midbrain domain), which in turn could partially rescue the wild-type phenotype. In addition, some of the conclusions reached may be reflective of the different ages that were analyzed.

Neural Shh is not required to specify the hypothalamus as a region Transcription factor *Nkx2-1* is specifically expressed in a ventral forebrain region encompassing the presumptive hypothalamus (Shimamura et al. 1995) and is a major determinant of hypothalamic fate and the first ventral forebrain marker to be expressed during development. The hypothalamus without neural Shh expresses *Nkx2-1*, implying that expression of *Nkx2-1* depends on prechordal *Shh* (see below, Sect. 2.6). The hypothalamus without neural Shh does not express (or only in extremely reduced domains) *Dlx2* and *Dbx1*. In both mutants, the identity of the hypothalamus is respected (Fig. 2.5d–f), that is, a dorsalizing influence did not overtake the presumptive hypothalamic region. This is based on the expression of *Foxg1*, which does not expand into diencephalon, and the expression of the transcription factor *Rax*, key for tuberal specification (Shimogori et al. 2010). In addition, the hypothalamic regions, as much as they can be recognized, are found in their appropriate anteroposterior (or alar–basal, see Sect. 2.1.2) positions.

2.5.2.2 Hypothalamic Regional Phenotypes (See Also Sect. 2.7)

In both the *Shh* cKO-*Foxb1* and the *Shh* cKO-*Nkx2-1* mutants, the mammillary region is the least affected region. At E18.5, it expresses appropriate markers such as *Sim1*, *Pitx2*, *Nkx2-1*, *Lhx1*, *Lhx5, and Emx2,* but not *Foxb1*, *Otp*. It is reduced in size, but possesses a small axonal bundle in the place of the mammillotegmental tract (Szabó et al. 2009) (Fig. 2.5j, k). Shimogori et al. (2010) report a reduction of the premammillary marker *Lef1* at E12.5, while Szabó et al. (2009) found that the premammillary region was present at E18.5 (Fig. 2.5j, k). The tuberal region is more affected in size and differentiation than the mammillary: characteristic markers of this region like *Nr5a1*, *Pomc*, and *Nkx6-2* were lost at E12.5 in the *Shh* cKO-*Nkx2-1* mutants. At E18.5, *Calb* expression identified a ventromedial nucleus reduced in size in the *Shh* cKO-*Foxb1* mutant. The hypothalamic anterior region is also reduced in these mutants. Only a few scattered cells are left that express neuropeptide *Avp* (a marker of the paraventricular nucleus, possibly the most recognizable nucleus of the hypothalamus). Also, there is attenuated expression of characteristic transcription factors *Otp* and *Sim1* (which label the paraventricular nucleus during development). None of the studies analyzes the preoptic region in depth, although in the *Shh* cKO-*Nkx2-1* it appears extremely reduced in size at E12.5.

2.5.2.3 General Growth and Dorsoventral Patterning in Neural *Shh* Mutants

Mutant analysis at E18.5 showed that an important function of neural *Shh* is to stimulate growth and, in this way, coordinate dorsoventral and mediolateral patterning (Cayuso et al. 2006; Ishibashi and McMahon 2002; Sagner and Briscoe 2017). Accordingly, the *Nkx2-1* territory (presumptive hypothalamus) is reduced in the *Shh* cKO-*Foxb1* mutant by E12.5, and both neural *Shh* mutants show a much smaller hypothalamus. At E18.5, growth has allowed for some recovery of early patterning defects (Szabó et al. 2009) although in this mutant, remnant telencephalic *Shh* expression may be contributing to this partial rescue (Sect. 2.5.2.1). The growth defect appears as a large gap (asterisk in Fig. 2.5e, f, k) in the mutant forebrain between thalamus and hypothalamus. Intriguingly, formation of the lateral *Shh* expression domain (Lateral Progenitor Domain) appears to depend on neural *Shh* and therefore that domain is absent in the mutant (black arrows in Fig. 2.5g–i). Additionally, a neuroepithelial region termed "intrahypothalamic diagonal" and expressing *Lhx6*, *Arx*, and *Nkx2-1* does not develop in the absence of neural *Shh* (Shimogori et al. 2010). This suggests that tissue growth stimulated from the lateral domain has a role in "stitching together" the alar and basal portions of the forebrain. Finally, the gap between alar and basal forebrain is larger in the *Shh* cKO-*Foxb1* mutant because of the disappearance of the *Shh* expression domain of the zona limitans, which is key for growth of the Prethalamus and Thalamic Eminence (Fig. 2.5d–f).

2.5.2.4 Mediolateral Patterning in Neural Shh Mutants

The Medial Progenitor Domain is probably present in these mutants because it can be partially rescued by prechordal *Shh* acting through *Gli2* (Sect. 2.4). The Lateral Progenitor Domain is absent in the neural Shh mutant (Fig. 2.5g–i). This leads to loss of a large part of the lateral hypothalamus; accordingly, marker genes of the lateral hypothalamic area are reduced to very few scattered cells (*Pmch*), or not present at all (*Hcrt*) (Szabó et al. 2009).

2.5.3 Hypophysis, Eye, and Holoprosencephaly Phenotypes (See Also Sects. 2.6.3.2 and 2.7.2)

Analyzing mutant mouse lines carrying mutations that decrease Shh in the neuroepithelium directly or indirectly, have noticed an important role of Shh in hypophysis development. Zhao et al. (2012) used an *SBE2-Cre* line to abolish *Shh* expression in the presumptive hypothalamus and observed an ectopic, dysmorphic, and small infundibulum (primordium of the posterior hypophysis) as well as a duplicated and not completely differentiated Rathke's pouch (primordium of the anterior hypophysis). Failure to form the infundibulum as well as a multiplicity of Rathke's pouches have been described in mice mutant for *Rax*, a transcription factor activated by *Shh* and, in turn, essential to maintain *Shh* expression in the presumptive hypothalamus (Orquera et al. 2016). In mice carrying null mutations of transcription factor *Tbx3*, *Shh* expression abnormally persists in the tuberal region beyond E9.5.

In these embryos, the infundibulum does not develop and Rathke's pouch shows decreased proliferation (Trowe et al. 2013). In conditional mutants of *Shh* using *Hex1* as a *Cre* driver, *Shh* expression in the anterior hypothalamus is lost. These mutants do not develop an infundibulum or Rathke's pouch (Carreno et al. 2017). *SBE2-Cre* × $Shh^{loxP/loxP}$ mice lack *Shh* in the presumptive hypothalamus and show lack of formation of the optic disk (the point of exit of the retinal axons from the eye) and a hypoplastic optic nerve (Zhao et al. 2012). Early in the development of the mammalian head, the prechordal plate is essential for splitting of the eye field and the primary (unpaired) telencephalic vesicle into left and right. Failure of this process leads to a major phenotype, holoprosencephaly (cyclopia). Some mouse mutants in transcription factor *Six3*, expressed in the rostral portion of the ventral telencephalon, show holoprosencephaly due to deficiency in *Shh* in that expression domain (Geng et al. 2008, 2016). However, none of the three neural *Shh* mutants (Shimogori et al. 2010; Szabó et al. 2009; Zhao et al. 2012) shows holoprosencephaly.

2.6 Interactions of *Shh* with Other Genes in the Developing Hypothalamus

2.6.1 Prechordal *Shh* and *Bmp* Genes

Experiments on chick embryos have shown that during gastrulation, the prechordal plate migrates rostrally after detaching from the node. At the same time, the presumptive hypothalamic floor plate cells migrate rostrally, at times in register with the prechordal plate, and are influenced by SHH and BMP ligands secreted by it. If SHH confers "floor plate" identity, the BMPs bestow the hypothalamic floor plate with "rostral" identity, so that they become "hypothalamic floor plate," characteristically expressing not only *Shh* but *Nkx2-1* and *Bmp7* (Dale et al. 1997; Patten et al. 2003; Placzek and Briscoe 2005) (summarized in Fig. 2.6a). In experimental conditions, SHH can by itself rostralize the forebrain floor plate (without the help of BMP), but only at very high, nonphysiological concentrations (Dale et al. 1997).

2.6.2 Wnt-Based Anteroposterior Patterning of the Neural Plate

Shh secreted by the prechordal plate acts on the overlying neural tube already patterned along the anteroposterior axis. This early patterning depends on a balance between the posteriorizing influence of Wnt ligands and the anteriorizing influence of Wnt antagonists (like Fgf8 from the anterior neural ridge (Shimamura and Rubenstein 1997)). As a consequence, the forebrain is divided into an anterior domain expressing transcription factor *Six3* and a posterior one expressing *Irx3*, meeting each other at the level of the zona limitans (Braun et al. 2003; Kobayashi et al. 2002). Repression of *Wnt1* by Six3 is required to maintain the forebrain territory and therefore the hypothalamus (Lagutin et al. 2003). Wnt antagonism is also essential for the ventral forebrain to acquire hypothalamic character (Kapsimali et al. 2004).The caudalizing Wnt signals (Braun et al. 2003) have to be consistently

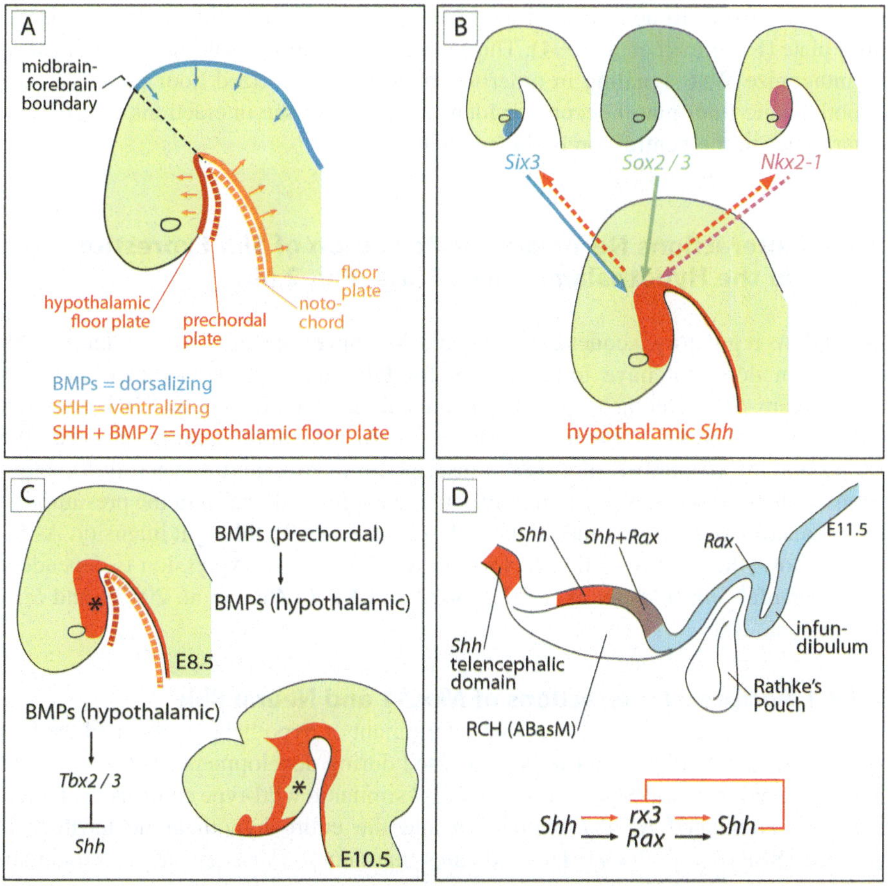

Fig. 2.6 Diagrams depicting *Shh* cooperating and interacting with other genes in the developing hypothalamus. (**a**) In most of the neural tube, SHH (orange-colored arrows) from the notochord (dashed orange-colored line) ventralizes the floor plate (solid orange-colored line), which then continues secreting SHH; BMP ligands (blue line and arrows) act as dorsalizers. Rostral to the midbrain–forebrain boundary (dashed black line), the prechordal plate (red dashed line) secretes SHH plus BMP7, which cooperate to induce the hypothalamic floor plate (red line) (data from Dale et al. 1997). (**b**) *Shh* expression by the hypothalamic floor plate (red) requires *Nkx2-1* expression (pink; at least indirectly, dashed pink arrow) as well as the direct influence (continuous arrows) of SIX3 (blue) and SOX2/3 (green) non-*Shh* regulatory sequences. SHH, in turn, maintains *Six3* and *Nkx2-1* expression (dashed red arrows) (data from Geng et al. 2008; Sussel et al. 1999; Zhao et al. 2012). (**c**) The striking change in Shh expression pattern (red) between the middle stage (top left) and the late stage (bottom right) depends on early secretion of BMP ligands from the prechordal plate (dashed red line), which activate expression of Bmp genes in the presumptive hypothalamus, leading to expression of Tbx transcription factors which directly repress Shh expression in a restricted area (asterisk) (data from Manning et al. 2006; Trowe et al. 2013). (**d**) Top: The *Shh* (red) and *Rax* (blue) expression domains in the midline of the presumptive hypothalamus by E11.5 overlap in a domain (brown) of the retrochiasmatic area (RCH or ABasM). Bottom: In the zebrafish, *Rax* homolog *rx3* is activated by *Shh* (red arrows) in order to confer specific neuroendocrine cell fates to progenitors; then rx3 is repressed by Shh influence (red lines). The interaction is partially

inhibited in order to achieve and maintain anterior character in the hypothalamic floor plate (Kapsimali et al. 2004). The early anteroposterior patterning and the need to antagonize Wnt signaling in order to obtain an anteriorized floor plate (i.e., the hypothalamic floor plate) become evident in several of the interactions of *Shh* with other genes in the ventral forebrain (see below).

2.6.3 Interactions Necessary for Activation of *Shh* Expression in the Hypothalamic Floor Plate (Fig. 2.6b)

The DNA regulatory sequence necessary to confer specific hypothalamic Shh expression does not have binding sites for Gli transcription factors (which act downstream *Shh*) (Jeong et al. 2006), indicating that prechordal SHH does not induce *Shh* expression by the overlying neuroepithelium directly. Instead, the expression of a number of other transcription factors in the ventral forebrain neuroepithelium is necessary to finally activate expression of *Shh* in the presumptive hypothalamus at the right time, in the right place, and with the right intensity. As far as we know, these transcription factors are *Nkx2-1* (whose expression is dependent on prechordal *Shh* (Dale et al. 1997)), *Sox2* and *Sox3* (Zhao et al. 2012), and *Six3* (Geng et al. 2008).

2.6.3.1 Reciprocal Interactions of *Nkx2-1* and Neural Shh

Transcription factor *Nkx2-1* is a major determinant of hypothalamic fate and the first specific hypothalamic marker to be expressed during development. At E9.5, mutant mice deficient in neural *Shh* expresses *Nkx2-1* similar to wild-type controls. At E11.5, however, attenuated *Nkx2-1* expression and the entire hypothalamic territory is reduced (Szabó et al. 2009). This indicates that neural *Shh* is essential to maintain appropriate levels of *Nkx2-1* expression in the presumptive hypothalamus, as well as for the growth of the region. Mice deficient in *Nkx2-1* do not express *Shh* in the forebrain at early stages (Sussel et al. 1999). Since the phenotype of *Nkx2-1* mutants (Kimura et al. 1996) is different and milder from that of neural *Shh* mutants (Sect. 2.5), we assume that *Nkx2-1* is not necessary for neural *Shh* expression, only for its timely onset. The complexity of the relation between these genes is illustrated by reports that in *Nkx2-1* mutants, the midline territory of expression of *Shh* in the presumptive hypothalamic neuroepithelium is expanded; in conditional mutants abolishing functional *Shh* expression in the *Hesx1* ventral forebrain expression domain, expression of *Nkx2-1* is reduced along the anteroposterior axis (Carreno et al. 2017); *Nkx2-1* expression is maintained by neural *Shh* in the telencephalic domain.

Fig. 2.6 (continued) conserved in the mouse (black arrows), where *Shh* is required to activate *Rax* expression to accomplish a similar role, and *Rax*, in turn, maintains *Shh* expression at a very specific point in time; repression of *Rax* by *Shh* is also conserved in the mouse (data from Lu et al. 2013; Muthu et al. 2016; Orquera et al. 2016)

2.6.3.2 Interaction of Neural Shh with *Six3*

One important determinant of anteroposterior patterning is transcription factor *Six3*. *Six3* is expressed in the presumptive ventral forebrain, forming a boundary with *Irx3* which represents posterior (Braun et al. 2003). *Six3* confers "anterior" character to the *Shh*-expressing forebrain ventral midline, so that *Nkx2-1* is expressed only there. *Six3,* however, is not upstream *Nkx2-1*, it just predisposes the cells to respond to prechordal *Shh* (and possibly BMPs)(Kobayashi et al. 2002). *Six3* is necessary to maintain neural *Shh* expression in the early rostral domain in the ventral forebrain. For this reason, *Six3* is (in certain genetical backgrounds) critical to prevent holoprosencephaly (Geng et al. 2008; Jeong et al. 2008). Neural *Shh* interacts with *Foxg1*, *Six3*, *Fgf8*, *Bmp4*, and *Wnt8b* for splitting the eye field and the early, unpaired telencephalic vesicle (Geng et al. 2016), and subtle alterations in gene dosage or allelic variations of one or more genes can greatly affect the general outcome. In agreement with this complexity, the cyclops phenotype appears only in *Six3* mutants of certain backgrounds and, in humans, this phenotype is not fully penetrant and shows clinical variability (Roessler and Muenke 2010).

2.6.3.3 *Sox2* and *Sox3*

Sox1, Sox2, and *Sox3* form the SoxB1 subfamily of transcription factors. The expression domain of *Sox2* and *Sox3* overlaps with that of *Shh* in the early presumptive hypothalamic neuroepithelium, and they are necessary for *Shh* expression to reach appropriate intensity. They act through direct interaction with the long-range Shh forebrain enhancer SBE2. This interaction is dose dependent, so that the dose of SoxB1 factors bound to SBE2 is key, rather than the identity of the factors (Zhao et al. 2012).

2.6.4 Interactions of Neural Shh with Other Transcription Factors and Signaling Molecules

After the expression of Shh is activated in the presumptive hypothalamus, it activates *Rax* expression (and is maintained by it). Repression of *Shh* by *Tbx2* and *Tbx3* in a restricted domain is necessary for correct development of the infundibulum. *Shh* maintains the expression of *Bmp7* and/or *Pax2* in the developing eye. Finally, mutual repression of *Fgf10* and *Shh* is emerging as crucial for the specification of hypothalamic regions.

2.6.4.1 *Tbx2* and *Tbx3* (Fig. 2.6c)

The interaction of neural *Shh* with T-box transcription factors *Tbx2* and *Tbx3* is particularly complex and interesting, because it helps explain the impressive change in expression pattern undergone by neural Shh in the tuberal and mammillary hypothalamus in the late stage (see Chap. 2) and illuminates the development of the hypophysis. In chick embryo explants (Manning et al. 2006), experimental blocking of *Tbx2* expression by means of chordin (a BMP inhibitor) prevents the repression of *Shh* expression; the resulting abnormal presence of Shh causes a cessation of cell proliferation and prevents the specific marker gene *Emx2* to be expressed in the presumptive mammillary region. In this model, it is not possible to examine subsequent alterations in hypophysis development. In the mouse, *Tbx2* is not required for hypothalamic development, but *Tbx3* shows the same interaction with Shh as in chicken. *Tbx2* and *Tbx3* could act redundantly, but *Tbx2* expression is lost in the *Tbx3* mutant. In the mouse, however, although Tbx3 is essential to repress *Shh* expression in the presumptive tubero-mammillary region, the knockout of *Tbx3* causes no obvious changes in differentiation (expression of *Emx2* and *Fgf10* are normal) but prevents the formation of the infundibulum (presumptive neurohypophysis, Trowe et al., 2013). In addition, these studies showed that *Tbx2* and *3* can repress *Shh* expression in this region directly, through two mechanisms. First, both TBX2 and TBX3 can bind to SOX2 (or SOX3) (necessary to activate Shh), in this way making it unavailable to bind to the required *Shh* long-range enhancer SBE2. Second, both can also specifically bind to long-range enhancer of Shh SFPE2 repressing *Shh* expression in this region again.

2.6.4.2 *Rax* (Fig. 2.6d)

"Retina and anterior neural fold homeobox" (*Rax*) is a transcription factor essential for eye development (Mathers et al. 1997) and is also expressed in the presumptive hypothalamic neuroepithelium. An interaction between this transcription factor and *Shh* is essential for the specification of certain hypothalamic neurons in the anterior and tuberal regions. This has been analyzed in detail in the mouse (Lu et al. 2013; Orquera et al. 2016) and in the zebrafish (Muthu et al. 2016).

In the zebrafish, *rx3* (*Rax* homolog) is first activated and then repressed by Shh signaling, important in the patterning of the neuroendocrine system. *Rx3* is essential for the specification of the progenitors of neurons expressing either *pomc*, *ff1b*, or *avp*. The *pomc*-expressing neurons are found in the periventricular tuberal nucleus, functionally analogous to the arcuate nucleus of the tuberal region in mammals; *ff1b* is a homolog of mammalian *SF1* (*Nr5a1*), expressed in the ventromedial nucleus of the tuberal region; *avp* is expressed in the neurosecretory preoptic area (functionally analogous to the periventricular nucleus of the anterior region in mammals). *Rx3* not only specifies cell fate for those specific neuronal populations of the anterior/tuberal region of the zebrafish hypothalamus, but is essential for their survival, proliferation, and migration. At the same time, *rx3* opposes the dorsomedial nucleus and the neurohypophysis fates in anterior/tuberal progenitors. The effect of *rx3* on the specification of *avp* progenitors could be mediated by activation of *otpb* (*Otp* homolog), which in zebrafish is essential for the specification of neurons in the anterior/tuberal regions (Muthu et al. 2016).

In the mouse, *Rax* expression is activated by neural Shh and, in turn, *Rax* is essential to maintain neural *Shh* expression in the most medial domain of the presumptive hypothalamus at a very precise time-point (E8.5). Timed-conditional inactivation experiments show that, after E8.5, *Rax* is dispensable for hypothalamic *Shh* expression in the mouse (Orquera et al. 2016). The same report indicates that there is an early requirement for *Rax* in the specification and proliferation of the presumptive tuberal neuroepithelium. The effects of *Rax* inactivation on hypothalamic development could be due to lack of neural *Shh* (since *Rax* is needed at E8.5 to keep *Shh* activated from then on): reduced proliferation leading to thinning of the midline forebrain neuroepithelium and/or reduced expression of proneural genes *Ascl1* and *Ngn3*. Additionally, upon losing neural *Shh* expression (in the *Rax* mutant), the rostral expression boundary of markers like *Otx2*, *Tbx3*, and *Fgf10* shifts rostrally. Perhaps as a consequence, ectopic Rathke's pouches (presumptive anterior hypophysis) are found in these mutants (same as in neural Shh mutants (Zhao et al. 2012)) and the infundibulum is missing (see Sect. 2.7.2).

About the role of *Rax* in progenitor specification there is some disagreement. One report including a very detailed description of the *Rax* expression domain finds that this gene is not expressed in the presumptive neuroepithelium of the anterior region (Lu et al. 2013). As a consequence, *Rax* does not specify neurons of the anterior region (i.e., *Avp*- or *Oxt*-expressing neurons, for instance). By other reports, *Rax* is indeed expressed in the presumptive neuroepithelium of the anterior region (Orquera et al. 2016; Shimogori et al. 2010). Accordingly, Orquera et al. (2016) find a reduction in the expression of *Six6*, *Lhx1*, *Nkx2-2*, which are markers of the suprachiasmatic nucleus. Both analyses of *Rax* mutants find that Rax is essential for the specification of some progenitors of the arcuate nucleus. Orquera et al. (2016) report a reduction in markers found in this nucleus of the tuberal region including *Isl1*, *Pomc*, and *Th* in neuroendocrine neurons as well as *Sst* (a marker of some non-neuroendocrine arcuate neurons; reviewed in (Alvarez-Bolado 2018)). According to Lu et al. (2013), *Rax* is essential to confer fate specification to the progenitors of *Pomc*-expressing neurons in the arcuate nucleus and of *SF1 (Nr5a1)*-expressing neurons in the ventromedial nucleus, both in the tuberal region.

Contrary to what happens in zebrafish, in the mouse, *Rax* is not involved in the development of the anterior region (i.e., not involved in the specification of *Avp*-expressing neurons of the paraventricular nucleus). This may be due to *Rax* having an ancestral developmental mechanism in fishes and a phylogenetically more recent and more restricted function in mammals. This is consistent with the fact that *rx3* acts through activation of *optb* (*Otp* homolog), a transcription factor essential for anterior/tuberal region in the fish. In the mouse, however, *Otp* is specific of anterior fates, particularly in the paraventricular and supraoptic nuclei (although it also regulates the development of *Sst*-expressing neurons of the arcuate nucleus (Acampora et al. 1999)). For the sake of completeness, it is worth mentioning that *Rax* has an important role in tanycyte differentiation (Miranda-Angulo et al. 2014; Salvatierra et al. 2014) and a tanycyte phenotype has not been reported in *Shh* mutants.

2.6.4.3 Interaction of Neural *Shh* with *Bmp7* and *Pax2* in the Development of the Optic Nerve

Neural *Shh* is required to maintain the expression of *Bmp7* and/or *Pax2* in the ventral region of the developing eye so that retinal axons can exit the eye and form the optic nerve (Zhao et al. 2012).

2.6.4.4 *Fgf10*

Both *Fgf10* and *Shh* are expressed in the presumptive hypothalamic neuroepithelium and the territories of expression of the two markers tend to be complementary (rather than overlapping), so that they seem to repress each other, in a direct or indirect way. This expression pattern is key to the development of the hypophysis in chick embryos (Fu et al. 2017) (see Sect. 2.7.2), and, at least in this last model, seems key to the specification and growth of the anterior region (see Sect. 2.7.3).

2.7 Shh in Hypothalamic Regions and Hypophysis

2.7.1 Neural Shh in the Tuberal Region

Together with the mammillary, the tuberal region forms the basal portion of the hypothalamus (Sect. 2.1.2) and most of it derives from Shh-expressing neuroepithelium (Sect. 2.3). The ventromedial, dorsomedial, and arcuate nuclei are anatomical landmarks of this region. The arcuate nucleus in particular is easy to recognize by its functionally important neuronal subpopulations expressing *Pomc* and *Carptp*, *Npy*, and *Agrp* or *Th* and *Ghrh*. The medial portions of the arcuate and ventromedial nuclei derive from the medial progenitor domain, and the rest derives from the Lateral Progenitor Domain (Sect. 2.3). In neural *Shh* mutants, the tuberal region is reduced in size and mostly undifferentiated (Sect. 2.5). Transcription factors essential for tuberal development and depending on neural Shh for activation or maintenance of expression are *Nkx2-1* and *Rax* (Sects. 2.6.3.1 and 2.6.4.2). Mice missing *Nkx2-1* lack differentiation of the tuberal and the mammillary regions, which appear to be, however, not much smaller than in wild-type brains (Kimura et al.1996). Of note, the mammillary region of the *Nkx2-1* mutant is more affected than that of neural Shh mutants, while the anterior region is not affected in *Nkx2-1* mutants but very much in neural *Shh* mutants (see below). In *Rax* mutant mice, the arcuate nucleus particularly, and also the ventromedial nucleus, are missing specific cell populations (Lu et al. 2013; Orquera et al. 2016). The *Rax* homolog *rx3* is essential for the development of the anterior/tuberal region in zebrafish (Muthu et al. 2016).

2.7.2 Neural Shh in the Developing Hypophysis

A series of elegant experiments on chick embryos, supported by analysis of specific mouse mutant phenotypes, has led to a description of how successive phases of activation and repression of *Shh,* tightly regulated in time and space, dynamically regulate the development of adenohypophysis and neurohypophysis. The role of Shh in the development of the hypophysis (infundibulum and Rathke's pouch) can be tentatively subdivided into three steps, of which the first and second are consecutive and the second and third approximately simultaneous.

First, *Shh* is expressed in the entire presumptive tuberal region (i.e., in the RDVM). In the chick, this expression it is required for the development of a group of anterior ventral midline cells called "collar cells." These are characterized by *Fgf3* and *Sox3* expression and have a key role in the formation of the infundibulum (Pearson et al. 2011). Next, *Shh* is repressed by *Tbx2* (*Tbx3* in mammals) (Manning et al. 2006; Trowe et al. 2013) in order for the differentiation of the tuberal and mammillary regions to proceed (Manning et al. 2006), as well as for infundibulum morphogenesis (Trowe et al. 2013). Therefore, without *Shh*, no infundibulum develops. But, if *Shh* is expressed in the presumptive tuberal–mammillary region for too long, then no infundibulum develops either. Since *Shh* expression is maintained by several factors, in the absence of these, no *Shh*, therefore no infundibulum: e.g., without *Sox3*, no *Shh*, no infundibulum (Zhao et al. 2012); without *Rax*, no *Shh*, no infundibulum (Orquera et al. 2016). *Shh* is necessary but not sufficient for hypophysis development, as other factors are also needed. For instance, as has been observed, in the *Nkx2-1* mutant, *Shh* is expanded, but this does not rescue the hypophysis phenotype in this mutant (Carreno et al. 2017). Finally, *Shh* is required to specify Rathke's pouch progenitors in the chick (Fu et al. 2017) as well as in the mouse (Carreno et al. 2017), identified by expression of *Lhx3* and *Lhx4*. Again, Shh is required but not sufficient to specify them, since BMP and FGF signaling are also required (Carreno et al. 2017).

2.7.3 Neural Shh in the Anterior Region

The anterior region is in the alar portion of the hypothalamus (Sect. 2.1.2) and its characteristic nuclei are the paraventricular, the supraoptic, and the suprachiasmatic. According to mouse mutant phenotypes (Sect. 2.5), the development of the anterior region is strongly dependent on neural *Shh* (Sect. 2.3). In zebrafish, transcription factor *rx3* patterns in this region (Muthu et al. 2016). In the mouse, *Rax* (*rx3* homolog) is expressed in the anterior region (Shimogori et al. 2010; Pak et al. 2014) and it could have a role in anterior patterning (Sect. 2.6.4.2). The anterior region does not derive from *Shh*-expressing neuroepithelium in the mouse, i.e., it does not originate in the basal hypothalamus (Alvarez-Bolado et al. 2012). In the chicken, however, anterior progenitors derive from *Fgf10*- and *Shh*-expressing midline progenitors present in the presumptive tuberal region (Fu et al. 2017). The most rostral of these progenitors downregulate *Fgf10* and move (or proliferate) forward (or in the alar direction) abandoning the presumptive tuberal region, downregulate *Shh* and proliferate intensively, generating the anterior region (Fu et al. 2017). In this way, the anterior region of the chick would indeed belong to the *Shh* lineage.

In both chicken and mouse, the rostral border (alar border) of the Shh-expressing domain in the presumptive hypothalamus abuts the *Foxg1* (*BF1*) domain that marks the telencephalon (Fu et al. 2017; Shimamura et al. 1995). Since in the mouse the anterior region is not Shh-derived, it could be *Foxg1*-derived, i.e., telencephalic, like the preoptic region. There is scarce information on the lineage of *Foxg1* in the forebrain, but at least part of the anterior region seems to belong to the *Foxg1* lineage, including part of the paraventricular and/or supraoptic nuclei (Kawaguchi et al. 2016). The *Foxg1* mutant mouse shows a phenotype in the magnocellular neuroendocrine system (Frullanti et al. 2016). Maybe a clue can be found in the fact that the radial glial fibers that presumably guide the migration of newborn neurons originate in telencephalon as well as diencephalon and they end in the preoptic and anterior regions (Tobet et al. 1995).

2.7.4 Neural Shh in the Preoptic Region

In terms of neuronal nuclei and connections, the preoptic region is the most complex and diverse. Furthermore, several of these nuclei are sexually dimorphic. The preoptic region is involved in the regulation of neuroendocrine secretion (through numerous efferents to neuroendocrine nuclei of the anterior and tuberal regions), cardiovascular responses, fluid homeostasis, sleep, body temperature, and sexual behavior (Simerly 2004). Regulation of preoptic region patterning and development is not well understood. The morphological complexity and the lack of neuroendocrine nuclei with obvious neuropeptide markers make it difficult to study. These problems are compounded by a dilemma. The preoptic region was considered a part of the hypothalamus since it was first defined (Le Gros Clark 1938) and certainly, the hypothalamus is a singular brain region with specific functions (Thompson and Swanson 2003). However, the preoptic region is now considered part of the telencephalon (Puelles et al. 2012). The problem of the "clash of nomenclatures" between development and adult has been commented before (Zhang and Alvarez-Bolado 2016). Unfortunately, when dealing with the hypothalamus, the preoptic region is considered telencephalic and therefore often left out of the scope of the paper (Shimogori et al. 2010; Fu et al. 2017), while papers dealing with telencephalic development do not mention the preoptic region since it is hypothalamic (Xu et al. 2008). The net result being that we do not know much about the patterning of this functionally important region.

2.7.4.1 *Shh* Expression in the Developing Preoptic Region of the Mouse

In the embryonic forebrain, the preoptic region is immediately rostral to the optic recess and ventral to the medial ganglionic eminence (Fig. 2.1). The onset of Shh expression in the ventral telencephalon takes place at E8.5, simultaneously with Nkx2-1. Later on, Shh is expressed in the subventricular zone and in the mantle (in the hypothalamic primordium *Shh* is expressed only by the neuroepithelium) (Shimamura et al. 1995; Sussel et al. 1999). Activation of *Shh* expression in this region is dependent on *Nkx2-1* (Sussel et al. 1999). The role of Shh in the basal telencephalon probably does not contribute to patterning but to the maintenance of appropriate gene expression (Machold et al. 2003). For instance, maintenance of proper levels of *Nkx2.1* expression in the progenitors of the medial ganglionic eminence depends on *Shh* (Xu et al. 2005). At E13.5, the preoptic region is characterized by coexpression of *Nkx2-1*, *Nkx2-2,* and *Shh*, and has two different progenitor domains (one of them co-expresses *Nkx6-2*, *Dbx1*, and *Lhx2* and the other does not) (Flames et al. 2007).

2.7.4.2 Shh-Derived Cells in the Preoptic Region

The neuroepithelium of the preoptic region is characterized by *Shh* expression but, despite this fact, the only *Shh*-derived cells found in this region at the end of development, are astrocytes and some very few scattered neurons (Alvarez-Bolado et al. 2012). The reason is that the preoptic progenitor domains produce interneurons for the amygdala (Hirata et al. 2009), the cortex (Gelman et al. 2009), and the lateral ganglionic eminence (Marin et al. 2000) as well as oligodendrocytes for the optic nerve and possibly for more caudal hypothalamic regions (Gao and Miller 2006; Ono et al. 2017) under the control of *Shh* (Merchán et al. 2007). Where do the preoptic neurons originate? As for the origin of the neurons of the anterior region (Sect. 2.7.3), the answer can be given by *Foxg1*-lineage analysis. The sparse data seem to indicate that the preoptic region and part of the anterior are of telencephalic origin (Kawaguchi et al. 2016).

2.7.4.3 Mutant Phenotypes and Experimental Approaches to the Preoptic Region

The role of neural *Shh* in the preoptic region is not known. In the *Shh* cKO-*Foxb1* mutant this region was not very affected (Szabó et al. 2009) possibly because expression in the telencephalic domain was not completely removed. The *Nkx2-1* mutant does not maintain the expression of *Shh*. For this reason, part of its phenotype could be due to lack of *Shh*. To judge from the phenotype of the *Nkx2-1* mouse (Kimura et al. 1996), the septal region including the nucleus of the diagonal band, derives from this region, together with the "preoptic areas." Sussel et al. (1999) found that the telencephalic domain of *Nkx2-1* gives rise to medial ganglionic eminence, septum and preoptic region. Phenotypical analyses of *Nkx2-1* mutants (Kimura et al. 1996; Sussel et al. 1999) do not include neuroanatomical descriptions of the mutant preoptic area. Analysis of the development of the Shh-expressing ventral midline in chick embryos indicate that the most rostral hypothalamus is the anterior region; the *Shh–Foxg1* boundary is considered the rostral boundary of the hypothalamus (Shimogori et al. 2010; Fu et al. 2017), in this way excluding the preoptic region.

2.7.5 Neural Shh in the Mammillary Region

The mammillary region is the most caudal (or the most basal) of the hypothalamic regions. Although it does not have the neuropeptide-expressing neuroendocrine nuclei that characterizes the anterior and tuberal regions, the mammillary region is morphologically very distinctive with the large mammillary body and its prominent axonal tree forming the mammillothalamic and mammillotegmental branches. In the mouse, the mammillary region starts receiving *Shh*-expressing medial progenitor domain cells at E8.0, peaks from E9.0 to E10.0 and reduced numbers of cells keep being added through E12.0 (Alvarez-Bolado et al. 2012). In the chicken, this process is different. *Fgf10*-expressing progenitors in the tuberal region proliferate beyond their caudal (or basal) boundary; then they downregulate *Fgf10*, upregulate specific mammillary marker *Emx2,* and resume proliferation to form the mammillary region (a similar process takes place in the anterior region; Sect. 2.7.3) (Fu et al. 2017).

The mammillary body and premammillary nuclei show differential needs for neural *Shh.* In the absence of neural *Shh* in this region, expression of a specific marker of the mammillary body, *Foxb1,* is not maintained. Yet, other specific markers of this nucleus (*Emx2, Lhx1, Nkx2-1, Sim1*) are still expressed (although in very reduced domains). The premammillary area, however, is more affected: it loses specific marker expression at E12.5 and by E18.5 the entire area seems to have disappeared. Note: in the *Nkx2-1* mutant brain, the mammillary region does not differentiate at all (Kimura et al. 1996), that is, it is much more affected than in neural *Shh* mutants. In chick embryos, repression of *Shh* by *Tbx2* in the tubero-mammillary region is necessary for *Emx2* to be expressed, that is, for the differentiation and further proliferation of specific mammillary progenitors (Manning et al. 2006). Analysis of *Tbx3* mutants shows that, in the mouse, inhibition of *Shh* expression in this region is not necessary for *Emx2* to be expressed (Trowe et al. 2013). Gli3A is specifically required for the proliferation of the mammillary body (Haddad-Tóvolli et al. 2015).

2.8 Future Goals

In the near future, it would be interesting to see progress in at least these four areas: the role of the prechordal plate in hypothalamic specification and patterning, the origin of hypothalamic regions, the relation between the Shh pathway and neuroepithelial cilia and the genomic networks regulating hypothalamic neuron specification.

Shh in the Prechordal Plate

The realization that the main function of neural Shh in the developing hypothalamus is to counteract Gli3R function leads back to the patterning role of Shh secreted by the prechordal plate, as well as to the patterning of the prechordal plate itself. Much remains to be done in the molecular genetic control, for instance, are there specific enhancers to govern Shh expression in the prechordal plate? Which proteins activate such enhancer(s)? What molecular interactions pattern the prechordal plate (and so pre-pattern the hypothalamus)?

The phenomenology of prechordal plate formation and migration as well as its changing influence on the overlying presumptive hypothalamic floor has been gleaned mostly from studies in chicken embryos. Therefore, an important question is how similar to each other are the prechordal plates of Birds and Mammals? Rabbit embryos show a flat gastrula (as opposed to the egg-cylinder gastrula of the mouse) accessible to manipulation. The mechanistic analysis of this model is in its beginnings but it is starting to yield unexpected and interesting results (Kremnyov et al. 2018).

Role of *Shh* in the Specification of the Four Hypothalamic Regions
How do the hypothalamic regions develop? Studies in the chick have shown how the regions expand rostrally and caudally from an initial tuberal region. This "anisotropic growth model" (Fu et al. 2017; see above Sect. 2.7.3) has explanatory power and is supported by careful experiments. Our Shh lineage data suggests that there could be some intriguing mechanistic differences, however, between birds and mammals. Again here, the rabbit flat gastrula could be very useful and has the potential to become a long-range research project. More specifically, we do not know near enough about the origin of the cells of the preoptic and anterior hypothalamic areas (location of progenitors and migration pathways) and what exactly is the role of Shh in their specification.

What Is the Role of the Neuroepithelial Cilia in Hypothalamic Development?
The Shh membrane receptor and downstream signaling components of the Shh pathway are located to the primary cilium (Huangfu et al. 2003). The importance of the primary cilium of neuroepithelial cells for the patterning function of Shh on the basal forebrain is only starting to be investigated (Andreu-Cervera et al. 2019) and has the marks of a fruitful research field for the future.

Shh and the Developmental Genomic Regulatory Networks of the Hypothalamus
The question about the complexity of the brain (including the hypothalamus) boils down to a great extent to the question of the regionalization and specification of the neuroepithelium. A great deal is already known about the regionalization of the neural plate, including the presumptive hypothalamus by Hh, Fgf, and Bmp ligands. In the mouse and fish, analysis of mutants guided by microsurgical and explant experiments in chicken will be needed, as well as expression resources, bioinformatics approaches and mutant analysis. Ultimately, the entire genomic regulatory network underpinning the regionalization and specification of the neuroepithelium and cascading from Shh and other factors (Wnt, BMP, Fgf in particular) will have to be revealed.

Key References

Alvarez-Bolado et al. (2012)—Fate-mapping shows that the Shh-expressing domains of the presumptive hypothalamus generate the tuberal and mammillary regions (but not the preoptic and anterior regions).

Chiang et al. (1996)—Inactivation of Shh in transgenic mice confirmed that Shh is required for the formation of ventral structures (including the entire hypothalamus) in the central nervous system.

Dale et al. (1997)—Experiments on chick embryos show that the hypothalamic floor plate is specified by the prechordal mesoderm.

Geng et al. (2008)—A network of essential regulators controls the partition of the telencephalon into left and right hemispheres. As a consequence, in certain genetic backgrounds, neural Shh is required for this process to be successful.

Haddad-Tóvolli et al. (2015)—Analysis of mouse mutant phenotypes shows how different combinations of Gli transcription factors, acting downstream Shh signalling, are involved in the specification of the four classical hypothalamic regions.

Manning L et al. (2006)—The mechanism behind the impressive change in the pattern of Shh expression in the hypothalamus, and its role, explained.

Szabó et al. (2009)—Thirteen years after the full Shh null mutant (Chiang et al. 1996), the role of neural Shh vs prechordal Shh in hypothalamic development were dissected by using conditional mouse mutants.

Shimogori et al. (2010)—A second conditional mutant analyzing the role of Shh from different sources (neural or prechordal) on the developing hypothalamus.

Zhao et al. (2012)—The authors use a novel conditional mutant to show that neural Shh is essential for the proper development of the eye and hypophysis, a role that had escaped previous analyses.

References

Acampora D, Postiglione MP, Avantaggiato V, Di Bonito M, Vaccarino FM, Michaud J, Simeone A (1999) Progressive impairment of developing neuroendocrine cell lineages in the hypothalamus of mice lacking the Orthopedia gene. Genes Dev 13:2787–2800

Altman J, Bayer SA (1986) The development of the rat hypothalamus. Adv Anat Embryol Cell Biol 100:1–178

Alvarez-Bolado G (2018) Development of neuroendocrine neurons in the mammalian hypothalamus. Cell Tissue Res. https://doi.org/10.1007/s00441-018-2859-1

Alvarez-Bolado G, Paul FA, Blaess S (2012) Sonic hedgehog lineage in the mouse hypothalamus: from progenitor domains to hypothalamic regions. Neural Dev 7:4. https://doi.org/10.1186/1749-8104-7-4

Andreu-Cervera A, Anselme I, Karam A, Laclef C, Catala M, Schneider-Maunoury S (2019) The ciliopathy gene *Ftm/Rpgrip1l* controls mouse forebrain patterning via region-specific modulation of Hedgehog/Gli signaling. J Neurosci 39(13):2398–2415. https://doi.org/10.1523/J.NEUROSCI.2199-18

Aoto K, Shikata Y, Imai H, Matsumaru D, Tokunaga T, Shioda S, Yamada G, Motoyama J (2009) Mouse Shh is required for prechordal plate maintenance during brain and craniofacial morphogenesis. Dev Biol 327:106–120. https://doi.org/10.1016/j.ydbio.2008.11.022

Blaess S, Szabó N, Haddad-Tovolli R, Zhou X, Alvarez-Bolado G (2014) Sonic hedgehog signaling in the development of the mouse hypothalamus. Front Neuroanat 8:156. https://doi.org/10.3389/fnana.2014.00156

Braun MM, Etheridge A, Bernard A, Robertson CP, Roelink H (2003) Wnt signaling is required at distinct stages of development for the induction of the posterior forebrain. Development 130:5579–5587

Briscoe J, Thérond PP (2013) The mechanisms of Hedgehog signalling and its roles in development and disease. Nat Rev Mol Cell Biol 14:416–429. https://doi.org/10.1038/nrm3598

Carreno G, Apps JR, Lodge EJ, Panousopoulos L, Haston S, Gonzalez-Meljem JM, Hahn H, Andoniadou CL, Martinez-Barbera JP (2017) Hypothalamic sonic hedgehog is required for cell specification and proliferation of LHX3/LHX4 pituitary embryonic precursors. Development 144:3289–3302. https://doi.org/10.1242/dev.153387

Cayuso J, Ulloa F, Cox B, Briscoe J, Marti E (2006) The Sonic hedgehog pathway independently controls the patterning, proliferation and survival of neuroepithelial cells by regulating Gli activity. Development 133:517–528

Chiang C, Litingtung Y, Lee E, Young KE, Corden JL, Westphal H, Beachy PA (1996) Cyclopia and defective axial patterning in mice lacking Sonic hedgehog gene function. Nature 383:407–413

Dale JK, Vesque C, Lints TJ, Sampath TK, Furley A, Dodd J, Placzek M (1997) Cooperation of BMP7 and SHH in the induction of forebrain ventral midline cells by prechordal mesoderm. Cell 90:257–269

Dassule HR, Lewis P, Bei M, Maas R, McMahon AP (2000) Sonic hedgehog regulates growth and morphogenesis of the tooth. Development 127:4775–4785

Dessaud E, McMahon AP, Briscoe J (2008) Pattern formation in the vertebrate neural tube: a sonic hedgehog morphogen-regulated transcriptional network. Development 135:2489–2503

Echelard Y, Epstein DJ, St-Jacques B, Shen L, Mohler J, McMahon JA, McMahon AP (1993) Sonic hedgehog, a member of a family of putative signaling molecules, is implicated in the regulation of CNS polarity. Cell 75:1417–1430

Ericson J, Muhr J, Placzek M, Lints T, Jessell TM, Edlund T (1995) Sonic hedgehog induces the differentiation of ventral forebrain neurons: a common signal for ventral patterning within the neural tube. Cell 81:747–756

Flames N, Pla R, Gelman DM, Rubenstein JL, Puelles L, Marin O (2007) Delineation of multiple subpallial progenitor domains by the combinatorial expression of transcriptional codes. J Neurosci 27:9682–9695. https://doi.org/10.1523/JNEUROSCI.2750-07.2007

Frullanti E, Amabile S, Lolli MG, Bartolini A, Livide G, Landucci E, Mari F, Vaccarino FM, Ariani F, Massimino L, Renieri A, Meloni I (2016) Altered expression of neuropeptides in FoxG1-null heterozygous mutant mice. Eur J Hum Genet 24:252–257. https://doi.org/10.1038/ejhg.2015.79

Fu T, Towers M, Placzek MA (2017) Fgf10+ progenitors give rise to the chick hypothalamus by rostral and caudal growth and differentiation. Development 144:3278–3288. https://doi.org/10.1242/dev.153379

Gao L, Miller RH (2006) Specification of optic nerve oligodendrocyte precursors by retinal ganglion cell axons. J Neurosci 26:7619–7628. https://doi.org/10.1523/JNEUROSCI.0855-06.2006

Gelman DM, Martini FJ, Nóbrega-Pereira S, Pierani A, Kessaris N, Marín O (2009) The embryonic preoptic area is a novel source of cortical GABAergic interneurons. J Neurosci 29:9380–9389. https://doi.org/10.1523/JNEUROSCI.0604-09.2009

Geng X, Speirs C, Lagutin O, Inbal A, Liu W, Solnica-Krezel L, Jeong Y, Epstein DJ, Oliver G (2008) Haploinsufficiency of Six3 fails to activate Sonic hedgehog expression in the ventral forebrain and causes holoprosencephaly. Dev Cell 15:236–247. https://doi.org/10.1016/j.devcel.2008.07.003

Geng X, Acosta S, Lagutin O, Gil HJ, Oliver G (2016) Six3 dosage mediates the pathogenesis of holoprosencephaly. Development 143:4462–4473. https://doi.org/10.1242/dev.132142

Haddad-Tóvolli R, Paul FA, Zhang Y, Zhou X, Theil T, Puelles L, Blaess S, Alvarez-Bolado G (2015) Differential requirements for Gli2 and Gli3 in the regional specification of the mouse hypothalamus. Front Neuroanat 9:34. https://doi.org/10.3389/fnana.2015.00034

Hirata T, Li P, Lanuza GM, Cocas LA, Huntsman MM, Corbin JG (2009) Identification of distinct telencephalic progenitor pools for neuronal diversity in the amygdala. Nat Neurosci 12:141–149. https://doi.org/10.1038/nn.2241

Huangfu D, Liu A, Rakeman AS, Murcia NS, Niswander L, Anderson KV (2003) Hedgehog signalling in the mouse requires intraflagellar transport proteins. Nature 426:83–87. https://doi.org/10.1038/nature02061

Hui CC, Joyner AL (1993) A mouse model of greig cephalopolysyndactyly syndrome: the extra-toesJ mutation contains an intragenic deletion of the Gli3 gene. Nat Genet 3:241–246

Hui CC, Slusarski D, Platt KA, Holmgren R, Joyner AL (1994) Expression of three mouse homologs of the Drosophila segment polarity gene cubitus interruptus, Gli, Gli-2, and Gli-3, in ectoderm- and mesoderm-derived tissues suggests multiple roles during postimplantation development. Dev Biol 162:402–413. https://doi.org/10.1006/dbio.1994.1097

Ishibashi M, McMahon AP (2002) A sonic hedgehog-dependent signaling relay regulates growth of diencephalic and mesencephalic primordia in the early mouse embryo. Development 129:4807–4819

Jeong Y, El-Jaick K, Roessler E, Muenke M, Epstein DJ (2006) A functional screen for sonic hedgehog regulatory elements across a 1 Mb interval identifies long-range ventral forebrain enhancers. Development 133:761–772. https://doi.org/10.1242/dev.02239

Jeong Y, Leskow FC, El-Jaick K, Roessler E, Muenke M, Yocum A, Dubourg C, Li X, Geng X, Oliver G, Epstein DJ (2008) Regulation of a remote Shh forebrain enhancer by the Six3 homeoprotein. Nat Genet 40:1348–1353. https://doi.org/10.1038/ng.230

Kapsimali M, Caneparo L, Houart C, Wilson SW (2004) Inhibition of Wnt/Axin/beta-catenin pathway activity promotes ventral CNS midline tissue to adopt hypothalamic rather than floorplate identity. Development 131:5923–5933

Kawaguchi D, Sahara S, Zembrzycki A, O'Leary DDM (2016) Generation and analysis of an improved Foxg1-IRES-Cre driver mouse line. Dev Biol 412:139–147. https://doi.org/10.1016/j.ydbio.2016.02.011

Kimura S, Hara Y, Pineau T, Fernandez-Salguero P, Fox CH, Ward JM, Gonzalez FJ (1996) The T/ebp null mouse: thyroid-specific enhancer-binding protein is essential for the organogenesis of the thyroid, lung, ventral forebrain, and pituitary. Genes Dev 10:60–69

Kobayashi D, Kobayashi M, Matsumoto K, Ogura T, Nakafuku M, Shimamura K (2002) Early subdivisions in the neural plate define distinct competence for inductive signals. Development 129:83–93

Krauss S, Concordet JP, Ingham PW (1993) A functionally conserved homolog of the Drosophila segment polarity gene hh is expressed in tissues with polarizing activity in zebrafish embryos. Cell 75:1431–1444. https://doi.org/10.1016/0092-8674(93)90628-4

Kremnyov S, Henningfeld K, Viebahn C, Tsikolia N (2018) Divergent axial morphogenesis and early Shh expression in vertebrate prospective floor plate. EvoDevo 9:4. https://doi.org/10.1186/s13227-017-0090-x

Lagutin OV, Zhu CC, Kobayashi D, Topczewski J, Shimamura K, Puelles L, Russell HR, McKinnon PJ, Solnica-Krezel L, Oliver G (2003) Six3 repression of Wnt signaling in the anterior neuroectoderm is essential for vertebrate forebrain development. Genes Dev 17:368–379

Le Gros Clark WE (1938) Morphological aspects of the hypothalamus. In: Le Gros Clark WE, Beattie J, Riddoch G, Dott NM (eds) The hypothalamus. Morphological, functional, clinical and surgical aspects. Oliver and Boyd, Edinburgh

Legué E, Joyner AL (2010) Genetic fate mapping using site-specific recombinases. In: Wassarman PM, Soriano PM (eds) Guide to techniques in mouse development, Part B: Mouse molecular genetics, 2nd edn. Elsevier, Amsterdam

Lu F, Kar D, Gruenig N, Zhang ZW, Cousins N, Rodgers HM, Swindell EC, Jamrich M, Schuurmans C, Mathers PH, Kurrasch DM (2013) Rax is a selector gene for mediobasal hypothalamic cell types. J Neurosci 33:259–272. https://doi.org/10.1523/JNEUROSCI.0913-12.2013

Machold R, Hayashi S, Rutlin M, Muzumdar MD, Nery S, Corbin JG, Gritli-Linde A, Dellovade T, Porter JA, Rubin LL, Dudek H, McMahon AP, Fishell G (2003) Sonic hedgehog is required for progenitor cell maintenance in telencephalic stem cell niches. Neuron 39:937–950

Manning L, Ohyama K, Saeger B, Hatano O, Wilson SA, Logan M, Placzek M (2006) Regional morphogenesis in the hypothalamus: a BMP-Tbx2 pathway coordinates fate and proliferation through Shh downregulation. Dev Cell 11:873–885

Marin O, Anderson SA, Rubenstein JL (2000) Origin and molecular specification of striatal interneurons. J Neurosci 20:6063–6076
Mathers PH, Grinberg A, Mahon KA, Jamrich M (1997) The Rx homeobox gene is essential for vertebrate eye development. Nature 387:603–607. https://doi.org/10.1038/42475
Merchán P, Bribián A, Sánchez-Camacho C, Lezameta M, Bovolenta P, de Castro F (2007) Sonic hedgehog promotes the migration and proliferation of optic nerve oligodendrocyte precursors. Mol Cell Neurosci 36:355–368. https://doi.org/10.1016/j.mcn.2007.07.012
Miranda-Angulo AL, Byerly MS, Mesa J, Wang H, Blackshaw S (2014) Rax regulates hypothalamic tanycyte differentiation and barrier function in mice. J Comp Neurol 522:876–899. https://doi.org/10.1002/cne.23451
Mo R, Freer AM, Zinyk DL, Crackower MA, Michaud J, Heng HH, Chik KW, Shi XM, Tsui LC, Cheng SH, Joyner AL, Hui C (1997) Specific and redundant functions of Gli2 and Gli3 zinc finger genes in skeletal patterning and development. Development 124:113–123
Muthu V, Eachus H, Ellis P, Brown S, Placzek M (2016) Rx3 and Shh direct anisotropic growth and specification in the zebrafish tuberal/anterior hypothalamus. Development 143:2651–2663. https://doi.org/10.1242/dev.138305
Nüsslein-Volhard C, Wieschaus E (1980) Mutations affecting segment number and polarity in Drosophila. Nature 287:795–801
Ono K, Yoshii K, Tominaga H, Gotoh H, Nomura T, Takebayashi H, Ikenaka K (2017) Oligodendrocyte precursor cells in the mouse optic nerve originate in the preoptic area. Brain Struct Funct 222:2441–2448. https://doi.org/10.1007/s00429-017-1394-2
Orquera DP, Nasif S, Low MJ, Rubinstein M, de Souza FSJ (2016) Essential function of the transcription factor Rax in the early patterning of the mammalian hypothalamus. Dev Biol 416:212–224. https://doi.org/10.1016/j.ydbio.2016.05.021
Pak T, Yoo S, Miranda-Angulo AL, Wang H, Blackshaw S (2014) Rax-CreERT2 knock-in mice: a tool for selective and conditional gene deletion in progenitor cells and radial glia of the retina and hypothalamus. PLoS One 9(4):e90381. https://doi.org/10.1371/journal.pone.0090381
Patten I, Kulesa P, Shen MM, Fraser S, Placzek M (2003) Distinct modes of floor plate induction in the chick embryo. Development 130:4809–4821. https://doi.org/10.1242/dev.00694
Pearson CA, Ohyama K, Manning L, Aghamohammadzadeh S, Sang H, Placzek M (2011) FGF-dependent midline-derived progenitor cells in hypothalamic infundibular development. Development 138:2613–2624. https://doi.org/10.1242/dev.062794
Placzek M, Briscoe J (2005) The floor plate: multiple cells, multiple signals. Nat Rev Neurosci 6:230–240
Puelles L, Martínez S, Martínez-de-la-Torre M, Rubenstein JLR (2004) Gene maps and related histogenetic domains in the forebrain and midbrain. In: Paxinos G (ed) The rat nervous system, 3rd edn. Academic, San Diego, CA
Puelles L, Martinez-de-la-Torre M, Bardet S, Rubenstein JLR (2012) Hypothalamus. In: Watson C, Paxinos G, Puelles L (eds) The mouse nervous system. Elsevier-Academic, San Diego
Rallu M, Machold R, Gaiano N, Corbin JG, McMahon AP, Fishell G (2002) Dorsoventral patterning is established in the telencephalon of mutants lacking both Gli3 and Hedgehog signaling. Development 129:4963–4974
Riddle RD, Johnson RL, Laufer E, Tabin C (1993) Sonic hedgehog mediates the polarizing activity of the ZPA. Cell 75:14011416. https://doi.org/10.1016/0092-8674(93)90626-2
Roelink H, Augsburger A, Heemskerk J, Korzh V, Norlin S, Ruiz i Altaba A, Tanabe Y, Placzek M, Edlund T, Jessell TM, Dodd J (1994) Floor plate and motor neuron induction by *vhh-1*, a vertebrate homolog of *hedgehog* expressed by the notochord. Cell 16:761–775. https://doi.org/10.1016/0092-8674(94)90514-2
Roessler E, Muenke M (2010) The molecular genetics of holoprosencephaly. Am J Med Genet C Semin Med Genet 154C:52–61. https://doi.org/10.1002/ajmg.c.30236
Ruiz i Altaba A, Jessell TM (1993) Midline cells and the organization of the vertebrate neuraxis. Curr Opin Genet Dev. 3:633–640. https://doi.org/10.1016/0959-437X(93)90100-4

Sagner A, Briscoe J (2017) Morphogen interpretation: concentration, time, competence, and signaling dynamics. Wiley Interdiscip Rev Dev Biol 6. https://doi.org/10.1002/wdev.271

Salvatierra J, Lee DA, Zibetti C, Duran-Moreno M, Yoo S, Newman EA, Wang H, Bedont JL, de Melo J, Miranda-Angulo AL, Gil-Perotin S, Garcia-Verdugo JM, Blackshaw S (2014) The LIM homeodomain factor Lhx2 is required for hypothalamic tanycyte specification and differentiation. J Neurosci 34:16809–16820. https://doi.org/10.1523/JNEUROSCI.1711-14.2014

Shimamura K, Rubenstein JL (1997) Inductive interactions direct early regionalization of the mouse forebrain. Development 124:2709–2718

Shimamura K, Hartigan DJ, Martinez S, Puelles L, Rubenstein JL (1995) Longitudinal organization of the anterior neural plate and neural tube. Development 121:3923–3933

Shimogori T, Lee DA, Miranda-Angulo A, Yang Y, Wang H, Jiang L, Yoshida AC, Kataoka A, Mashiko H, Avetisyan M, Qi L, Qian J, Blackshaw S (2010) A genomic atlas of mouse hypothalamic development. Nat Neurosci 13:767–775. https://doi.org/10.1038/nn.2545

Simerly RB (2004) Anatomical substrates of hypothalamic integration. In: Paxinos G (ed) The rat nervous system, 3rd edn. Elsevier, Amsterdam

Sussel L, Marin O, Kimura S, Rubenstein JL (1999) Loss of Nkx2.1 homeobox gene function results in a ventral to dorsal molecular respecification within the basal telencephalon: evidence for a transformation of the pallidum into the striatum. Development 126:3359–3370

Szabó NE, Zhao T, Cankaya M, Theil T, Zhou X, Alvarez-Bolado G (2009) Role of neuroepithelial Sonic hedgehog in hypothalamic patterning. J Neurosci 29:6989–7002. https://doi.org/10.1523/JNEUROSCI.1089-09.2009

Thompson RH, Swanson LW (2003) Structural characterization of a hypothalamic visceromotor pattern generator network. Brain Res Brain Res Rev 41:153–202

Tobet SA, Paredes RG, Chickering TW, Baum MJ (1995) Telencephalic and diencephalic origin of radial glial processes in the developing preoptic area/anterior hypothalamus. J Neurobiol 26:75–86. https://doi.org/10.1002/neu.480260107

Trowe MO, Zhao L, Weiss AC, Christoffels V, Epstein DJ, Kispert A (2013) Inhibition of Sox2-dependent activation of Shh in the ventral diencephalon by Tbx3 is required for formation of the neurohypophysis. Development 140:2299–2309. https://doi.org/10.1242/dev.094524

Xu Q, Wonders CP, Anderson SA (2005) Sonic hedgehog maintains the identity of cortical interneuron progenitors in the ventral telencephalon. Development 132:4987–4998. https://doi.org/10.1242/dev.02090

Xu Q, Tam M, Anderson SA (2008) Fate mapping Nkx2.1-lineage cells in the mouse telencephalon. J Comp Neurol 506:16–29. https://doi.org/10.1002/cne.21529

Xuan S, Baptista CA, Balas G, Tao W, Soares VC, Lai E (1995) Winged helix transcription factor BF-1 is essential for the development of the cerebral hemispheres. Neuron 14:1141–1152. https://doi.org/10.1016/0896-6273(95)90262-7

Yu K, McGlynn S, Matise MP (2013) Floor plate-derived sonic hedgehog regulates glial and ependymal cell fates in the developing spinal cord. Development 140:1594–1604. https://doi.org/10.1242/dev.090845

Zhang Y, Alvarez-Bolado G (2016) Differential developmental strategies by Sonic hedgehog in thalamus and hypothalamus. J Chem Neuroanat 75:20–27. https://doi.org/10.1016/j.jchemneu.2015.11.008

Zhao L, Zevallos SE, Rizzoti K, Jeong Y, Lovell-Badge R, Epstein DJ (2012) Disruption of SoxB1-dependent Sonic hedgehog expression in the hypothalamus causes septo-optic dysplasia. Dev Cell 22:585–596. https://doi.org/10.1016/j.devcel.2011.12.023

Development of the Hypothalamus in *Xenopus laevis*

3

Nerea Moreno and Agustín González

Abstract

Xenopus laevis (African clawed frog) is a model organism for the study of early embryonic development of vertebrates. It is especially suited to analyze the neuroendocrine system because it possesses hormonal systems and mechanisms of gene regulation comparable to those in rodents and humans. The hypothalamus is the structure of the forebrain that plays a fundamental role in the neuroendocrine and homeostatic regulation of the body, representing a central hub for the neural networks involved in these functions. As such the hypothalamus is the nexus between the autonomic and limbic systems, which condition the behavioral and regulatory responses of the animal. During the development of *Xenopus*, the hypothalamus is located rostral to the diencephalon, possesses derivatives of the alar and basal plates, and includes molecular compartments, comparable to amniotes (reptiles, birds, and mammals), expressing transcription factors such as Nkx2.1, Islet 1, and Otp in unique subpopulations.

Keywords

Prosencephalon · Forebrain patterning · Expression patterns · Evolution

Abbreviations

BH	Basal hypothalamus
CT	Caudal tuberal region
Dien	Diencephalon
hp1	Hypothalamic prosomeric domain 1
hp2	Hypothalamic prosomeric domain 2

N. Moreno (✉) · A. González
Department of Cell Biology, Faculty of Biology, University Complutense, Madrid, Spain
e-mail: nerea@bio.ucm.es

S. Wray, S. Blackshaw (eds.), *Developmental Neuroendocrinology*, Masterclass in Neuroendocrinology 9, https://doi.org/10.1007/978-3-030-40002-6_3

Hyp	Hypophysis
M	Mammillary region
Ma	Mammillary area proper
oc	Optic chiasm
P1	Prosomere 1
P2	Prosomere 2
P3	Prosomere 3
P3b	Basal plate of P3
Pa	Paraventricular nucleus
PA	Pallium
Pac	Caudal paraventricular nucleus
Par	Rostral paraventricular nucleus
PO	Preoptic region
PT	Pretectum
PTh	Prethalamus
RM	Retromammillary region
RT	Rostral tuberal region
Sag	Sagittal
SC	Suprachiasmatic region
SCc	Caudal suprachiasmatic region
SCr	Rostral suprachiasmatic region
SPa	Subparaventricular area
SPA	Subpallium
Th	Thalamus
Tub	Tuberal region

3.1 Why Is the *Xenopus laevis* Hypothalamus a Model of Interest in Developmental Analysis?

From the late 1940s to the 1970s, when modern pregnancy tests were not available, *Xenopus laevis* frogs were injected with the urine of possibly pregnant women and a positive result was indicated when the presence of human chorionic gonadotropin (hCG) induced the frogs to lay a large number of eggs (Lee-Liu et al. 2017). At the same time, in many laboratories, this ease with which female frogs could be induced to lay eggs, which could subsequently be fertilized in vitro, along with the relatively large size of the eggs and embryos, and their development *ex utero*, made *Xenopus laevis* an excellent model organism to study the early embryonic development of vertebrates. In addition, due to its metamorphosis phase, this model has provided insight into the developing neuroendocrine systems, as well as a model to study endocrine disrupters. In fact, in recent years evidence has accumulated that persistent organic pollutants, heavy metals, pharmaceutical and personal care products, agricultural chemicals, and aerospace products are acting as endocrine disrupters that may alter the activity of hypothalamus–pituitary–thyroid axis and subsequently interfere with the endocrine systems of both wildlife and humans by affecting reproductive biology and the thyroid system. In this context, it was also demonstrated that hormonal control of development during the human perinatal period is critically important and complex, with multiple hormones that regulate fetal growth, brain development, and organ maturation in preparation for birth. Notably, many significant parallels exist between metamorphosis of frogs and the birth of mammals, and underlying these analogous developmental changes are conserved neuroendocrine components (Buchholz 2017). Frogs and mammals have identical hormones that reach their peak at metamorphosis and birth, have conserved hormone receptors and mechanisms of gene regulation, and share comparable roles for hormones in many target organs. Indeed, in both frogs and mammals, genetic and environmental perturbations of hormonal control can cause irreversible morphological and physiological changes and can predispose individuals to diseases of adulthood.

The hypothalamic development in *Xenopus* has been the focus of thorough studies in recent years (Domínguez et al. 2013, 2014, 2015). In these studies, *Xenopus* was selected mainly due to the convenience of "hormone induced" breeding that makes embryos available for experimentation at all times of the year and that they are easy to maintain under laboratory conditions. In addition, an accurate timetable of development exists (Nieuwkoop and Faber 1967) that allows comparative interpretation of developmental processes. Notably, analysis of the amphibian hypothalamus serves as a model for understanding perinatal events in mammals due to the analogies in development of these structures.

3.1.1 The Hypothalamus in the Current Prosomeric Model of Brain Regionalization

In all vertebrates, the hypothalamus plays a key role in survival of the animal, acting as the bridge between the external (visual, olfactory, and gustatory) and internal stimuli and endocrine, autonomic and limbic responses. Thus, the hypothalamus is considered the forebrain territory par excellence for controlling homeostatic processes. The hypothalamus has been studied to a great extent, however the interpretation of its regionalization during development has been debated throughout the last decades and additional studies may shed further light on this fundamental, anatomical question.

From an anatomical point of view, the strong curvature of the neural tube flexure at the junction of the midbrain and forebrain forms a sharp cephalic flexure involving the hypothalamus (see Fig. 3.1). This developmental event has made it hard to identify the basic units or subdivisions in the mature hypothalamus and understand the topological relationships between them. Thus, following mere topographical positions and interpreting "classical" transverse sections, the hypothalamus was considered a diencephalic region beneath the thalamus and different nomenclatures have been used for nuclei and hypothalamic regions in different vertebrates. The

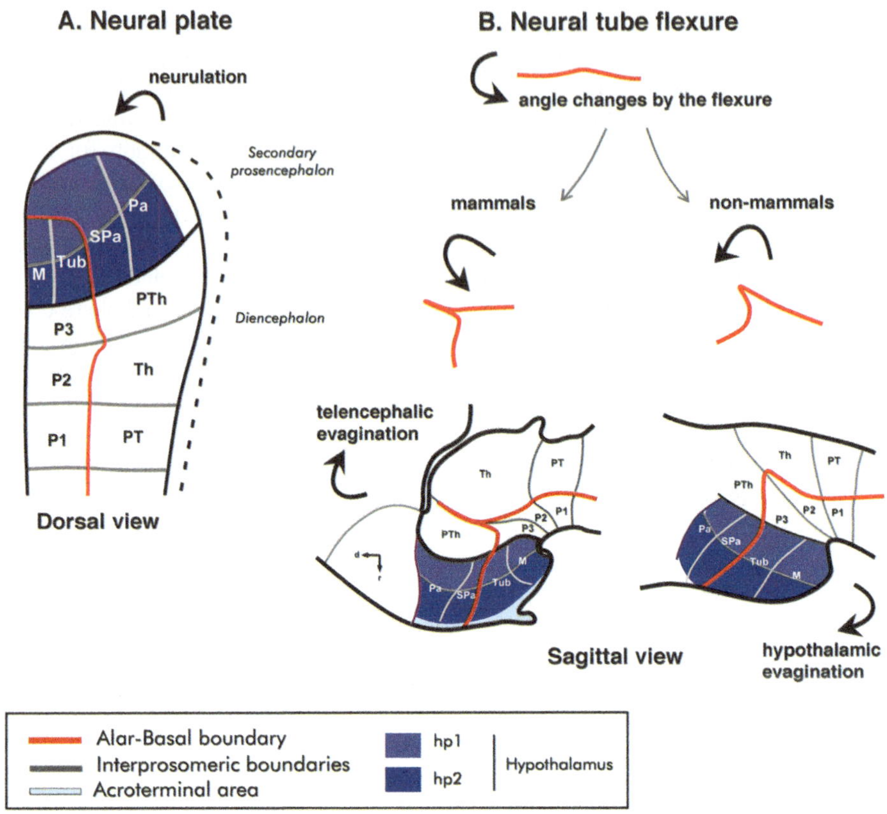

Fig. 3.1 Schematic comparison of the hypothalamic development between mammals and non-mammals. At the neural plate stage (**a**), the brain undergoes neurulation forming the neural tube (**b**). The hypothalamic anatomy is influenced by the flexure of the neural tube. In mammals the longitudinal axis bends almost 90° forming a sharp flexure and the rostral tube is moved to a "ventral" position, whereas in non-mammals this angle is less pronounced. Additionally, the telencephalic evagination in the case of mammals is huge and proportional over hp1 and hp2, flattening the hypothalamic territory, whereas in non-mammals, this event pushes hp2, turning the hypothalamus more "ventral." In contrast to non-mammals, the hypothalamic evagination in mammals contributes to the elongation of the hypothalamic territory

different nomenclature established in the past has made it challenging to compare results within the research community.

During the last 25 years, the "prosomeric model" of forebrain organization has been reinforced by numerous neuroanatomical studies (Puelles and Rubenstein 2003). This model is based on data from developmental expression patterns of many different regulatory genes (such as Nkx2, Six3, Dlx, OTx2, Otp, and Isl1) that allowed investigators to conceptualize and explain the organization of different brain regions, including the hypothalamus (Puelles et al. 2012; Domínguez et al. 2015) in many different species, including *Xenopus*. In this model, the forebrain is organized into transverse segments (prosomeres, see Fig. 3.1) and longitudinal zones that are also defined histogenetic domains based on regulatory gene expression patterns. According to the original prosomeric model and its subsequent revisions, the hypothalamus is excluded from the diencephalon, which is composed of three neuromeres (prosomeres P1–P3, Fig. 3.1). The rostral-most forebrain is designated the secondary prosencephalon that contains the hypothalamus (rostral to diencephalic P3), the telencephalon impar, and the telencephalic hemispheres (Puelles and Rubenstein 2003). The interpretation of the parts of the secondary prosencephalon is fraught with difficulties, mainly derived from the early optic and hemispheric evaginations and the different degree of development shown across vertebrates that disturb the primary pattern of this region. However, morphological, molecular, and hodological (connectivity) results have continued to highlight the organization of the main parts of the secondary prosencephalon and its subdivisions, particularly in mice, although comparable results have been obtained in other amniotes, amphibians, and fishes (reviewed in Domínguez et al. 2015; Ferran et al. 2015).

Based on gene expression patterns and other molecular markers, interpreted under the current prosomeric model (Fig. 3.1), such as FoxG1, the preoptic region (PO, see Fig. 3.2) is part of the subpallium of the telencephalon (SPA, Fig. 3.2), instead of the hypothalamus, where it was traditionally included (Puelles et al. 2012). It forms the dorsal boundary of the hypothalamic territory. In the proper hypothalamus, a transverse and complete limit, called intrahypothalamic boundary, has been described marking anteroposterior patterning and defining subdivisions. Thus, terminal (rostral, hp2: hypothalamic prosomeric domain 2, Fig. 3.1 dark blue area) and peduncular (caudal, hp1: hypothalamic prosomeric domain 1, Fig. 3.1 lighter blue area) regions were distinct on the basis of the differential expression of transcription factors in the early neural tube. In addition, a band of Nkx2.2 expressing cells becomes bent at the cephalic flexure and highlights the continuity of the dorsoventral patterning across the hypothalamus, revealing the existence of the longitudinal organization in the forebrain, and indicating alar versus basal territories. As such, four main domains can be considered in the hypothalamus, because both the hp1 and hp2 segments contain alar and basal parts (see Fig. 3.1). Furthermore, two longitudinal subdivisions are also formed in the alar and basal hypothalamic regions. The paraventricular area (Pa), topologically the dorsalmost part of the hypothalamus, is located adjacent to the preoptic region and mainly expresses the transcription factors Tbr1, Sim1, and Otp, and produces the magnocellular and parvocellular neurosecretory cell populations of the paraventricular, supraoptic, and periventricular nuclei. The second domain is the subparaventricular area (SPa), which forms the ventral part of the alar hypothalamus and is characterized by the expression of Dlx genes and gives rise to the suprachiasmatic, the anterior hypothalamic nucleus, and the subparaventricular zone. Sonic hedgehog and the transcription factor Nkx2.1 specifically characterize the basal hypothalamus formed by two longitudinal subdivisions: tuberal (Tub) and mammillary (M) regions. The tuberal region produces the anterobasal and posterobasal areas, and the ventromedial, dorsomedial, and arcuate nuclei. Rostrally to all those areas, in the terminal region a distinct acroterminal territory has been defined (see Fig. 3.1, yellow dots, Ferran et al. 2015).

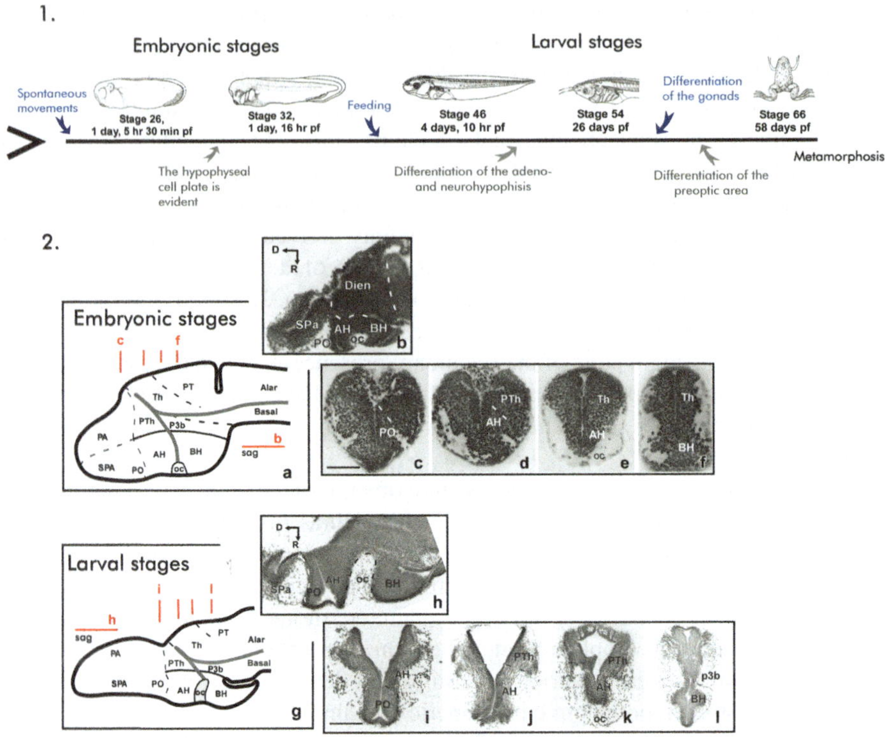

Fig. 3.2 (**1**) Schematic representation of the timing of the main developmental events during *Xenopus laevis* development (based on Nieuwkoop and Faber 1967) pf: post-fertilization. (**2**) Brain region changes from embryonic to larval. Photomicrographs of Nissl-stained sagittal (**b**, **h**) and transverse (**c–f**; **i–l**) sections through the hypothalamus of an embryonic (**b–f**) and a larval (**h–l**) *Xenopus*; from anterior to posterior, at the approximate levels indicated on the schemes (**a**, **g**). These photomicrographs illustrate the segregated regions recognized in the hypothalamus using techniques of the time (Nieuwkoop and Faber 1967). The scale bar in **c** is valid for **b–f** = 100 μm. The scale bar in **i** is valid for **h–l** = 200μm

It corresponds to the rostral-most portion of the neural tube and is responsible for the development of structures such as the lamina terminalis and the optic chiasm in the alar hypothalamus, whereas in the basal hypothalamus it gives rise to the median eminence and the associated arcuate nucleus, the infundibulum, and the neurohypophysis (Morales-Delgado et al. 2014; Puelles et al. 2012).

3.1.2 How to Study the Developmental Hypothalamic Region in *Xenopus*?

3.1.2.1 Experimental Approaches

Many aspects of developmental and cellular neurobiology have been addressed using *Xenopus* as the animal model (Gurdon and Hopwood 2000). Analysis of the development of the *Xenopus* brain, including the hypothalamus, has been the basis of numerous studies using a variety of experimental techniques and protocols. As with any laboratory model organism, *Xenopus* possesses standard protocols for laboratory breeding and husbandry, techniques to modify gene expression (e.g., antisense morpholino oligonucleotides), the generation of transgenic animals, and genome editing (e.g., CRISPR/Cas9 system; Bhattacharya et al. 2015). In addition, there are currently genetic and genomic data available from *Xenopus* to analyze genes, gene families, and gene networks (see Tandon et al. 2017).

3.1.2.2 Initial Neuroanatomical Studies

Using classic histochemical techniques provided abundant and useful information, but they are not currently considered sufficiently accurate given that the anatomical references within a particular region, such as the ventricular sulci, the position of the ventricles, and even the boundaries between regions, were not clear. Fortunately, new resources (Pearl et al. 2012) have facilitated experiments that have provided details of gene regulation and patterning via the identification of morphogens, transcription factors, neuropeptides, and neurotransmitters, which allow identification of developing hypothalamic areas and their limits.

3.1.2.3 High Resolution Analysis of Cells

Immunohistochemistry and/or in situ hybridization show the exact position of different molecules or genes at different developing stages. These techniques, together with RTPCR, have identified genes that specify different hypothalamic regions at very early developmental stages, even before the presence of the neuropeptides and neurotransmitters that functionally define the region. This new information, together with classic data, such as ^{3}H-thymidine autoradiography birth-dating techniques (Tay and Straznicky 1982) are making it possible to delineate the transition from the neural plate to formation of hypothalamic nuclei. Analyses of neural circuits using the isolated *Xenopus* nervous system (Kelley et al. 2017), together with emerging technologies for whole brain imaging (Tomer et al. 2015), and behavioral tests will significantly advance our understanding of which nuclei integrate into the circuits that generate behavioral patterns.

3.2 Hypothalamic Organization During the Development of *Xenopus laevis*

Throughout development, brain changes can easily be correlated with the striking body transformations that take place through embryonic and larval development. These include changes in the hypothalamus that progressively lead to specific developmental events related to hormonal control, sexual maturity and behavioral events which functionally are related to maturation of the hypothalamus.

3.2.1 Developmental Stages

For *Xenopus laevis*, Nieuwkoop and Faber (1967) described and illustrated in precise tables, the different stages of embryonic and larval development mainly based on external anatomy. This staging is still used today (Fig. 3.2, 1). *Xenopus laevis* development encompasses embryos (stages 1–45) and larvae (stages 46–66), at stage 66 metamorphosis is completed and the tadpole becomes a froglet. According to these tables, from stage 24 the animal begins to show initial motor reactions to external stimulation, and by stage 26 such movements are spontaneous. But it is only from stage 28 that changes in brain morphology related to the hypothalamus are detected, with the hypophyseal primordium (future pituitary) penetrating beyond the optic stalk, and the infundibulum rapidly developing by progressive thinning of the ventral wall. From stage 29/30, the chiasmatic ridge is rostrally separated from the lamina terminalis, and it is just later, at stage 31, that the hypophyseal cell plate reaches the caudal border of the chiasmatic ridge, but it will be connected to the ectoderm until stage 32. All of these events occur at early embryonic developmental stages. Thus, during late embryonic stages, the hypothalamus is already present as an anatomical entity, but its maturity is likely linked to subsequent changes in the animal as a whole. The next main event that takes place with respect to behavioral responses is the beginning of independent feeding at stage 45, when larval stages begin. At stage 47, the pars intermedia of the hypophysis begins to differentiate, and at the same stage, the genital ridges begin to protrude into the coelomic cavity. At stage 49 the infundibular extension of the neurohypophysis begins to develop and the first anlage of tuberal part of the adenohypophysis can be observed. From stage 50, the gonadal rudiments form and the thyroid follicles can be observed, but it is not until stage 52 that sexual differentiation of the gonads begins, and at stage 55 the first meiotic divisions are observed.

Thus, in *Xenopus*, the development of the hypothalamus actually takes place over a relatively long period, throughout embryonic and larval development (Fig. 3.2, 2). This makes this system a good model to dissect apart the signals that lead to a mature, functional hypothalamus. Of note, it is known that from the cell blastomere of stage 16, hypothalamic dopaminergic cells can already be detected (Huang and Moody 1992). The development of the anterior hypothalamus occurs mainly around embryonic stages 20–25, while the posterior part develops later, through the late embryonic and early larval stages 40–52. In contrast, the preoptic area which is mainly related to reproductive and parental behaviors, and now defined as part of the telencephalon, develops almost at the end of development, from stage 58 (Tay and Straznicky 1982).

3.2.2 The Main Neuroendocrine Hypothalamus: The Paraventricular Region

3.2.2.1 Developmental Gene Expression Pattern

Neuroendocrine cells that release their neuropeptides directly into the general circulation in the neurohypophysis or into the portal system supplying the adenohypophysis, are primarily, but not exclusively, situated in the most dorsal region of the hypothalamus (Figs. 3.1 and 3.3). In this region, groups of cells in the paraventricular area (Pa) that secrete neuropeptides have been described (Domínguez et al. 2015). From the earliest stages of development, this region is defined molecularly by the expression of the gene orthopedia (Otp, Fig. 3.3a). Otp in the *Xenopus* embryonic hypothalamus is involved in the specification of this region. During development, the Otp-positive territory is distinguished from the adjacent prethalamic (prosomere p3. Fig. 3.3, PTh) and suprachiasmatic (Fig. 3.3, SC) territories by the lack of expression of Isl1, Nkx2.1, and Shh (Fig. 3.3b, c, f–h and r). In addition, the expression of Nkx2.2 in the rostral portion of this area subdivides it into two regions (Pac and Par; Fig. 3.3i–m, Domínguez et al. 2013).

3.2.2.2 Neuropeptides

Neurochemically different cell populations in the Pa of the hypothalamus in *Xenopus* include neurons containing several neuropeptides such as somatostatin (SOM; Fig. 3.3k), mesotocin (MST; Fig. 3.3j; an oxytocin-like peptide), arginine-vasotocin (AVT), corticotropin-releasing hormone (CRH), thyrotropin-releasing hormone (TRH), or orexins (González et al. 1995; Domínguez et al. 2015; López et al. 2017). These cell populations project to the pars nervosa of the pituitary gland (Tuinhof et al. 1994), and a profuse connection exists between this region and the tuberal hypothalamus (Tub, Figs. 3.1 and 3.4; Domínguez et al. 2013) that may be involved in regulating ingestion of food. Numerous opsin-positive neurons have been found in the previously named "hypothalamic magnocellular preoptic nucleus," and most of them co-express MST. Axonal processes from these cells reach the neural lobe of the pituitary gland, suggesting that in *Xenopus*, mesotocinergic hypothalamic neurons, which are directly involved in photoreception and could be regulating seasonal adaptation changes as well. (Álvarez-Viejo et al. 2003). Independent of function, the early emergence of cells expressing TRH or MST and SOM, and the presence of Otp has been correlated (Fig. 3.3j, k; Domínguez et al. 2013, 2015), suggesting that the transcription factor Otp may participate in the acquisition of their phenotypes. Thus, in both Xenopus and mouse, transcription factors are involved in the specification of hypothalamic postmitotic populations and in the differentiation of independent nuclei within this territory.

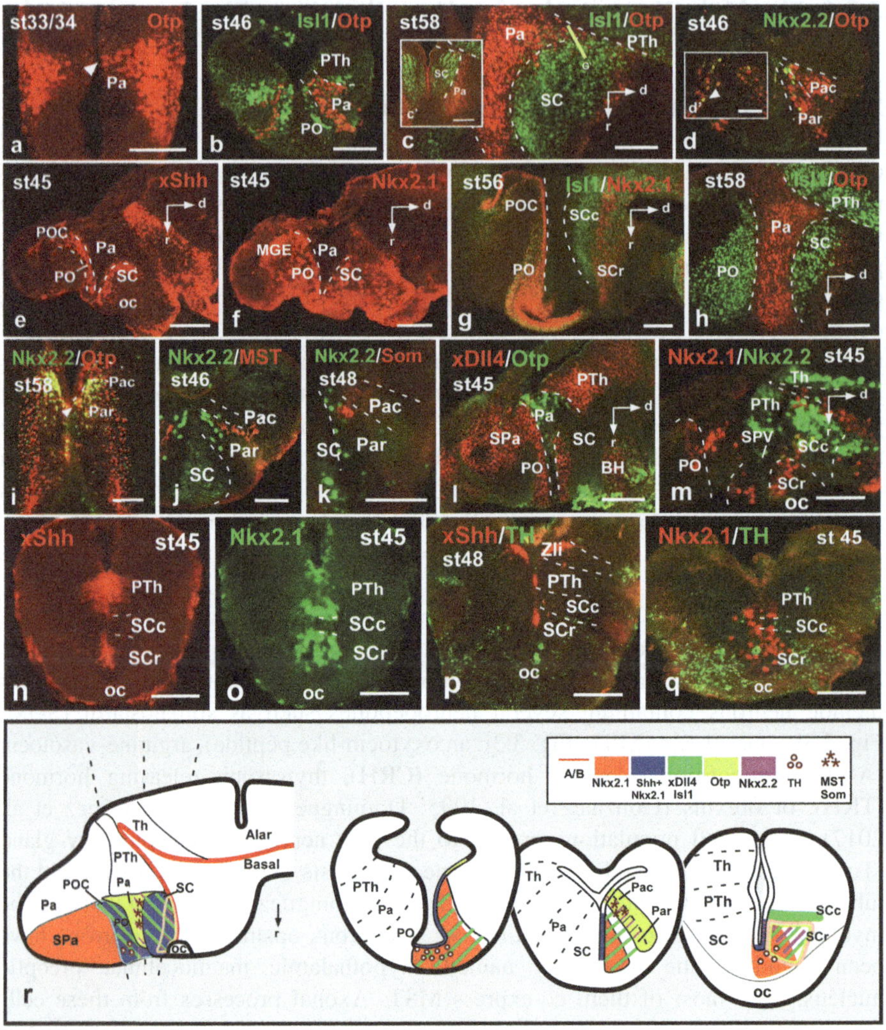

Fig. 3.3 Alar hypothalamic regions identified by molecular markers. Upper panel: Photomicrographs of sagittal (**c**, **e–h**, **l**, **m**) and transverse sections (**a–d′**, **i–k**, **n–g**) through the developing *Xenopus* alar hypothalamic region showing the single expressions of Otp (**a**), xShh (**e**, **n**), and Nkx2.1 (**f**, **o**), and the combined expressions of Isl1 and Nkx2.1 (**g**), Isl1/Otp (**b**, **c**, **h**), Nkx2.2/Otp (**d**, **i**), Nkx2.2/MST (**j**), Nkx2.2/SOM (**k**), xDll4/Otp (**l**), Nkx2.1/Nkx2.2 (**m**), xShh/TH (**p**), and Nkx2.1/TH (**q**). The arrowheads in **d′** and **i** indicate double labeled cells, and in (**a**) indicate Otp ventricular expressing cells. The yellow line in c indicates the level of the section shown in **c′**. Developmental stages are indicated in each photograph. Lower panel: Schematic representation (**r**) summarizes the combinatorial code of markers present in the alar hypothalamus and its adjacent domains in lateral and transverse views (**r**). MST, mesotocin; TH, tyrosine hydroxylase. Scale bars in **a**, **b**, **e**, **h**, **j–m** = 50 μm; in **c**, **d**, **g**, **i**, **n–q** = 100 μm. All photomicrographs are modified from Domínguez et al. (2013)

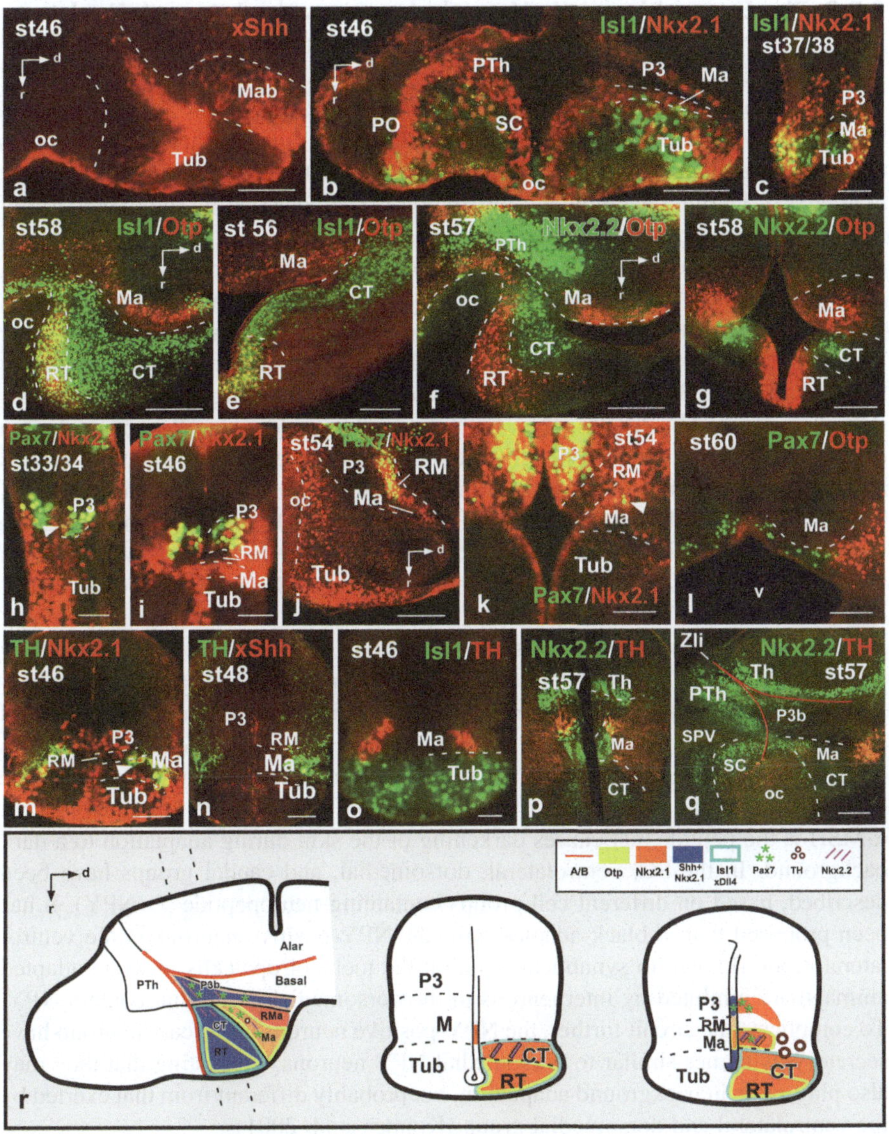

Fig. 3.4 Basal hypothalamic regions identified by molecular markers. Upper panel: Photomicrographs: transverse (**c**, **e**, **g**, **h**, **i**, **k–p**) and sagittal (**a**, **b**, **d**, **f**, **j**, **q**) sections through the developing *Xenopus* basal hypothalamic region showing the single expressions of xShh (**a**) and the combined expressions of Isl1 and Nkx2.1 (**b**, **c**), Isl1 and Otp (**d**, **e**), Nkx2.2 and Otp (**f**, **g**), Pax7 and Nkx2.1 (**h–k**), Pax7 and Otp (**l**), TH and Nkx2.1 (**m**), Shh (**n**), Isl1 (**o**), Nkx2.2 (**p**, **q**). The arrowheads in **h**, **k**, and **m** indicate double labeled cells. The red line in **q** indicates the alar/basal boundary. Developmental stages (st) are indicated in each photograph. Lower panel: schematic representation summarizes the combinatorial code of markers present in the basal hypothalamus and its adjacent domains in lateral and transverse views. TH: tyrosine hydroxylase. Scale bars in **a–i**; **k–q** = 100 μm; in **j** = 50 μm. All the photomicrographs are modified from Domínguez et al. (2014)

3.2.3 The Suprachiasmatic Hypothalamus as the Internal Clock During Development: The Subparaventricular Hypothalamus

3.2.3.1 Gene Expression Patterns and Connections

The subparaventricular region (Spa; Figs. 3.1 and 3.3) of the alar hypothalamus is the most ventral part and contains the suprachiasmatic nucleus (SC; Domínguez et al. 2013). During development, this region is characterized by the expression of Dlx genes, which control the expression of GABAergic populations, and Isl1 (Fig. 3.3l; Domínguez et al. 2013). The expression of transcription factors Nkx2.1 and Nkx2.2 subdivides the suprachiasmatic nucleus into rostro-caudal subdomains (Fig. 3.3m; SCr and SCc). In addition, the expression of the morphogen sonic hedgehog (Shh) in the rostral portion (SCr; Fig. 3.3n) overlaps with the expression of Nkx2.1 (Fig. 3.3o), suggesting a regulatory role of Shh through Nkx2.1 actions (Domínguez et al. 2013). The suprachiasmatic region in *Xenopus* is also characterized during development by the presence of catecholaminergic cell groups restricted to the rostral domain (identified by expression of tyrosine hydroxylase, TH, Fig. 3.3p, q; Moreno et al. 2012; Domínguez et al. 2013). Furthermore, the suprachiasmatic region has important connections with the tuberal hypothalamus (Tub, Fig. 3.1) and the mammillary complex (M, Fig. 3.1; Domínguez et al. 2013).

3.2.3.2 Suprachiasmatic Nucleus Functions

Functionally, the SC nucleus is involved in adaptation of an animal to its surroundings, a very important mechanism for wildlife survival. The development and structure of this nucleus has been studied in detail in *Xenopus*. Specially, a group of SC neurons inhibits the release of the α-melanophore stimulating hormone (α-MSH) from the neuroendocrine melanotropic cells of the pituitary gland. α-MSH is the peptide that causes darkening of the skin during adaptation to a dark background. In the SC, ventrolateral, dorsomedial, and caudal groups have been described, based on different cell groups containing neuropeptide Y (NPY). It has been proposed that in black-adapted animals, NPY-positive neurons in the ventrolateral group (known to synaptically inhibit the melanotrope cells in white-adapted animals) are inhibited by interneurons in the dorsomedial group that contain NPY. To complicate the circuit further, the NPY-positive neurons in the caudal group have secretory dynamics similar to dorsomedial NPY neurons, suggesting that they may also play a role in background adaptation, but probably different from that exerted by the ventrolateral and dorsomedial group (Kramer et al. 2001).

These SC NPY inhibitory neurons, designated suprachiasmatic melanotrope-inhibiting neurons (SMINs), possess more and larger synapses on melanotrope cells in white adapted compared with black-background adapted animals. In addition, in the latter animals, the melanotropes are larger and produce more proopiomelanocortin, the precursor of α-MSH. On a white background, presynaptic SMIN plasticity is reflected by a higher expression of inhibitory NPY and is closely associated with postsynaptic melanotrope plasticity, namely with a higher expression of the NPY Y1 receptor. Interestingly, the melanotrope cells in these animals also show a higher expression of receptors for TRH and urocortin 1, two neuropeptides that stimulate α-MSH secretion. Possibly, in white-adapted animals melanotropes are sensitized to neuropeptide stimulation, so that when the frog moves to a black background, they can immediately initiate the secretion of α-MSH to achieve rapid adaptation to the new background condition.

3.2.4 The Basal Hypothalamus: The Tuberal and Mammillary Territories

3.2.4.1 Tuberal Region Molecular Expression Pattern

In the basal hypothalamus (Fig. 3.4), the tuberal region is the most rostral portion. From early embryonic stages, it is molecularly characterized by the expression of Shh and Nkx2.1 close to the ventricles and by Isl1 expression in the adjacent zone (Fig. 3.4a–c; Domínguez et al. 2014). Furthermore, starting at embryonic stages, the transcription factor Otp is located exclusively in the rostral tuberal portion (RT; Fig. 3.4d–e), which constitutes the anterobasal/retrochiasmatic region that gives rise to the arcuate nucleus. In addition, the caudal domain is characterized by a lack of Otp and the expression of Nkx2.2 (CT; Fig. 3.4f, g; Domínguez et al. 2014). In the caudal domain, both Isl1 and Nkx2.2 have been related to the development of the ventromedial nucleus, including specification of the basal phenotype, specification of the ventromedial fate, and regulation of reproductive behavior through the estrogen receptor α (Davis et al. 2004; Kurrasch et al. 2007). In the tuberal hypothalamus, Otp in SOM cell specification has also been described, as in the Pa area located dorsally. These tuberal SOM cells probably migrate into the arcuate nucleus. Similarly, TRH cells have been found in the tuberal region of anurans, which probably originated in Pa of the alar region (Domínguez et al. 2015). Finally, expression of Dll4 in the more caudal part of the tuberal region in *Xenopus* is closely related to the positive GABAergic cell population detected in this location, suggesting an involvement of Dll4 in the specification of the GABAergic phenotype in the tuberal region, as in other areas (Domínguez et al. 2014).

3.2.4.2 Mammillary Region Expression Pattern

The boundary between the tuberal and the mammillary territories in amphibians is mainly defined by the lack of Isl1 expression within the continuous Nkx2.1 positive tuberal/mammillary region (Fig. 3.4b–e, r; Domínguez et al. 2014). In addition, both regions can also be distinguished by differential expression of Otp, which is expressed in the mammillary region and is lacking in the caudal-most tuberal domain (Fig. 3.4d–g; Domínguez et al. 2014). The mammillary band has been divided into two different regions: the proper mammillary region (Ma), which occupies a most rostral position and is Nkx2.1+/Shh-, and the retromammillary region (RM) located more caudally, where the expression patterns of Shh and Nkx2.1 are inverted being Nkx2.1-/Shh+ (Fig. 3.4a–c, i, j). This Ma/RM subdivision is demarcated by the expression of Nkx2.2, Otp, Dll4, and Lhx1 in Ma (Fig. 3.4a–l, p, q). In addition, some scattered Pax7+ cells are present in the subventricular zone of Ma. The cells likely originate from the adjacent Pax7 expressing population of the P3 basal plate (Fig. 3.4h–l), which indicates that part of the mammillary region arises in the prethalamic tegmentum. Fate map analysis is needed to confirm this. In *Xenopus*, the Lhx7 expression in the alar hypothalamus continues ventrally reaching the mammillary territory (Domínguez et al. 2013), suggesting the existence of a tuberomammillary terminal zone in anurans as in mammals (Domínguez et al. 2014). The presence of a catecholaminergic cell population (TH positive, González et al. 1994; Domínguez et al. 2014) that coexpresses Nkx2.1 and Otp (Fig. 3.4n–q) also characterizes the mammillary region. As for the specification of this cell population, there is currently controversy about its origin, but several data support a contribution of diencephalic areas to the mammillary territory since Pax7 expressing cells likely originated in P3 and subsequently colonized the mammillary region (Domínguez et al. 2014).

3.3 Concluding Remarks and Perspectives: *Xenopus*, a Good Model for Vertebrates or a Distinct Amphibian Condition?

In multiple species analyzed in recent years, it has been described that the hypothalamus includes comparable molecular compartments, and each compartment shows a tendency to a common organization with respect to the profile of molecular expression, but at the same time, the hypothalamic organization is subject to evolutive/adaptive changes/modifications during the anamnio-amniotic transition. In this context, the evolutionary leap of *Xenopus*, an amphibian (amphibia are the only group of anamniote tetrapods), to the rest of the tetrapods, implies relevant adaptive changes for the conquest of the water-independence of the new environment, and therefore it has some obvious consequences in the organization of the brain. The main common features of the hypothalamus are its position and regionalization, since in all models it is topographically located rostral to the diencephalon and it is divided into alar and basal regions. In addition, each of them is rostro-caudally subdivided into two different domains according to molecular criteria (reviewed in Domínguez et al. 2015). Moreover, the chemoarchitectonic profiles of the territories functionally define them by the presence of multiple postmitotic cell populations occupying neuroendocrine territories. In contrast, in amniotes a preoptohypothalamic boundary is evident, which is likely related to the higher complexity of the paraventricular area that starts to express transcription factors such as Pax6. In addition, in this alar territory, the restricted expression of Nkx2.1 in the subparaventricular region observed in *Xenopus* is gradually more restricted reaching a total lack in mammals, which could be related to the gradual pallial and thalamic expansion that take place in amniotes.

Key References

Bhattacharya et al. (2015)—This study demonstrates that the CRISPR/Cas9 system in Xenopus can efficiently modify both alleles in the F0 generation.

Buchholz (2017)—The study, describing the roles of thyroid hormones and their receptors in the tissue transformations comprising development, highlights the key experimental findings and endocrine, molecular, and genomic resources available in Xenopus.

Domínguez et al. (2015)—This article summarizes the main data on the expression patterns of different transcription factors and neuroactive substances in the developing hypothalamus of Xenopus laevis and the turtle Pseudemys scripta

Ferran et al. (2015)—Review of the main distinct dorsoventral progenitor domains detected in the mouse developing hypothalamus based on gene expression markers.

Puelles et al. (2012)—This chapter constitutes one of the most complete revision of the mouse hypothalamic organization.

Tandon et al. (2017)—This article reports the main advantages in Xenopus to study multiple aspects of cell and developmental biology and demonstrates its usefulness for genetic manipulation approaches.

References

Álvarez-Viejo M, Cernuda-Cernuda R, DeGrip WJ, Álvarez-Lopez C, García-Fernández JM (2003) Co-localization of mesotocin and opsin immunoreactivity in the hypothalamic preoptic nucleus of *Xenopus laevis*. Brain Res 969(1–2):36–43. https://doi.org/10.1016/S0006-8993(03)02273-X

Bhattacharya D, Marfo CA, Li D, Lane M, Khokha MK (2015) CRISPR/Cas9: an inexpensive, efficient loss of function tool to screen human disease genes in *Xenopus*. Dev Biol 408 (2):196–204. https://doi.org/10.1016/j.ydbio.2015.11.003

Buchholz DR (2017) *Xenopus* metamorphosis as a model to study thyroid hormone receptor function during vertebrate developmental transitions. Mol Cell Endocrinol 459:64–70. https://doi.org/10.1016/j.mce.2017.03.020

Davis AM, Seney ML, Walker HJ, Tobet SA (2004) Differential colocalization of Islet-1 and estrogen receptor alpha in the murine preoptic area and hypothalamus during development. Endocrinology 145(1):360–366. https://doi.org/10.1210/en.2003-0996

Domínguez L, Morona R, González A, Moreno N (2013) Characterization of the hypothalamus of *Xenopus laevis* during development. I. The alar regions. J Comp Neurol 521(4):725–759. https://doi.org/10.1002/cne.23222

Domínguez L, González A, Moreno N (2014) Characterization of the hypothalamus of *Xenopus laevis* during development. II. The basal regions. J Comp Neurol 522(5):1102–1131. https://doi.org/10.1002/cne.23471

Domínguez L, González A, Moreno N (2015) Patterns of hypothalamic regionalization in amphibians and reptiles: common traits revealed by a genoarchitectonic approach. Front Neuroanat 9:3. https://doi.org/10.3389/fnana.2015.00003

Ferran JL, Puelles L, Rubenstein JL (2015) Molecular codes defining rostrocaudal domains in the embryonic mouse hypothalamus. Front Neuroanat 9:46. https://doi.org/10.3389/fnana.2015.00046

González A, Marin O, Tuinhof R, Smeets WJ (1994) Ontogeny of catecholamine systems in the central nervous system of anuran amphibians: an immunohistochemical study with antibodies against tyrosine hydroxylase and dopamine. J Comp Neurol 346(1):63–79. https://doi.org/10.1002/cne.903460105

González A, Muñoz A, Muñoz M, Marín O, Smeets WJ (1995) Ontogeny of vasotocinergic and mesotocinergic systems in the brain of the South African clawed frog *Xenopus laevis*. J Chem Neuroanat 9(1):27–40. https://doi.org/10.1016/0891-0618(95)00063-d

Gurdon JB, Hopwood N (2000) The introduction of *Xenopus laevis* into developmental biology: of empire, pregnancy testing and ribosomal genes. Int J Dev Biol 44(1):43–50

Huang S, Moody SA (1992) Does lineage determine the dopamine phenotype in the tadpole hypothalamus?: a quantitative analysis. J Neurosci 12(4):1351–1362

Kelley DB, Elliott TM, Evans BJ, Hall IC, Leininger EC, Rhodes HJ et al (2017) Probing forebrain to hindbrain circuit functions in *Xenopus*. Genesis 55(1–2). https://doi.org/10.1002/dvg.22999

Kramer BM, Welting J, Berghs CA, Jenks BG, Roubos EW (2001) Functional organization of the suprachiasmatic nucleus of *Xenopus laevis* in relation to background adaptation. J Comp Neurol 432(3):346–355

Kurrasch DM, Cheung CC, Lee FY, Tran PV, Hata K, Ingraham HA (2007) The neonatal ventromedial hypothalamus transcriptome reveals novel markers with spatially distinct patterning. J Neurosci 27(50):13624–13634. https://doi.org/10.1523/Jneurosci.2858-07.2007

Lee-Liu D, Mendez-Olivos EE, Munoz R, Larrain J (2017) The African clawed frog *Xenopus laevis*: a model organism to study regeneration of the central nervous system. Neurosci Lett 652:82–93. https://doi.org/10.1016/j.neulet.2016.09.054

López JM, Morales L, González A (2017) Spatiotemporal development of the orexinergic (hypocretinergic) system in the central nervous system of *Xenopus laevis*. Brain Behav Evol 88(2):127–146. https://doi.org/10.1159/000449278

Morales-Delgado N, Castro-Robles B, Ferran JL, Martínez-de-la-Torre M, Puelles L, Díaz C (2014) Regionalized differentiation of CRH, TRH, and GHRH peptidergic neurons in the mouse hypothalamus. Brain Struct Funct 219(3):1083–1111. https://doi.org/10.1007/s00429-013-0554-2

Moreno N, Domínguez L, Morona R, González A (2012) Subdivisions of the turtle *Pseudemys scripta* hypothalamus based on the expression of regulatory genes and neuronal markers. J Comp Neurol 520(3):453–478. https://doi.org/10.1002/cne.22762

Nieuwkoop PD, Faber J (1967) Normal table of *Xenopus laevis* (Daudin). North Holland, Amsterdam

Pearl EJ, Grainger RM, Guille M, Horb ME (2012) Development of *Xenopus* resource centers: the National Xenopus Resource and the European Xenopus Resource Center. Genesis 50 (3):155–163. https://doi.org/10.1002/dvg.22013

Puelles L, Rubenstein JL (2003) Forebrain gene expression domains and the evolving prosomeric model. Trends Neurosci 26(9):469–476. https://doi.org/10.1016/S0166-2236(03)00234-0

Puelles L, Martínez de la Torre M, Bardet S, Rubenstein JLR (2012) Hypothalamus. In: Watson C, Paxinos G, Puelles L (eds) The mouse nervous system. Academic-Elsevier, San Diego, pp 221–313

Tandon P, Conlon F, Furlow JD, Horb ME (2017) Expanding the genetic toolkit in *Xenopus*: approaches and opportunities for human disease modeling. Dev Biol 426(2):325–335. https://doi.org/10.1016/j.ydbio.2016.04.009

Tay D, Straznicky C (1982) The development of the diencephalon in *Xenopus*. An autoradiographic study. Anat Embryol (Berl) 163(4):371–388

Tomer R, Khairy K, Amat F, Keller PJ (2015) Quantitative high-speed imaging of entire developing embryos with simultaneous multiview light-sheet microscopy. Nat Methods 9(7):755–763. https://doi.org/10.1038/nmeth.2062

Tuinhof R, González A, Smeets WJ, Scheenen WJ, Roubos EW (1994) Central control of melanotrope cells of *Xenopus laevis*. Eur J Morphol 32(2–4):307–310

Gene Regulatory Programs in the Development of Hypothalamic Arcuate Nucleus Neurons

4

Jae W. Lee, Christian Huisman, and Seunghee Lee

Abstract

Neurons in the hypothalamic arcuate nucleus (ARC) relay and translate important cues from the periphery into the central nervous system to maintain critical homeostatic processes such as energy balance, growth, and reproductive behaviors. Despite extensive studies dedicated to unraveling the physiological functions of various neuronal types in the ARC, the gene regulatory networks that drive their development are still poorly understood. This chapter provides a concise overview of the transcription factors that control the development of ARC neurons, followed by discussions of several outstanding challenges in the field and future directions to address those issues using the recent technological advances in genome-wide studies.

Keywords

Arcuate nucleus · Gene regulation · Transcription factor · POMC · AgRP · NPY · GHRH

J. W. Lee (✉)
Department of Biologial Sciences, University at Buffalo, Buffalo, NY, USA
e-mail: jlee269@buffalo.edu

C. Huisman
Department of Pediatrics, Oregon Health & Science University, Portland, OR, USA

S. Lee
College of Pharmacy, Research Institute of Pharmaceutical Sciences, Seoul National University, Seoul, Korea

S. Wray, S. Blackshaw (eds.), *Developmental Neuroendocrinology*, Masterclass in Neuroendocrinology 9, https://doi.org/10.1007/978-3-030-40002-6_4

4.1 Introduction

The arcuate nucleus of the hypothalamus (ARC) is located in the mediobasal hypothalamus. It is adjacent to the third ventricle and the median eminence, and it contains neuronal cell types that play critical roles in diverse physiological homeostatic processes, such as energy balance, growth, and reproductive behaviors (Biebermann et al. 2012; Bluet-Pajot et al. 2001; Hill et al. 2008; Hrabovszky 2014). In particular, five types of arcuate neurons have been extensively characterized over the past few decades (Fig. 4.1). Arcuate neurons expressing growth hormone releasing hormone (GHRH) trigger secretion of growth hormone (GH) in the anterior pituitary gland (Bluet-Pajot et al. 2001). GH induces the hepatic expression of insulin-like growth factor 1 (IGF1), which controls bone epiphyses, growth plate development, muscle and adipose tissue development and glucose homeostasis (Cohen and Rosenfeld 1994). Neurons expressing the orexigenic neuropeptides "agouti-related peptide (AgRP)" and neuropeptide Y (NPY), herein termed AgRP-neurons, enhance food intake and reduce energy expenditure, while neurons expressing the anorexigenic neuropeptides α-melanocyte stimulating hormone (αMSH, cleaved from pro-opiomelanocortin, POMC) and "cocaine and amphetamine regulated transcript" (CART) inhibit food intake and increase energy consumption (Biebermann et al. 2012). Neurons expressing Kiss1 activate secretion from gonadotropin releasing hormone (GnRH)-neurons, thereby playing essential roles in reproduction (Hill et al. 2008). Tuberoinfundibular dopamine (TIDA)-neurons in the ARC regulate the secretion of prolactin from the pituitary gland (Voogt et al. 2001).

Recently, 24 neuronal cell types were cataloged in the adult ARC using the *Drop-seq* technology that can simultaneously perform genome-wide analysis of RNA expression in thousands of individual cells (Campbell et al. 2017) (Box 4.1). This study also revealed heterogeneity of arcuate neurons that had been previously considered a homogeneous population of cells, including two types of AgRP-neurons, three of POMC-neurons, and six of tyrosine hydroxylase (TH)-expressing

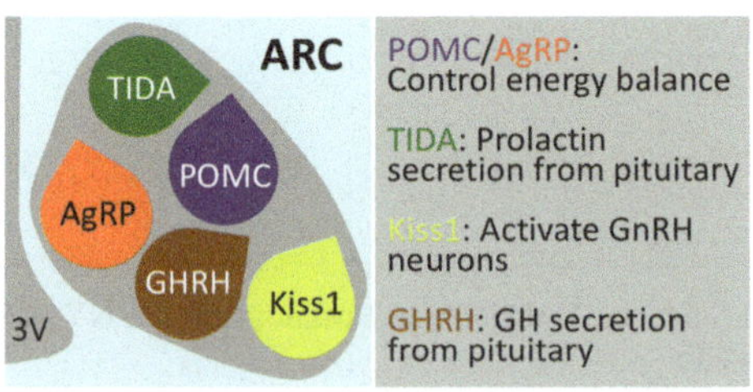

Fig. 4.1 Different neurons populate the arcuate nucleus (ARC). The ARC contains as many as 24 neuronal cell types according to the recent scRNA-seq results (Campbell et al. 2017). Only a few types of neurons have been extensively characterized, including tuberoinfundibular dopaminergic (TIDA)-, agouti-related peptide (AgRP)-, pro-opiomelanocortin (POMC)-, growth hormone relasing hormone (GHRH)- and Kiss1-neurons. *GH* growth hormone, *GnRH* gonadotropin releasing hormone

neurons (Campbell et al. 2017). Specific roles for each of these ARC neuronal populations are just beginning to be addressed. For instance, in female mice, AgRP-neurons that co-express somatostatin (Sst) were shown to specifically mediate corticotropin-releasing factor (CRF)-driven thermogenesis to cope with environmental challenges (Kuperman et al. 2016). Unfortunately, the gene regulatory networks that direct the development of arcuate neurons, which are the subject of this chapter, are poorly understood. Clearly, this information is needed to enable our understanding of the developmental trajectories through which common progenitors differentiate to distinct types of arcuate neurons (Biebermann et al. 2012; Bluet-Pajot et al. 2001; Hill et al. 2008; Hrabovszky 2014) and provide crucial insight into how and when the seemingly homogeneous populations of neurons segregate to multiple, functional subpopulations (Campbell et al. 2017).

Development of the hypothalamus involves multiple organizing signals and transcription factors. In mice, prior to embryonic day 10 (E10), the diencephalic region is separated from surrounding regions by the influence of organizing signals such as Wnts (for wingless/integrins), Shh (for sonic hedgehog), Bmps (for bone morphogenetic proteins, and Fgfs (for fibroblast growth factors). These organizing signals eventually induce the expression of the homeodomain (HD) transcription factor (TF) Nkx2.1, which is a key marker of hypothalamic tissue (Martinez-Ferre and Martinez 2012). During the subsequent neurogenic period (E10–16), neural stem cells in the ventricular zone give rise to neural precursors (Ifft 1972; Ishii and Bouret 2012). Because neurons, astrocytes, and oligodendrocytes all derive from the same progenitor pool, the proneural TFs that characterize the neurogenic period simultaneously drive neurogenesis and suppress gliogenesis (Rowitch and Kriegstein 2010). Notably, neurogenesis in the hypothalamus proceeds in an outside-in fashion (Altman and Bayer 1978). Neurons in the most lateral nuclei, such as the lateral hypothalamus area (LHA), are generated first. This neurogenic period is followed by birth of cells that will reside in more medially located nuclei such as the dorsomedial nucleus (DMH) and the ventromedial nucleus (VMH), and finally by neurons in nuclei along the midline such as the ARC (Byerly and Blackshaw 2009).

In this chapter, we give an overview of the findings related to the gene regulatory programs in the development of various arcuate neuronal types, which have been elucidated over the past few decades. We then highlight several outstanding challenges in the field, together with future studies to address those issues using the recent technological breakthroughs in genome-wide studies.

4.2 TFs in Progenitors and Fate Specification of ARC Neurons

A number of TFs, including HD and basic-helix-loop-helix (bHLH), are expressed in spatially and temporally distinct patterns in the developing arcuate neurons and play crucial roles in their development (Table 4.1; Fig. 4.2). It should be emphasized that many of these TFs are also involved in the development of other regions of the brain, although our discussion in this chapter is mostly focused on their developmental function of ARC neurons.

Table 4.1 Transcription factors in the development of ARC neurons

Transcription factors	Type	ARC phenotypes in KO models
Nkx2.1	HD	No ARC
Rax	HD	POMC↓
Dbx1	HD	AgRP↓
Isl1	HD	POMC/AgRP/GHRH↓
Bsx	HD	AgRP↓
Gsx1	HD	GHRH↓
Otp	HD	AgRP↓
Dlx1/2	HD	GHRH↓
Ngn3	bHLH	POMC↓
Nhlh2	bHLH	POMC↓
Mash1	bHLH	GHRH↓ POMC↓
Rbpjk		POMC↑ AgRP↑
Ikaros	Zinc-finger	GHRH↓

AgRP agouti-related peptide, *Arc* arcuate nucleus, *bHLH* basic-helix-loop-helix, *GHRH* growth hormone releasing hormone, *HD* homeodomain, *KO* gene knockout, *POMC* pro-opiomelanocortin

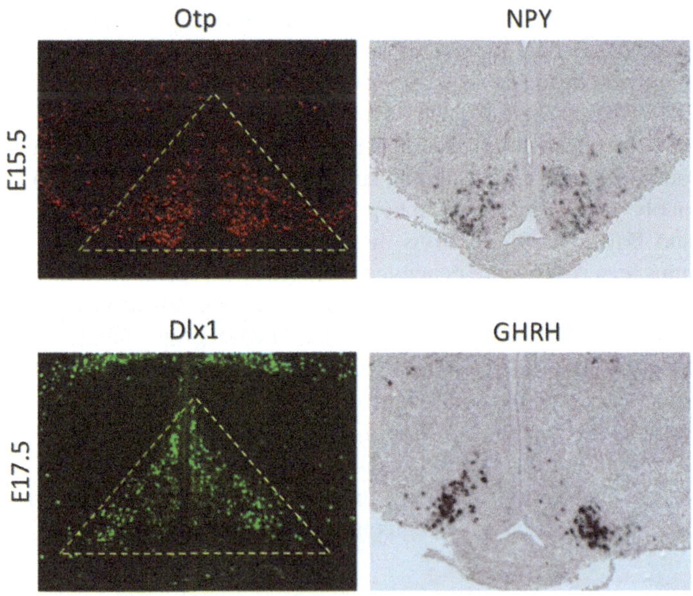

Fig. 4.2 Expression of Otp, NPY, Dlx1, GHRH in the developing arcuate nucleus (ARC). The expression in transverse brain sections of (left) the transcription factors (TFs) Otp (red) and Dlx1 (green) detected with immuno-fluorescent histochemistry (IHC); and (right) mRNAs for neuropeptides neuropeptide Y (NPY (a marker of agouti-related peptide (AgRP)-neurons, which derive from Otp^{+} cells) and growth hormone releasing hormone (GHRH, a marker of GHRH-neurons, which derive from $Dlx1^{+}$ cells) using in situ hybridization (ISH). The ARC region is denoted, left, as a yellow (dashed) triangle, and right, by autoradiographic sliver grains (black). E15.5, 17.5: embryonic days of development

4.2.1 Nkx2.1 (aka TTF-1 for Thyroid Transcription Factor-1)

The HD TF Nkx2.1 marks hypothalamic tissue (Martinez-Ferre and Martinez 2012). By E10.5, Nkx2.1 mRNA is detected in the progenitor zones of the rostroventral diencephalon and rostroventral telencephalon. Nkx2.1 mRNA and protein are expressed in both neuroepithelial progenitors and subsets of their postmitotic neurons (Lazzaro et al. 1991; Marin et al. 2000; Price et al. 1992; Sussel et al. 1999). By E14.5, Nkx2.1 is detectable in the ARC and persists in this region postnatally (Marin et al. 2000). Accordingly, transgenic mice, in which LacZ expression is driven by approximately a 4 kb fragment of potential regulatory DNA from the 5′ flanking region of *Nkx2.1*, show robust expression of LacZ in the ARC (Pan et al. 2004). In Nkx2.1-deficient mice, the development of the ARC is absent, while other hypothalamic nuclei are present (Kimura et al. 1996; Marin et al. 2002; Sussel et al. 1999), suggesting that Nkx2.1 is essential for the formation of ARC during development.

4.2.2 Retina and Anterior Neural Fold Homeobox (Rax, aka Rx)

The HD TF Rax, a key regulator of early retinal development (Zhang et al. 2000), also plays crucial roles in the formation of the ventral neuroendocrine hypothalamus (Muranishi et al. 2012). Nkx2.1-expressing neurons, which include many developing neurons in the ARC, appear to be derived from Rax-expressing cells. Elimination of Rax throughout ARC progenitors in mice results in loss of ventral hypothalamic expression of Nkx2.1 and POMC (Lu et al. 2013). These results suggest that Rax controls the production of POMC and possibly other ARC neuronal cell types, given the finding that POMC-derived progenitor cells differentiate into multiple ARC neuronal types (Padilla et al. 2010; Sanz et al. 2015).

4.2.3 Developing Brain Homeobox 1 (Dbx1)

Standard and inducible fate-mapping revealed that progenitors expressing the HD TF Dbx1 produce some populations of both AgRP- and POMC-neurons in the developing ARC (Sokolowski et al. 2016). Studies using *Nkx2.1-Cre* mice (Xu et al. 2008) to produce Dbx1 hypothalamic-specific conditional knockout (cKO) mice showed a significant decrease in ARC NPY expression at E13.5 while no changes in POMC expression were detected (Sokolowski et al. 2015). These results suggest that, while $Dbx1^{+}$ progenitors generate both AgRP- and POMC-neurons in the ARC, Dbx1 only plays a crucial role in the production of AgRP-neurons (NPY positive). In contrast, Dbx1-independent mechanisms are required for specification of the POMC-neurons in the ARC.

4.2.4 Isl1

The HD TF Isl1 is required for the development of POMC-neurons (Nasif et al. 2015). Multiple changes are found in mice lacking Isl1 in Nkx2.1^{+} expressing cells (using *Nkx2.1-Cre* mice). These mice show loss of the orexigenic neuropeptides AgRP and NPY, the anorexigenic neuropeptide αMSH, and the growth-stimulatory peptide GHRH, accompanied by severe deficits in both feeding and linear growth (Lee et al. 2016). Isl1 can directly enhance the expression of AgRP by cooperating with two key orexigenic TFs, glucocorticoid receptor (GR) and brain-specific homeobox factor (Bsx) (Lee et al. 2016). Isl1 also activates the expression of POMC in the ARC by directly binding to the neuronal *Pomc* enhancers nPE1 and nPE2 (Lam et al. 2015; Nasif et al. 2015). These data indicate that Isl1 is involved in the specification of AgRP-neurons, POMC-neurons, and GHRH-neurons within the ARC (Lee et al. 2016).

4.2.5 Bsx

The HD TF Bsx is expressed in the ARC as early as E15.5 (Cremona et al. 2004). In the adult mouse brain, Bsx expression is restricted to the hypothalamus with prominent expression in the ARC and the DMH as well as some cells in the LHA (Sakkou et al. 2007). In Bsx mutant brains, the expression of AgRP and NPY is downregulated but not abolished (Sakkou et al. 2007). Although it is unclear if Bsx is required for the early fate specification of AgRP-neurons, it is possible that Bsx and Isl1 have essential but redundant roles in the fate specification of AgRP-neurons as well as their upregulation of AgRP (Lee et al. 2016).

4.2.6 Gsx1 (aka Gsh1)

Mice lacking the HD TF Gsx1 show severe deficits in pituitary development, accompanied by extreme dwarfism (Li et al. 1996). Interestingly, Gsx1 is expressed in the regions of the diencephalon, which develops into the thalamus and hypothalamus but not in embryonic precursors of the pituitary (Valerius et al. 1995). These results suggest that the deficits in pituitary development likely reflect the role(s) of Gsx1 in the hypothalamus. Indeed, Gsx1 directly controls the expression of GHRH in the ARC (Li et al. 1996).

4.2.7 Orthopedia (Otp)

In the mouse (from E12.5), the HD TF Otp is expressed in neurons that give rise to the PVN, supraoptic (SON), anterior periventricular (aPV), and ARC nuclei (Simeone et al. 1994). Otp-null mice die soon after birth and display progressive impairment of crucial developmental events, including reduced cell proliferation, abnormal cell migration. In these mice, terminal differentiation of the parvocellular and magnocellular neurons of the aPV, PVN, SON, and ARC is disrupted (Acampora et al. 1999). In the ARC of Otp-null brains, the expression of Sst but not GHRH is abolished (Acampora et al. 1999; Wang and Lufkin 2000). We recently reported that Opt is highly enriched in AgRP-neurons and plays a critical role in their development (Lee et al. 2018). Together, these data suggest a role of Otp in the production of AgRP-neurons and other Sst^{+} neuronal types in the ARC during development.

4.2.8 Distal-Less Homeobox-1 (Dlx1) and Dlx2

The HD TFs Dlx1 and Dlx2 are expressed in the region of the developing and adult ARC (Eisenstat et al. 1999; Saino-Saito et al. 2003). Dlx1/2 plays important and redundant roles in the fate specification of $GABA^{+}$ and TH^{+} neurons in the forebrain (Andrews et al. 2003; Cobos et al. 2007; Long et al. 2007; Petryniak et al. 2007; Qiu et al. 1995). The immunochemical analyses with a pan-Dlx antibody detecting multiple Dlx factors (including Dlx1/2) reveal that Dlx factors are also expressed in TH^{+} neurons and non-AgRP-neuronal type $GABA^{+}$ neurons in the ARC (Yee et al. 2009). Adult Dlx1-null mice show a two-fold reduction in the number of TH^{+} neurons in the ARC (Yee et al. 2009), whereas the role of Dlx2 in the hypothalamus remains unexplored. We recently reported that Dlx1 and Dlx2 are highly expressed in TH^{+} neurons in the ARC, including GHRH-neurons and play critical roles in the production of GHRH-neurons (Lee et al. 2018).

4.2.9 Neurogenin-3 (Ngn3)

Proneural bHLH TF Ngn3 expression begins as early as E9.5 and disappears by E17.5 (Pelling et al. 2011). Fate mapping studies show that many of the earliest born POMC-neurons, around E10.5, derive from Ngn3-expressing progenitors, while later born POMC-neurons develop from non-Ngn3 expressing progenitors (Pelling et al. 2011). A large fraction of AgRP-neurons also appears to derive from Ngn3-expressing progenitors (Pelling et al. 2011). Ngn3-null mice show a substantial reduction in the number of POMC-neurons born between E10.5 and 13.5, although this number rebounds somewhat by E17.5 (Pelling et al. 2011). Other cell types in the ARC, such as GHRH-neurons, remained unaffected by deletion of Ngn3. When Ngn3 is knocked out only in $Nkx2.1^{+}$ cells using the inducible *Nkx2.1-iCre* mice (Tsai et al. 2012), the resulting cKO mice show a reduced number of POMC-neurons beginning at E10.5 that continues into the adulthood, thereby developing early-onset obesity and hyperphagia (Anthwal et al. 2013). These results demonstrate critical roles of Ngn3 in the development of POMC-neurons.

4.2.10 Nhlh2 (aka NSCL-2)

The bHLH TF Nhlh2 is expressed in the developing nervous system as early as E9.5 (Begley et al. 1992; Gobel et al. 1992). Nhlh2-null mice show hyperphagia and obesity as well as infertility (Good et al. 1997). These mice show no alteration in expression of POMC mRNA, although their αMSH levels are reduced, which is due to a reduction in the expression and activity of prohormone convertases (Pc1/3) that cleave POMC to adrenocorticotropic hormone (ACTH) and αMSH (Jing et al. 2004). The infertility of Nhlh2-null mice is explained by the finding that migration of GnRH-neurons to their final hypothalamic sites is significantly reduced in *Nhlh1/Nhlh2*-double KO mice (Kruger et al. 2004). Interestingly, POMC-neurons are reduced in mice when *Nhlh2* is deleted specifically in GnRH-expressing neurons (Schmid et al. 2013), suggesting a critical role for GnRH cells in the development of POMC-neurons.

4.2.11 Mash1 (aka Ascl1 for Achaete-Scute Complex 1)

The proneural bHLH TF Mash1 is expressed throughout the basal retrochiasmatic neuroepithelium and loss of Mash1 results in hypoplasia of the ARC (McNay et al. 2006). This is due to a failure of neurogenesis and an increase in apoptosis, defects that are rescued by ectopic expression of Ngn2 (another proneural bHLH TF) under the control of the Mash1 promoter (McNay et al. 2006). In addition to its role in neurogenesis, Mash1 is specifically required for the expression of Gsx1 (McNay et al. 2006), which drives the expression of GHRH (Li et al. 1996). Although Mash1 is not required for POMC expression, it is required for normal development of POMC-neurons (McNay et al. 2006). These data demonstrate that Mash1 is both required for the generation of ventral neuroendocrine neurons as well as for specification of these neurons.

4.2.12 Rbpjk

Rbpjk is a key downstream effector TF of Notch signaling (Lasky and Wu 2005). When the Notch intracellular domain translocates to the nucleus, it binds to Rbpjk (a transcriptional repressor), converting it into an activator, and inducing the expression of downstream target genes. Mice lacking Rbpjk in Nkx2.1-expressing cells show an expansion of the ARC and an increased number of POMC-neurons (Aujla et al. 2013). Mash1 expression is also increased in these animals, and this could account for the aberrant POMC expression, as Mash1 is an important factor in the differentiation of POMC-neurons (McNay et al. 2006). The number of NPY-expressing AgRP-neurons is also increased (Aujla et al. 2013). Mice with a constitutively active Notch1 intracellular domain show a reduction in ARC area and a complete absence of POMC- and AgRP-neurons (Aujla et al. 2013), mirroring the phenotypes of *Mash1*-KO mice (McNay et al. 2006).

4.2.13 Ikaros

Ikaros is a zinc finger TF that is critical for lymphohematopoiesis and immune regulation (Hu et al. 2017). Ikaros is also expressed in GHRH-neurons in the mouse embryonic ARC. Overexpression of Ikaros enhances GHRH promoter activity and induces endogenous GHRH gene expression in N3 hypothalamic cells (Ezzat et al. 2006). Notably, Ikaros-null mice exhibit dwarfism with characteristics of pituitary GH deficiency, and GHRH immunoreactivity was undetectable in the hypothalamus, consistent with Ikaros being essential for GHRH expression.

4.3 Additional TFs in Arcuate Neuronal Development

From the public databases such as GenePaint (Visel et al. 2004) and Allen Brain Atlas (Jones et al. 2009; Lein et al. 2007) as well as the recent expression studies of Shimogori et al. (2010) (Box 4.2), it is clear that more TFs are expressed in the developing ARC, which include the TFs Pou2f2, Sox14, Tbx2, Tbx19, Arx, Pbx1, Insm1, and Lhx5. Future investigation of the role of these TFs in arcuate neuronal development will allow us to establish which TFs, alone or together, play crucial roles in regulating the production of the many arcuate neuronal types during embryogenesis.

4.4 Coordinate Development of Different Arcuate Neurons

It is important to note that what appear to be distinct classes of arcuate neurons are indeed, functionally interconnected among themselves. For instance, reproduction, a highly energy-demanding process, is coupled to energy balance, and leptin serves as a key mediator of this coupling by controlling both food intake via AgRP-/POMC-neurons and reproduction via Kiss1-neurons (Chehab 2014). In addition, AgRP-neurons project onto GnRH and Kiss1-neurons while POMC-neurons project onto Kiss1-neurons, forming a circuit that allows for cross talk between food intake and reproduction (Chehab 2014). Another example is the well-known correlation between nutritional status and linear growth. As such, GH/GHRH levels are inversely correlated with circulating levels of the anorexic hormone leptin, which in turn targets arcuate neurons that control energy balance (Considine et al. 1996; Dieguez and Casanueva 1995; Grinspoon et al. 1996). Clearly, the generation of all of the different ARC neuronal subtypes is critical for homeostasis. Currently, evidence suggests that production of multiple arcuate neurons during embryogenesis involves common progenitors. These progenitors then differentiate into multiple types of neurons via gene regulatory programs that involve distinct fate-specifying TFs. How these programs are initiated/specified is unclear, but these programs allow concise regulation of each developmental lineage. In this context, it is interesting to note that POMC^{+} precursors generate several arcuate neuronal types, including POMC-, AgRP-, and Kiss1-neurons (Padilla et al. 2010; Sanz et al. 2015). It is also possible that the production of GHRH-neurons is coordinated with the generation of AgRP- and/or POMC-neurons via common progenitors.

4.5 Challenges to Define Gene Regulatory Programs and the Genome-Wide Solutions

In addition to identifying and characterizing the TFs that drive the development of arcuate neuronal subtypes, several challenges remain to be addressed to fully understand the gene regulatory programs that direct this developmental process. In particular, three issues are noteworthy.

First, for most TFs that are known to control the development of arcuate neurons (Table 4.1), information on their critical downstream target genes is largely missing. Determining the binding sites of these TFs in the different neuronal cell types of the ARC will help unravel the detailed molecular basis of their action in fate specification and development of arcuate neurons. This information should also facilitate our understanding of how different arcuate neurons coordinately develop during embryogenesis.

Second, precise assessment of the whole spectrum of actions that each TF plays in fate specification and development of the different arcuate neuronal types is seriously hampered by the fact that we currently have only a limited number of early fate markers for each arcuate neuronal population. Thus, we are lacking tools to identify lineage specification pathways.

Third, directly following is that *Cre* lines that can delete genes of interest in a specific population of arcuate neurons during earlier fate specification stages of development are missing. Such *Cre* lines will allow us to delineate the developmental action of each TF in specific populations of arcuate neurons independent of its action in other cell types. Potential resolutions to these issues, using recent advances in genome-wide studies, are discussed below.

4.5.1 Connecting Individual TFs in the Gene Regulatory Networks That Coordinately Integrate Development of Different Arcuate Neurons

The first step to reconstruct the gene regulatory programs for arcuate neuronal development is to identify TFs that play critical roles in arcuate neuronal development (Table 4.1). The next step is to identify the genome-wide target genes of each of these TFs in the developing ARC. This will also facilitate our effort to decipher how these TFs function, alone and/or in combination, to coordinate the production of different populations of arcuate neurons. A strategy used in our recent study provides a template for such work. We found that the HD TFs Otp and Dlx1 (and its homolog Dlx2) play essential roles in the development of AgRP- and GHRH-neurons, respectively (Lee et al. 2018). In addition to a loss of GHRH-neurons, we found that mice with *Nkx2.1-Cre*-directed cKO for the adjoining *Dlx1* and *Dlx2* genes

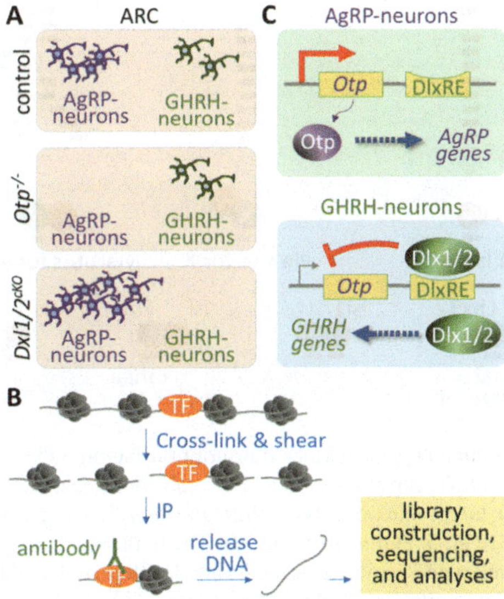

Fig. 4.3 The Dl1/2-Otp regulatory axis in the development of GHRH- and AgRP-neurons. (**a**) Summary of deficits in the development of growth hormone releasing hormone (GHRH)- and agouti-related peptide (AgRP)-neurons in $Otp^{-/-}$ and *Dlx1/2*-conditional knockout (cKO) mice. While $Otp^{-/-}$ mice lose AgRP-neurons, *Dlx1/2*-cKO mice lose GHRH-neurons but gain AgRP-neurons. (**b**) CHIP-seq schematic, including cross-linking of tissues, preparation of lysed chromatin and sonification, immunoprecipitation of chromatin fragments bound by transcription factor (TF) using antibody against the TF, and reverse cross-linking and isolation of DNA, followed by library constructiuon, sequencing and analyses. (**c**) Our CHIP-seq experiments using anti-Dlx1 antibody as well as the ensuing functional assays revealed that Dlx1 suppresses the expression of Otp by directly binding the silencer region located downstream of the *Otp* gene in developing GHRH-neurons (Lee et al. 2018). Direct target genes of Otp in developing AgRP-neurons and of Dlx1/2 in developing GHRH-neurons remain to be investigated.

show a significant increase in AgRP-neurons (Fig. 4.3a). Consistent with these results, these mice exhibit not only dwarfism, likely reflecting the loss of GHRH-neurons, but also many postnatal phenotypes that appear to reflect the increased number of AgRP-neurons, including reduced levels of energy expenditure and brown fat thermogenesis (Lee et al. 2018). Our subsequent genome-wide experiments revealed a surprising link between Otp and Dlx1/2.

First, we performed chromatin immunoprecipitation (CHIP) of E16.5 hypothalamus using an anti-Dlx1 antibody followed by sequencing, a technology named CHIP-seq that enables the genome-wide mapping of binding sites for TFs with extremely high resolution (Barski and Zhao 2009) (Fig. 4.3b). Notably, a functional Dlx1-binding site in the downstream region of the *Otp* gene was identified. Occupation of this site by Dlx1 resulted in suppression of Otp expression (Fig. 4.3c). We genetically showed that upregulation of Otp increased the number of AgRP-neurons in *Dlx1/2*-cKO mice (Lee et al. 2018). Given the finding that Dlx1/2 is expressed in developing GHRH-neurons, we concluded that GHRH- and AgRP-neurons share

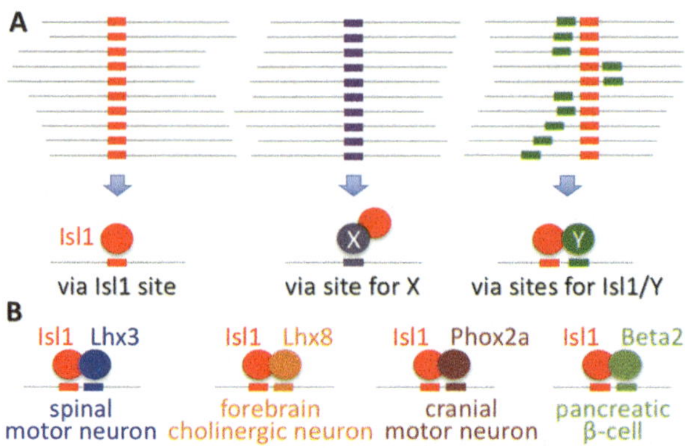

Fig. 4.4 De novo motif analysis and partner transcription factors (TFs). (**a**) Unbiased search of CHIP-seq peaks for motifs that recruit the target TF reveals multiple modes of DNA binding by the TF. TFs often bind their target sites indirectly through other TFs (e.g., X in this diagram) or in synergy with another TF (e.g., Y in this diagram). (**b**) The homeodomain (HD) TF Isl1 is known to pair with other TFs in a cell type-specific manner during development, and this combinatorial mode of actions allows one TF to control multiple different fates

common progenitors and that developing GHRH-neurons are still plastic with the potential to differentiate to AgRP-neurons. Thus, suppression of Otp expression by Dlx1/2 forces the progenitors to take the lineage to GHRH-neurons rather than the erroneous pathway to AgRP-neurons. These results demonstrate that the Dlx1/2-Otp gene regulatory axis directs a balanced production of GHRH- and AgRP-neurons during embryogenesis, contributing to the postnatal homeostatic functions of hypothalamic arcuate neurons that control energy balance and growth.

In addition to directly binding their cognate binding motifs, TFs often bind DNA either indirectly through protein–protein interactions with other DNA-bound TFs (e.g., X in Fig. 4.3a) or synergistically through adjacent motifs for other TFs (e.g., Y in Fig. 4.4a). Once the genome-wide occupancy of a TF is determined using the CHIP-seq technology, de novo motif analysis of CHIP-seq peaks (Bailey et al. 2009; Machanick and Bailey 2011) will allow us to predict its different modes of DNA binding as well as partner TFs. Hence, de novo motif analyses, coupled with the ensuing validation experiments, also provide crucial insights into the identity of partner TFs that function together with the original TF (e.g., X and Y in Fig. 4.4a), contributing to our understanding of how distinct fate-determining TFs are integrated into the gene regulatory programs for arcuate neuronal development. It is possible that some of the TF in Table 4.1 may be engaged in such protein–protein interactions, resulting in an additional layer of cross talk among key TFs. Such cross talk may direct the development of neuronal heterogeneity now known to exist in the ARC. In this regard, it is interesting to note that Isl1 directs fate determination of "specific" cell types by forming cell type-specific complexes with other TFs (Cho et al. 2014; Habener et al. 2005; Lee and Pfaff 2001; Mazzoni et al. 2013; Peng et al. 2005) (Fig. 4.4b). As Isl1 is required for the generation of several distinct types of arcuate neurons (Lee et al. 2016), Isl1 may partner with other TFs in specifying cell-fate of each arcuate neuronal subtype.

4.5.2 The Need for Early Fate Markers of Arcuate Neurons

The lack of early fate markers has prohibited us from deciphering the full spectrum of activities that each TF exerts in the production of arcuate neuronal populations. Most fate markers that are currently being used in the field, such as GHRH for GHRH-neurons and AgRP for AgRP-neurons, represent late fate markers that identify already committed or differentiated neurons. In addition, markers currently used as early fate markers, label multiple neuronal types. For example, POMC marks early ARC cells but lacks specificity given POMC-progenitors produce POMC-neurons, AgRP-neurons as well as Kiss1-neurons (Padilla et al. 2010; Sanz et al. 2015). Similarly, we recently found that NPY^{+} progenitors likely give rise to AgRP- and GHRH-neurons (Lee et al. 2018). The multiple lineages from POMC^{+} and NPY^{+} precursors highlight the need for additional early fate markers, which will enable us to understand the exact roles of each TF in specifying different arcuate neuronal cell types. Future investigation into the ontogeny of arcuate neurons, such as the relationship among POMC^{+} and NPY^{+} precursors and their progenies, coupled with our understanding of the gene regulatory programs for arcuate neuronal development, will reveal how different types of arcuate neurons are produced in the right ratio during development to perform highly interconnected functions at post-natal stages.

The recent advances in single cell RNA-seq (scRNA-seq) technology (Fig. 4.5) will greatly facilitate the efforts to identify additional early fate markers of developing arcuate neurons, as this technology provides the transcriptome of individual developing arcuate neurons. Using this technology in the hypothalamus at various embryonic stages, we can identify additional early fate markers for each cell type. In addition, developmental trajectories of each arcuate neuronal population and when common progenitors begin to diverge to different arcuate neuronal lineages during embryogenesis may be unraveled. Single cell RNA-seq can also be performed for control and KO/cKO models for each TF, and their comparative analyses will

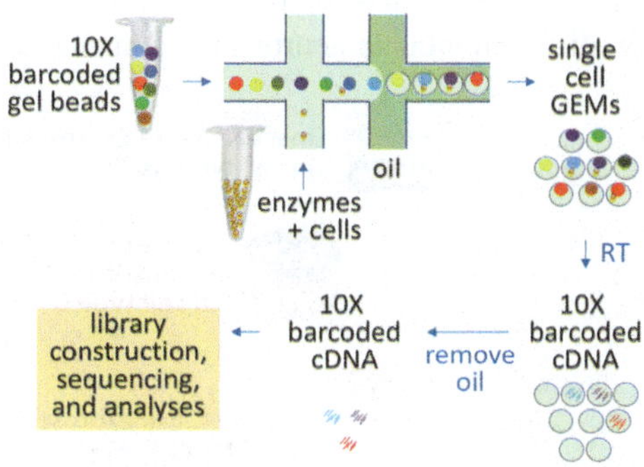

Fig. 4.5 Schematics for scRNA-seq. Dissociated cells are highly diluted so that only a single cell mixed with reagents is encapsulated into nanoliter-sized GEMs (gel bead in emulsion) together with a bead containing a barcoded oligonucleotide. GEMs are subjected to reverse transcription, and the resulting cDNA is isolated and constructed into a library, followed by sequencing and analyses

identify target genes of each TF. Integration of these results with the above described CHIP-seq results will identify target genes that are directly controlled by the TF, ultimately allowing us to fully reconstruct the gene regulatory programs that direct the development of individual arcuate neurons.

4.5.3 The Need for More Specific Cre Lines

Cre lines that can delete genes of interests only in a specific population of developing arcuate neurons will allow us to investigate the developmental action of each TF in a specific arcuate neuronal type independent of its action in other arcuate and non-arcuate neuronal types. Notably, while *Nkx2.1-Cre* transgenic line has been widely used for studying roles of genes of interests in the developing ARC, it deletes genes in most neuronal types in the developing ARC (Xu et al. 2008). Moreover, Nkx2.1 is also expressed in the thyroid, lung, and other parts of the brain (Lazzaro et al. 1991). This specificity issue also extends to other *Cre* lines that delete genes of interests in developing arcuate neurons, such as *Npy-Cre* (Milstein et al. 2015) and *Pomc-Cre* (Balthasar et al. 2004). The promoters of early fate markers to be identified from the aforementioned single cell RNA-seq approaches may allow us to create more specific *Cre* lines. However, many of these early markers may still be expressed in more than a single cell type. To help resolve this issue, an additional layer of specificity can be achieved using the split-Cre system, in which co-expression of catalytically inactive N-terminal and C-terminal fragments of Cre reactivates Cre activity (Casanova et al. 2003; Hirrlinger et al. 2009) (Fig. 4.6). In this system, for instance, two different promoters from early fate marker genes that are expressed in arcuate cell types A and B (promoter 1) and B, C, and D (promoter 2), respectively, will drive the expression of reconstituted active Cre in the overlapping cell type B (Fig. 4.5).

Demonstrating a postnatal homeostatic phenotype that correlates with a developmental deficit in an arcuate neuronal subtype is an excellent way to validate cell specification. Notably, different arcuate neurons are connected to each other as well

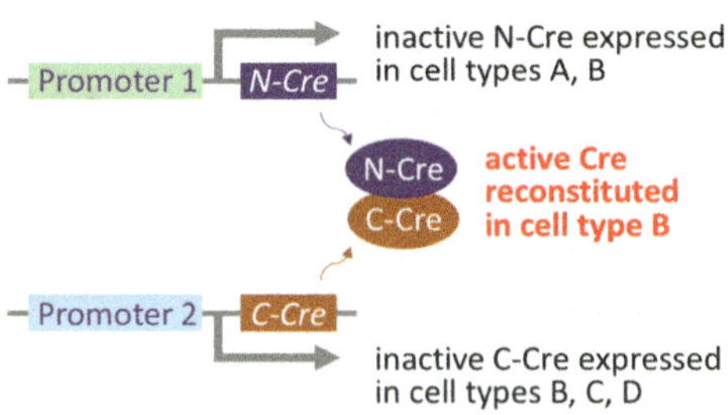

Fig. 4.6 Schematics for the split-Cre system. To obtain higher specificity, expression of the catalytically inactive N-terminal and C-terminal fragments of Cre is driven by two less-specific promoters, respectively. Only in cell types in which both promoters are active, N-terminal and C-terminal fragments of Cre are coexpressed restoring Cre activity

as with neurons in other brain regions, forming complex neural circuitries that mediate homeostatic processes. Therefore, less-specific *Cre* lines, such as the *Nkx2.1-Cre* transgenic line (Xu et al. 2008), pose a challenge for deciphering such complex circuitry. For instance, mice with conditional deletions of *Isl1* or *Dlx1/2* in developing arcuate neurons display developmental deficits in arcuate neurons that control energy homeostasis and linear growth. Postnatally, these mice show dwarfism as well as reduced energy expenditure and late-onset obesity (Lee et al. 2016, 2018). However, the specificity issues of *Nkx2.1-Cre* line make it difficult to conclude that the postnatal phenotypes observed with *Isl1*- or *Dlx1/2*-cKO mice are mainly due to the developmental deficits in arcuate neurons (Lee et al. 2016, 2018). For instance, Isl1 may also play a role in the development of the thyroid gland (Westerlund et al. 2008). Thus, potential hypothyroidism resulting from the lack of Isl1 may be the major culprit or have also contributed to the metabolic phenotypes observed in the *Isl1*-cKO mice (Lee et al. 2016). These experiments highlight the necessity for the development of more specific *Cre* lines that will enable us to successfully link developmental deficits to adult ARC-related phenotypes.

4.6 Future Perspectives

The technological breakthroughs, such as CHIP-seq and scRNA-seq, combined with the publicly available databases including GenePaint (Visel et al. 2004) and Allen Brain Atlas (Jones et al. 2009; Lein et al. 2007) as well as the pioneering expression studies of Dr. Blackshaw's group (Shimogori et al. 2010) provide an unprecedented opportunity to fully reconstruct the gene regulatory programs that direct the balanced development of diverse neuronal types and subtypes in the developing ARC. Future studies are needed to identify transcriptional coregulators that enable or disable the action of TFs in the development of arcuate neurons (Mannervik et al. 1999). In particular, roles of epigenetic chromatin regulators in arcuate neuronal development is an interesting subject, given their broad effect on brain development and autism spectrum disorders (De Rubeis et al. 2014; Iossifov et al. 2014; Michaelson et al. 2012; O'Roak et al. 2012). Moreover, some patients with inactivated chromatin regulators show phenotypes that may involve deficit in arcuate neuronal development, such as dwarfism (Lintas and Persico 2018; Niikawa et al. 1981). With the arrival of more specific *Cre* lines, we may be able to connect prominent metabolic features of those patients to deficits in the development of specific types of arcuate neurons. For the next few years, we will begin to witness the emergence of fully decoded gene regulatory programs that direct the development of arcuate neurons.

Acknowledgments This work was supported by grants from NIH/NIDDK (R01 DK064678 and R01 DK103661 to J.W.L.), the Korea Health Technology R&D Project through the Korea Health Industry Development Institute, funded by the Ministry of Health & Welfare, Republic of Korea (HI17C0447 to S.L.), and National Research Foundation of Korea (NRF-2012M3A9D1054705 to S.L.).

Key References

Campbell et al. (2017)—This article reports different types of arcuate nucleus neurons in the adult for the first time using scRNA-seq approach.

Lee et al. (2016)—This article reports crucial roles of the TF Isl1 in the development of multiple arcuate nucleus neuronal types.

Lee et al. (2018)—This article shows how different TFs involved in the production of GHRH- and AgRP-neurons are integrated into a regulatory axis that prevents developing GHRH-neurons from erroneously taking AgRP-neuronal fate during embryogenesis.

Nasif et al. (2015)—This article describes the developmental role of the TF Isl1 in specifying POMC-neurons.

Shimogori et al. (2010)—Using microarray analysis at different developmental time points, and developmental in situ hybridization, for genes expressed over the course of hypothalamic neurogenesis, this article identifies markers that stably label each major hypothalamic nucleus over the entire course of neurogenesis and constructs a molecular atlas of the developing hypothalamus.

Box 4.1 Identification of the Adult Arcuate Nucleus Neuronal Types

This box contains interesting information on the discovery of hypothalamic arcuate nucleus neurons in the adult. Despite extensive studies on the physiological roles of many arcuate nucleus neurons, their exact composition has not been well-defined. Using the recent advances in scRNA-seq technology, the group in Boston cataloged arcuate nucleus neuronal types in an unbiased way for the first time [7]. This paradigm-shifting study identified many transcriptionally distinct cell populations, including 24 arcuate nucleus neurons. One unexpected finding from this study was the presence of multiple subtypes for some neurons, such as three subtypes of POMC-neurons, two subtypes of AgRP-neurons and six subtypes of TH-neurons (GHRH- and five other TH-subtypes). This study also serves as an invaluable resource for genes specifically expressed in each adult arcuate neuronal type.

Box 4.2 A Molecular Atlas of the Developing Hypothalamus

Understanding of the developmental processes for various hypothalamus neurons has been seriously hampered by the lack of information on genes specifically expressed in each developing cell type, including upstream signaling molecules and downstream gene regulators such as transcription factors and coregulators. Dr. Seth Blackshaw's group at Johns Hopkins University performed microarray analysis at 12 different developmental time points, and then in situ hybridization for 1045 genes that were dynamically expressed during hypothalamic neurogenesis [66]. This paradigm-shifting study

(continued)

Box 4.2 (continued)
identified markers that stably labeled each major hypothalamic nucleus during the course of neurogenesis, which will serve as a fundamental platform to study the molecular pathways that mediate hypothalamic development.

References

Acampora D, Postiglione MP, Avantaggiato V, Di Bonito M, Vaccarino FM, Michaud J, Simeone A (1999) Progressive impairment of developing neuroendocrine cell lineages in the hypothalamus of mice lacking the Orthopedia gene. Genes Dev 13:2787–2800

Altman J, Bayer SA (1978) Development of the diencephalon in the rat. II. Correlation of the embryonic development of the hypothalamus with the time of origin of its neurons. J Comp Neurol 182:973–993

Andrews GL, Yun K, Rubenstein JL, Mastick GS (2003) Dlx transcription factors regulate differentiation of dopaminergic neurons of the ventral thalamus. Mol Cell Neurosci 23:107–120

Anthwal N, Pelling M, Claxton S, Mellitzer G, Collin C, Kessaris N, Richardson WD, Gradwohl G, Ang SL (2013) Conditional deletion of neurogenin-3 using Nkx2.1iCre results in a mouse model for the central control of feeding, activity and obesity. Dis Model Mech 6:1133–1145

Aujla PK, Naratadam GT, Xu L, Raetzman LT (2013) Notch/Rbpjkappa signaling regulates progenitor maintenance and differentiation of hypothalamic arcuate neurons. Development 140:3511–3521

Bailey TL, Boden M, Buske FA, Frith M, Grant CE, Clementi L, Ren J, Li WW, Noble WS (2009) MEME SUITE: tools for motif discovery and searching. Nucleic Acids Res 37:W202–W208

Balthasar N, Coppari R, McMinn J, Liu SM, Lee CE, Tang V, Kenny CD, McGovern RA, Chua SC Jr, Elmquist JK et al (2004) Leptin receptor signaling in POMC neurons is required for normal body weight homeostasis. Neuron 42:983–991

Barski A, Zhao K (2009) Genomic location analysis by ChIP-Seq. J Cell Biochem 107:11–18

Begley CG, Lipkowitz S, Gobel V, Mahon KA, Bertness V, Green AR, Gough NM, Kirsch IR (1992) Molecular characterization of NSCL, a gene encoding a helix-loop-helix protein expressed in the developing nervous system. Proc Natl Acad Sci USA 89:38–42

Biebermann H, Kuhnen P, Kleinau G, Krude H (2012) The neuroendocrine circuitry controlled by POMC, MSH, and AGRP. Handb Exp Pharmacol 209:47–75

Bluet-Pajot MT, Tolle V, Zizzari P, Robert C, Hammond C, Mitchell V, Beauvillain JC, Viollet C, Epelbaum J, Kordon C (2001) Growth hormone secretagogues and hypothalamic networks. Endocrine 14:1–8

Byerly MS, Blackshaw S (2009) Vertebrate retina and hypothalamus development. Wiley Interdiscip Rev Syst Biol Med 1:380–389

Campbell JN, Macosko EZ, Fenselau H, Pers TH, Lyubetskaya A, Tenen D, Goldman M, Verstegen AM, Resch JM, McCarroll SA et al (2017) A molecular census of arcuate hypothalamus and median eminence cell types. Nat Neurosci 20:484–496

Casanova E, Lemberger T, Fehsenfeld S, Mantamadiotis T, Schutz G (2003) Alpha complementation in the Cre recombinase enzyme. Genesis 37:25–29

Chehab FF (2014) 20 years of leptin: leptin and reproduction: past milestones, present undertakings, and future endeavors. J Endocrinol 223:T37–T48

Cho HH, Cargnin F, Kim Y, Lee B, Kwon RJ, Nam H, Shen R, Barnes AP, Lee JW, Lee S et al (2014) Isl1 directly controls a cholinergic neuronal identity in the developing forebrain and spinal cord by forming cell type-specific complexes. PLoS Genet 10:e1004280

Cobos I, Borello U, Rubenstein JL (2007) Dlx transcription factors promote migration through repression of axon and dendrite growth. Neuron 54:873–888
Cohen P, Rosenfeld RG (1994) Physiologic and clinical relevance of the insulin-like growth factor binding proteins. Curr Opin Pediatr 6:462–467
Considine RV, Sinha MK, Heiman ML, Kriauciunas A, Stephens TW, Nyce MR, Ohannesian JP, Marco CC, McKee LJ, Bauer TL et al (1996) Serum immunoreactive-leptin concentrations in normal-weight and obese humans. N Engl J Med 334:292–295
Cremona M, Colombo E, Andreazzoli M, Cossu G, Broccoli V (2004) Bsx, an evolutionary conserved brain specific homeoboX gene expressed in the septum, epiphysis, mammillary bodies and arcuate nucleus. Gene Expr Patterns 4:47–51
De Rubeis S, He X, Goldberg AP, Poultney CS, Samocha K, Cicek AE, Kou Y, Liu L, Fromer M, Walker S et al (2014) Synaptic, transcriptional and chromatin genes disrupted in autism. Nature 515:209–215
Dieguez C, Casanueva FF (1995) Influence of metabolic substrates and obesity on growth hormone secretion. Trends Endocrinol Metab 6:55–59
Eisenstat DD, Liu JK, Mione M, Zhong W, Yu G, Anderson SA, Ghattas I, Puelles L, Rubenstein JL (1999) DLX-1, DLX-2, and DLX-5 expression define distinct stages of basal forebrain differentiation. J Comp Neurol 414:217–237
Ezzat S, Mader R, Fischer S, Yu S, Ackerley C, Asa SL (2006) An essential role for the hematopoietic transcription factor Ikaros in hypothalamic-pituitary-mediated somatic growth. Proc Natl Acad Sci USA 103:2214–2219
Gobel V, Lipkowitz S, Kozak CA, Kirsch IR (1992) NSCL-2: a basic domain helix-loop-helix gene expressed in early neurogenesis. Cell Growth Differ 3:143–148
Good DJ, Porter FD, Mahon KA, Parlow AF, Westphal H, Kirsch IR (1997) Hypogonadism and obesity in mice with a targeted deletion of the Nhlh2 gene. Nat Genet 15:397–401
Grinspoon S, Gulick T, Askari H, Landt M, Lee K, Anderson E, Ma Z, Vignati L, Bowsher R, Herzog D et al (1996) Serum leptin levels in women with anorexia nervosa. J Clin Endocrinol Metab 81:3861–3863
Habener JF, Kemp DM, Thomas MK (2005) Minireview: transcriptional regulation in pancreatic development. Endocrinology 146:1025–1034
Hill JW, Elmquist JK, Elias CF (2008) Hypothalamic pathways linking energy balance and reproduction. Am J Phys Endocrinol Metab 294:E827–E832
Hirrlinger J, Scheller A, Hirrlinger PG, Kellert B, Tang W, Wehr MC, Goebbels S, Reichenbach A, Sprengel R, Rossner MJ et al (2009) Split-cre complementation indicates coincident activity of different genes in vivo. PLoS One 4:e4286
Hrabovszky E (2014) Neuroanatomy of the human hypothalamic kisspeptin system. Neuroendocrinology 99:33–48
Hu Y, Yoshida T, Georgopoulos K (2017) Transcriptional circuits in B cell transformation. Curr Opin Hematol 24:345–352
Ifft JD (1972) An autoradiographic study of the time of final division of neurons in rat hypothalamic nuclei. J Comp Neurol 144:193–204
Iossifov I, O'Roak BJ, Sanders SJ, Ronemus M, Krumm N, Levy D, Stessman HA, Witherspoon KT, Vives L, Patterson KE et al (2014) The contribution of de novo coding mutations to autism spectrum disorder. Nature 515:216–221
Ishii Y, Bouret SG (2012) Embryonic birthdate of hypothalamic leptin-activated neurons in mice. Endocrinology 153:3657–3667
Jing E, Nillni EA, Sanchez VC, Stuart RC, Good DJ (2004) Deletion of the Nhlh2 transcription factor decreases the levels of the anorexigenic peptides alpha melanocyte-stimulating hormone and thyrotropin-releasing hormone and implicates prohormone convertases I and II in obesity. Endocrinology 145:1503–1513
Jones AR, Overly CC, Sunkin SM (2009) The Allen Brain Atlas: 5 years and beyond. Nat Rev Neurosci 10:821–828

Kimura S, Hara Y, Pineau T, Fernandez-Salguero P, Fox CH, Ward JM, Gonzalez FJ (1996) The T/ebp null mouse: thyroid-specific enhancer-binding protein is essential for the organogenesis of the thyroid, lung, ventral forebrain, and pituitary. Genes Dev 10:60–69

Kruger M, Ruschke K, Braun T (2004) NSCL-1 and NSCL-2 synergistically determine the fate of GnRH-1 neurons and control necdin gene expression. EMBO J 23:4353–4364

Kuperman Y, Weiss M, Dine J, Staikin K, Golani O, Ramot A, Nahum T, Kuhne C, Shemesh Y, Wurst W et al (2016) CRFR1 in AgRP neurons modulates sympathetic nervous system activity to adapt to cold stress and fasting. Cell Metab 23:1185–1199

Lam DD, de Souza FS, Nasif S, Yamashita M, Lopez-Leal R, Otero-Corchon V, Meece K, Sampath H, Mercer AJ, Wardlaw SL et al (2015) Partially redundant enhancers cooperatively maintain mammalian pomc expression above a critical functional threshold. PLoS Genet 11: e1004935

Lasky JL, Wu H (2005) Notch signaling, brain development, and human disease. Pediatr Res 57:104r–109r

Lazzaro D, Price M, de Felice M, Di Lauro R (1991) The transcription factor TTF-1 is expressed at the onset of thyroid and lung morphogenesis and in restricted regions of the foetal brain. Development 113:1093–1104

Lee SK, Pfaff SL (2001) Transcriptional networks regulating neuronal identity in the developing spinal cord. Nat Neurosci 4(Suppl):1183–1191

Lee B, Lee S, Lee SK, Lee JW (2016) The LIM-homeobox transcription factor Isl1 plays crucial roles in the development of multiple arcuate nucleus neurons. Development 143:3763–3773

Lee B, Kim J, An T, Kim S, Patel EM, Raber J, Lee S-K, Lee S, Lee JW (2018) Dlx1/2 and Otp coordinate the production of hypothalamic GHRH- and AgRP-neurons. Nat Commun 9:2026

Lein ES, Hawrylycz MJ, Ao N, Ayres M, Bensinger A, Bernard A, Boe AF, Boguski MS, Brockway KS, Byrnes EJ et al (2007) Genome-wide atlas of gene expression in the adult mouse brain. Nature 445:168–176

Li H, Zeitler PS, Valerius MT, Small K, Potter SS (1996) Gsh-1, an orphan Hox gene, is required for normal pituitary development. EMBO J 15:714–724

Lintas C, Persico AM (2018) Unraveling molecular pathways shared by Kabuki and Kabuki-like syndromes. Clin Genet 94(3–4):283-295

Long JE, Garel S, Alvarez-Dolado M, Yoshikawa K, Osumi N, Alvarez-Buylla A, Rubenstein JL (2007) Dlx-dependent and -independent regulation of olfactory bulb interneuron differentiation. J Neurosci Off J Soc Neurosci 27:3230–3243

Lu F, Kar D, Gruenig N, Zhang ZW, Cousins N, Rodgers HM, Swindell EC, Jamrich M, Schuurmans C, Mathers PH et al (2013) Rax is a selector gene for mediobasal hypothalamic cell types. J Neurosci Off J Soc Neurosci 33:259–272

Machanick P, Bailey TL (2011) MEME-ChIP: motif analysis of large DNA datasets. Bioinformatics 27:1696–1697

Mannervik M, Nibu Y, Zhang H, Levine M (1999) Transcriptional coregulators in development. Science 284:606–609

Marin O, Anderson SA, Rubenstein JL (2000) Origin and molecular specification of striatal interneurons. J Neurosci Off J Soc Neurosci 20:6063–6076

Marin O, Baker J, Puelles L, Rubenstein JL (2002) Patterning of the basal telencephalon and hypothalamus is essential for guidance of cortical projections. Development (Cambridge, England) 129:761–773

Martinez-Ferre A, Martinez S (2012) Molecular regionalization of the diencephalon. Front Neurosci 6:73

Mazzoni EO, Mahony S, Closser M, Morrison CA, Nedelec S, Williams DJ, An D, Gifford DK, Wichterle H (2013) Synergistic binding of transcription factors to cell-specific enhancers programs motor neuron identity. Nat Neurosci 16:1219–1227

McNay DE, Pelling M, Claxton S, Guillemot F, Ang SL (2006) Mash1 is required for generic and subtype differentiation of hypothalamic neuroendocrine cells. Mol Endocrinol 20:1623–1632

Michaelson JJ, Shi Y, Gujral M, Zheng H, Malhotra D, Jin X, Jian M, Liu G, Greer D, Bhandari A et al (2012) Whole-genome sequencing in autism identifies hot spots for de novo germline mutation. Cell 151:1431–1442

Milstein AD, Bloss EB, Apostolides PF, Vaidya SP, Dilly GA, Zemelman BV, Magee JC (2015) Inhibitory gating of input comparison in the CA1 microcircuit. Neuron 87:1274–1289

Muranishi Y, Terada K, Furukawa T (2012) An essential role for Rax in retina and neuroendocrine system development. Develop Growth Differ 54:341–348

Nasif S, de Souza FS, Gonzalez LE, Yamashita M, Orquera DP, Low MJ, Rubinstein M (2015) Islet 1 specifies the identity of hypothalamic melanocortin neurons and is critical for normal food intake and adiposity in adulthood. Proc Natl Acad Sci USA 112:E1861–E1870

Niikawa N, Matsuura N, Fukushima Y, Ohsawa T, Kajii T (1981) Kabuki make-up syndrome: a syndrome of mental retardation, unusual facies, large and protruding ears, and postnatal growth deficiency. J Pediatr 99:565–569

O'Roak BJ, Vives L, Girirajan S, Karakoc E, Krumm N, Coe BP, Levy R, Ko A, Lee C, Smith JD et al (2012) Sporadic autism exomes reveal a highly interconnected protein network of de novo mutations. Nature 485:246–250

Padilla SL, Carmody JS, Zeltser LM (2010) Pomc-expressing progenitors give rise to antagonistic neuronal populations in hypothalamic feeding circuits. Nat Med 16:403–405

Pan Q, Li C, Xiao J, Kimura S, Rubenstein J, Puelles L, Minoo P (2004) In vivo characterization of the Nkx2.1 promoter/enhancer elements in transgenic mice. Gene 331:73–82

Pelling M, Anthwal N, McNay D, Gradwohl G, Leiter AB, Guillemot F, Ang SL (2011) Differential requirements for neurogenin 3 in the development of POMC and NPY neurons in the hypothalamus. Dev Biol 349:406–416

Peng SY, Wang WP, Meng J, Li T, Zhang H, Li YM, Chen P, Ma KT, Zhou CY (2005) ISL1 physically interacts with BETA2 to promote insulin gene transcriptional synergy in non-beta cells. Biochim Biophys Acta 1731:154–159

Petryniak MA, Potter GB, Rowitch DH, Rubenstein JL (2007) Dlx1 and Dlx2 control neuronal versus oligodendroglial cell fate acquisition in the developing forebrain. Neuron 55:417–433

Price M, Lazzaro D, Pohl T, Mattei MG, Ruther U, Olivo JC, Duboule D, Di Lauro R (1992) Regional expression of the homeobox gene Nkx-2.2 in the developing mammalian forebrain. Neuron 8:241–255

Qiu M, Bulfone A, Martinez S, Meneses JJ, Shimamura K, Pedersen RA, Rubenstein JL (1995) Null mutation of Dlx-2 results in abnormal morphogenesis of proximal first and second branchial arch derivatives and abnormal differentiation in the forebrain. Genes Dev 9:2523–2538

Rowitch DH, Kriegstein AR (2010) Developmental genetics of vertebrate glial-cell specification. Nature 468:214–222

Saino-Saito S, Berlin R, Baker H (2003) Dlx-1 and Dlx-2 expression in the adult mouse brain: relationship to dopaminergic phenotypic regulation. J Comp Neurol 461:18–30

Sakkou M, Wiedmer P, Anlag K, Hamm A, Seuntjens E, Ettwiller L, Tschop MH, Treier M (2007) A role for brain-specific homeobox factor Bsx in the control of hyperphagia and locomotory behavior. Cell Metab 5:450–463

Sanz E, Quintana A, Deem JD, Steiner RA, Palmiter RD, McKnight GS (2015) Fertility-regulating Kiss1 neurons arise from hypothalamic POMC-expressing progenitors. J Neurosci Off J Soc Neurosci 35:5549–5556

Schmid T, Gunther S, Mendler L, Braun T (2013) Loss of NSCL-2 in gonadotropin releasing hormone neurons leads to reduction of pro-opiomelanocortin neurons in specific hypothalamic nuclei and causes visceral obesity. J Neurosci Off J Soc Neurosci 33:10459–10470

Shimogori T, Lee DA, Miranda-Angulo A, Yang Y, Wang H, Jiang L, Yoshida AC, Kataoka A, Mashiko H, Avetisyan M et al (2010) A genomic atlas of mouse hypothalamic development. Nat Neurosci 13:767–775

Simeone A, D'Apice MR, Nigro V, Casanova J, Graziani F, Acampora D, Avantaggiato V (1994) Orthopedia, a novel homeobox-containing gene expressed in the developing CNS of both mouse and Drosophila. Neuron 13:83–101

Sokolowski K, Esumi S, Hirata T, Kamal Y, Tran T, Lam A, Oboti L, Brighthaupt SC, Zaghlula M, Martinez J et al (2015) Specification of select hypothalamic circuits and innate behaviors by the embryonic patterning gene dbx1. Neuron 86:403–416

Sokolowski K, Tran T, Esumi S, Kamal Y, Oboti L, Lischinsky J, Goodrich M, Lam A, Carter M, Nakagawa Y et al (2016) Molecular and behavioral profiling of Dbx1-derived neurons in the arcuate, lateral and ventromedial hypothalamic nuclei. Neural Dev 11:12

Sussel L, Marin O, Kimura S, Rubenstein JL (1999) Loss of Nkx2.1 homeobox gene function results in a ventral to dorsal molecular respecification within the basal telencephalon: evidence for a transformation of the pallidum into the striatum. Development 126:3359–3370

Tsai HH, Li H, Fuentealba LC, Molofsky AV, Taveira-Marques R, Zhuang H, Tenney A, Murnen AT, Fancy SP, Merkle F et al (2012) Regional astrocyte allocation regulates CNS synaptogenesis and repair. Science (New York, NY) 337:358–362

Valerius MT, Li H, Stock JL, Weinstein M, Kaur S, Singh G, Potter SS (1995) Gsh-1: a novel murine homeobox gene expressed in the central nervous system. Dev Dyn 203:337–351

Visel A, Thaller C, Eichele G (2004) GenePaint.org: an atlas of gene expression patterns in the mouse embryo. Nucleic Acids Res 32:D552–D556

Voogt JL, Lee Y, Yang S, Arbogast L (2001) Regulation of prolactin secretion during pregnancy and lactation. Prog Brain Res 133:173–185

Wang W, Lufkin T (2000) The murine Otp homeobox gene plays an essential role in the specification of neuronal cell lineages in the developing hypothalamus. Dev Biol 227:432–449

Westerlund J, Andersson L, Carlsson T, Zoppoli P, Fagman H, Nilsson M (2008) Expression of Islet1 in thyroid development related to budding, migration, and fusion of primordia. Dev Dyn 237:3820–3829

Xu Q, Tam M, Anderson SA (2008) Fate mapping Nkx2.1-lineage cells in the mouse telencephalon. J Comp Neurol 506:16–29

Yee CL, Wang Y, Anderson S, Ekker M, Rubenstein JL (2009) Arcuate nucleus expression of NKX2.1 and DLX and lineages expressing these transcription factors in neuropeptide Y(+), proopiomelanocortin(+), and tyrosine hydroxylase(+) neurons in neonatal and adult mice. J Comp Neurol 517:37–50

Zhang L, Mathers PH, Jamrich M (2000) Function of Rx, but not Pax6, is essential for the formation of retinal progenitor cells in mice. Genesis 28:135–142

Winding the Clock: Development of Hypothalamic Structures Controlling Biological Timing and Sleep

5

Dong Won Thomas Kim and Seth Blackshaw

Abstract

The daily cycle of sleep and wake governs all aspects of behavior and is under tight homeostatic regulation. The core circadian oscillator in turn restricts sleep to specific intervals of the solar day. The neurons that regulate both sleep and circadian timing reside largely in the hypothalamus, and are linked in a complex, reciprocally connected, and still incompletely characterized network. While our knowledge of the neural circuitry controlling circadian rhythms and sleep has advanced considerably, our understanding of the mechanisms by which they are assembled during embryonic and postnatal development lag far behind. In this chapter, we review advances in the understanding of hypothalamic circuitry controlling circadian rhythms and sleep and our current knowledge of the gene regulatory networks that guide their development. In addition, we discuss the

D. W. T. Kim
Solomon H. Snyder Department of Neuroscience, Johns Hopkins University School of Medicine, Baltimore, MD, USA

S. Blackshaw (✉)
Solomon H. Snyder Department of Neuroscience, Johns Hopkins University School of Medicine, Baltimore, MD, USA

Department of Ophthalmology, Johns Hopkins University School of Medicine, Baltimore, MD, USA

Department of Neurology, Johns Hopkins University School of Medicine, Baltimore, MD, USA

Center for Human Systems Biology, Johns Hopkins University School of Medicine, Baltimore, MD, USA

Institute for Cell Engineering, Johns Hopkins University School of Medicine, Baltimore, MD, USA
e-mail: sblack@jhmi.edu

S. Wray, S. Blackshaw (eds.), *Developmental Neuroendocrinology*, Masterclass in Neuroendocrinology 9, https://doi.org/10.1007/978-3-030-40002-6_5

potential for future studies of hypothalamic development to provide new insights into the mechanisms by which animals regulate the timing and duration of sleep.

Keywords
Circadian · Development · Hypocretin · Sleep · Suprachiasmatic · Timing · Transcription factor · Zona incerta

5.1 Fast, Slow, and Timely Hypothalamic Processes Controlling Sleep

5.1.1 Introduction

The hypothalamus has long been known to be a central and critical regulator of sleep. Focal lesions to the anterior hypothalamus and preoptic area often result in insomnia, while lesions to the posterior hypothalamus result in lethargy and hypersomnia (Scammell et al. 2017; Weber and Dan 2016). Several distinct neural processes interact to regulate sleep. First of these is the neural circuitry that directly regulates sleep/wake transitions. Sleep and wakefulness are mutually exclusive states, and transitions between these states occur rapidly, with "half asleep" being a state that lasts only seconds. Reciprocal transitions between wake and sleep can occur frequently, particularly in species such as mice and human infants. An analogous process in turn mediates rapid, biphasic, and bidirectional transitions between non-rapid eye movement (NREM) and rapid eye movement (REM) sleep phases. The neural circuitry that controls sleep and wake involves three major components. The first and most extensively studied directly regulates the transition between wake, NREM, and REM sleep. The second and third, which are much less well understood, regulate homeostatic sleep pressure, and circadian changes in sleep pressure, respectively. Sensory signals and changes in internal states can in turn modulate sleep/wake transitions. An overview of hypothalamic cell types that control sleep, wakefulness, and biological timing is shown in Fig. 5.1.

5.1.2 Neural Circuitry Directly Regulating Sleep/Wake Transitions

Recent studies, particularly since the advent of optogenetics, have identified several hypothalamic neuronal subtypes whose activity is sleep regulated, and which rapidly induce either wakefulness or sleep following activation or inhibition (Fig. 5.2a). Primary wake-promoting neurons act by sending widespread, and typically excitatory, projections throughout the forebrain. In the hypothalamus, these include the histaminergic neurons of the tuberomammillary nucleus (TMN), whose wake-promoting signals are blunted by conventional antihistamines; orexinergic neurons of the lateral hypothalamus (LH), which degenerate in narcoleptic patients; and other populations such as glutamatergic neurons of the supramammilary nucleus (Pedersen

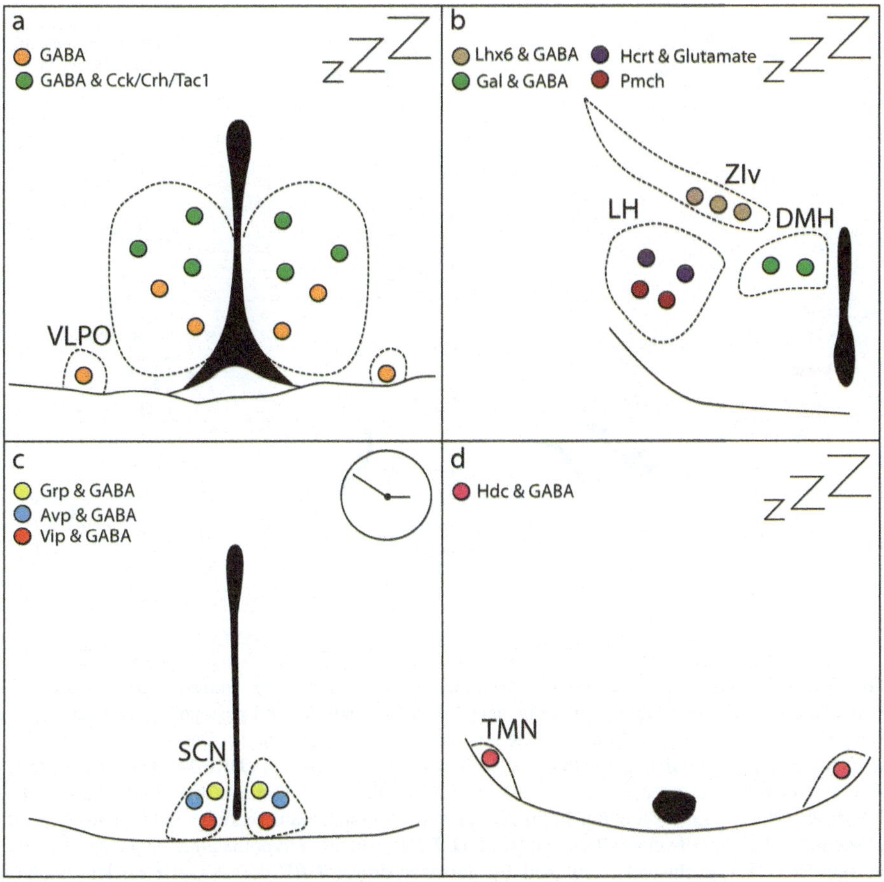

Fig. 5.1 Locations of hypothalamic sleep and wake-promoting neurons. (**a**) Sleep-promoting neurons of the preoptic area. (**b**) Sleep and wake-promoting neurons of the dorsolateral hypothalamus. (**c**) Neuronal subtypes of the suprachiasmatic nucleus regulating biological timing. (**d**) Wake-promoting neurons of the posteroventral hypothalamus. Diagrams of transverse sections through hyopothalamus, centered on third ventricle (black). Abbreviations: *Avp* arginine vasopressin, *Cck* cholecystokinin, *Crh* corticotropin-releasing hormone, *DMH* dorsomedial hypothalamic nucleus, *GABA* gamma-amino butyric acid, *Gal* galanin, *Grp* gastrin-related peptide, *Hdc* histamine decarboxylase, *LH* lateral hypothalamus, *Pmch* pro-melanin concentrating hormone, *SCN* suprachiasmatic nucleus, *Tac1* tachykinin 1, *TMN* tuberomammillary nucleus, *Vip* vasoactive intestinal peptide, *VLPO* ventrolateral preoptic area, *ZIv* zona incerta, ventral region

et al. 2017). Alternatively, wake-promoting neurons can be inhibitory and often function by directly inhibiting the activity of sleep-promoting neurons. Examples of such inhibitory wake-promoting hypothalamic neurons include subpopulations of GABAergic neurons in the LH (Venner et al. 2016) and in the preoptic area (POA) (Chung et al. 2017) (Fig. 5.2a).

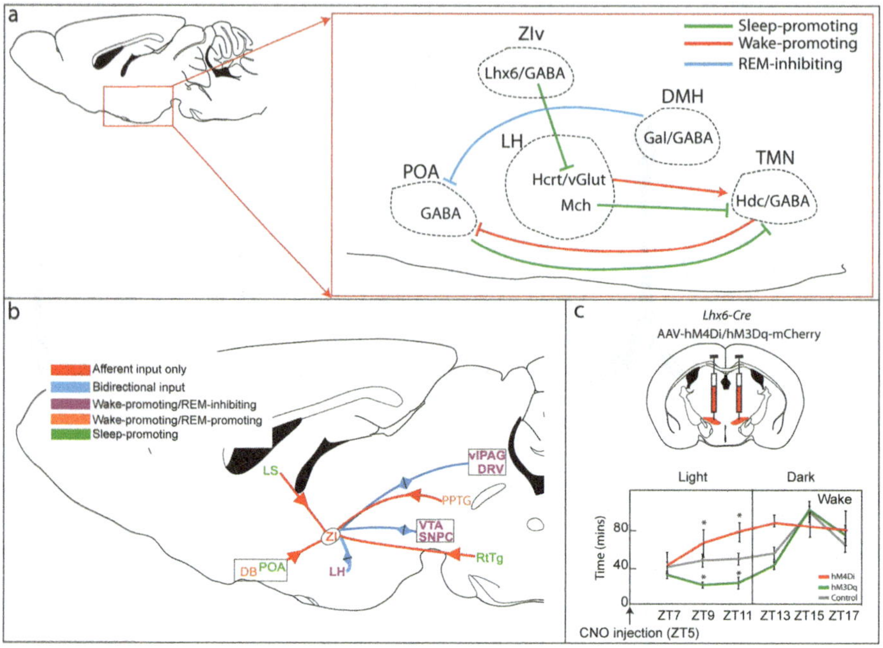

Fig. 5.2 Connections and actions of hypothalamic sleep-regulating neurons. (**a**) Schematic of hypothalamic neurons directly regulating sleep and wake. (**b**) Pre- and postsynaptic connections of sleep pressure-regulating Lhx6-expressing neurons of the zona incerta. (**c**) Delayed effects on sleep of chemogenetic-dependent activation and inhibition of Cre-dependent DREADD constructs delivered Lhx6-expressing neurons of the zona incerta by bilateral stereotaxic injection [adapted from Liu et al. (2017)]. (**a**, **b**) Sagittal brain sections; (**c**) coronal brain section. Abbreviations: *CNO* clozapine-*N*-oxide, *DB* ventral diagonal band, *DMH* dorsomedial hypothalamic nucleus, *DREADD* designer receptor exclusively activated by designer drugs, *DRV* dorsal raphe nucleus, ventral region, *LH* lateral hypothalamus, *LS* lateral septum, *POA* preoptic area, *PPTG* pedunculopontine tegmental nucleus, *REM* rapid eye movement, *RtTg* rostral tegmental nucleus, *SNPC* substantia nigra, pars compacta, *TMN* tuberomammillary nucleus, *vlPAG* ventrolateral periaqueductal grey, *VTA* ventral tegmental area, *ZIv* zona incerta, ventral region, *ZT* zeitgeber time

Conversely, primary sleep-promoting neurons identified thus far are invariably GABAergic, and send direct, inhibitory inputs to wake-promoting neurons. Examples include sleep-active GABAergic POA neurons, which directly suppress the activity of histaminergic and orexinergic neurons (Chung et al. 2017; Saito et al. 2013). Other sleep-promoting hypothalamic neurons, such as MCH-positive LH neurons, do so by directly inhibiting wake-promoting neurons in the midbrain and brainstem (Ferreira et al. 2017). Sleep-promoting hypothalamic neurons, in turn, generally directly promote either NREM or REM sleep, with GABAergic POA neurons selectively promoting NREM and galaninergic DMH neurons promoting REM (Chen et al. 2018; Chung et al. 2017). Wake-promoting neurons send reciprocal projections to sleep-promoting neurons, and in turn inhibit these cells, usually by activation of Gi-coupled GPCRs. This biphasic nature of sleep and wakefulness is thus encoded in the core neural circuitry that controls the initiation and maintenance of sleep.

5.1.3 Neural Circuitry Regulating Homeostatic Sleep Pressure

A second, and much less well understood, component of sleep-regulatory circuitry are the neural mechanisms that regulate homeostatic sleep pressure (Fig. 5.1). Homeostatic sleep pressure accumulates continuously during time spent awake, is measured separately for NREM and REM sleep phases, and is gradually dissipated during sleep (Borbely 1982). Sleep deprivation, once relieved, leads to a compensatory rebound in sleep time (Brunner et al. 1990). The molecular signals of sleep pressure are still mostly obscure but are known to include byproducts of normal cellular metabolism such as adenosine, which activates GPCRs that are blocked by caffeine (Elmenhorst et al. 2007, 2017). The neural circuitry that encodes sleep pressure, and relays these signals to the neural circuitry that directly regulates sleep/wake transition, is even less well characterized. In contrast to the fast-acting core circuitry that directly regulates sleep/wake transitions, it is both slow acting and highly stable in the face of sensory input and changes in internal state. Although nNOS-positive cortical interneurons are activated by elevated sleep pressure (Dittrich et al. 2015; Morairty et al. 2013), little is still known about how sleep pressure is detected by hypothalamic sleep-regulatory circuitry.

Insight into how such circuitry may be organized comes from the recent identification of Lhx6-positive GABAergic neurons of the hypothalamic zona incerta (ZI) as candidate regulators of homeostatic sleep pressure (Liu et al. 2017). These neurons are activated in response to accumulating sleep pressure, and are both directly presynaptic to multiple populations of wake-promoting neurons and extensively interconnected with another (Fig. 5.2b). Direct activation and inhibition of these cells produce robust but slow-onset changes in sleep/wake patterns (Fig. 5.2c), likely reflecting the effect of inputs from these, which counteract and modulate direct inhibitory inputs onto directly wake-promoting neurons. Other extensively interconnected populations of GABAergic neurons are found in both the hypothalamus and other brain regions and may also play a similar role in signaling homeostatic sleep pressure.

5.1.4 Neural Circuitry Regulating Circadian Sleep Pressure

A third component is controlled by the core circadian clock located in the suprachiasmatic nucleus (SCN) of the hypothalamus and modulates sleep pressure depending on time of day (Fig. 5.1) (Daan et al. 1984). It signals elevated sleep pressure during the dark phase for diurnal species and during the light phase for nocturnal species. This master clock regulates not only the sleep/wake cycle but also controls virtually all aspects of mammalian physiology (Reddy and O'Neill 2009). Much like the circuitry regulating homeostatic sleep pressure, changes in the activity of this component of the sleep-regulatory circuit occur slowly, though cyclically. Moreover, the core circadian oscillator makes extensive use of negative feedback at both the cellular level and in local circuitry to encode these slow but robust changes in activity levels.

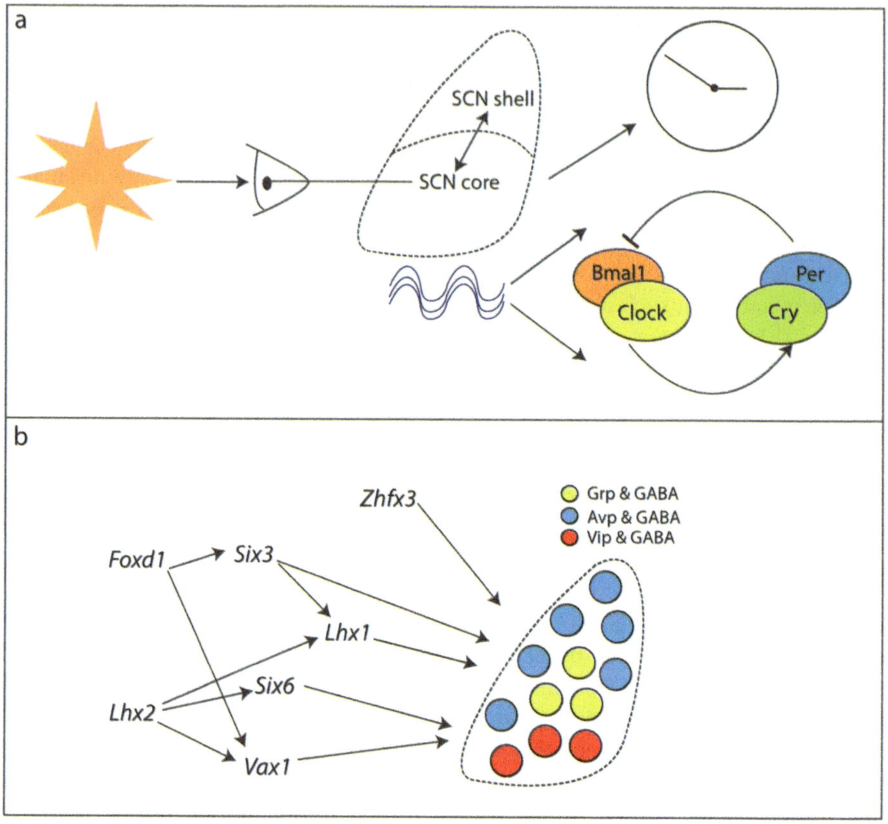

Fig. 5.3 Suprachiasmatic nucleus (SCN): circadian rhythms and neuronal differentiation. (**a**) Overview of light-dependent regulation of the core circadian oscillator in the SCN. Light is relayed to the SCN via intrinsically photosensitive retinal ganglion cells to regulate rhythmic gene expression. SCN neurons. (**b**) Transcriptional regulatory networks (left) controlling SCN differentiation, (right) and controlling expression of clock-regulating neuropeptides (*Avp* arginine vasopressin, *Grp* gastrin releasing peptide, *Vip* vasoactive intestinal peptide)

Like all cells in the adult, a biochemical circadian oscillator is active in neurons of the SCN (Fig. 5.3). This takes the form of a negative feedback loop, with a heterodimeric transcriptional complex consisting of the transcription factors *Bmal1* and *Clock*, which drive the expression of *Per* and *Cry* genes, which in turn feedback to inhibit their own transcription. This core oscillator is reinforced and stabilized by linked transcriptional negative feedback loops, involving the nuclear hormone receptor superfamily members *Rora*, *Rorb*, and *Nr1d1* (Partch et al. 2014). In sharp contrast to virtually all other tissues, however, the cellular clocks of SCN neurons undergo spontaneous synchronization when grown in dispersed cell culture and remain robustly synchronized when maintained ex vivo in intact organ culture (Mohawk and Takahashi 2011). Moreover, while cellular clocks in other tissues are readily reset and synchronized by a range of external signals—including temperature, metabolites, and growth factors—cells in the SCN robustly retain their phase and synchrony under a broad range of treatments. These unique features arise at least in part as a result of the organization of local neural circuitry in the SCN. Like the ZI, the SCN consists of interconnected GABAergic neurons, and this extensive collateral inhibition buffers the SCN against perturbation. Inhibition of neuronal excitability in the SCN reverses this stability, allowing cellular oscillators to be readily reset by light and temperature signals that would normally produce no change in phase (Buhr et al. 2010; Yamaguchi et al. 2013). SCN neurons also release a diverse array of neuropeptides—including Vip, Avp, Grp, Penk, and Vgf—which likewise contribute to phase synchrony and stabilization. While the molecular mechanisms and, to a lesser extent, the local circuitry that controls circadian timekeeping in the SCN, are now fairly well understood, the neural circuitry that relays these signals from the SCN to other brain regions is not.

5.1.5 Sensory and Internal Signals Regulating Sleep/Wake Transitions

Finally, a range of sensory and internal signals can modulate sleep, usually by directly promoting wakefulness without altering either homeostatic or circadian sleep pressure. Sensory signals include loud noises, temperature extremes, and pain, with the most complex of these in its effects being light. Light directly inhibits (in the case of diurnal mammals) or promotes (in the case of nocturnal mammals) the onset of sleep. Moreover, while circadian sleep pressure is highly robust in the face of most sensory and internal signals, it can be reset by light signals relayed directly from intrinsically photosensitive retinal ganglion cells (ipRGCs) (Hattar et al. 2002). These ipRGCs directly activate neurons in the ventral SCN, which in turn leads to a gradual and progressive resetting of cellular SCN clocks. This phase shift occurs at the rate of approximately 1 h per day, with misalignment of circadian and clock time leading to the phenomenon of jet lag (Sack et al. 2007). Some examples of internal states that acutely modulate sleep (though not sleep pressure) include sympathetic and behavior arousal, such as hunger and thirst, or psychotropic drugs that directly target arousal circuitry (Saper et al. 2010). Despite the strong and rapid effects of these signals on sleep, surprisingly little is known about the circuitry mediating clock-dependent regulation of sleep. In addition, virtually nothing is known about the mechanisms by which changes in sleep state in turn induce sleep-dependent changes in internal states and sensory processing.

5.1.6 Ontogeny of Sleep–Wake Cycles and Behavioral Rhythmicity

As any parent of a newborn knows, light and circadian-dependent regulation of sleep/wake cycle develops gradually during the postnatal period. Using standard EEG recordings, wake, NREM, and REM sleep can be clearly distinguished in premature human infants at 6 months gestational age (Foreman et al. 2008), and in rodents between postnatal day 12 and 20 (Daszuta and Gambarelli 1985; Frank and Heller 1997; Nelson et al. 2013). However, analysis of other sleep-regulated behaviors, such as muscle tone measured by electromyogram (EMG), suggest that sleep/wake transitions may be evident in rodents as early as postnatal day 2 (Gall et al. 2008). Light-dependent regulation of sleep–wake cycles is evident by weaning in rodents (Nelson et al. 2013; Sack et al. 2007) and begin to emerge by 10–12 weeks postnatal in humans (Davis et al. 2004). Circadian control of circulating cortisol and core body temperature also becomes evident around this age (Joseph et al. 2015), and this is likely the point at which sleep/wake cycles also come under circadian regulation (Blumberg et al. 2014; Galland et al. 2012). This process occurs long after molecular rhythms are first detected in the SCN. These are first observed by E15, coincident with expression of the first SCN-enriched neuropeptides, and become strong and well consolidated by postnatal day 2 (Carmona-Alcocer et al. 2018), implying a substantial delay in the development of efferent neural circuitry relaying SCN output. There is a gradual shift in sleep homeostasis through life, with progressively less time spent in sleep, with major reductions observed first in REM, then in deep NREM (Dzierzewski et al. 2018; Miles and Dement 1980). Sleep/wake phases are relatively phase-advanced in childhood, phase-delayed in adolescence and early adulthood, and phase-advanced and less well consolidated in the elderly. Both Mendelian and multifactorial genetic variation control the extent to which sleep/wake cycles are phase-advanced or phase-delayed relative to the solar day, giving rise to what are known as chronotypes (Roenneberg et al. 2003). This chronotypic variation in sleep/wake timing peaks in early adulthood and declines slowly throughout the lifespan (Fischer et al. 2017).

5.2 Development of Sleep-Regulatory Circuitry

5.2.1 Introduction

Like all CNS structures, the developing hypothalamus is initially patterned by gradients of growth and guidance factors, which then induce expression of transcription factors in a concentration-dependent manner. These broadly expressed transcription factors then act in a combinatorial fashion to specify individual subregions prior to the onset of hypothalamic neurogenesis. Once neurogenesis begins, many of these same transcription factors have additional roles in controlling the specification and differentiation of individual neuronal subtypes that regulate biological timing and sleep, among other physiological behaviors. Finally, a largely uncharacterized set of molecular cues control the formation of axonal projections and synaptic connections that underpin these circuits. We will review our current knowledge of how these processes guide the formation of the sleep-regulating cells for which this is best understood: the neurons of the SCN and the dorsolateral hypothalamus.

5.2.2 Early Hypothalamic Patterning and Role of Extrinsic Factors

The POA derives from *Foxg1*-positive telencephalic neuroepithelium, while the hypothalamus proper arises from *Foxd1*-positive diencephalic neuroepithelium. The extrinsic factors discussed here arise from different sources and have substantially different effects on the patterning of these structures.

5.2.2.1 Shh

Shh (sonic hedgehog) has been studied more extensively than any other extrinsic factor in the context of hypothalamic development. Shh expression in the notochord, and its anterior derivative the prechordal plate, leads to induction of Shh expression in the ventral floor plate all along the neuraxis, including in hypothalamic neuroepithelium. Loss of function of Shh at this stage blocks development of all hypothalamic structures (Blaess et al. 2014). Soon thereafter, Shh expression is induced in a more dorsally located region known at the basal plate domain, and floorplate expression is suppressed. Selective loss of function of Shh in the basal plate domain prior to the onset of neurogenesis leads to the loss of all anterior and tuberal hypothalamic markers, along with severe defects in neural progenitor proliferation. This prevents development of all sleep-associated neurons, with the exception of supramammilary nucleus neurons (Shimogori et al. 2010).

5.2.2.2 Canonical Wnt Signaling

Canonical Wnt signaling also plays a critical role in hypothalamic patterning and the development of sleep-regulatory neurons. Selective loss of function of beta-catenin—which is essential for mediating canonical Wnt signaling—in hypothalamic neuroepithelium leads to a reduction in the number of multiple sleep-regulatory neurons, including GABAergic neural precursors in the SCN and LH, as well as orexinergic neural precursors (Newman et al. 2018b). Overexpression of constitutively active beta-catenin blocks specification of all hypothalamic GABAergic neurons but increases generation of LH orexinergic neurons (Newman et al. 2018b). Milder phenotypes are seen in mice and zebra fish showing altered expression of *Lef1*, a major transcriptional effector for canonical Wnt signaling (Lee et al. 2006; Xie et al. 2017). Loss of function of *Lef1* selectively disrupts development of LH MCH neurons, while the gain of function increases their number.

5.2.2.3 Other Extrinsic Factors

Other extrinsic factors are known to regulate hypothalamic development. In chick, a temporal gradient of FGF signaling plays a critical role in anteroposterior patterning and proliferation of early hypothalamic progenitors (Fu et al. 2017). At later stages of chick development, BMP signaling specifies anterioventral identity and drives proliferation of ventral hypothalamic progenitors (Manning et al. 2006; Ohyama et al. 2008). Finally, Notch/Delta signaling has been shown to regulate proliferation and neurogenesis in both anterior and tuberal hypothalamic progenitors (Aujla et al. 2013; Biehl et al. 2018). Although it is likely that all of these pathways play an

a

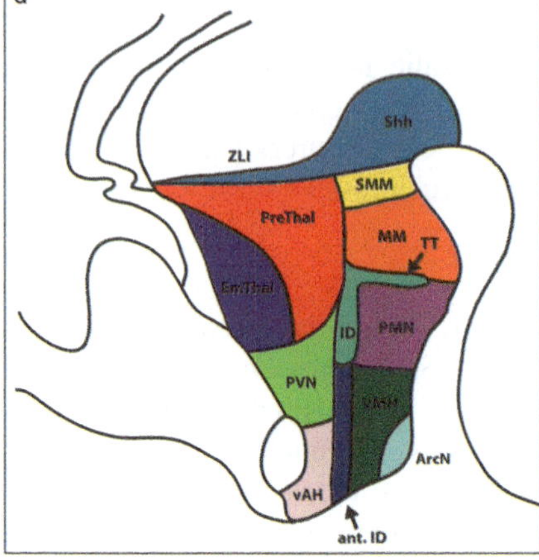

b

VMH	PVN	ID/TT	PMN	MM	SMM
Fezf1	*Otp*	*Lhx6*	*Hmx2*	*Pcp4*	*Pitx2*
Nr0b1	*Sim1*	*Dlx1/2*	*Hmx3*	*Foxb1*	*Barhl1*
Nr5a1	*Neurod1*	*Arx*	*Lef1*	*Lhx1*	*Irx5*
Sox14	*Neurod6*	*Lhx1*	*Bub3*	*Pitx2*	*Nr4a2*
S100a10	*Sim2*	*Lhx8*	*Cited1*	*Nrsn1*	*Lmx1a*

Fig. 5.4 Molecular organization of the hypothalamus and prethalamus in the embryonic (day E12) mouse brain. (**a**) Parasagittal schematic (adapted from Shimogori et al. (2010)). (**b**) Selected molecular markers for the indicated regions. Abbreviations: from top to bottom, left group: *ZLI* zona limitans intrathalamica, *PreThal* prethalamus, *EmThal* thalamic eminence, *PVN* hypothalamic paraventricular nucleus, *vAH* ventral anterior hypothalamus, *ID* intrahypothalamic diagonal, *TT* tuberomammillary terminal, *VMH* ventromedial hypothalamus, *ArcN* arcuate nucleus, *PMN* premammillary nucleus, *MM* mammillary nucleus, *SMM* supramammillary nucleus, *Shh* sonic hedgehog

important role in development of sleep-regulatory neurons in mammals, their specific function has not been investigated.

5.2.3 A Temporal Cascade of Intrinsic Factors Controlling SCN Development

Despite its small size (~20,000 neurons total), the physiological importance and compact, easily identifiable structure of the SCN has resulted in the transcription factors that control its development being better characterized than any other structure in the hypothalamus. Nearly a decade of work has made considerable progress in identifying a hierarchical network of transcription factors that controls specification, differentiation, and adult function of the SCN (Fig. 5.4).

5.2.3.1 Dual Regulation of SCN Identity by *Lhx2*

The LIM homeodomain transcription factor *Lhx2* is expressed during early stages of development in both optic neuroepithelium and adjacent neuroepithelium that give rise to the anterioventral hypothalamus (Porter et al. 1997; Shimogori et al. 2010). Loss of function of *Lhx2* blocks expression of genes specific to both the eye field and the SCN. Strikingly, this also leads to an ectopic upregulation of genes specific to the adjacent neuroendocrine paraventricular nucleus (PvN) and thalamic eminence

(EmThal) (Roy et al. 2013). In the neuroretina, the repressive function of *Lhx2* persists even after neurogenesis has begun (de Melo et al. 2016; Roy et al. 2013).

5.2.3.2 Promotion of SCN Identity and Nucleogenesis by *Six3*, *Six6*, and *Vax1*

The homeodomain transcription factors *Six3* is broadly expressed in anterior hypothalamic neuroepithelium, while *Six6* and *Vax1* show a more restricted expression that is confined to anterioventral hypothalamus and eye field. *Lhx2* is necessary to maintain expression of *Six6* and *Vax1*, though not *Six3*, in this region (Roy et al. 2013). Loss of function of *Six3* in hypothalamic neuroepithelium leads to a loss of all SCN-specific molecular markers, along with the absence of a histologically defined SCN (VanDunk et al. 2011). *Six6* loss of function results in milder and more variable defects, with some persistence of SCN-specific markers, and moderate to severe defects in SCN nucleogenesis (Clark et al. 2013). *Vax1* defects are milder yet and exhibit only moderate defects in expression of SCN-specific genes (Hanne et al. 2016). Although both *Six3* and *Six6* expression persist in adult SCN (Clark et al. 2013; VanDunk et al. 2011), loss of *Six3* expression in SCN neural precursors produced no obvious developmental phenotype.

5.2.3.3 Delayed Action of Foxd1 in Maintenance of SCN Differentiation

The Forkhead domain transcription factor *Foxd1* is broadly, transiently, and selectively expressed in hypothalamic neuroepithelial cells, although expression is maintained longest in anterior hypothalamus (Salvatierra et al. 2014; Shimogori et al. 2010). The loss of function of *Foxd1* results in a severe, although puzzlingly delayed, defect in SCN differentiation. *Foxd1* mutants initially show normal expression of early SCN markers, although they display modest reductions in expression of both *Six3* and *Vax1* expression (Newman et al. 2018a). However, beginning at E16 in mouse, once neurogenesis is completed, progressive defects in expression of SCN-specific genes are observed. By P0, virtually all SCN-specific markers are lost, along with all histological evidence for a distinct SCN (Newman et al. 2018a). The mechanism mediating these effects is unclear, and the long delay between termination of *Foxd1* expression and phenotype onset is striking. It is possible that loss of function of *Foxd1* triggers epigenetic changes that selectively inactivate cis-regulatory elements that maintain expression of SCN-specific genes or other transcription factors required for SCN differentiation at later developmental ages.

5.2.3.4 *Lhx1* Controls SCN Terminal Differentiation

The first selective marker of postmitotic SCN neural precursors is the LIM homeodomain factor *Lhx1*, which is detected beginning at E11 in mouse. *Lhx1* expression in SCN is lost in *Lhx2*- and *Six3*-mutants and, while initially normal in *Foxd1*-mutants, is lost by P0 (Table 5.1). Selective loss of function of *Lhx1* in anterior hypothalamic neuroepithelium does not disrupt either initial SCN specification or nucleogenesis. Instead, mutant animals lose expression of genes that are selectively expressed in mature SCN neurons, including virtually all SCN-enriched neuropeptides, including *Vip*, *Avp*, and *Grp* (Bedont et al. 2014). Selective loss of

Table 5.1 Behavioral defects in mouse mutants that disrupt development of hypothalamic neurons controlling sleep and circadian timing

Mutant line	Cellular phenotype	Behavioral phenotype	References
Six3-Cre: Lhx1$^{lox/lox}$	Broadly decreased expression of SCN-enriched neuropeptides.	Loss of SCN-enriched neuropeptides. Disruption of activity, sleep, and body temperature rhythms. Loss of resistance to temperature-dependent resetting of SCN clock. Loss of light-dependent regulation of sleep.	Bedont et al. (2014, 2017)
Rora-Cre: Lhx1$^{lox/lox}$	Decreased expression of a subset of SCN-enriched neuropeptides.	Moderate defects in circadian activity rhythms. Enhanced light-dependent resetting of circadian phase.	Hatori et al. (2014)
Zfhx3$^{sci/sci}$	Broadly reduced expression of SCN-enriched neuropeptides.	Defective circadian activity rhythms. Altered sleep and interval timing.	Balzani et al. (2016) and Parsons et al. (2015)
Ubc-CreER; Zfhx3$^{lox/lox}$	Not determined.	Disrupted circadian activity rhythms.	Wilcox et al. (2017)
Foxd1-Cre; Lhx6$^{lox/lox}$	Loss of hypothalamic Lhx6+ neurons.	Reduced sleep time, increased wake time.	Liu et al. (2017)
Lhx9$^{-/-}$	Reduction in Hcrt+ neuron number.	Reduced locomotor activity and increased wake time.	Dalal et al. (2013)

function in postmitotic SCN neural precursors results in a milder but broadly similar phenotype (Hatori et al. 2014). *Lhx1* expression is maintained in adult SCN (VanDunk et al. 2011) and is regulated by both the circadian clock and by light (Hatori et al. 2014).

Changes in gene expression in *Lhx1*-deficient mice appear to reflect both the direct and indirect effects of *Lhx1*. *Lhx1* directly regulates *Vip* expression (Hatori et al. 2014) and *Vip* shows the strongest effects of all SCN-enriched neuropeptides on regulation of cellular and behavioral rhythms (Maywood et al. 2011). Loss of function of *Vip* leads to defects in circadian rhythmicity, and loss of expression of a large fraction of *Lhx1*-dependent genes, including neuropeptides such as *Avp* and *Penk*, implying that *Lhx1*-dependent regulation of *Vip* expression accounts for these effects (Bedont et al. 2017). However, expression of other SCN-enriched neuropeptides such as *Grp*, *Nms*, and *Prok2* is unaffected by *Vip* loss of function, implying that *Lhx1* may act as a master regulator of SCN neuropeptide expression (Table 5.1).

5.2.3.5 *Zfhx3* Regulates Expression of SCN-Enriched Neuropeptides

The zinc finger homeodomain factor *Zfhx3* is expressed in both neural progenitors and postmitotic precursor cells of the anterioventral hypothalamus (Parsons et al.

2015). *Zfhx3* hypomorphic mutants resemble *Lhx1* mutants in showing lower expression levels of SCN-enriched neuropeptides—including *Vip*, *Grp*, *Nms*, and *Prok2* (though not *Avp*)—along with a limited subset of other genes. Both mutants show normal SCN specification and nucleogenesis are maintained. Like *Lhx1*, *Zfhx3* directly regulates transcription of SCN-enriched neuropeptides (Parsons et al. 2015). However, the total number of SCN neurons expressing these neuropeptides is unchanged, implying that *Zfhx3* is not necessary for the expression of these genes but rather regulates their expression levels. Also like *Lhx1*, *Zfhx3* is expressed in mature SCN neurons and is required in adult to maintain normal expression of many of these same SCN-enriched genes (Wilcox et al. 2017) (Table 5.1).

5.2.3.6 Other Candidate Regulators of SCN Differentiation and Function

Gene expression analysis has identified other transcription factors expressed in anterioventral hypothalamus and/or SCN neural precursors that are strong candidates for controlling SCN development. These include homeodomain transcription factors such as *Dlx* family members and *Arx*, which are also necessary for normal development of telencephalic interneurons (Shimogori et al. 2010). Two *Zfhx3* paralogues—*Zfhx2* and *Zfhx4*—are also expressed in SCN and associated structures. Like *Zfhx3*, they show reduced expression in *Foxd1* mutant mice (Newman et al. 2018a) and may play a similar role in control of SCN differentiation and/or function.

5.2.4 Intrinsic Factors Controlling Specification of Wake and Sleep-Promoting Neurons in the Posterolateral Hypothalamus

Much less is known about transcription factors that control development of sleep-regulating neurons in the posterior hypothalamus, with work on this topic having mostly focused on the function of the LIM homeodomain factors *Lhx6* and *Lhx9*. *Lhx6* and *Lhx9* are expressed in adjacent, nonoverlapping regions of the posterolateral hypothalamus from E11 into adulthood in mouse (Shimogori et al. 2010). *Lhx6* expression is ultimately restricted to GABAergic neurons in the ventral ZI, DMH, and LH (Liu et al. 2017), while *Lhx9* is restricted to excitatory orexinergic neurons (Dalal et al. 2013). Although loss of function of *Lhx6* disrupts the function of sleep-promoting ZI neurons, exactly how this gene regulates the specification and/or function of these cells is still unclear. *Lhx9* expression is detected in orexinergic cells long before orexin itself is expressed (Shimogori et al. 2010), and in zebra fish, *Lhx9* directly regulates orexin expression (Liu et al. 2017). In mice, loss of function of *Lhx9* leads to a reduction in both the levels of orexin expression and a 30% reduction in the number of orexigenic neurons (Dalal et al. 2013) (Table 5.1). *Lhx9*-deficient zebra fish shows a complete loss of orexin expression (Liu et al. 2017), In contrast, overexpression of *Lhx9* in both species increases the number of orexinergic neurons, but in mice did so only in posterior hypothalamus (Liu et al. 2015). Moreover, viral overexpression of *Lhx9* in adult *Lhx9*-deficient mice does not rescue defects in hypocretin expression (Liu et al. 2015). Gene expression profiling analysis of orexinergic neurons has identified other transcription factors that are selectively expressed in orexinergic neurons, although their role in differentiation and function of these cells has not been investigated.

5.2.5 Development of Other Hypothalamic Cell Types Regulating Sleep and Wake

Little is known about the transcriptional regulatory network controlling development of other sleep-regulatory circuitry, although multiple candidate genes have been identified. These include the homeodomain factors *Hmx2/3*, which are expressed in both the POA and posterior hypothalamic neural progenitors that give rise to both histaminergic and orexinergic neurons (Liu et al. 2015). *Dlx* family genes are also expressed in developing GABAergic neurons throughout the hypothalamus and POA and may play important roles in specification and differentiation of multiple wake and sleep-promoting neuronal subtypes.

5.2.6 Control of Nucleogenesis, Axonal Targeting, and Synaptogenesis

Virtually nothing is known about the molecular mechanisms that control nucleogenesis and circuit formation, with most of what is known being negative data. Despite the relative ease in visualizing the retinohypothalamic tract, no genes are known to selectively regulate innervation of the SCN by ipRGCs. Despite its prominent role in SCN development, retinohypothalamic tract formation occurs normally in *Lhx1* mice (Bedont et al. 2014). Likewise, nothing is known about the mechanisms guiding the extensive reciprocal synaptic connectivity of the SCN or ZI, which are likely to be critical for their circadian and/or sleep-regulating function. Further studies on this topic are urgently needed.

5.3 Insights into Hypothalamic Circuitry and Physiology Gained from Studies of Hypothalamic Development

5.3.1 *Lhx1* and Control of SCN Clock Function

Mice mutant for *Lhx1* have provided important insights into the molecular mechanisms regulating circadian synchrony and behavioral output of the SCN. SCN neurons express a diverse collection of neuropeptides, including Vip, Avp, Grp, Prok2, Nms, Penk, and Vgf (Abrahamson and Moore 2001). Expression of many of these is regulated by both circadian time and light, and they have been proposed to both regulate the synchrony of cellular oscillators within the SCN and mediate output from the SCN to other brain regions, allowing them to keep time with the master clock (Aton and Herzog 2005). Genetic and pharmacological analysis have identified a hierarchy of function among these neuropeptides, at least with respect to synchronization of cellular oscillators, with Vip being dominant over Avp, and Avp over Grp (Maywood et al. 2011). The sheer number of candidate neuropeptides, however, has not made it feasible to investigate the role of neuropeptides in central clock function more generally, as opposed to investigating the function of individual neuropeptides.

Disruption of *Lhx1* function in hypothalamic neuroepithelial cells leads to a near-complete loss of expression of all SCN-enriched neuropeptides without grossly affecting either regional organization or GABAergic neurotransmission in the SCN (Bedont et al. 2014). Cellular rhythms within the SCN remain locally synchronized in explants from *Lhx1*-deficient mice, but desynchrony is observed among different regions within the SCN, and this becomes more pronounced over time (Bedont et al. 2014). This is reflected in the behavioral phenotype of mutant mice, which though grossly normal, show either phase splitting or lack of circadian rhythmicity in locomotor and body temperature rhythms (Bedont et al. 2014).

Interestingly, the loss of neuropeptide expression in *Lhx1*-deficient mice renders the core behavioral oscillator highly susceptible to perturbation and resetting. SCN explants from *Lhx1*-deficient mice lose their resistance to resetting by temperature pulses, essentially phenocopying samples in which neuronal excitability is suppressed using tetrodotoxin (Bedont et al. 2017). Moreover, induction of fever by rhythmic injections of lipopolysaccharide is sufficient to reset both core body temperature and locomotor rhythms of these mutant mice (Bedont et al. 2017). These same animals are also rendered susceptible to phase shifts induced by intracerebroventricular canulation of SCN-enriched neuropeptides that do not normally influence behavioral rhythms (Bedont et al. 2014). Mice showing later loss of function of *Lhx1* show milder phenotypes that also reflect reduced coupling of cellular oscillators in the SCN (Hatori et al. 2014). Taken together, these data show that *Lhx1*-regulated neuropeptide gene expression plays a central role in the stabilization of cellular oscillators within the SCN, buffering them against resetting by internal signals and also in mediating efferent signaling from the SCN to other brain regions.

5.3.2 *Lhx1* and Light-Dependent Regulation of Sleep and Circadian Rhythms

Lhx1-deficient mice also show severe defects in light-dependent regulation of both behavioral rhythms and sleep. Although projections from ipRGCs to the SCN remain intact following early deletion of *Lhx1* (Bedont et al. 2014), these animals do not show increases in c-fos expression in the SCN in response to 30 min light pulses (Bedont et al. 2017). Moreover, while activity rhythms of *Lhx1*-deficient mice synchronize with 12 h light-dark cycles, light periods shorter than 6 h are not effective at doing so. This stands in sharp contrast to wild-type mice, which robustly synchronize and entrain to skeleton photoperiods where light phases are as short as

1 h (Jud et al. 2005). This implies that SCN-enriched neuropeptides are essential for relaying light signals from ipRGCs to cellular clocks in the SCN. These phenotypes are not observed when *Lhx1* is selectively deleted in SCN neural precursors, with these mutants showing enhanced resetting of circadian activity rhythms by light pulse (Hatori et al. 2014).

Early deletion of *Lhx1* leads to an unexpected loss of acute light-dependent regulation of sleep (Bedont et al. 2017). This phenotype is distinct from light-dependent regulation of the core circadian clock, which is completely disrupted in these mutants, and identifies an unexpected role for the SCN in direct regulation of sleep/wake transitions. These mutants do not show any changes in sleep homeostasis. The neural circuitry downstream of the SCN that controls acute light-dependent regulation of sleep remains unclear, however.

5.3.3 *Zfhx3* and Control of Circadian Rhythms, Interval Timing, and Sleep

Like *Lhx1*, *Zfhx3* has an essential role in control of circadian locomotor rhythms and sleep. Dominant ENU-induced missense mutations in *Zfhx3* show shortened biochemical and circadian activity periods, while these were lengthened by *Zfhx3* knockdown (Parsons et al. 2015). In contrast, selective knockout of *Zhfx3* in adult animals results in either shortened rhythms or else led to complete arrhythmicity. Animals that retained rhythmicity rapidly entrained to phase-advanced (but not phase-delayed) light cycles, implying that coupling among cellular SCN oscillators was attenuated (Parsons et al. 2015).

Missense mutations in *Zfhx3* mutants also showed evidence for shortened interval timing, when trained to respond with a 10s delay to a conditioned stimulus (Balzani et al. 2016), suggesting that they may show a more general defect in biological timing. *Zfhx3* missense mutants also show a reduction in rebound NREM sleep following 6 h. of sleep deprivation but do not show any significant differences in sleep homeostasis under baseline conditions (Balzani et al. 2016). Light-dependent regulation of sleep is also intact in these mutants (Balzani et al. 2016). Since these data were obtained using constitutive mutants, and *Zfhx3* is expressed broadly in the forebrain, it is unclear whether defects in interval timing and sleep homeostasis are a direct result of loss of *Zfhx3* in the SCN or other hypothalamic structures.

5.3.4 *Lhx6* and Detection of Sleep Pressure

Lhx6-expressing neurons in the ventral ZI are directly activated by sleep pressure and send inhibitory GABAergic projections to wake-promoting neurons in both the LH and midbrain (Liu et al. 2017). Slice recordings show that optogenetic activation of *Lhx6* ZI neurons directly inhibits wake-promoting hypocretin-expressing and GABAergic cells in the LH (Liu et al. 2017). In addition, *Lhx6*-positive ZI neurons receive presynaptic inputs from multiple sleep/wake-regulating neurons.

Conditional deletion of *Lhx6* from the developing diencephalon leads to decreases in both NREM and REM sleep. Furthermore, selective activation and inhibition of *Lhx6*-positive ZI neurons bidirectionally regulate sleep time in adult animals, in part through hypocretin-dependent mechanisms (Liu et al. 2017). There is a striking contrast between the very rapid inhibition of wake-promoting cells seen using ex vivo optogenetic analysis in slice preparations and the much slower changes in sleep/wake patterns seen following in vivo chemogenetic modulation of the activity of ZI *Lhx6*-positive neurons. This likely reflects the fact that *Lhx6*-positive ZI neurons are in turn functionally interconnected through inhibitory collateral projections, resulting in slow global shifts in activity levels even following direct stimulation. This pattern of connectivity may underpin the role of these neurons in sensing and signaling sleep pressure to neuronal subtypes that directly regulate sleep–wake transitions.

5.3.5 *Lhx9* and Orexinergic Neurons

Lhx9-deficient mice show reduced locomotor activity and increased rest time, although sleep was not directly investigated (Dalal et al. 2013). Unfortunately, in zebra fish, the effects of gain and loss of function of *Lhx9* on sleep behavior have also not been investigated (Liu et al. 2015)—but since *Lhx9* is crucial for Hcrt-expressing neuronal differentiation, we expect there is a defect in sleep behavior.

5.4 General Conclusions and Future Directions

5.4.1 The Urgent Need for Unambiguous Identification of Sleep-Regulating Neurons

Aside from the transcriptional regulatory network controlling specification of the SCN, little is known about the molecular mechanisms controlling development of the central circadian pacemaker and the sleep/wake cycle. Since congenital defects in these neural circuits are of considerable medical importance, this is somewhat surprising, and needs to be prioritized in future. A key problem here is the lack of unambiguous molecular markers for sleep-regulating neuronal subtypes. Most physiological studies rely on the use of fast neurotransmitters or neuropeptides to identify individual cell types linked to sleep regulation. These are generally fairly poor markers of cell identity, except in the rare cases where the cell that expresses genes show limited and highly restricted expression (e.g., Hcrt, MCH, histaminergic neurons). ScRNA-Seq studies, which have barely scratched the surface in characterizing hypothalamic cell diversity, have already identified hundreds of distinct GABAergic and glutamatergic neuronal subtypes, and considerable diversity among neurons expressing individual neuropeptides or calcium binding proteins (Campbell et al. 2017; Chen et al. 2017; Romanov et al. 2017).

Knowledge of a cell's transcription factor expression profile and, in particular, the expression profile of that cell immediately following terminal differentiation, can greatly assist in accurate classification of major cell types, avoiding potentially confounding effects of both aging and the effects of experience and environmental-driven variation in gene expression that accumulate over the lifespan (Hrvatin et al. 2018; Martinez-Jimenez et al. 2017) Developmental studies can thus be of great help in helping identify a unique molecular address for each main subtype of sleep-regulating cells—an essential step for unambiguous analysis of their function.

5.4.2 Comparative Studies: What Can Be Learned from Nonmammalian Vertebrates?

It is unclear how evolutionarily conserved this circuitry is likely to be. The location of the central circadian clock in nonmammalian vertebrates is still unclear, with the location and function of SCN being ambiguous even in birds. Some conservation of key sleep-regulatory neurons (e.g., Hcrt, Lhx6, Pmch) is observed in zebra fish (Liu et al. 2015; Wolf and Ryu 2013; Xie et al. 2017). In some cases, these have also been shown to be functionally homologous (Liu et al. 2015). Studies of nonmammalian vertebrates will help resolve this question—in particular, a systematic and unbiased analysis of gene expression in developing hypothalamus like that performed in developing mouse hypothalamus (Shimogori et al. 2010).

5.4.3 Human Studies and ES/iPS-Based Approaches to Understanding Sleep

Recent years have seen rapid improvement in protocols for directed differentiation of human ES and iPS cells into a broad range of neuronal subtypes, including a number of mediobasal hypothalamic cell types (Merkle et al. 2015; Ogawa et al. 2018; Wang et al. 2015; Wataya et al. 2008; Yao et al. 2017). The more recent development of hypothalamic organoid preparations (Qian et al. 2016, 2018) raises the possibility of potentially analyzing the formation and function of sleep-regulating structures such as the SCN or ZI in vitro.

Before this can come about, however, we will need to know much more about the basic molecular mechanisms controlling hypothalamic patterning and neurogenesis. In particular, we lack a clear understanding of general principles guiding organization and function of even the SCN. For instance, we have no idea how individual transcription factors that have been found to be necessary to regulate rhythmicity actually do so. What are their genomic targets? Is expression of these factors sufficient for initiation of rhythmicity and synchrony of cellular clocks, as well as necessary? The ability to perform gain and loss of function analysis in organoid preparations (Eldred et al. 2018; Mariani et al. 2015), in combination with techniques for globally profiling transcription factor target sites such as ATAC-

Seq-based footprinting, offers the possibility of identifying the gene regulatory networks that control the formation of key components of human sleep-regulatory circuitry. Ultimately, this may allow us to build a sleeping brain in a dish.

Key References

Bedont et al. (2014)—Identified Lhx1 as a master regulation of neuropeptide expression in neurons of the suprachiasmatic nucleus, and demonstrated the necessity of these neuropeptides in stabilizing the master circadian clock.

Carmona-Alcocer et al. (2018)—Demonstrated that synchronized activity rhythms first emerge in the mouse suprachiasmatic nucleus by embryonic day 15, and show mature organization by postnatal day 2.

Dalal et al. (2013)—Identified Lhx9 as a critical regulator of development of hypocretin neurons.

Liu et al. (2017)—Identified Lhx6-positive neurons of the zona incerta as responsive to sleep pressure, and demonstrated that they are necessary and sufficient for promotion of sleep.

Parsons et al. (2015)—Identified Zfhx3 as a master regulator of the development of the suprachiasmatic nucleus.

References

Abrahamson EE, Moore RY (2001) Suprachiasmatic nucleus in the mouse: retinal innervation, intrinsic organization and efferent projections. Brain Res 916:172–191

Aton SJ, Herzog ED (2005) Come together, right...now: synchronization of rhythms in a mammalian circadian clock. Neuron 48:531–534

Aujla PK, Naratadam GT, Xu L, Raetzman LT (2013) Notch/Rbpjkappa signaling regulates progenitor maintenance and differentiation of hypothalamic arcuate neurons. Development 140:3511–3521

Balzani E, Lassi G, Maggi S, Sethi S, Parsons MJ, Simon M, Nolan PM, Tucci V (2016) The Zfhx3-mediated axis regulates sleep and interval timing in mice. Cell Rep 16:615–621

Bedont JL, LeGates TA, Slat EA, Byerly MS, Wang H, Hu J, Rupp AC, Qian J, Wong GW, Herzog ED et al (2014) Lhx1 controls terminal differentiation and circadian function of the suprachiasmatic nucleus. Cell Rep 7:609–622

Bedont JL, LeGates TA, Buhr E, Bathini A, Ling JP, Bell B, Wu MN, Wong PC, Van Gelder RN, Mongrain V et al (2017) An LHX1-regulated transcriptional network controls sleep/wake coupling and thermal resistance of the central circadian clockworks. Curr Biol 27:128–136

Biehl MJ, Kaylan KB, Thompson RJ, Gonzalez RV, Weis KE, Underhill GH, Raetzman LT (2018) Cellular fate decisions in the developing female anteroventral periventricular nucleus are regulated by canonical Notch signaling. Dev Biol 442:87–100

Blaess S, Szabo N, Haddad-Tovolli R, Zhou X, Alvarez-Bolado G (2014) Sonic hedgehog signaling in the development of the mouse hypothalamus. Front Neuroanat 8:156

Blumberg MS, Gall AJ, Todd WD (2014) The development of sleep-wake rhythms and the search for elemental circuits in the infant brain. Behav Neurosci 128:250–263

Borbely AA (1982) A two process model of sleep regulation. Hum Neurobiol 1:195–204

Brunner DP, Dijk DJ, Tobler I, Borbely AA (1990) Effect of partial sleep deprivation on sleep stages and EEG power spectra: evidence for non-REM and REM sleep homeostasis. Electroencephalogr Clin Neurophysiol 75:492–499

Buhr ED, Yoo SH, Takahashi JS (2010) Temperature as a universal resetting cue for mammalian circadian oscillators. Science 330:379–385

Campbell JN, Macosko EZ, Fenselau H, Pers TH, Lyubetskaya A, Tenen D, Goldman M, Verstegen AM, Resch JM, McCarroll SA et al (2017) A molecular census of arcuate hypothalamus and median eminence cell types. Nat Neurosci 20:484–496

Carmona-Alcocer V, Abel JH, Sun TC, Petzold LR, Doyle FJ 3rd, Simms CL, Herzog ED (2018) Ontogeny of circadian rhythms and synchrony in the suprachiasmatic nucleus. J Neurosci 38:1326–1334

Chen R, Wu X, Jiang L, Zhang Y (2017) Single-cell RNA-Seq reveals hypothalamic cell diversity. Cell Rep 18:3227–3241

Chen KS, Xu M, Zhang Z, Chang WC, Gaj T, Schaffer DV, Dan Y (2018) A hypothalamic switch for REM and non-REM sleep. Neuron 97(1168–1176):e1164

Chung S, Weber F, Zhong P, Tan CL, Nguyen TN, Beier KT, Hormann N, Chang WC, Zhang Z, Do JP et al (2017) Identification of preoptic sleep neurons using retrograde labelling and gene profiling. Nature 545:477–481

Clark DD, Gorman MR, Hatori M, Meadows JD, Panda S, Mellon PL (2013) Aberrant development of the suprachiasmatic nucleus and circadian rhythms in mice lacking the homeodomain protein Six6. J Biol Rhythm 28:15–25

Daan S, Beersma DG, Borbely AA (1984) Timing of human sleep: recovery process gated by a circadian pacemaker. Am J Phys 246:R161–R183

Dalal J, Roh JH, Maloney SE, Akuffo A, Shah S, Yuan H, Wamsley B, Jones WB, de Guzman Strong C, Gray PA et al (2013) Translational profiling of hypocretin neurons identifies candidate molecules for sleep regulation. Genes Dev 27:565–578

Daszuta A, Gambarelli F (1985) Early postnatal development of EEG and sleep-waking cycle in two inbred mouse strains. Brain Res 354:39–47

Davis KF, Parker KP, Montgomery GL (2004) Sleep in infants and young children: part one: normal sleep. J Pediatr Health Care 18:65–71

de Melo J, Zibetti C, Clark BS, Hwang W, Miranda-Angulo AL, Qian J, Blackshaw S (2016) Lhx2 is an essential factor for retinal gliogenesis and Notch signaling. J Neurosci 36:2391–2405

Dittrich L, Morairty SR, Warrier DR, Kilduff TS (2015) Homeostatic sleep pressure is the primary factor for activation of cortical nNOS/NK1 neurons. Neuropsychopharmacology 40:632–639

Dzierzewski JM, Dautovich N, Ravyts S (2018) Sleep and cognition in older adults. Sleep Med Clin 13:93–106

Eldred KC, Hadyniak SE, Hussey KA, Brenerman B, Zhang PW, Chamling X, Sluch VM, Welsbie DS, Hattar S, Taylor J et al (2018) Thyroid hormone signaling specifies cone subtypes in human retinal organoids. Science 362:eaau6348

Elmenhorst D, Meyer PT, Winz OH, Matusch A, Ermert J, Coenen HH, Basheer R, Haas HL, Zilles K, Bauer A (2007) Sleep deprivation increases A1 adenosine receptor binding in the human brain: a positron emission tomography study. J Neurosci 27:2410–2415

Elmenhorst D, Elmenhorst EM, Hennecke E, Kroll T, Matusch A, Aeschbach D, Bauer A (2017) Recovery sleep after extended wakefulness restores elevated A1 adenosine receptor availability in the human brain. Proc Natl Acad Sci USA 114:4243–4248

Ferreira JGP, Bittencourt JC, Adamantidis A (2017) Melanin-concentrating hormone and sleep. Curr Opin Neurobiol 44:152–158

Fischer D, Lombardi DA, Marucci-Wellman H, Roenneberg T (2017) Chronotypes in the US—influence of age and sex. PLoS One 12:e0178782

Foreman SW, Thomas KA, Blackburn ST (2008) Individual and gender differences matter in preterm infant state development. J Obstet Gynecol Neonatal Nurs 37:657–665

Frank MG, Heller HC (1997) Development of diurnal organization of EEG slow-wave activity and slow-wave sleep in the rat. Am J Phys 273:R472–R478

Fu T, Towers M, Placzek MA (2017) Fgf10(+) progenitors give rise to the chick hypothalamus by rostral and caudal growth and differentiation. Development 144:3278–3288
Gall AJ, Todd WD, Ray B, Coleman CM, Blumberg MS (2008) The development of day-night differences in sleep and wakefulness in Norway rats and the effect of bilateral enucleation. J Biol Rhythm 23:232–241
Galland BC, Taylor BJ, Elder DE, Herbison P (2012) Normal sleep patterns in infants and children: a systematic review of observational studies. Sleep Med Rev 16:213–222
Hanne HM, Meadows JD, Trang C, Hereford B, Bharti K, Gorman MR, Mellon PL (2016) Deletion of SIX3 or VAX1 in the SCN impairs circadian rhythms and fertility. In Endocrine Society's 98th Annual Meeting and Expo, April 1–4, 2016, Boston
Hatori M, Gill S, Mure LS, Goulding M, O'Leary DD, Panda S (2014) Lhx1 maintains synchrony among circadian oscillator neurons of the SCN. Elife 3:e03357
Hattar S, Liao HW, Takao M, Berson DM, Yau KW (2002) Melanopsin-containing retinal ganglion cells: architecture, projections, and intrinsic photosensitivity. Science 295:1065–1070
Hrvatin S, Hochbaum DR, Nagy MA, Cicconet M, Robertson K, Cheadle L, Zilionis R, Ratner A, Borges-Monroy R, Klein AM et al (2018) Single-cell analysis of experience-dependent transcriptomic states in the mouse visual cortex. Nat Neurosci 21:120–129
Joseph D, Chong NW, Shanks ME, Rosato E, Taub NA, Petersen SA, Symonds ME, Whitehouse WP, Wailoo M (2015) Getting rhythm: how do babies do it? Arch Dis Child Fetal Neonatal Ed 100:F50–F54
Jud C, Schmutz I, Hampp G, Oster H, Albrecht U (2005) A guideline for analyzing circadian wheel-running behavior in rodents under different lighting conditions. Biol Proced Online 7:101–116
Lee JE, Wu SF, Goering LM, Dorsky RI (2006) Canonical Wnt signaling through Lef1 is required for hypothalamic neurogenesis. Development 133:4451–4461
Liu J, Merkle FT, Gandhi AV, Gagnon JA, Woods IG, Chiu CN, Shimogori T, Schier AF, Prober DA (2015) Evolutionarily conserved regulation of hypocretin neuron specification by Lhx9. Development 142:1113–1124
Liu K, Kim J, Kim DW, Zhang YS, Bao H, Denaxa M, Lim SA, Kim E, Liu C, Wickersham IR et al (2017) Lhx6-positive GABA-releasing neurons of the zona incerta promote sleep. Nature 548:582–587
Manning L, Ohyama K, Saeger B, Hatano O, Wilson SA, Logan M, Placzek M (2006) Regional morphogenesis in the hypothalamus: a BMP-Tbx2 pathway coordinates fate and proliferation through Shh downregulation. Dev Cell 11:873–885
Mariani J, Coppola G, Zhang P, Abyzov A, Provini L, Tomasini L, Amenduni M, Szekely A, Palejev D, Wilson M et al (2015) FOXG1-dependent dysregulation of GABA/glutamate neuron differentiation in autism spectrum disorders. Cell 162:375–390
Martinez-Jimenez CP, Eling N, Chen HC, Vallejos CA, Kolodziejczyk AA, Connor F, Stojic L, Rayner TF, Stubbington MJT, Teichmann SA et al (2017) Aging increases cell-to-cell transcriptional variability upon immune stimulation. Science 355:1433–1436
Maywood ES, Chesham JE, O'Brien JA, Hastings MH (2011) A diversity of paracrine signals sustains molecular circadian cycling in suprachiasmatic nucleus circuits. Proc Natl Acad Sci USA 108:14306–14311
Merkle FT, Maroof A, Wataya T, Sasai Y, Studer L, Eggan K, Schier AF (2015) Generation of neuropeptidergic hypothalamic neurons from human pluripotent stem cells. Development 142:633–643
Miles LE, Dement WC (1980) Sleep and aging. Sleep 3:1–220
Mohawk JA, Takahashi JS (2011) Cell autonomy and synchrony of suprachiasmatic nucleus circadian oscillators. Trends Neurosci 34(7):349–358
Morairty SR, Dittrich L, Pasumarthi RK, Valladao D, Heiss JE, Gerashchenko D, Kilduff TS (2013) A role for cortical nNOS/NK1 neurons in coupling homeostatic sleep drive to EEG slow wave activity. Proc Natl Acad Sci USA 110:20272–20277
Nelson AB, Faraguna U, Zoltan JT, Tononi G, Cirelli C (2013) Sleep patterns and homeostatic mechanisms in adolescent mice. Brain Sci 3:318–343

Newman EA, Kim DW, Wan J, Wang J, Qian J, Blackshaw S (2018a) Foxd1 is required for terminal differentiation of anterior hypothalamic neuronal subtypes. Dev Biol 439:102–111

Newman EA, Wu D, Taketo MM, Zhang J, Blackshaw S (2018b) Canonical Wnt signaling regulates patterning, differentiation and nucleogenesis in mouse hypothalamus and prethalamus. Dev Biol 442:236–248

Ogawa K, Suga H, Ozone C, Sakakibara M, Yamada T, Kano M, Mitsumoto K, Kasai T, Kodani Y, Nagasaki H et al (2018) Vasopressin-secreting neurons derived from human embryonic stem cells through specific induction of dorsal hypothalamic progenitors. Sci Rep 8:3615

Ohyama K, Das R, Placzek M (2008) Temporal progression of hypothalamic patterning by a dual action of BMP. Development 135:3325–3331

Parsons MJ, Brancaccio M, Sethi S, Maywood ES, Satija R, Edwards JK, Jagannath A, Couch Y, Finelli MJ, Smyllie NJ et al (2015) The regulatory factor ZFHX3 modifies circadian function in SCN via an AT motif-driven Axis. Cell 162:607–621

Partch CL, Green CB, Takahashi JS (2014) Molecular architecture of the mammalian circadian clock. Trends Cell Biol 24:90–99

Pedersen NP, Ferrari L, Venner A, Wang JL, Abbott SBG, Vujovic N, Arrigoni E, Saper CB, Fuller PM (2017) Supramammillary glutamate neurons are a key node of the arousal system. Nat Commun 8:1405

Porter FD, Drago J, Xu Y, Cheema SS, Wassif C, Huang SP, Lee E, Grinberg A, Massalas JS, Bodine D et al (1997) Lhx2, a LIM homeobox gene, is required for eye, forebrain, and definitive erythrocyte development. Development 124:2935–2944

Qian X, Nguyen HN, Song MM, Hadiono C, Ogden SC, Hammack C, Yao B, Hamersky GR, Jacob F, Zhong C et al (2016) Brain-region-specific organoids using mini-bioreactors for modeling ZIKV exposure. Cell 165:1238–1254

Qian X, Jacob F, Song MM, Nguyen HN, Song H, Ming GL (2018) Generation of human brain region-specific organoids using a miniaturized spinning bioreactor. Nat Protoc 13:565–580

Reddy AB, O'Neill JS (2009) Healthy clocks, healthy body, healthy mind. Trends Cell Biol 20:36–44

Roenneberg T, Wirz-Justice A, Merrow M (2003) Life between clocks: daily temporal patterns of human chronotypes. J Biol Rhythm 18:80–90

Romanov RA, Zeisel A, Bakker J, Girach F, Hellysaz A, Tomer R, Alpar A, Mulder J, Clotman F, Keimpema E et al (2017) Molecular interrogation of hypothalamic organization reveals distinct dopamine neuronal subtypes. Nat Neurosci 20:176–188

Roy A, de Melo J, Chaturvedi D, Thein T, Cabrera-Socorro A, Houart C, Meyer G, Blackshaw S, Tole S (2013) LHX2 is necessary for the maintenance of optic identity and for the progression of optic morphogenesis. J Neurosci 33:6877–6884

Sack RL, Auckley D, Auger RR, Carskadon MA, Wright KP Jr, Vitiello MV, Zhdanova IV, American Academy of Sleep, M (2007) Circadian rhythm sleep disorders: part I, basic principles, shift work and jet lag disorders. An American Academy of Sleep Medicine review. Sleep 30:1460–1483

Saito YC, Tsujino N, Hasegawa E, Akashi K, Abe M, Mieda M, Sakimura K, Sakurai T (2013) GABAergic neurons in the preoptic area send direct inhibitory projections to orexin neurons. Front Neural Circuits 7:192

Salvatierra J, Lee DA, Zibetti C, Duran-Moreno M, Yoo S, Newman EA, Wang H, Bedont JL, de Melo J, Miranda-Angulo AL et al (2014) The LIM homeodomain factor Lhx2 is required for hypothalamic Tanycyte specification and differentiation. J Neurosci 34:16809–16820

Saper CB, Fuller PM, Pedersen NP, Lu J, Scammell TE (2010) Sleep state switching. Neuron 68:1023–1042

Scammell TE, Arrigoni E, Lipton JO (2017) Neural circuitry of wakefulness and sleep. Neuron 93:747–765

Shimogori T, Lee DA, Miranda-Angulo A, Yang Y, Wang H, Jiang L, Yoshida AC, Kataoka A, Mashiko H, Avetisyan M et al (2010) A genomic atlas of mouse hypothalamic development. Nat Neurosci 13:767–775

VanDunk C, Hunter LA, Gray PA (2011) Development, maturation, and necessity of transcription factors in the mouse suprachiasmatic nucleus. J Neurosci 31:6457–6467

Venner A, Anaclet C, Broadhurst RY, Saper CB, Fuller PM (2016) A novel population of wake-promoting GABAergic neurons in the ventral lateral hypothalamus. Curr Biol 26:2137–2143

Wang L, Meece K, Williams DJ, Lo KA, Zimmer M, Heinrich G, Martin Carli J, Leduc CA, Sun L, Zeltser LM et al (2015) Differentiation of hypothalamic-like neurons from human pluripotent stem cells. J Clin Invest 125:796–808

Wataya T, Ando S, Muguruma K, Ikeda H, Watanabe K, Eiraku M, Kawada M, Takahashi J, Hashimoto N, Sasai Y (2008) Minimization of exogenous signals in ES cell culture induces rostral hypothalamic differentiation. Proc Natl Acad Sci USA 105:11796–11801

Weber F, Dan Y (2016) Circuit-based interrogation of sleep control. Nature 538:51–59

Wilcox AG, Vizor L, Parsons MJ, Banks G, Nolan PM (2017) Inducible knockout of mouse Zfhx3 emphasizes its key role in setting the pace and amplitude of the adult circadian clock. J Biol Rhythm 32:433–443

Wolf A, Ryu S (2013) Specification of posterior hypothalamic neurons requires coordinated activities of Fezf2, Otp, Sim1a and Foxb1.2. Development 140:1762–1773

Xie Y, Kaufmann D, Moulton MJ, Panahi S, Gaynes JA, Watters HN, Zhou D, Xue HH, Fung CM, Levine EM et al (2017) Lef1-dependent hypothalamic neurogenesis inhibits anxiety. PLoS Biol 15:e2002257

Yamaguchi Y, Suzuki T, Mizoro Y, Kori H, Okada K, Chen Y, Fustin JM, Yamazaki F, Mizuguchi N, Zhang J et al (2013) Mice genetically deficient in vasopressin V1a and V1b receptors are resistant to jet lag. Science 342:85–90

Yao L, Liu Y, Qiu Z, Kumar S, Curran JE, Blangero J, Chen Y, Lehman DM (2017) Molecular profiling of human induced pluripotent stem cell-derived hypothalamic neurones provides developmental insights into genetic loci for body weight regulation. J Neuroendocrinol 29(2)

Pituitary Development and Organogenesis: Transcription Factors in Development and Disease

6

Alexandre Z. Daly and Sally A. Camper

Abstract

Many transcription factors that regulate pituitary gland development and specification of hormone-producing cell types have been discovered. Here we focus on 13 transcription factors that play key roles in these processes, many of which are mutated in patients with congenital pituitary insufficiency disorders. We present the methods by which each of these transcription factors were identified. Each of the genes we discuss has been well studied in genetically engineered mice, facilitating an understanding of the impact of each single gene defect on the expression of downstream targets. Finally, we highlight technical advances that promise to yield future insights into epigenetic regulation of cell fate.

Keywords

Cell specification · Stem cell · Hypopituitarism

6.1 Introduction

The pituitary gland is located at the base of the brain, and it is a small organ, approximately the size of a pea in humans. It is often referred to as the master gland because it is a key regulator of multiple organ systems, controlling growth, fertility, stress response, and homeostasis. The pituitary gland develops between 5 and 9 weeks of gestation in humans and embryonic day 10–19 in mice (Xue et al. 2013). An invagination of oral ectoderm, called Rathke's pouch, forms the anterior and intermediate lobes of the pituitary gland, and evagination of the neural ectoderm forms the posterior lobe of the pituitary gland and pituitary stalk. The posterior lobe contains the axon terminals for oxytocin and vasopressin and pituicytes. The anterior

A. Z. Daly · S. A. Camper (✉)
Department of Human Genetics, University of Michigan, Ann Arbor, MI, USA
e-mail: scamper@med.umich.edu

S. Wray, S. Blackshaw (eds.), *Developmental Neuroendocrinology*, Masterclass in Neuroendocrinology 9, https://doi.org/10.1007/978-3-030-40002-6_6

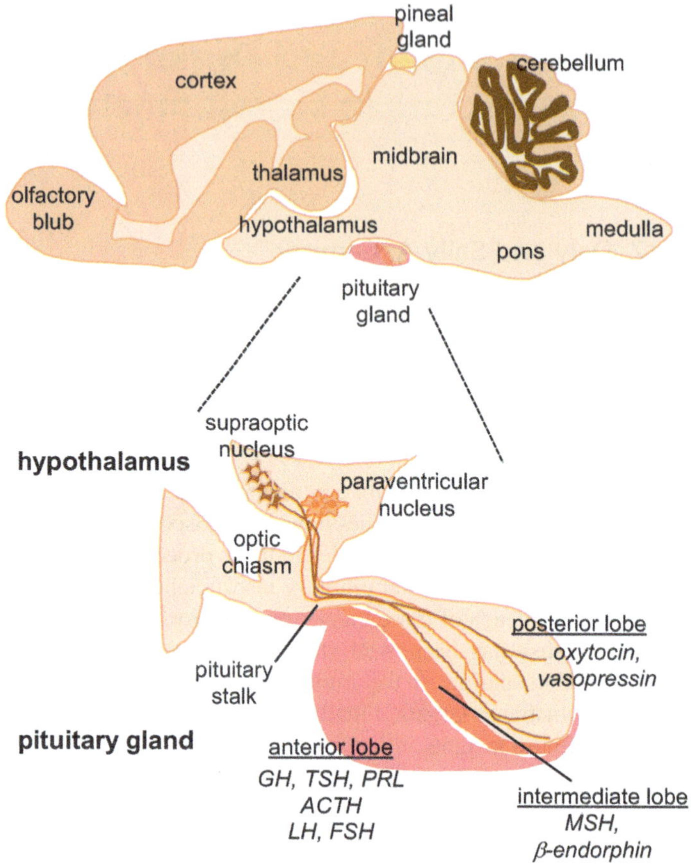

Fig. 6.1 Anatomy of the pituitary gland and base of the hypothalamus. The pituitary gland contains anterior (ant) and intermediate (int) lobes derived from the oral ectoderm that forms Rathke's pouch. The posterior lobe (post) and pituitary stalk are derived from neural ectoderm of the ventral diencephalon, and they contain the axon terminals of vasopressin and oxytocin neurons that project from both the supraoptic nucleus and the paraventricular nucleus of the hypothalamus. Hypothalamic peptides regulate the release of anterior pituitary hormones into the hypophyseal portal system [adapted from Rappaport and Amselem (2001)]

lobe is composed of cells specialized in the secretion of polypeptide hormones (Fig. 6.1). Hypothalamic neurons secrete factors into the hypophyseal portal system that regulate pituitary function. In the past three decades, a great deal of progress has been made in understanding the mechanisms that drive the differentiation of hormone-producing cells of the anterior pituitary gland during development. Multiple transcription factors and signaling pathways are involved, and defects in the genes that encode these factors and pathways are an important contributor to congenital pituitary hormone deficiency (Castinetti et al. 2016; Fang et al. 2016; Kelberman et al. 2009; Rizzoti 2015). The focus of this review will be on the pituitary transcription factors that drive the differentiation of pituitary hormone-producing cells with an emphasis on the history of discovery for those that are most relevant to human pituitary insufficiency. Other chapters in this volume review the development of the hypothalamus and its regulation of pituitary function (See the reviews by Placzek and Towers on the development of the hypothalamus, the review from Alvarez-Bolado on SHH and GLI in the hypothalamus and Lee's review on the arcuate nucleus.)

6.1.1 The Hypophyseal Portal System, Pituitary Vasculature, and Accessory Cells of the Pituitary Gland

The hypophyseal portal system refers to the blood vessels that provide connections between the hypothalamus and the anterior pituitary gland. This vasculature develops e14.5–e18.5 in mice and appears complete by 12 weeks of gestation in humans (Xue et al. 2013). The capillaries are highly fenestrated, facilitating rapid molecular exchange between the hypothalamus and pituitary lobes. Oxygen levels and blood flow are regulated and have an effect on pituitary hormone production and secretion (Lafont et al. 2010; Schaeffer et al. 2010).

The mature pituitary gland has six distinct hormone-producing cell types, as well as folliculo-stellate cells and pituicytes (Fig. 6.2). The hormone-producing cell types in the anterior lobe include corticotropes that produce adrenocorticotropin (ACTH) from the pro-hormone pro-opiomelanocortin (POMC), somatotropes that produce GH, lactotropes that produce prolactin, gonadotropes that produce the heterodimeric glycoprotein hormones luteinizing hormone (LH) and follicle-stimulating hormone (FSH) and thyrotropes that produce the heterodimeric glycoprotein hormone thyroid-stimulating hormone (TSH). The intermediate lobe, which remains distinct in rodents, contains melanotropes that produce melanocyte-stimulating hormone from POMC. Folliculo-stellate (FS) cells are non-endocrine cells that were named because of the star-like morphology of cytoplasmic processes (Rinehart and Farquhar 1953). FS cells appear between postnatal days 10 and 20 in the rat, throughout the anterior lobe parenchyma, directly adjacent to the hormone-producing cells (Soji et al. 1997). FS cells serve a support function for hormone-producing cells through the release of cytokines and growth factors. FS cells are excitatory and are involved in the coordinated, pulsatile release of hormones from the hormone-producing cells, which themselves form homotypic networks (Fauquier et al. 2001; Mollard et al. 2012). Pituicytes are glial-like cells located within the posterior lobe. The larger parenchymatous pituicytes regulate hormone output by completely enveloping the neurosecretory processes of the oxytocin and vasopressin axons when hormone secretion is low, and they recede as hormone secretion increases. They also regulate hormone output by secretion of taurine (Rosso and Mienville 2009). The smaller fibrous pituicytes do not appear to be involved in this regulatory function and are less well understood (Pow et al. 1989).

The vasculature is important for pituitary gland function because it transports molecules from pituitary target organs that feedback at the level of the pituitary gland

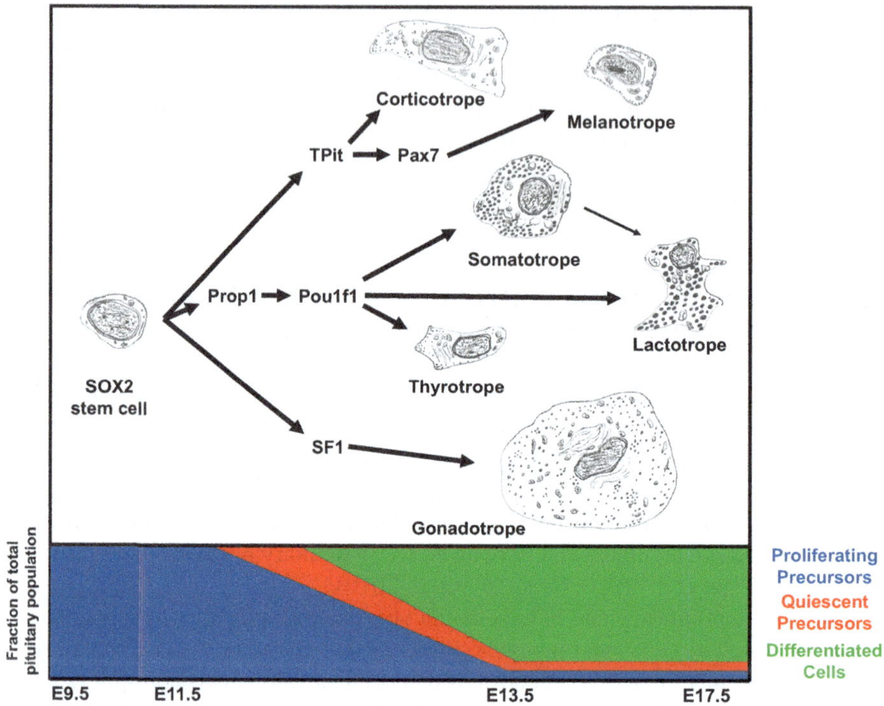

Fig. 6.2 Critical transcription factors control the development of specialized hormone-producing cells. The major hormone-producing cell types of the anterior and intermediate lobes of the pituitary gland are derived from SOX2-expressing progenitor cells. Combinations of transcription factors drive specific cell fates and antagonize differentiation into alternate fates. Each cell type has distinctive secretory granules and shape (Matsumoto and Ishii 1992). Proliferating precursors leave the cell cycle and differentiate into hormone-producing cells during gestation

and hypothalamus, maintaining homeostasis. This multi-organ feedback regulation is called an axis. The best known of these axes are the hypothalamic–pituitary–gonadal axis, hypothalamic–pituitary–adrenal axis, and the hypothalamic–pituitary–thyroid axis, although axes involving bone, liver, and other organs also exist (Fekete and Lechan 2014; Imam et al. 2009; Jin and Yang 2014; Keller-Wood 2015; Simon et al. 2002).

6.1.2 Cell Signaling During Pituitary Gland Development

Some of the major signaling pathways that have been implicated in pituitary development are SHH, FGF, BMP, Wnt, Notch, Hippo, EGF, and retinoic acid (Cohen et al. 1999; Djakoure et al. 1996; Lodge et al. 2019; Zhu et al. 2007). Hypothalamic SHH is necessary for inducing expression of the pituitary transcription factors LHX3 and LHX4, which drive progenitor proliferation and prevent cell death (Carreno et al.

2017). FGF8 and FGFR1 mutations cause pituitary hormone deficiency in humans (Fang et al. 2016), and mouse studies show reduced cell proliferation and enhanced cell death in the pituitary primordia of *Fgf8*, *Fgf10*, and *Fgfr2* mutants (Ohuchi et al. 2000). BMP and the antagonist noggin affect pituitary growth and shape during development, and altered BMP signaling is associated with pituitary adenomas (Davis and Camper 2007; Ericson et al. 1998; Giacomini et al. 2006; Treier et al. 2001). Similarly, WNT signaling regulates pituitary organogenesis and acts in a paracrine manner to stimulate excess cell proliferation in adenomas (Chambers et al. 2013). The ligand, beta-catenin, is a cofactor for several key pituitary transcription factors including PITX2, TCF7L2 (TCF4), LEF1, NR5A1, and PROP1. Notch signaling regulates the timing of pituitary progenitor cell cycle exit and differentiation (Cheung et al. 2018a; Raetzman et al. 2004; Zhu et al. 2006). YAP and TAZ are transcriptional regulators downstream of the Hippo signaling pathway, which suppresses their function by phosphorylation. This pathway plays an essential role in the regulation of pituitary progenitor expansion and normal organ size (Lodge et al. 2019). There is a great deal of cross talk between these pathways, in that perturbation of one signaling pathway typically affects the other signaling pathways. EGF and retinoic acid signaling are recognized as important signaling pathways but their precise roles during organogenesis have not been established (Rhodes et al. 1993; Roh et al. 2001).

6.1.3 Hypopituitarism

The major pathologies associated with the pituitary are adenomas, which can be associated with either over- or underproduction of hormones, and congenital or acquired hypopituitarism (Higham et al. 2016; Melmed 2011). The main emphasis of this review will be on congenital hypopituitarism, which had an estimated prevalence of 45.5 per 100,000 people (Regal et al. 2001). This is a genetically heterogeneous condition that can present with a single pituitary hormone deficiency or with multiple deficiencies. About 50% of cases that originally present with isolated growth hormone deficiency (IGHD) progress to combined pituitary hormone deficiency (CPHD). CPHD is defined as a reduction of at least two pituitary hormones, and usually, GH is one of them. It has a prevalence of 1 in 8000 individuals, and more than 30 genes have been implicated as causal factors for CPHD (Fang et al. 2016).

6.1.4 Overview

Here we review the transcription factors that are critical for developing the specialized cells of the anterior and intermediate lobes of the pituitary gland, emphasizing the history of scientific discovery surrounding each factor, from its discovery through to its most recent characterization. This reveals the continuing evolution and diversity of effective technical approaches. These include identification of cis-acting sequences important for regulation of hormone gene expression and identification of the trans-acting factors that bind those sites, positional cloning of mutations in mice and human patients, and exploration of epigenetic regulation of chromatin accessibility. We also highlight areas for future discovery that may take advantage of high-throughput sequencing and high-efficiency, targeted germline disruption tools.

6.2 SOX2 (LCC, YSB)

6.2.1 Gene Discovery

Sox2 was first discovered in a whole embryo cDNA screen in search of genes similar to the male, sex-determining gene *Sry*, which binds DNA through a motif or box characteristic of high mobility group proteins (Gubbay et al. 1990). *Sox2* is expressed in progenitor cell populations in many tissues, and its expression is negatively correlated with differentiation (Avilion et al. 2003; Li et al. 1998; Zappone et al. 2000). SOX2 is one of the factors that, along with OCT3/4, c-MYC, and KLF4, were demonstrated by Yamanaka and colleagues to be sufficient to induce pluripotency in differentiated cells from mice and humans (Takahashi et al. 2007; Takahashi and Yamanaka 2006).

6.2.2 Knockout Mouse Phenotype, Downstream Targets, and Interacting Factors

Sox2 is expressed in the developing pituitary and hypothalamus, and it has a role in pituitary development (Takahashi et al. 2007; Takahashi and Yamanaka 2006). Embryos homozygous for *Sox2* loss of function alleles die around implantation, and heterozygous mice have genetic background-dependent reduction in body size and male infertility (Avilion et al. 2003). The developing pituitaries of heterozygous mutants are bifurcated, similar to the dysmorphology observed in *Wnt5a* mutants, and both males and females exhibit reduced differentiation into somatotropes and gonadotropes (Fig. 6.3) (Cha et al. 2004; Rizzoti and Lovell-Badge 2005). About two-thirds of the heterozygotes survive to adulthood, and in these survivors, the reduction in pituitary size is modest, and GH and LH content is only significantly reduced in males.

Martinez-Barbera and colleagues generated a conditional deletion of *Sox2* using the *Hesx1-cre* strain, deleting *Sox2* in the pituitary gland and areas of the brain between e10.5 and e12.5 (Rizzoti and Lovell-Badge 2005). This resulted in severe pituitary hypoplasia and greatly reduced the expressions of *Pou1f1*, *Gh*, and *Tshb*. Gonadotropin deficiency appeared to be secondary to the reduction in GnRH neurons in these mice. Cell proliferation is significantly decreased in the conditionally deleted *Sox2* pituitary glands, and it appears that the progenitors have not migrated from the progenitor zone of Rathke's pouch to colonize the anterior lobe. This suggests that *Sox2* has an important role in pituitary progenitor proliferation and transition to differentiation.

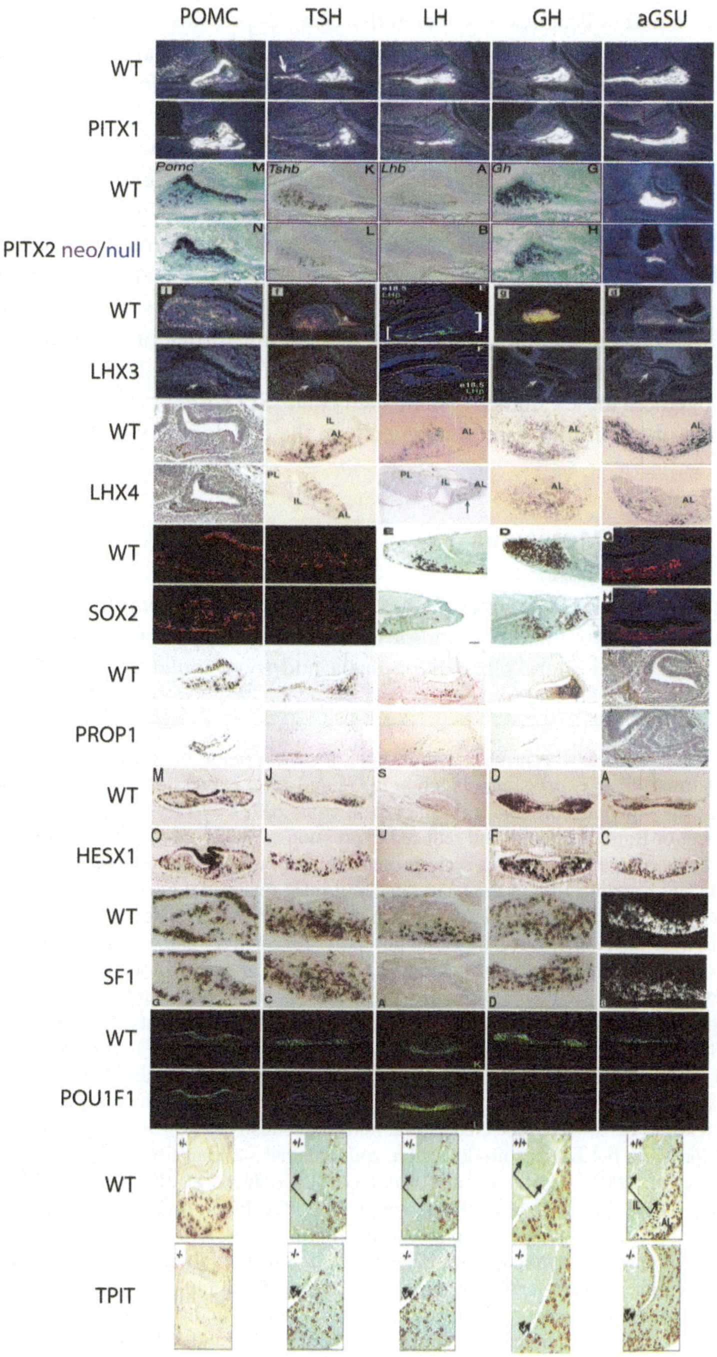

Fig. 6.3 The effect of transcription factor mutations on pituitary development has been revealed using genetically engineered and spontaneous mutant mice. Deletion of *Pitx1* has modest effects including slight reduction of TSH and LH, slight increase in POMC and no change in GH and αGSU, as observed by in situ hybridization in P0 animals (Szeto et al. 1999). In situ hybridization in

Robinson and colleagues discovered that *Sox2*-expressing cells in the adult mouse pituitary were still mitotically active and that, given appropriate culture conditions, these SOX2-positive cells can self-renew and differentiate in vitro into all of the hormone-producing cells of the anterior lobe of the pituitary gland (Fauquier et al. 2008). Both Lovell-Badge and Martinez-Barbera and colleagues used lineage tracing to demonstrate that SOX2-positive cells give rise to all of the different hormone-producing cells in the pituitary gland in vivo (Andoniadou et al. 2013; Rizzoti et al. 2013). SOX2-positive cells contribute to tissue homeostasis, and they are involved in responding to end-organ ablation, such as adrenalectomy and gonadectomy. Mice with a constitutively active allele of β-catenin exhibit clusters of pituitary cells that overexpress β-catenin, express *Sox2*, and induce transformation of neighboring cells that mimic craniopharyngioma, a childhood tumor (Gaston-Massuet et al. 2011). Taken together these studies established that SOX2-expressing cells represent progenitors of the hormone-producing cells in the developing pituitary.

6.2.3 Patient Mutations

Most patients with *SOX2* mutations have serious eye abnormalities that can include anophthalmia, microphthalmia, and/or optic nerve hypoplasia, and hypopituitarism characterized by pituitary gland hypoplasia and hypogonadotropic hypogonadism

---

Fig. 6.3 (continued) newborn mice homozygous for a hypomorphic allele of *Pitx2*, $Pitx2^{neo/neo}$, revealed normal *Pomc* expression, reduction in *Gh* and *Tshb* expression, and no detectable *Lhb* or *Fshb* transcripts (panels bordered in purple) (Suh et al. 2002). $Pitx2^{-/-}$ embryos had arrested the development of Rathke's pouch and reduced expression of αGSU at e13.5 mice (Lin et al. 1999). Trace amounts of *Pomc* transcripts were present, but all other hormones were not detectable (not shown). In situ hybridization in the *Lhx3* knockout mice revealed a near-total loss of *Pomc*, *Tshb*, *Gh*, and *Cga* (αGSU) transcripts, and immunohistochemistry failed to detect significant amounts of LH. In situ hybridization of *Lhx4* knockout pituitaries revealed a significant reduction in *Tshb*, *Lhb*, *Gh*, and *Cga* transcripts at e18.5 (Sheng et al. 1997), and immunohistochemistry showed largely similar POMC expression in e14.5 mice (Sheng et al. 1997). Immunohistochemistry of *Sox2* heterozygous null mice revealed reduced staining for GH and LH at e18.5 (Kelberman et al. 2006). There is no detectable *Pou1f1* staining in $Hesx^{cre/+}$; $Sox2^{flox/flox}$ mice at e14.5 or later in gestation (not shown) (Jayakody et al. 2012). The staining for αGSU and POMC was somewhat normal at e15.5, considering the hypoplasia of the organ. Only *Pou1f1* independent TSH-expressing cells were detected. Immunohistochemistry of $Prop1^{-/-}$ pituitaries revealed POMC expression but little or no detectable caudomedial staining for TSH, LH, or GH at e18.5 (Nasonkin et al. 2004). αGSU staining was modestly reduced at e14.5 in $Prop1^{df/df}$ mice, but NR5A1 cells were normal at birth (Raetzman et al. 2002). In situ hybridization analysis of e17.5 $Hesx1^{R160C/R160C}$ mutants revealed approximately normal levels of *Pomc*, *Tshb*, *Lhb*, *Gh*, and *Cga* expression, although the pituitaries were dysmorphic (Sajedi et al. 2008). These missense mutant mice have similar pituitary features to mice homozygous for a null allele, but the forebrain is less affected. Analysis of newborn SF1-mutant mice revealed undetectable LH, diminished *Cga* transcripts and normal immunostaining for ACTH, TSH, and GH (Ingraham et al. 1994). $Pou1f1^{dw/dw}$ mutant newborns have normal POMC immunostaining, undetectable TSH and GH, and increased LH (Mortensen et al. 2011). *Cga* appeared reduced. Immunohistochemical analysis of *Tpit* knockouts revealed undetectable POMC, no effect on caudomedial staining of GH, TSH, LH, or αGSU in the anterior lobe, but ectopic expression of TSH, LH, and αGSU in the intermediate lobe (Pulichino et al. 2003b).

(Gaston-Massuet et al. 2011; Gubbay et al. 1990; Kelberman et al. 2006; Ragge et al. 2005). While all of the patients with *SOX2* mutations exhibited reduced gonadotrope function, the degree of GH deficiency varied. The response of patients to GnRH stimulation was also variable.

6.3 PITX Factors (PITX1: Ptx1, Bft, P-OTX; PITX2: Brx, Munc30, Otlx2, Rieg)

6.3.1 Gene Discovery

The PITX transcription factors, PITX1 and PITX2, are expressed early in pituitary gland development and persist through adulthood. Their names derive from their pituitary expression and paired-like homeobox DNA-binding domains, and they have important functions in the hind limb, heart, eye, and other tissues (Gage et al. 1999; Kitamura et al. 1999; Lin et al. 1999; Lu et al. 1999; Szeto et al. 1999). These factors were found in very different ways, and they have overlapping functions within the pituitary gland (Charles et al. 2005).

Pitx1 was discovered by searching for factors that bind the regulatory elements of the pro-opiomelanocortin gene (*Pomc*). POMC is a pro-hormone that is expressed in both corticotropes of the anterior lobe and melanotropes or the intermediate lobe. POMC undergoes differential cleavage to produce adrenocorticotropin (ACTH) in corticotropes and melanocyte-stimulating hormone (MSH) and β-endorphin in melanotropes. In vitro analysis revealed that the major regulatory region of the *Pomc* promoter-proximal region was between 166 and 480 bp upstream of the transcription start site (Therrien and Drouin 1991). Drouin and colleagues found that deleting a small section from the middle of this element nearly ablated *Pomc* expression (Lamonerie et al. 1996). They screened a cDNA expression library from AtT-20 cells, an immortalized corticotrope-like cell, for factors that bind this element, and identified *Pitx1*, which has high homology with the mouse *Otx* genes. *Pitx1* was identified independently by screening the same cDNA expression library for factors that interact with POU1F1 (Szeto et al. 1996), a pituitary transcription factor described below.

PITX2 was discovered as one cause of Rieger syndrome, which was first described in the 1800s as a genetic disorder characterized by eye malformations, dental hypoplasia, craniofacial dysmorphism, and umbilical stump abnormalities. The causal gene was mapped to a 50-kb region on chromosome 4 (Datson et al. 1996). Genomic DNA from CpG islands in this region was used to screen human craniofacial and fetal brain cDNA libraries, and a contig of *RIEG* cDNAs was obtained (Semina et al. 1996). *RIEG* was most closely related to *Pitx1*, and it was later renamed *PITX2*. Murray and colleagues reported mutations in six families with Rieger syndrome. *Pitx2* was discovered independently by screening an adult mouse pituitary cDNA library for homeobox genes (Gage and Camper 1997). It was predicted to be a regulator of anterior structure formation in mice based on its expression pattern and its location on mouse chromosome 3 within a region of synteny homology to the region of human 4q25 where Rieger syndrome had been mapped.

6.3.2 Gene Expression

The expression patterns of *Pitx1* and *Pitx2* differ, but there are some regions of striking overlap. *Pitx1* is expressed in the pituitary, intestine, tongue, oral epithelium, the mandible, salivary glands, duodenum, nasal epithelium, and the hind limb (Lamonerie et al. 1996; Lanctot et al. 1999; Szeto et al. 1996). *Pitx1* is expressed in the pituitary primordia by embryonic day 9.5 (e9.5) (Treier et al. 1998). *Pitx2* is expressed in the eye, the dental lamina, limb mesenchyme, umbilical vessels, and in Rathke's pouch as early as e8.5 (Semina et al. 1996). The early expression of both PITX factors suggested their importance for pituitary development.

6.3.3 Downstream Targets

PITX1 binds the promoters of several genes that encode pituitary hormones, including *Cga* (chorionic gonadotropin alpha, the common subunit of the heterodimeric hormones LH, FSH, and TSH), *Gh*, and *Prl*, and induces their transcription (Szeto et al. 1996). PITX1 was also implicated in the transcription of *Gnrhr*, *Pou1f1*, *Tshb*, *Nr5a1*, and *Lhx3* using cell transfection assays (Tremblay et al. 1998). However, immunostaining suggests that gonadotropes and thyrotropes have the most abundant expression of PITX1 and PITX2 in the adult pituitary gland (Charles et al. 2006). PITX2 activity is enhanced by β-catenin, and the complex regulates Cyclin D2 expression (Kioussi et al. 2002).

6.3.4 Knockout Mice

Mice homozygous for *Pitx1* deletion had disrupted hind limb formation and a very mild pituitary phenotype with a slight reduction of gonadotropes and thyrotropes (Szeto et al. 1999). All of the major transcription factors were largely unaffected, and *Pomc* expression was normal.

By comparison, *Pitx2* had a much stronger effect on the developing pituitary. Four independent *Pitx2* knockout mouse lines were published in 1999 (Gage et al. 1999; Kitamura et al. 1999; Lin et al. 1999; Lu et al. 1999). Camper and colleagues developed several alleles, which included a hypomorphic or reduced function allele, $Pitx2^{neo}$, a null allele, $Pitx2^{-}$, and an inducible null allele (Suh et al. 2002). Various combinations of these alleles were used to assess dosage-sensitive effects of *Pitx2* loss of function, i.e., an allelic series. $Pitx2^{-/-}$ and $Pitx2^{neo/neo}$ embryos die in utero, at e13.5 and e18.5, respectively. Both exhibit right lung isomerism, and dosage-sensitive effects on eye, pituitary gland, and heart development. The ventral body wall of $Pitx2^{-/-}$ embryos fails to close, resulting in the externalization of the heart, liver, and abdominal organs. *Pitx2* heterozygotes accurately model the ocular features of Rieger syndrome (Chen and Gage 2016).

The initial induction of Rathke's pouch occurs at e10.5 in $Pitx2^{-/-}$ embryos, but there are extensive cell death and reduced cell proliferation, resulting in a very small pituitary primordium at e12.5 (Charles et al. 2005; Gage et al. 1999). *Hesx1*, *Prop1*, and *Lhx4* are not expressed in these mutants, but *Lhx3* is activated. Little or no hormone cell specification is detected (Lin et al. 1999). In contrast to the small pituitaries in *Pitx2*$^{-/-}$ and *Pitx2*$^{neo/-}$ embryos, there was no obvious reduction in the size of *Pitx2*$^{neo/neo}$ pituitaries (Suh et al. 2002). Somatotropes and thyrotropes were reduced, and gonadotropes were ablated. *Pitx2*$^{neo/neo}$ pituitaries had reduced expression of the transcription factors *Gata2*, *Egr1*, *Nr5a1*, and *Pou1f1*.

The overlapping expression patterns of *Pitx1* and *Pitx2* suggested the possibility of compensatory function. Indeed, double heterozygotes (*Pitx1*$^{+/-}$, *Pitx2*$^{+/-}$ and *Pitx1*$^{+/-}$, *Pitx2*$^{+/neo}$) have very poor viability, and hypoplastic pituitaries (Charles et al. 2005). *Lhx3* expression was not detectable in rare double mutants, suggesting that *Pitx* genes are required for *Lhx3* expression.

6.3.5 Human Mutations

Heterozygous *PITX1* loss of function mutations have been reported in two families with lower extremity abnormalities (Alvarado et al. 2011; Gurnett et al. 2008; Klopocki et al. 2012). Genomic rearrangements that result in ectopic expression of PITX1 in the developing forelimbs cause Liebenberg syndrome, a homeotic transformation of the arms to a leg morphology (Kragesteen et al. 2019; Spielmann et al. 2012).

Rieger syndrome is a rare, autosomal dominant disorder affecting the development of the eyes, teeth, and abdominal wall. It can be caused by mutations in *PITX2* or *FOXC1*. The majority of patients with Rieger syndrome have normal height, although there are some anecdotal reports of Rieger syndrome patients with growth insufficiency. For example, a three-generation family with Rieger syndrome and either short stature or growth hormone deficiency was reported (Sadeghi-Nejad and Senior 1974). The proband responded to GH replacement therapy. The nature of the mutation in this family and other historical cases is unknown. Many Rieger syndrome patients have been screened for *PITX2* mutations, but no systematic screening of CPHD patients without eye defects has been reported.

6.4 LIM Homeodomain Proteins: LHX2 (ap, apterous, Lh-2, LH2A), LHX3 (P-LIM, Lim3), and LHX4 (Gsh-4)

6.4.1 Gene Discovery

LHX proteins contain both *L*IM-domains, composed of two contiguous zinc fingers and a *H*omeobo*x*-domain. Three LHX factors involved in pituitary development, *Lhx2*, *Lhx3*, and *Lhx4*, and they were discovered independently. *Lhx2* was discovered in a screen of cell lines for VDJ recombinase activity (Xu et al. 1993). *Lhx3* was discovered in a screen for *Lhx* factors in a cDNA library generated from *Xenopus* gastrula (called Xlim3 in *Xenopus*) (Taira et al. 1992). *Lhx4* (previously known as Gsh-4) was discovered in a screen for homeobox genes in mouse genomic DNA (Singh et al. 1991). *Lhx2* is expressed in developing B cells, telencephalon, retina, ventral diencephalon, posterior pituitary gland, and infundibulum (Porter et al. 1997). *Lhx3* is expressed in the pituitary and pineal glands, the retina, spinal cord, and hindbrain. *Lhx4* is expressed in the developing pituitary, neural tube, hindbrain, and spinal cord (Li et al. 1994; Sheng et al. 1997).

6.4.2 Knockout Mice

Mouse knockouts of *Lhx2*, *Lhx3*, and *Lhx4* revealed the importance of each gene for pituitary development (Ellsworth et al. 2008; Li et al. 1994; Raetzman et al. 2002; Sheng et al. 1996, 1997; Zhao et al. 2010).

Lhx2 is expressed in the developing ventral diencephalon from e9.5 to e12.5 in regions that become the pituitary infundibulum and posterior lobe (Zhao et al. 2010). Embryos homozygous for *Lhx2* null alleles have pituitary dysmorphology including bifurcation of Rathke's pouch, excess proliferation in the infundibular area, and failure of the neural ectoderm to evaginate. Despite this, all hormone-producing cell types of the anterior lobe are detectable, consistent with the expression of critical signaling molecules such as BMP and FGFs by the abnormal infundibular structure. The mice are not viable and also have anophthalmia, anemia, and malformations of the cortex (Porter et al. 1997). A cohort of 59 patients with pituitary deficiency and eye abnormalities were negative for *LHX2* mutations (Perez et al. 2012).

Lhx3 is expressed in the pituitary placode at e9.5 and persists through adulthood (Sheng et al. 1996). *Lhx3*$^{+/-}$ mice are phenotypically normal and fertile, but mice homozygous for the null allele were either stillborn or died within 24 h after birth. Rathke's pouch formed in *Lhx3*$^{-/-}$ embryos but failed to expand. The thin, underdeveloped pouch fails to separate from the oral ectoderm. These mutant pouches exhibited reduced cell proliferation and enhanced cell death, and there is evidence of abnormal dorsal–ventral polarity (Ellsworth et al. 2008). The cyclin-dependent kinase inhibitor *Cdkn1a* (p21) is expressed more dorsally than normal, and *Cdkn1c* (p57) is absent in the caudal region of the pouch (Gergics et al. 2015). *Hesx1* and *Isl1* expression were initiated at e10.5 but failed to be sustained at e12.5. Induction of lineage-specific transcription factor gene expression—*Tbx19*, *Neurod1*, *Nr5a1*, and *Pou1f1*—were very poor. *Lhx3*$^{-/-}$ embryos contained a few differentiated corticotropes but no other differentiated hormone-producing cells were detected.

Lhx4 is expressed in the pituitary primordium between e9.5 and e12.5, and it is restricted to the prospective anterior lobe at e12.5 (Sheng et al. 1997). Expression diminishes at e15.5, but it is detectable in adult pituitary anterior and intermediate lobes. *Lhx4*$^{-/-}$ newborns appear normal but die shortly after birth due to failure of the lungs to inflate (Li et al. 1994). Dexamethasone treatment, used to accelerate lung development, improved but did not completely rescue survival. Pituitary insufficiency can result in poor lung development in certain genetic backgrounds (Nasonkin et al. 2004). *Lhx4*$^{-/-}$ embryos have anterior pituitary lobe hypoplasia (Sheng et al. 1997). Enhanced cell death is evident at e12.5, and *Lhx3* expression is significantly delayed (Gergics et al. 2015; Raetzman et al. 2002). *Isl1* expression does not become ventralized at e11.5 and it is not detected at e12.5. *Pitx2* expression is also not maintained. Expression of pituitary hormone genes is reduced but detectable.

The functions of *Lhx3* and *Lhx4* overlap (Sheng et al. 1997). Both $Lhx3^{+/-}$, $Lhx4^{-/-}$ and $Lhx3^{-/-}$, $Lhx4^{+/-}$ embryos can form Rathke's pouch, but it fails to expand and is more severely affected than in either single mutant. $Lhx3^{-/-}$, $Lhx4^{-/-}$ double mutants have a very severe pituitary phenotype (Sheng et al. 1997). The double mutants form a pouch rudiment, and invagination of the oral ectoderm occurs, but the rudiment fails to separate from the oral ectoderm and remains at the pharyngeal cavity beneath the palate.

6.4.3 Human Mutations

Following characterization of the $Lhx3^{-/-}$ mice, patients with Combined *P*ituitary *H*uman *D*eficiency (CPHD) were screened for *LHX3* mutations. Loss of function mutations were found in several unrelated patients (Bechtold-Dalla Pozza et al. 2012; Bhangoo et al. 2006; Bonfig et al. 2011; Gregory et al. 2015; Jullien et al. 2019; Kristrom et al. 2009; Netchine et al. 2000; Pfaeffle et al. 2007; Rajab et al. 2008; Ramzan et al. 2017; Sobrier et al. 2012). Generally, the mutations were recessive and associated with multiple pituitary hormone deficiencies. Limited neck rotation, hyperplastic anterior pituitary gland, sensorineural hearing impairment, respiratory problems, and skeletal abnormalities were found in some cases. Heterogeneity in the presentation of pituitary size is observed in CPHD patients with mutations in other genes including *PROP1* (Riepe et al. 2001).

Mutations in *LHX4* have been identified in multiple unrelated patients with CPHD. One is a homozygous lethal missense variant (p.T126M) (Gregory et al. 2015), but the rest are heterozygous loss of function mutations with incomplete penetrance (Dateki et al. 2010; Filges et al. 2012; Pfaeffle et al. 2008; Rochette et al. 2015; Tajima et al. 2007, 2010). This indicates that the dosage sensitivity to *LHX4* loss of function is more extreme in humans and mice.

In conclusion, LHX3 and LHX4 have essential and overlapping roles in anterior pituitary development in mouse and man, while LHX2 has a role in posterior pituitary lobe development.

6.5 HESX1 (Rpx)

6.5.1 Gene Discovery and Gene Expression

Hesx1 was discovered by a screen for *H*omeobox factors in an *E*mbryonic *S*tem cell cDNA library, and independently, in a screen for homeobox factors expressed during gastrulation, at e7.5 (Hermesz et al. 1996; Thomas et al. 1995; Thomas and Rathjen 1992). *Hesx1* is expressed sequentially in the anterior visceral endoderm, anterior definitive endoderm, and the cephalic neural plate. While officially known as *Hesx1*, it was also named *R*athke's *P*ouch homeobox (*Rpx1*) because of its robust expression in the developing pituitary primoridia at e9.5. Its expression in the developing pituitary gland is extinguished by PROP1 at e14.5 (Dasen et al. 2001; Gage et al. 1996).

6.5.2 Knockout Mouse Phenotype, Downstream Targets, Upstream Regulators

Robinson and colleagues generated a *Hesx1* deletion in mice and demonstrated that 93% of homozygous mutants died before weaning (Dattani et al. 1998). The phenotype was variable and included developmental abnormalities of the forebrain, eye, and pituitary gland. The pituitary glands were missing (5%), misplaced, dysmorphic, and/or small (Dasen et al. 2001). *Hesx1* deficiency permits excessive FGF signaling, resulting in multiple invaginations of oral ectoderm instead of a single Rathke's pouch (Dasen et al. 2001). Most mice heterozygous for the deletion were normal, but 1% displayed mild abnormalities of the same organs affected in homozygous mice.

Rosenfeld and colleagues demonstrated that HESX1 acts as a repressor of *Pou1f1* expression by binding the *Pou1f1* early enhancer (Dasen et al. 2001; Olson et al. 2006). HESX1 binds the co-repressor transducin-like enhancer of split 1 (TLE1) through its engrailed homology domain and also interacts with a corepressor, histone deacetylase complex containing Nuclear Receptor Co-Repressor (NCoR). Forced expression of HESX1 and TLE1 during pituitary development is sufficient to suppress POU1F1-dependent cell lineages. Both HESX1 and PROP1 are paired-type homeodomain transcription factors that can compete for the same DNA binding sites. In the presence of beta-catenin, PROP1 binds the early enhancer of *Pou1f1* and initiates *Pou1f1* gene expression.

Both *Lhx3* and *Pitx2* are necessary upstream regulators of *Hesx1* in the developing pituitary gland (Gage et al. 1999; Sheng et al. 1996). A 100-bp element that directs spatial expression of *Hesx1* in Rathke's pouch lies ~3.3 kb 3′ of the gene and is bound by PITX2 and GATA2 (Chou et al. 2006). The promoter proximal −570 bp, exons 1 and 2 and intron 1 of *Hesx1* are sufficient for transgene reporter expression in the anterior visceral endoderm, anterior neural ectoderm, and anterior neural plate. A negative element that suppresses *Hesx1* expression in the hypothalamus, and responds to inductive signals from Rathke's pouch, has been mapped −568 to −532 bp upstream of the gene.

6.5.3 Patient Mutations

The phenotypes of *Hesx1* knockout mice suggested that *HESX1* mutations might cause pituitary and anterior structure malformations in humans. A screen of 61 patients with holoprosencephaly, Septo-Optic Dysplasia (SOD), or pituitary insufficiency uncovered two siblings with agenesis of the corpus callosum and panhypopituitarism that were homozygous for a missense mutation in *HESX1*, p. R53C, which abrogates DNA binding (Dattani et al. 1998). This disease association was confirmed by screening a larger cohort of 228 patients with congenital pituitary defects for mutations in *HESX1* (Thomas et al. 2001). Three heterozygous missense mutations were identified, p.S170L, p.C509T, and p.A541G, which were associated with incomplete penetrance and variable phenotypic presentation from SOD to hypopituitarism, or Isolated Growth Hormone Deficiency (IGHD).

6.5.4 Summary

Hesx1 was discovered in two independent cDNA screens for homeobox transcription factors. LHX1, LHX3, and LMX1B activate its expression in early head development, and PITX2 and GATA2 activate its expression in Rathke's pouch. Pituitary expression is extinguished by PROP1, which also releases HESX1-mediated repression of *Pou1f1*, activating the development of somatotropes, lactotropes, and thyrotropes. Homozygous mutations in mice and heterozygous mutations in humans cause variable phenotypes that include reduced forebrain, anophthalmia, and pituitary dysfunction. The cause of variable presentation is not known.

6.6 PROP1

6.6.1 Gene Discovery

Prophet of Pit1 or PROP1 is a pituitary-specific transcription factor that was identified by positionally cloning a spontaneous dwarf mouse mutant, Ames dwarf, df (Sornson et al. 1996). These mice were first described phenotypically 58 years ago by Drs. Schaible and Gowen at Iowa State University in Ames, Iowa, as they studied the descendants of an irradiation experiment (Schaible and Gowen 1961). Homozygous mutant mice have anterior pituitary hypoplasia, growth insufficiency, hypothyroidism, and infertility, which are attributed to lack of growth hormone, prolactin, and thyroid-stimulating hormone (Bartke 1965; Cheng et al. 1983; Gage et al. 1995). *Hesx1* expression was not properly repressed in the df/df mutant pituitaries, placing the df locus downstream of *Hesx1* (Gage et al. 1995). No expression of *Pou1f1* could be detected in the pituitaries of developing *df/df* mutants by in situ hybridization, indicating that *Prop1* is required for activation of *Pou1f1* (Andersen et al. 1995; Gage et al. 1996). However, small clusters of cells expressing *Pou1f1* and GH, PRL or TSH could be detected in adult *df/df* mutant pituitaries by immunohistochemistry (Gage et al. 1995). These committed somatotropes did not express detectable levels of *Ghrhr* and were insensitive to increased stimulation by growth hormone-releasing hormone (Gage et al. 1995).

Genetic mapping placed the *df* locus on mouse chromosome 11 in a region of synteny homology with human chromosome 5q (Buckwalter et al. 1991). Using genetically directed representational difference analysis, the critical region was narrowed to a 400–600 kb window (Sornson et al. 1996). Genomic clones from this region were probed with cDNA from the developing pituitary, resulting in the identification of *Prop1*, which encodes a 233 amino acid paired homeodomain-containing protein. The Ames dwarf mice had a missense mutation in the homeodomain, p.S83P, that abrogates DNA binding. A null allele of *Prop1* was genetically engineered by inserting a LacZ/Neo cassette into the first exon of *Prop1* (Nasonkin et al. 2004). These *Prop1*$^{-/-}$ mice essentially phenocopy the *Prop1*$^{df/df}$ mice when compared to the same genetic background. *Prop1* expression is highly restricted to the pituitary, beginning at embryonic day 10.5, reaching a peak at e12.5, declining through development, and remaining detectable in the postnatal period (Perez Millan et al. 2016; Sornson et al. 1996).

6.6.2 Downstream Targets

PROP1 binds β-catenin and competes with the HESX1-corepressor complex for binding at the early enhancer of *Pou1f1* to induce *Pou1f1* expression (Dasen et al. 2001; Olson et al. 2006; Sornson et al. 1996). Deletion of β-catenin before the expression of *Pou1f1* blocks *Pou1f1* expression, while deletion of β-catenin after the onset of *Pou1f1* expression had no effect, suggesting a narrow critical window of β-catenin expression in lineage determination. This β-catenin phenotype occurs independently of LEF1, TCF3, or TCF4 action. PROP1 interacts directly with the corepressors TLE, Reptin, and HDAC1 to bind and repress the expression of *Hesx1*.

Lineage tracing reveals that all hormone-producing cell types of the anterior and intermediate lobes of the pituitary descend from *Prop1*-expressing cells (Davis et al. 2016). This places *Prop1* just downstream of *Sox2* expression in pituitary progenitors (Perez Millan et al. 2016) (Fig. 6.2). *Prop1*$^{df/df}$ mice have a vascularization defect, reduced expression of cyclin D1 and cyclin D2 in pituitary progenitors, and a failure of proliferating cells to migrate away from the stem cell niche into the parenchyma, which appears to be a failed epithelial to mesenchymal-like transition (Ward et al. 2005). Many mutant cells undergo apoptosis. Notch signaling is normally active in the transitional zone between the stem cell niche and the differentiating cells, but *Prop1* mutants do not express *Notch2* (Himes and Raetzman 2009; Raetzman et al. 2004). Colony forming assays and RNA-Seq were used to assess the effect of *Prop1* deficiency on stem cells. Mutants had a reduced stem cell pool, abnormal colony morphology, and changes in gene expression consistent with a failed epithelial-to-mesenchymal transition. To identify additional downstream targets of *Prop1* that might underlie the features of mutant pituitaries, Pérez Millán and colleagues developed a biotin-tagged allele of PROP1 in an immortalized pituitary cell line, facilitating direct identification of PROP1-binding sites in chromatin (Perez Millan et al. 2016). Binding sites were identified in or near *Notch2*, *Gli2*, and *Zeb2*. *Zeb2* activation appeared to be an essential step in the process of engaging progenitors to undergo EMT.

6.6.3 Upstream Regulators

Multiple species alignment of PROP1 genomic sequences from a variety of mammals revealed three areas of high conservation: upstream, downstream, and within intron 1 of *Prop1* (Ward et al. 2007). In the context of the *Cga* promoter, the intronic enhancer is sufficient to drive expression in the stem cell niche, located in the dorsal aspect of the pituitary gland in transgenic mice. The *Cga* promoter alone drives expression in the more ventral aspect of the organ, consistent with the location of differentiating gonadotropes and thyrotropes. This enhancer sequence is bound by RBpJK, indicating that Notch signaling feeds back to upregulate *Prop1* expression (Zhu et al. 2006).

6.6.4 Patient Mutations

Quickly after the discovery of mouse *Prop1* by positional cloning, the human ortholog was identified and recessive loss of function mutations was found in four unrelated families with CPHD (Wu et al. 1998). All three mutations either reduced or completely eliminated PROP1's ability to bind DNA, convincingly implicating these lesions as the cause of CPHD. Mutations in *PROP1* are the most common known cause of CPHD (Cogan et al. 1998; Deladoey et al. 1999). The most frequent mutations are frameshifts causing premature termination; many patients are homozygous for c.[301-302delAG] or c.[150delA]. Portuguese (Lemos et al. 2006), Lithuanian (Navardauskaite et al. 2014), and Czech (Lebl et al. 2005) cohorts have a high incidence of the 2-bp deletion, and the 1-bp deletion is common in Croatia (Dusatkova et al. 2016). A thorough investigation of the haplotype revealed that the 2-bp deletion founder mutation arose approximately 101 generations ago in central eastern Europe, and it either reoccurred or was transferred to the Iberian peninsula approximately 23 generations ago (Dusatkova et al. 2016). The Iberian haplotype and mutation were subsequently transferred to Latin America. Combining all population groups analyzed so far, *PROP1* mutations are estimated to cause ~11% of familial and sporadic cases of CPHD (Fang et al. 2016). All mutations discovered to date are recessive, and the phenotype includes deficiency of GH, TSH, gonadotropins, and progressive loss of ACTH. We speculate that the progressive hormone loss may result from depletion of the pituitary stem cell pool.

6.7 POU1F1 (PIT-1, GHF-1)

6.7.1 Gene Discovery

Karin and Rosenfeld independently discovered the first pituitary-specific transcription factor, demonstrated that it regulated expression of the *Gh* and *Prl* genes, and named it GHF1 and PIT1, respectively (Bodner et al. 1988; Ingraham et al. 1988). It was later renamed POU1F1. Rosenfeld's group identified cis-acting sequences in the

Gh and *Prl* genes that were necessary for their expression in pituitary cell cultures and used them as probes to screen a pituitary cDNA expression library (Ingraham et al. 1988; Nelson et al. 1986). They identified the *Pou1f1* cDNA, detected transcripts only in the pituitary gland, and showed that *Pou1f1* expression vectors were sufficient to activate *Gh* and *Prl* reporter genes in transfected, heterologous cells. Karin's group took the approach of biochemically purifying the protein that bound the cis-acting sequences in the *Gh* promoter and obtaining the amino acid sequence of a peptide from the binding protein (Bodner et al. 1988; Bodner and Karin 1987). Probes complementary to the peptide sequence were used to screen pituitary cDNA libraries and GHF1 was identified. Loss of GHF1 resulted in the loss of *Gh* expression (McCormick et al. 1988). Subsequent structure–function analyses revealed the presence of the nuclear localization signal and the roles of the homeodomain and POU-specific domain in specificity of DNA binding (Theill et al. 1989).

6.7.2 Knockout Mice

George D. Snell described the phenotype of an autosomal recessive, spontaneous dwarf mouse mutant in 1929, and it was named dw (Snell 1929). The growth insufficiency of the mutant mice is evident 14 days postnatally, and adult mutants also exhibit hypothyroidism and hypogonadism. Tissue transplantation experiments demonstrated that the multi-organ defects were rescued by pituitary tissue, and mutant pituitaries did not improve function when implanted in the sella of normal mice, pinpointing the pituitary as the root cause of the hormone deficiencies (Carsner and Reynolds 1960; Smith and MacDowell 1930).

Finally, 61 years after the dw Snell dwarf mutation was described phenotypically, lesions in *Pou1f1* were implicated as causal (Camper et al. 1990; Li et al. 1990). *Pou1f1* was tightly linked genetically to the *dw* locus, and the gene was rearranged in an allelic variant, dw^J (Eicher and Beamer 1980). DNA sequencing revealed a G → T missense mutation, resulting in a p.W251C change of a highly conserved residue in the homeodomain that abrogates DNA binding. These findings extended the role of *Pou1f1* from direct activation of *Gh* and *Prl* gene expression to include stimulating the development and specification of cells that express *Gh*, *Prl*, and *Tshb*. These three specialized cell types became known as the *Pou1f1* lineage.

6.7.3 Patient Mutations

Two years after linking the mouse dwarfism phenotype to lesions in *Pou1f1*, Khono and colleagues identified the first *POU1F1* mutation found in a patient with TSH, GH, and prolactin deficiencies (Tatsumi et al. 1992). This c.C638T nonsense mutation resulted in a truncation, p.R172X, N-terminal to the homeodomain of POU1F1. Numerous *POU1F1* mutations have been described in patients with CPHD (reviewed in Fang et al. 2016). The majority are recessive loss of function mutations that present with deficiency of GH, PRL, and TSH. However, there are a few notable exceptions. The p.R271W mutation is dominant negative and incompletely penetrant, possibly due to variability in mono-allelic expression of the mutant versus normal allele (Ohta et al. 1992; Okamoto et al. 1994; Radovick et al. 1992). The p.R271W mutation interferes with binding to C/EBPα, and it permits recruitment to centromeric heterochromatin (Enwright et al. 2003). The mutation also interferes with tethering of POU1F1 dimers to the nuclear matrix via interactions with β-catenin, SATB1, and matrin-3 (Skowronska-Krawczyk et al. 2014). A p. K216E mutation in the homeodomain is also a dominant cause of CPHD (Cohen et al. 1999). This mutation acts by interfering with the ability of POU1F1 to autoactivate its late enhancer in response to retinoic acid stimulation (see below). A p.P76L mutation in the transactivation domain causes dominant IGHD in humans and recessive growth insufficiency in mice (Sobrier et al. 2016). TSH and PRL were not measured in the mouse model. The amino acid substitution increases DNA binding at the human *GH1* locus control region and increases binding to several transcription factors, including PITX1, LHX3a, and ELK1. The p.P76L change decreases transactivation of *Gh*, but it has no effect on the *Prl* promoter. Two other proline substitutions in the transactivation domain have been reported in children with GH, PRL, and TSH deficiency: p.P14L and p.P24L (Fofanova et al. 1998; Ohta et al. 1992). Detailed functional studies were not carried out. The p.P14L variant appears dominant with incomplete penetrance, and no inheritance data were presented for the p.P24L allele. In sum, mutations in *POU1F1* can cause CPHD or IGHD and be recessive or dominant, and dominant mutations are informative about the normal mechanism of action.

6.7.4 Upstream Regulators

Pou1f1 expression is selectively high in the anterior lobe of the pituitary. The cis-acting sequences necessary for *Pou1f1* expression were mapped extensively in transgenic mice (Rhodes et al. 1993). A 390-bp enhancer located ~10.4–10.2 kb upstream of the *Pou1f1* transcription start site is sufficient for *Pou1f1* expression postnatally in the context of the minimal 327 bp *Pou1f1* promoter (DiMattia et al. 1997). This distal or "late enhancer" is bound by POU1F1 in an autoregulatory loop necessary for maintaining expression. Another enhancer element located between ~8 kb upstream of *Pou1f1* is bound by PROP1 and β-catenin, inducing initial expression of *Pou1f1*, but this proximal or "early enhancer" is not sufficient for the sustained expression of *Pou1f1* into adulthood (DiMattia et al. 1997; Olson et al. 2006; Sornson et al. 1996).

6.7.5 Downstream Targets

The role of POU1F1 in establishing thyrotropes and *Tshb* expression was discovered later (Charles et al. 2006; Dasen et al. 1999; Gordon et al. 1997). Ridgway and colleagues mapped cis-acting sequences necessary for *Tshb* expression in thyrotropic tumor cells and identified binding sites for POU1F1 and GATA2 within −133 to −88 of the *Tshb* transcription start site. Neither transcription factor had a major impact on *Tshb* transcription individually, but together they synergistically activated *Tshb*. Using dominant-negative and ectopic gene expression transgenic approaches, Rosenfeld and colleagues discovered that POU1F1 and GATA2 interact to promote thyrotrope development, while GATA2 alone promotes gonadotrope development. Specifically, they showed that broadly expressing *Gata2* throughout the developing pituitary blocks *Pou1f1* expression, reduces thyrotrope and somatotrope differentiation, and increases gonadotrope specification. Conversely, expressing *Pou1f1* broadly throughout the developing pituitary drives high levels of thyrotrope and somatotrope differentiation, and blocks gonadotrope specification. Camper and Ridgway collaborated to show that a pituitary-specific knockout of GATA2 diminished thyrotrope and gonadotrope differentiation. At birth, the mice have very few thyrotropes, but they recover as the mice age, likely through sensitive feedback loops that maintain thyroid hormone production at the proper level. These mice also became fertile. Marked upregulation of *Gata3* was observed, which may compensate for loss of *Gata2*. A pituitary-specific *Gata2*, *Gata3* double knockout would be necessary to assess whether loss of both GATA factors is sufficient to block gonadotrope and thyrotrope development. Nevertheless, these studies established an important role for both *Pou1f1* and *Gata2* in pituitary development.

POU1F1-mediated regulation of gene expression and cell fate requires interaction with chromatin-modifying factors. POU1F1 binds to the *Gh* promoter by e13.5–e14.5, but *Gh* transcription is not activated until a complex containing the histone lysine demethylase, LSD1 is recruited (Wang et al. 2007). A pituitary-specific deletion of *Lsd1* has little effect on *Pou1f1* expression, but there is little or no expression of *Gh*, *Tshb*, or *Prl*. Differentiation appears to be suppressed because Notch signaling fails to be silenced. LSD1 binds and activates the *Gh* promoter through interaction with an MLL1 coactivator complex. A region of the *Gh* promoter located at −161 to −146 is necessary for silencing *Gh* expression in lactotropes postnatally. ZEB1, a homeodomain transcription factor with seven Zn fingers, binds this element with a constellation of cofactors including LSD1 and corepressors. Estrogen, an activator of *Prl* expression, enhances the recruitment of LSD1- and ZEB1-mediated repression of *Gh* in lactotropes during the postnatal period in mice.

Recently, other POU1F1 binding partners were identified, shedding light on its mechanism of action (Skowronska-Krawczyk et al. 2014). Using immunoprecipitation and mass spectrometry, POU1F1 was discovered to interact with matrin-3, β-catenin, and SATB1. The dominant-negative *POU1F1* p.R271W mutation interferes with POU1F1's interaction with β-catenin and SATB1 which, in turn, results in the failure to interact with matrin-3 and reduced expression of POU1F1 target genes. Gene expression is rescued by fusing POU1F1 (p.R271W) with matrin-3. This shows convincingly that the multi-protein complex must be tethered to the nuclear matrix for optimal activity.

6.8 NR5A1 (Steroidogenic Factor 1, SF1, Ftz-f1, Adrenal 4-Binding Protein)

6.8.1 Gene Discovery

NR5A1 is considered to be the signature transcription factor for driving pituitary gonadotrope cell fate. It was initially discovered in the search for transcription factors regulating steroidogenic enzyme biosynthesis in the adrenal cortex. Parker and Morohashi independently made the observation that multiple cis-regulatory elements driving expression of steroidogenic enzyme genes and cytochrome P450 genes in the adrenal cortex had highly similar AGGTC motifs, suggesting that a single factor might regulate the expression of these proteins (Morohashi et al. 1992; Rice et al. 1991). In support of this, the factor(s) that bound all six of these elements were the same molecular weight and had identical chromatographic properties. They took a biochemical purification approach to identify the transcription factor binding these elements. Y1 adrenocortical tumor cells did not produce sufficient protein for purification and characterization. To circumvent this limitation, they took advantage of the evolutionary conservation of regulatory networks among mammals. The promoters of these three genes have high sequence similarity in cows and mice (Honda et al. 1990), and the bovine factor that binds the mouse regulatory elements is the same size as the mouse factor (Honda et al. 1993; Lala et al. 1992). They collected bovine adrenal cortex tissues, purified the binding protein, and named it steroidogenic factor 1 or SF1. This 53-kD protein could bind all six regulatory elements. Because the AGGTC motif is similar to the binding sites of nuclear hormone receptors, they predicted that SF1 would have a DNA-binding domain similar to the nuclear hormone receptor family. They screened a Y1 cDNA library with a probe designed to hybridize to the consensus sequence encoding a nuclear hormone receptor DNA binding domain. One of the cDNAs fit the expected expression profile of SF1, namely expression in the adrenal glands, the corpus luteum, and the testis. They proved that they had cloned the correct cDNA by expressing the protein in bacteria and demonstrating that it had the same binding affinity as that of SF1 purified from the bovine adrenal gland. The SF1 cDNA sequence encoded an orphan nuclear hormone receptor related to the *Drosophila* segmentation gene, fushi tarazu or ftz, which also specifies neuronal identity in flies. The gene was renamed as *NR5A1* according to the nomenclature for orphan nuclear receptors.

6.8.2 Gene Expression

In mice, *Nr5a1* transcripts were detected by in situ hybridization in the developing hypothalamus and the urogenital ridge, which is the precursor to the adrenal glands and gonads (Ikeda et al. 1994). *Nr5a1* sustains high expression in the adrenal glands through adulthood. Conversely, it is expressed highly in the developing bipotential gonad, and it is maintained in testes, particularly in Sertoli cells. Meanwhile, its expression in ovaries is reduced when the ovary first becomes morphologically distinct from the testis. *Nr5a1* expression in the bipotential gonad suggested a function in sexual development. *Nr5a1* was expressed in the ventromedial hypothalamus (VMH), which is involved in feeding, fear, thermoregulation, and sexual activity (Shima et al., 2005). These studies indicated that *Nr5a1* was expressed early in organ development and could, therefore, be a driver of cell fate in mammals.

6.8.3 Downstream Targets

Mellon and colleagues identified a DNA element that regulates gonadotrope expression of *Cga*, which encodes the alpha-subunit of the heterodimeric glycoprotein hormones, luteinizing hormone (LH), follicle-stimulating hormone (FSH), and thyroid-stimulating hormone (TSH) (Horn et al. 1992). Mellon showed that this element (which they called GSE for *G*onadotrope-*S*pecific *E*lement) was bound by a 54-kD protein, which they labeled GSEB-1 (GSE-*B*inding factor 1). They demonstrated that GSEB-1 is NR5A1 by showing that NR5A1 binds specifically to the GSE in vitro and that NR5A1-specific antibodies antagonize GSEB-1 function (Barnhart and Mellon 1994). This was the first evidence that NR5A1 had a role in the pituitary gland.

Mellon discovered that NR5A1 regulates the transcription of *Lhb*. This represented the second known direct target of NR5A1 in the pituitary gland. NR5A1 binds a GSE in the promoter region of *Lhb* and increases its expression (Halvorson et al. 1996; Horn et al. 1992). Nilson and colleagues showed that 776 bp of the bovine LHB proximal promoter region was sufficient to drive gonadotrope-specific expression in the transgenic mice (Keri et al. 1994). This sequence contained the GSE, was responsive to GnRH regulation but was not androgen or estrogen responsive. A mutation in the GSE element reduced transgene expression ten-fold (Keri and Nilson 1996). This suggested that NR5A1 stimulates LHB expression by binding the promoter-proximal GSE in cell culture and in vivo. Morohashi and colleagues found two regulatory elements of *Nr5a1*, showing that NR5A1 binds one of them to stimulate its own expression in adrenal cells (Honda et al. 1993), while Clay and colleagues demonstrated that NR5A1 directly regulates expression of *Gnrhr* via a GSE in the promoter-proximal region (Duval et al. 1997).

6.8.4 Interacting Factors

NR5A1 interacts with other transcription factors to promote gene expression. The Drouin lab demonstrated that PITX1 and NR5A1 interact directly and synergize to activate *Lhb* promoter but not *Cga* gene expression (Tremblay et al. 1998, 1999). The PITX1 binding site is not required for this activity. NR5A1 also interacts with the zinc-finger transcription factor EGR1 to regulate the expression of *Lhb* (Lee et al. 1996; Tremblay and Drouin 1999). Both NR5A1 and EGR1 binding sites are required for the synergism. Milbrandt and Charnay independently demonstrated that EGR1 is necessary for *Lhb* expression in mice (Lee et al. 1996; Topilko et al. 1998). There are no exact obvious NR5A1 binding sites in the *Fshb* promoter, but Mellon and colleagues found two sites bound by NR5A1 (Jacobs et al. 2003). NR5A1 interacts with Nuclear Factor Y (NFYA) on the *Fshb* promoter to synergistically activate *Fshb* transcription.

6.8.5 Knockout Mouse Phenotype

Nr5a1 knockout mice were generated by homologous recombination in embryonic stem (ES) cells (Luo et al. 1994). Homozygous mutant mice lack adrenal glands and gonads, revealing that NR5A1 is not only critical for expression of steroidogenic enzymes in these tissues, but it has a critical role early in adrenal and gonadal organogenesis. Moreover, the pituitaries of *Nr5a1*-null mice lack detectable transcripts for *Lhb*, *Fshb*, and the gonadotropin-releasing hormone receptor, *Gnrhr* (Ingraham et al. 1994). Expression of other pituitary hormones was normal. *Nr5a1* transcripts were detected at e13.5–e14.5, prior to the expression of other gonadotrope markers, suggesting that NR5A1 could be involved in gonadotrope differentiation, but it was still unclear whether *Nr5a1* was involved in differentiation or maintenance.

Nr5a1 mutants had structural abnormalities in the VMH, implicating NR5A1 in three levels of the reproductive axis: hypothalamus–pituitary–gonad (Ikeda et al. 1995). However, the GnRH neurons were normal. A 24-h treatment regime of GnRH injections can rescue expressions of *Lhb* and *Fshb* in hypogonadal mutants, *Gnrh*$^{hpg/hpg}$, and *Nr5a1* knockout mutants, suggesting that the gonadotrope progenitors are present in *Nr5a1* mutants, and that NR5A1 is not strictly required for gonadotrope fate (Ikeda et al. 1995). GnRH induces expression of many transcription factors in the pituitary gland, including *Egr1*, *Atf3*, c-Jun, and TCF/LEF, and increased expression of these factors may compensate for NR5A1 deficiency (Kaiser et al. 2000; Melamed et al. 2006; Salisbury et al. 2009; Tremblay and Drouin 1999).

To distinguish direct effects of NR5A1 on the pituitary gland from indirect effects resulting from altered hypothalamic and/or gonadal inputs to the pituitary gland, researchers developed a floxed allele of *Nr5a1* and induced deletion in the pituitary with a *Cga-cre* transgene (Cushman et al. 2000; Zhao et al. 2001). This ablated expression of *Fshb*, *Lhb*, and *Gnrhr*. These deficiencies caused failure to develop mature reproductive organs and infertility. Injection of these pituitary-specific knockout mice with pregnant mare serum gonadotropin (PMSG), a hormone produced by the equine placenta that has luteinizing- and follicle-stimulating activity, was sufficient to induce pituitary expression of *Lhb* and *Fshb*. This demonstrated that *Nr5a1* deficiency impacts pituitary development and function directly, but it can be overcome by supraphysiological hormonal stimulation.

6.8.6 Patient Mutations

The first patient with an *NR5A1* mutation was reported in 1999 (Achermann et al. 1999). The patient presented with primary adrenal failure, complete XY sex reversal, and streak-like gonads. Her pituitary gland responded to GnRH stimulation. She was heterozygous for a 2-bp mutation that causes a p.G35A missense in the first zinc finger of NR5A1 that abrogates DNA binding. Her phenotype is consistent with NR5A1's role in adrenal function and stimulation of LH and FSH expression. The sex reversal was surprising because mice heterozygous for a null allele of *Nr5a1* are phenotypically normal. Loss-of-function mutations have been associated with other cases of 46XY sex reversal (Achermann et al. 2002; Correa et al. 2004; Lin et al. 2007; Mallet et al. 2004), 46XX sex reversal (Hasegawa et al. 2004; Swartz et al. 2017b), undervirilization (Swartz et al. 2017a), adrenocortical insufficiency (Biason-Lauber and Schoenle 2000; Guran et al. 2016), premature ovarian failure (Lourenco et al. 2009), and spermatogenic failure (Bashamboo et al. 2010). Most mutations were dominant, some were associated with incomplete penetrance, and some were recessive (Baetens et al. 2017; Bashamboo et al. 2016; Igarashi et al. 2017; Sekido and Lovell-Badge 2008).

6.8.7 Upstream Regulators

Mice homozygous for a hypomorphic *Pitx2*^{neo} allele have no *Nr5a1* expression and, consequently, no *Lhb*, *Fshb*, or *Gnrhr* expression (Suh et al. 2002). This suggests that *Pitx2* is upstream of *Nr5a1* in the transcriptional hierarchy regulating pituitary development. Later, PITX2 was shown to bind a pituitary gonadotrope-specific enhancer within intron 6 of the *Nr5a1* gene (Shima et al. 2008). Several groups used transgenic mice to localize elements necessary for *Nr5a1* expression in various tissues (Karpova et al. 2005; Shima et al. 2005; Stallings et al. 2002). The action of *Nr5a1* in promoting gonadotrope fate is antagonized by TBX19 (Pulichino et al. 2003b).

6.9 POMC Transcription in Corticotropes (TBX19, NEUROD1)

6.9.1 Factor Discovery

Pro-opiomelanocortin (POMC) is a prohormone that is cleaved in anterior pituitary corticotropes to become adrenocorticotropic hormone (ACTH) and in melanotropes of the intermediate lobe into melanocyte-stimulating hormone (MSH) and beta-endorphin. Drouin used the AtT-20 mouse corticotrope tumor cell line to identify cis-acting elements that regulate *Pomc* expression. One element contained an E-box motif, which is typically bound by helix-loop-helix (HLH) transcription factors (Therrien and Drouin 1993). The beta HLH factor NeuroD1 is expressed in corticotrope cells and binds this element, acting together with PITX1 to activate

Pomc expression (Poulin et al. 1997). The detection of *Neurod1* expression in the developing pituitary gland from e12 to e16 is consistent with a role in initiating *Pomc* expression. *Neurod1* is transiently expressed in corticotropes, suggesting that there are other factors involved in sustained *Pomc* transcription (Poulin et al. 2000).

PITX1 was identified in an effort to identify a trans-acting factor that bound another cis-acting sequence within the promoter-proximal region of *Pomc* (Lamonerie et al. 1996). This element was sufficient to confer reporter gene expression in AtT-20 cells but not heterologous cells. Drouin and colleagues used this sequence to screen a cDNA expression library for DNA binding proteins, which led to the discovery of a novel homeodomain transcription factor PTX, now known as *Pitx1*. *Pitx1* is expressed in the pituitary primordium, and is detected in all pituitary cell types, but it is enriched in adult gonadotropes and thyrotropes (Charles et al. 2005; Lanctot et al. 1999). Mice homozygous for *Pitx1* knockout die at birth and have severe defects in mandible and hind limb development, but there is little effect on pituitary development due to functional overlap with the related *Pitx2* gene (Charles et al. 2005; Lanctot et al. 1999; Szeto et al. 1999).

A cis-acting sequence adjacent to the PITX1 binding site is required for *Pomc* expression, and the sequence motif, TCACACCA, is a T box transcription factor-binding site (Lamolet et al. 2001). PCR amplification of *Tbox* cDNAs in AtT-20 cells identified a novel factor, TPIT (officially known as TBX19), which is expressed early in pituitary development and is enriched in corticotropes and melanotropes. Ectopic expression of *Tpit* under the control of the *Cga* promoter in mice is sufficient to drive *Pomc* expression, but not *Neurod1* expression. Patients with isolated ACTH deficiency were screened and two unrelated individuals were identified with mutations in *TPIT*, confirming the critical role of this gene in POMC expression. Homozygous deletion of *Tpit* results in almost complete loss of POMC in both the anterior and the intermediate lobes, and the intermediate lobe is hypoplastic (Pulichino et al. 2003b). *Neurod1* expression is not obviously affected, suggesting that the commitment to corticotrope and melanotrope fate has been initiated. In the absence of TPIT, the progenitors in the intermediate lobe express *Nr5a1* and the alpha and beta subunits of TSH, FSH, and LH, suggesting that TPIT has a repressive effect on gonadotrope and thyrotrope cell fates. TPIT interacts directly with NR5A1 to inhibit its transcriptional activity. It is noteworthy that *Pou1f1*, *Gh* and *Prl* are not ectopically expressed in the intermediate lobes of *Tpit* mutants. Overexpressing *Tpit* under the control of the *Cga* promoter had little effect on *Tshb* expression, but it is sufficient to suppress expression of *Nr5a1*, *Lhb*, and *Fshb*, consistent with the idea that TPIT drives corticotrope and melanotrope fate while suppressing gonadotrope fate.

6.9.2 Patient Mutations

TPIT mutations are the most common, known cause of congenital, isolated ACTH deficiency (Couture et al. 2012; Metherell et al. 2004; Pulichino et al. 2003a; Vallette-Kasic et al. 2005). Patients with homozygous loss-of-function mutations typically present with undetectable plasma ACTH and corticosterone, and hypoplastic adrenal glands. Affected babies may have cholestatic jaundice and potentially fatal hypoglycemia. *Tpit* null mice have more yellow pigment than wild-type mice, due to the lack of MSH, but no obvious pigment defects are present in human patients.

6.9.3 Summary

Pomc transcription and the corticotrope and melanotrope cell fates are promoted by several transcription factors including TPIT, NeuroD1, PITX1, and PITX2. TPIT is a T box transcription factor that binds promoter-proximal region of POMC and induces its expression in concert with PITX factors. It is selectively expressed in corticotropes and melanotropes, activating the corticotrope and melanotrope lineage while repressing the gonadotrope lineage by antagonizing NR5A1 function in a DNA binding-independent manner. Loss of TPIT results in neonatal-onset isolated ACTH deficiency, but not juvenile onset.

6.10 PAX7

6.10.1 Gene Discovery

Pax7$^{-/-}$ mice were generated more than 20 years ago and revealed the importance of *Pax7* for muscle and brain development (Jostes et al. 1990; Kassar-Duchossoy et al. 2005; Mansouri and Gruss 1998; Oustanina et al. 2004; Relaix et al. 2005, 2006; Seale et al. 2000). The homozygous mutants die 3 weeks after birth, but the pituitary gland was not characterized (Mansouri et al. 1996). Karlstrom and colleagues showed that the SHH-responsive gene *Pax7* is expressed in the melanotropes of the developing zebra fish pituitary gland (Guner et al. 2008). Spatial and temporal gene expression profiling revealed that *Pax7* is expressed in progenitors that give rise to intermediate lobe melanotropes in mice, and it was hypothesized to be a critical factor for intermediate lobe cell fate through direct regulation of Notch signaling (Budry et al. 2012; Goldberg et al. 2011; Hosoyama et al. 2010).

6.10.2 Downstream Targets, Pioneering Activity

Drouin and colleagues showed that *Pax7*$^{-/-}$ mice lack intermediate lobe melanotropes, and the deficiency of *Pax7* permits ectopic differentiation of progenitors into corticotropes (Budry et al. 2012). Moreover, ectopic expression of *Pax7* in the immortalized, corticotrope-like AtT-20 cells is sufficient to convert them to a melanotrope identity associated with both reduced expression of corticotrope-enriched transcripts such as the *NeuroD1*, glucocorticoid receptor, CRH receptor, and vasopressin receptor 1b and activation of melanotrope-enriched transcripts such as the dopamine receptor *Drd2* and prohormone convertase 2, *Pcsk2*, which cleaves *Pomc* to produce MSH. Thus, *Pax7* is a selector of intermediate lobe melanotrope cell identity.

To understand the underlying mechanism of PAX7 action in reprogramming AtT-20 cells to the melanotrope fate, binding sites for TPIT, PAX7, and H3K4me1 were identified by chromatin immunoprecipitation (CHIP-Seq) and open chromatin was identified using FAIRE-Seq (Formaldehyde-Assisted Isolation of Regulatory Elements). Methylation of histone H3 at lysine 4 (H3K4me1) is associated with enhancers. Drouin and colleagues demonstrated that PAX7 can bind heterochromatin and act as pioneer factors to open the chromatin, yielding a FAIRE-Seq signal and permitting the binding of additional transcription factors such as TPIT (Mayran et al. 2018).

To further characterize PAX7 action, the corticotrope and melanotrope transcriptomes and epigenomes were compared. In so doing, these authors confirmed much of what they had shown previously, that PAX7 is necessary and sufficient for establishing a melanotrope transcriptome and epigenome. These authors made the interesting discovery that the majority of PAX7 action on the epigenome occurs at regions far from gene promoters, as the promoters of most genes in corticotropes and melanotropes are similar. They went on to show that the areas of the genome that PAX7 opens and activates represents a minor, but sizable fraction of PAX7 binding sites. Interestingly, they revealed a large fraction of sites containing a PAX7-binding motifs that are impervious to PAX7 binding. These sites were characterized by CTCF binding, and constitutive heterochromatin. While constitutive heterochromatin was resistant to PAX7 remodeling, facultative heterochromatin was readily bound by PAX7. Using inducible nuclear localization of PAX7 in AtT20− PAX7+ cells, these authors showed that PAX7 binds to open DNA very quickly (within 30 min), while it takes much longer (nearly 24 h) for PAX7 to bind areas of closed DNA. Unsurprisingly, it takes even longer for PAX7 to remodel bound, heterochromatin sites, requiring more than 3 days to fully open these areas of DNA. Finally, areas of PAX7 binding and opening were subject to reduction in CpG methylation, a very stable form of epigenetic memory, and most sites opened by PAX7 pioneering activity remain open long after PAX7 has been removed from the nucleus. With this information, the authors put forth the model that involves PAX7 binding to both euchromatin and facultative heterochromatin, remodeling of the areas of heterochromatin into areas of euchromatin, and allowing factors like TBX19 and STAT3 to bind and drive expression. This clearly defines the function of a pioneering transcription factor, and leaves us with the question of whether this mechanism of pioneering action represents a broader pattern of pioneer action, or is specific to PAX7.

6.10.3 Patient Mutations

PAX7 translocations resulting in fusion with the forkhead in rhabdomyosarcoma gene, *FKHR*, can cause pediatric alveolar rhabdomyosarcoma (Barr et al. 1996). Loss-of-function mutations would be expected to be lethal in the homozygous state, and none have been reported in humans (Mansouri et al. 1996).

6.10.4 Summary

PAX7 is the first validated pioneer transcription factor acting within the pituitary. Important for both muscle and brain development, it was not implicated in pituitary function until 2010. Melano/corticotrope precursor cells can become either cell type in development, and the presence of PAX7 railroads the development to a melanotrope fate whereas the absence of PAX7 results in a corticotrope fate. PAX7 does this by making melanotrope-specific TPIT-binding sites available to TPIT by binding to heterochromatin and converting it to euchromatin. The Drouin lab has shown a very thorough mechanism by which PAX7 does this, revealing the amount of time this chromatin-changing process takes, and showing its stability through repeated DNA replication cycles. However, the exact complex required for this change in the epigenetic landscape is unclear. Furthermore, all of the work done to date has focused on the "opening" of DNA that is specific to melanotropes, while it is yet unclear whether PAX7 "closes" corticotrope-specific DNA.

6.11 Current Model for the Transcription Factors of the Pituitary Gland and Open Questions

A complex cascade of transcription factors is responsible for the development and maintenance of the pituitary gland. PITX and LHX proteins are some of the earliest factors in promoting pituitary formation. *Pitx1* and *Pitx2* compensate for one another and are responsible for turning on *Lhx3*. LHX3 and LHX4 also have overlapping functions, and together, they are necessary for the pouch rudiment to expand. During these early stages, the majority of cells are undifferentiated, proliferating progenitor cells, defined by their expression of SOX2. Then, *Lhx3* activates *Prop1* throughout Rathke's pouch from e10.5 to e12.5, and *Prop1* quickly becomes restricted to the progenitors that are undergoing an epithelial to mesenchymal-like transition, involving activation of *Zeb2*, and have silenced *Sox2* expression. PROP1 represses *Hesx1* expression and induces *Pou1f1*, the major lineage determining transcription factor for somatotropes, lactotropes, and thyrotropes. POU1F1 interacts with epigenetic regulators to activate and repress expression of target genes necessary for the sub-specialization into three different hormone-producing cell fates. LSD1, ZEB1, estrogen receptor, and GATA2 are some of the factors involved in sub-specialization. During the last few days of embryogenesis, GATA2, NR5A1, and EGR1 drive gonadotrope fate, and TPIT and NEUROD1 drive corticotrope fate. PAX7 and TPIT act together to drive melanotrope fate.

This attractive model represents decades of sophisticated, pioneering work, and explains many basic steps in pituitary development. There are, however, questions that remain unanswered. How is the switch from proliferation to differentiation regulated? What additional transcription factors and signaling pathways are involved in cell specification?

SOX2 is firmly established as the marker for cycling progenitor cells in the developing mouse pituitary, and it is also expressed in the quiescent stem cell reserve in adult pituitary. The steps involved in switching cells from their SOX2 expressing, cycling state to their specific lineage is unclear. A thorough analysis of cell cycle regulators from e11.5 to e16.5 helped clarify this, as it revealed three distinct populations (Bilodeau et al. 2009). There is a cycling precursor population (defined as Ki67-positive, and presumably SOX2-positive), a non-cycling precursor population (defined as p57-positive and p27-negative), and a non-cycling differentiated population (p57-negative and p27-positive). Drouin and colleagues showed that the deletion of the major transcription factor TPIT resulted in an unusual population of non-cycling, undifferentiated progenitors that were p57+ and p27+. This suggests that p57 marks cell cycle exit, and major factors (PROP1, TPIT, NR5A1) silence p57, permitting differentiation, and locking cells into their fate. Unlike the temporal phasing of neuronal differentiation in the brain and retina, nearly all cells in the pituitary leave the cell cycle at approximately the same time between e11.5 and e13.5 (Davis et al. 2011). Two major questions remain: what triggers the transition from the proliferating to non-proliferating state, and how are the major transcription factors exclusively turned on in different progenitors? Notch, WNT, and YAP/TAZ signaling play important roles in regulating progenitor proliferation (Cheung et al. 2018a; Gaston-Massuet et al. 2011; Lodge et al. 2019) The second question concerning how each factor is turned on is almost certainly specific to the factor, as common regulatory networks across cell lineages would fail to explain the highly unique transcription network in the different cell types.

The prevailing model of cell-type differentiation in the pituitary involves an exit from the cell cycle with a major transcription factor that inexorably pushes it down a single path to induce a specific and exclusive fate. There is evidence to suggest, however, that the differentiation is not as linear as once thought, and the fate may be more plastic than the model suggests. Early analyses of the pituitary led Romeis to propose a "one cell one hormone" model (von Möllendorff and Bargmann 1940). However, a single gonadotrope can express and secrete both LH and FSH (Phifer et al. 1973; Robyn et al. 1973) and GH and TSH are co-expressed in hypothyroid mice (Horvath et al. 1990). Cells that express both GH and PRL may be a transient population that represents a precursor of both somatotropes and lactotropes (Borrelli et al. 1989; Frawley et al. 1985). Or, they may represent a transition from a GH-expressing cell to PRL-expressing through the action of POU1F1, LSD1, and ZEB1 (Wang et al. 2007). Boehm and colleagues discovered that at e17.5, nearly all cells expressing FSH also express TSH, and that many of these cells express NR5A1 suggesting a close relationship between gonadotropes and thyrotropes (Wen et al. 2010). These studies highlight how closely related the various cell types are, challenging the idea that major transcription factors direct an inflexible cell fate.

6.12 Technical Developments and Future Directions

Four major technological innovations are now available to gain insight into cell fate decisions during development: innovative DNA library preparation methods coupled with high-throughput sequencing, single-cell sequencing, CRISPR, and organoids.

6.12.1 High-Throughput Sequencing

Advances in sequencing technology have made genome-wide analyses of gene expression feasible. Sequencing mRNA from genetically marked pituitary cells has been invaluable to the pituitary field. Subpopulations of pituitary cells have been purified from transgenic mice with fluorescently labeled cells and fluorescence-activated cell sorting (FACS). This approach revealed heterogeneity of gonadotrope gene expression in different sexes at different times (Qiao et al. 2016) and the similarity between corticotropes and melanotropes, with only ~500 differentially expressed genes (Mayran et al. 2018). This method also revealed novel regulators of cell identity for somatotropes and lactotropes (Peel et al. 2018). There are limitations, however, to applying this technology to small numbers of cells collected from developing pituitaries, as they may contain different factors that drive cell fate from those that are employed to maintain it in adult animals.

Identification of genome-wide regions of open chromatin and transcription factor binding sites in specialized cells is valuable for understanding cell fate decisions. DNase hypersensitivity mapping has been largely replaced by *A*ssay for *T*ransposase-*A*ccessible *C*hromatin with sequencing (ATAC-seq) for the identification of putative regulatory elements (Buenrostro et al. 2013). These powerful methods have successfully been used in the pituitary, finding regulatory elements in melanotropes and corticotropes (Budry et al. 2012), implicating PAX7 as a pioneering transcription factor, and further characterizing the speed at which PAX7 euchromatinizes DNA and the longevity of the subsequent epigenetic marks (Mayran et al. 2018). Similarly, they have shown the chromatin accessibility at STAT3 and Glucocorticoid receptor binding sites in corticotropes (Langlais et al. 2012). Software programs are available to identify footprints of transcription factors in ATAC-seq data and DNA-binding motifs that underlie transcription factor binding (Baek and Sung 2016). Genome-wide binding sites for PROP1 (Perez Millan et al. 2016) were identified using *Ch*romatin *I*mmuno*p*recipitation with sequencing (CHIP-seq) (Barski et al. 2007). POU1F1 binding at active enhancers with Matrin3 have been reported also (Skowronska-Krawczyk et al. 2014).

CHIP-seq for epigenetic marks on histones can identify putative poised enhancers (histone H3 lysine 4 methylation, H3K4me1 alone) and active enhancers (H3K4me1 coupled with histone H3 lysine 27 acetylation, H3K27ac) (Creyghton et al. 2010). H3K27me3 marking is a little more nuanced, as the presence of this mark across the gene body is correlated with repression, whereas enrichment at the transcription start site or the promoter may be a sign of bivalency or even active transcription (Young et al. 2011).

Chromatin Conformation Capture (3C) enables researchers to determine which regulatory elements and promoters are interacting. This is important because enhancer elements may not regulate the closest gene, and they may lie at a substantial distance from the gene they regulate. Liebhaber and colleagues initially used traditional techniques of DNase I hypersensitivity mapping and transgenic animal models to discover and characterize the Locus Control Region (LCR) for the human growth hormone gene cluster (Jones et al. 1995). There are several different enhancers within the LCR that selectively regulate the expression of *GH1* and the surrounding *GH*-related genes in the pituitary and the placenta, respectively. The 3C method was used to discover the interaction of different enhancers within the LCR with the promoters of *GH1* and the neighboring related genes (Ho et al. 2008, 2013; Tsai et al. 2016). This method was also used to show that the early and distal enhancers of *Pou1f1* interact with the *Pou1f1* promoter sequentially, and this shift requires POU1F1, establishing an autoregulatory network, that results in the enrichment of H3K27ac at the *Pou1f1* promoter (Ho et al. 2015). Further application of this method will link enhancers found using CHIP-seq to the genes they regulate. A remaining challenge for the field is scaling the technology to small numbers of cells available during pituitary development and the tendency for individual enhancers to be redundant (Sarro et al. 2018).

6.12.2 Single Cell Sequencing

RNA sequencing of single cells (scRNA-seq) is now possible (Navin et al. 2011). Various methods can be used to capture single cells including fluidics, manual capture, and FACS (Shapiro et al. 2013), and there are different approaches for obtaining the RNA sequences from individual cells, including bar coding RNA library preparations from single cells followed by next-generation sequencing (Zheng et al. 2017; Ziegenhain et al. 2017). scRNA-seq analysis algorithms allow researchers to map out a shared differentiation pathway in an asynchronous population of cells (Trapnell et al. 2014). Temporal changes in gene expression that occur as differentiation proceeds can be inferred, revealing candidate genes that may drive these differentiation pathways. The scRNA-seq methods are particularly valuable for analysis of rare cell types or situations in which the relative contribution of individual cell types to the whole is shifting. Currently, very few publications containing single-cell sequencing data from pituitary exist (Cheung et al. 2018b; Ruf-Zamojski et al. 2018). But the work that has been done already shows that this approach is powerful and can reveal previously unappreciated subpopulations of cells and novel markers of specialized cell types.

6.12.3 CRISPR

Genetically engineered mice have been invaluable for understanding the role of individual genes in pituitary development (Camper et al. 1995). Transgenic mice were used to identify the regulatory elements of the *Gh* gene (Jones et al. 1995) and *Nr5a1* (Karpova et al. 2005; Stallings et al. 2002) and to tease apart the differentiation of the somatolactotrope (Borrelli et al. 1989). Homologous recombination in embryonic stem cells was used to characterize the function of *Hesx1* (Dasen et al. 2001) and numerous other pituitary genes, as described earlier. Despite the importance of these technologies, there are significant shortcomings, including expense and inefficiency, and position effects associated with random integration of transgenes. CRISPR-Cas9 (clustered regularly interspersed short palindromic repeats-CRISPR-associated protein 9) technology has reduced the time and expense needed to make genetic modifications in mammalian genomes (Cong et al. 2013). Suddenly, knockouts, knockins, conditional alleles, and even more complex tissue-specific CRISPR mice were developed. This technology will have a high impact on the field as it is more broadly implemented.

6.12.4 Organoids

Organoids offer the opportunity to study developmental processes ex vivo, which has helped in understanding the differentiation the gut (Sato et al. 2009), retina (Nakano et al. 2012), kidneys (Xia et al. 2013), and pituitary gland (Ozone et al. 2016; Suga et al. 2011). Pituitary *Sox2*-expressing and verapamil-sensitive cells can be cultured as pituispheres and differentiate into each of the anterior pituitary hormone-producing cell types (Chen et al. 2005; Fauquier et al. 2008). Subsequently, Sasai and colleagues demonstrated an incredible ability of mouse and human embryonic stem cells to self-organize and differentiate into functional pituitary hormone-producing cells that could rescue hormone deficiency, giving promise for future cell-based therapies for hypopituitarism (Ozone et al. 2016; Suga et al. 2011).

6.13 Future Directions

These powerful technological advances will help us clarify many questions that remain in understanding pituitary development, namely understanding the switch from proliferation to differentiation and the factors responsible for the differentiation of the cell types. Single-cell sequencing will reveal more nuanced populations of cells that will clarify the exact pathway from a *Sox2*-expressing progenitor cell, through a non-cycling intermediate precursor, to a cell type that has some combination of major transcription factors and hormones, to finally, a differentiated cell type that expresses high levels of hormones. Knowing exactly which factor is in which cell will rigorously generate a model for differentiation that will be less prone to revision. Highly enriched motifs and footprints in ATAC-seq data will implicate transcription factors that are critical to this process, shortening the long list of factors that will be generated by single-cell RNA-seq. ATAC-seq will also (in concert with CHIP-seq for histone marks) find the regulatory elements for these major factors, revealing transcriptional networks that are active at different timepoints during pituitary development. Using CRISPR, each factor's necessity and function will be directly assessed in vivo. Finally, once a larger suite of factors has been implicated, whose expression profile is clear, and whose function has been characterized, we will be able to generate highly accurate organoids, knowing which ingredients are necessary and sufficient for creating a pituitary gland.

One such area that requires further investigations are the thyrotropes. These represent ~5% of the total number of cells in the pituitary gland (Kulig et al. 1998), making them difficult to study. It is clear from the *Pou1f1*-dwarfed mice that POU1F1 is required for the adult population of thyrotropes (Fig. 6.3), and it interacts with GATA2 to drive thyrotrope fate (Dasen et al. 1999). There could be additional factors that are necessary to drive thyrotrope differentiation (Charles et al. 2006). The regulatory elements of *Tshb* are not known. POU1F1, GATA2, and MED220 all synergize on the *Tshb* promoter to activate its expression (Gordon et al. 1997, 2006), but these sequences are insufficient for expression in transgenic mice (Camper et al. 1995). Using single-cell sequencing, ATAC-seq and CHIP-seq for POU1F1 and histone marks, the factors responsible for thyrotropes, and the cis-regulatory elements to which they bind will become clear.

Acknowledgments We thank Karine Rizzoti (Crick, London) for images of *Sox2* mutant embryos, Hironori Bando for critical reading of the manuscript, and the National Institutes of Health for funding (T32GM007544 and T32HG000040 to AZD and HD034283 and HD097096 to SAC).

Key References

Sox2 deficiency impairs the differentiation of multiple pituitary cell types in humans and mice (Kelberman et al. 2006). *Sox2* expressing pituitary cells have the potential to self-renew and generate all other hormone producing cells of the pituitary gland (Fauquier et al. 2008). *Sox2* expressing progenitors are recruited to generate new hormone producing cells in adult animals challenged to increase hormone production (Andoniadou et al. 2013).

Pitx1 was cloned as a transcription factor that binds a critical element of the *Pomc* gene (Lamonerie et al. 1996). *Pitx2* was cloned as a cause of Rieger syndrome and in a search for homeobox genes expressed in the pituitary gland (Gage and Camper 1997; Semina et al. 1996). *Pitx1* deletion has a modest effect on pituitary gonadotropes and thyrotropes (Szeto et al. 1999). *Pitx2* has multiple roles in pituitary development including early induction of Rathke's pouch, differentiation of gonadotropes and the *Pou1f1* lineage, and overlapping function with *Pitx1* (Charles et al. 2005; Gage et al. 1999).

Lhx3 knockout affects early pituitary development and causes apoptosis of the pituitary primordium (Ellsworth et al. 2008; Sheng et al. 1996). *Lhx4* knockout affects early pituitary development and causes apoptosis in the pituitary primordium (Raetzman et al. 2002; Sheng et al. 1997). *Lhx2* knockout affects pituitary posterior lobe development (Zhao et al. 2010).

Hesx1 was identified as a transcription factor expressed in early development and then concentrated in Rathke's pouch (Hermesz et al. 1996). *Hesx1* deficient mice have multiple anomalies including pituitary hypoplasia, and severely affected individuals have septo-optic dysplasia (Dattani et al. 1998). HESX1 suppresses *Prop1* expression through interactions with co-repressors, and in the presence of beta-catenin, *Prop1* activites *Pou1f1* expression (Olson et al. 2006).

PROP1 is a pituitary-specific transcription factor that is necessary for activation of *Pou1f1* and for differentiation of thyrotropes, somatotropes and lactotropes (Sornson et al. 1996). *Prop1* is necessary for developing a normal pool of pituitary stem cells, for driving an epithelial to mesenchymal-like transition process through the activation of *Zeb2* and suppression of cell adhesion (Perez Millan et al. 2016). All hormone producing cells of the anterior and intermediate lobes of the pituitary gland are derived from a *Prop1* expressing progenitor (Davis et al. 2016).

POU1F1 was cloned based on its ability to bind critical elements of the GH and PRL genes (Bodner et al. 1988; Ingraham et al. 1988). *Pou1f1* is mutated in mice that lack somatotropes, lactotropes and thyrotropes (Camper et al. 1990; Li et al. 1990). POU1F1 interacts with GATA2 to drive thyrotrope fate, and *Gata2* may have overlapping function with *Gata3* (Charles et al. 2006; Dasen et al. 1999). Somatotropes and lactotropes are very similar cell types derived from POU1F1 expressing progenitors. Epigentic regulation involving LSD1 and ZEB1 affects the specialization of progenitors to somatotropes and lactotropes (Wang et al. 2007). POU1F1 acts at the nuclear matrix to regulate gene expression (Skowronska-Krawczyk et al. 2014).

NR5A1 is essential for the development of adrenal and multiple organs of the reproductive axis including the pituitary gland gonadotropes (Ingraham et al. 1994; Zhao et al. 2010). NR5A1 interacts with GATA2 to promote gonadotrope fate (Charles et al. 2006; Dasen et al. 1999).

TBX19 is critical for melanotrope and corticotrope fate and it antagonizes the action of NR5A1 in promoting gonadotrope fate (Lamolet et al. 2001; Pulichino et al. 2003). Homozygous deletion of *Tpit* results in almost complete loss of POMC in both the anterior and intermediate lobes, and the intermediate lobe is hypoplastic (Pulichino et al. 2003).

Pax7$^{-/-}$ mice lack intermediate lobe melanotropes, and the deficiency of *Pax7* permits ectopic differentiation of progenitors into corticotropes (Budry et al. 2012). PAX7 is a pioneering transcription factor that opens chromatin permitting access of transcription factors such as TPIT.

Single cell sequencing is a powerful technique for determining gene expression in individual pituitary cell types without the need for genetic labeling or cell sorting (Cheung et al. 2018b).

Pituitary organoids hold promise for studying pituitary development ex vivo (Ozone et al. 2016; Suga et al. 2011).

References

Achermann JC, Ito M, Ito M, Hindmarsh PC, Jameson JL (1999) A mutation in the gene encoding steroidogenic factor-1 causes XY sex reversal and adrenal failure in humans. Nat Genet 22:125–126

Achermann JC, Ozisik G, Ito M, Orun UA, Harmanci K, Gurakan B, Jameson JL (2002) Gonadal determination and adrenal development are regulated by the orphan nuclear receptor steroidogenic factor-1, in a dose-dependent manner. J Clin Endocrinol Metab 87:1829–1833

Alvarado DM, McCall K, Aferol H, Silva MJ, Garbow JR, Spees WM, Patel T, Siegel M, Dobbs MB, Gurnett CA (2011) Pitx1 haploinsufficiency causes clubfoot in humans and a clubfoot-like phenotype in mice. Hum Mol Genet 20:3943–3952

Andersen B, Pearse RV 2nd, Jenne K, Sornson M, Lin SC, Bartke A, Rosenfeld MG (1995) The Ames dwarf gene is required for Pit-1 gene activation. Dev Biol 172:495–503

Andoniadou CL, Matsushima D, Mousavy Gharavy SN, Signore M, Mackintosh AI, Schaeffer M, Gaston-Massuet C, Mollard P, Jacques TS, Le Tissier P et al (2013) Sox2(+) stem/progenitor cells in the adult mouse pituitary support organ homeostasis and have tumor-inducing potential. Cell Stem Cell 13:433–445

Avilion AA, Nicolis SK, Pevny LH, Perez L, Vivian N, Lovell-Badge R (2003) Multipotent cell lineages in early mouse development depend on SOX2 function. Genes Dev 17:126–140

Baek S, Sung MH (2016) Genome-scale analysis of cell-specific regulatory codes using nuclear enzymes. Methods Mol Biol 1418:225–240

Baetens D, Stoop H, Peelman F, Todeschini AL, Rosseel T, Coppieters F, Veitia RA, Looijenga LH, De Baere E, Cools M (2017) NR5A1 is a novel disease gene for 46,XX testicular and ovotesticular disorders of sex development. Genet Med 19:367–376

Barnhart KM, Mellon PL (1994) The orphan nuclear receptor, steroidogenic factor-1, regulates the glycoprotein hormone alpha-subunit gene in pituitary gonadotropes. Mol Endocrinol 8:878–885

Barr FG, Nauta LE, Davis RJ, Schafer BW, Nycum LM, Biegel JA (1996) In vivo amplification of the PAX3-FKHR and PAX7-FKHR fusion genes in alveolar rhabdomyosarcoma. Hum Mol Genet 5:15–21

Barski A, Cuddapah S, Cui K, Roh TY, Schones DE, Wang Z, Wei G, Chepelev I, Zhao K (2007) High-resolution profiling of histone methylations in the human genome. Cell 129:823–837

Bartke A (1965) The response of two types of dwarf mice to growth hormone, thyrotropin, and thyroxine. Gen Comp Endocrinol 5:418–426

Bashamboo A, Ferraz-de-Souza B, Lourenco D, Lin L, Sebire NJ, Montjean D, Bignon-Topalovic J, Mandelbaum J, Siffroi JP, Christin-Maitre S et al (2010) Human male infertility associated with mutations in NR5A1 encoding steroidogenic factor 1. Am J Hum Genet 87:505–512

Bashamboo A, Donohoue PA, Vilain E, Rojo S, Calvel P, Seneviratne SN, Buonocore F, Barseghyan H, Bingham N, Rosenfeld JA et al (2016) A recurrent p.Arg92Trp variant in steroidogenic factor-1 (NR5A1) can act as a molecular switch in human sex development. Hum Mol Genet 25:5286

Bechtold-Dalla Pozza S, Hiedl S, Roeb J, Lohse P, Malik RE, Park S, Duran-Prado M, Rhodes SJ (2012) A recessive mutation resulting in a disabling amino acid substitution (T194R) in the LHX3 homeodomain causes combined pituitary hormone deficiency. Horm Res Paediatr 77:41–51

Bhangoo AP, Hunter CS, Savage JJ, Anhalt H, Pavlakis S, Walvoord EC, Ten S, Rhodes SJ (2006) Clinical case seminar: a novel LHX3 mutation presenting as combined pituitary hormonal deficiency. J Clin Endocrinol Metab 91:747–753

Biason-Lauber A, Schoenle EJ (2000) Apparently normal ovarian differentiation in a prepubertal girl with transcriptionally inactive steroidogenic factor 1 (NR5A1/SF-1) and adrenocortical insufficiency. Am J Hum Genet 67:1563–1568

Bilodeau S, Roussel-Gervais A, Drouin J (2009) Distinct developmental roles of cell cycle inhibitors p57Kip2 and p27Kip1 distinguish pituitary progenitor cell cycle exit from cell cycle reentry of differentiated cells. Mol Cell Biol 29:1895–1908

Bodner M, Karin M (1987) A pituitary-specific trans-acting factor can stimulate transcription from the growth hormone promoter in extracts of nonexpressing cells. Cell 50:267–275
Bodner M, Castrillo JL, Theill LE, Deerinck T, Ellisman M, Karin M (1988) The pituitary-specific transcription factor GHF-1 is a homeobox-containing protein. Cell 55:505–518
Bonfig W, Krude H, Schmidt H (2011) A novel mutation of LHX3 is associated with combined pituitary hormone deficiency including ACTH deficiency, sensorineural hearing loss, and short neck-a case report and review of the literature. Eur J Pediatr 170:1017–1021
Borrelli E, Heyman RA, Arias C, Sawchenko PE, Evans RM (1989) Transgenic mice with inducible dwarfism. Nature 339:538–541
Buckwalter MS, Katz RW, Camper SA (1991) Localization of the panhypopituitary dwarf mutation (df) on mouse chromosome 11 in an intersubspecific backcross. Genomics 10:515–526
Budry L, Balsalobre A, Gauthier Y, Khetchoumian K, L'Honore A, Vallette S, Brue T, Figarella-Branger D, Meij B, Drouin J (2012) The selector gene Pax7 dictates alternate pituitary cell fates through its pioneer action on chromatin remodeling. Genes Dev 26:2299–2310
Buenrostro JD, Giresi PG, Zaba LC, Chang HY, Greenleaf WJ (2013) Transposition of native chromatin for fast and sensitive epigenomic profiling of open chromatin, DNA-binding proteins and nucleosome position. Nat Methods 10:1213–1218
Camper SA, Saunders TL, Katz RW, Reeves RH (1990) The Pit-1 transcription factor gene is a candidate for the murine Snell dwarf mutation. Genomics 8:586–590
Camper SA, Saunders TL, Kendall SK, Keri RA, Seasholtz AF, Gordon DF, Birkmeier TS, Keegan CE, Karolyi IJ, Roller ML et al (1995) Implementing transgenic and embryonic stem cell technology to study gene expression, cell-cell interactions and gene function. Biol Reprod 52:246–257
Carreno G, Apps JR, Lodge EJ, Panousopoulos L, Haston S, Gonzalez-Meljem JM, Hahn H, Andoniadou CL, Martinez-Barbera JP (2017) Hypothalamic sonic hedgehog is required for cell specification and proliferation of LHX3/LHX4 pituitary embryonic precursors. Development 144:3289–3302
Carsner RL, Reynolds EG (1960) Primary site of gene action in anterior pituitary dwarf mice. Science 131:829
Castinetti F, Reynaud R, Saveanu A, Jullien N, Quentien MH, Rochette C, Barlier A, Enjalbert A, Brue T (2016) Mechanisms in endocrinology: an update in the genetic aetiologies of combined pituitary hormone deficiency. Eur J Endocrinol 174(6):R239–R247
Cha KB, Douglas KR, Potok MA, Liang H, Jones SN, Camper SA (2004) WNT5A signaling affects pituitary gland shape. Mech Dev 121:183–194
Chambers TJ, Giles A, Brabant G, Davis JR (2013) Wnt signalling in pituitary development and tumorigenesis. Endocr Relat Cancer 20:R101–R111
Charles MA, Suh H, Hjalt TA, Drouin J, Camper SA, Gage PJ (2005) PITX genes are required for cell survival and Lhx3 activation. Mol Endocrinol 19:1893–1903
Charles MA, Saunders TL, Wood WM, Owens K, Parlow AF, Camper SA, Ridgway EC, Gordon DF (2006) Pituitary-specific Gata2 knockout: effects on gonadotrope and thyrotrope function. Mol Endocrinol 20:1366–1377
Chen L, Gage PJ (2016) Heterozygous Pitx2 null mice accurately recapitulate the ocular features of Axenfeld-Rieger syndrome and congenital glaucoma. Invest Ophthalmol Vis Sci 57:5023–5030
Chen J, Hersmus N, Van Duppen V, Caesens P, Denef C, Vankelecom H (2005) The adult pituitary contains a cell population displaying stem/progenitor cell and early embryonic characteristics. Endocrinology 146:3985–3998
Cheng TC, Beamer WG, Phillips JA 3rd, Bartke A, Mallonee RL, Dowling C (1983) Etiology of growth hormone deficiency in little, Ames, and Snell dwarf mice. Endocrinology 113:1669–1678
Cheung L, Le Tissier P, Goldsmith SG, Treier M, Lovell-Badge R, Rizzoti K (2018a) NOTCH activity differentially affects alternative cell fate acquisition and maintenance. Elife 7:e33318

Cheung LYM, George AS, McGee SR, Daly AZ, Brinkmeier ML, Ellsworth BS, Camper SA (2018b) Single-cell RNA sequencing reveals novel markers of male pituitary stem cells and hormone-producing cell types. Endocrinology 159:3910–3924

Chou SJ, Hermesz E, Hatta T, Feltner D, El-Hodiri HM, Jamrich M, Mahon K (2006) Conserved regulatory elements establish the dynamic expression of Rpx/HesxI in early vertebrate development. Dev Biol 292:533–545

Cogan JD, Wu W, Phillips JA 3rd, Arnhold IJ, Agapito A, Fofanova OV, Osorio MG, Bircan I, Moreno A, Mendonca BB (1998) The PROP1 2-base pair deletion is a common cause of combined pituitary hormone deficiency. J Clin Endocrinol Metab 83:3346–3349

Cohen LE, Zanger K, Brue T, Wondisford FE, Radovick S (1999) Defective retinoic acid regulation of the Pit-1 gene enhancer: a novel mechanism of combined pituitary hormone deficiency. Mol Endocrinol 13:476–484

Cong L, Ran FA, Cox D, Lin S, Barretto R, Habib N, Hsu PD, Wu X, Jiang W, Marraffini LA et al (2013) Multiplex genome engineering using CRISPR/Cas systems. Science 339:819–823

Correa RV, Domenice S, Bingham NC, Billerbeck AE, Rainey WE, Parker KL, Mendonca BB (2004) A microdeletion in the ligand binding domain of human steroidogenic factor 1 causes XY sex reversal without adrenal insufficiency. J Clin Endocrinol Metab 89:1767–1772

Couture C, Saveanu A, Barlier A, Carel JC, Fassnacht M, Fluck CE, Houang M, Maes M, Phan-Hug F, Enjalbert A et al (2012) Phenotypic homogeneity and genotypic variability in a large series of congenital isolated ACTH-deficiency patients with TPIT gene mutations. J Clin Endocrinol Metab 97:E486–E495

Creyghton MP, Cheng AW, Welstead GG, Kooistra T, Carey BW, Steine EJ, Hanna J, Lodato MA, Frampton GM, Sharp PA et al (2010) Histone H3K27ac separates active from poised enhancers and predicts developmental state. Proc Natl Acad Sci USA 107:21931–21936

Cushman LJ, Burrows HL, Seasholtz AF, Lewandoski M, Muzyczka N, Camper SA (2000) Cre-mediated recombination in the pituitary gland. Genesis 28:167–174

Dasen JS, O'Connell SM, Flynn SE, Treier M, Gleiberman AS, Szeto DP, Hooshmand F, Aggarwal AK, Rosenfeld MG (1999) Reciprocal interactions of Pit1 and GATA2 mediate signaling gradient-induced determination of pituitary cell types. Cell 97:587–598

Dasen JS, Martinez Barbera JP, Herman TS, Connell SO, Olson L, Ju B, Tollkuhn J, Baek SH, Rose DW, Rosenfeld MG (2001) Temporal regulation of a paired-like homeodomain repressor/TLE corepressor complex and a related activator is required for pituitary organogenesis. Genes Dev 15:3193–3207

Dateki S, Fukami M, Uematsu A, Kaji M, Iso M, Ono M, Mizota M, Yokoya S, Motomura K, Kinoshita E et al (2010) Mutation and gene copy number analyses of six pituitary transcription factor genes in 71 patients with combined pituitary hormone deficiency: identification of a single patient with LHX4 deletion. J Clin Endocrinol Metab 95:4043–4047

Datson NA, Semina E, van Staalduinen AA, Dauwerse HG, Meershoek EJ, Heus JJ, Frants RR, den Dunnen JT, Murray JC, van Ommen GJ (1996) Closing in on the Rieger syndrome gene on 4q25: mapping translocation breakpoints within a 50-kb region. Am J Hum Genet 59:1297–1305

Dattani MT, Martinez-Barbera JP, Thomas PQ, Brickman JM, Gupta R, Martensson IL, Toresson H, Fox M, Wales JK, Hindmarsh PC et al (1998) Mutations in the homeobox gene HESX1/Hesx1 associated with septo-optic dysplasia in human and mouse. Nat Genet 19:125–133

Davis SW, Camper SA (2007) Noggin regulates Bmp4 activity during pituitary induction. Dev Biol 305:145–160

Davis SW, Mortensen AH, Camper SA (2011) Birthdating studies reshape models for pituitary gland cell specification. Dev Biol 352:215–227

Davis SW, Keisler JL, Perez-Millan MI, Schade V, Camper SA (2016) All hormone-producing cell types of the pituitary intermediate and anterior lobes derive from Prop1-expressing progenitors. Endocrinology 157:1385–1396

Deladoey J, Fluck C, Buyukgebiz A, Kuhlmann BV, Eble A, Hindmarsh PC, Wu W, Mullis PE (1999) "Hot spot" in the PROP1 gene responsible for combined pituitary hormone deficiency. J Clin Endocrinol Metab 84:1645–1650

DiMattia GE, Rhodes SJ, Krones A, Carriere C, O'Connell S, Kalla K, Arias C, Sawchenko P, Rosenfeld MG (1997) The Pit-1 gene is regulated by distinct early and late pituitary-specific enhancers. Dev Biol 182:180–190

Djakoure C, Guibourdenche J, Porquet D, Pagesy P, Peillon F, Li JY, Evain-Brion D (1996) Vitamin A and retinoic acid stimulate within minutes cAMP release and growth hormone secretion in human pituitary cells. J Clin Endocrinol Metab 81:3123–3126

Dusatkova P, Pfaffle R, Brown MR, Akulevich N, Arnhold IJ, Kalina MA, Kot K, Krzisnik C, Lemos MC, Malikova J et al (2016) Genesis of two most prevalent PROP1 gene variants causing combined pituitary hormone deficiency in 21 populations. Eur J Hum Genet 24:415–420

Duval DL, Nelson SE, Clay CM (1997) A binding site for steroidogenic factor-1 is part of a complex enhancer that mediates expression of the murine gonadotropin-releasing hormone receptor gene. Biol Reprod 56:160–168

Eicher EM, Beamer WG (1980) New mouse dw allele: genetic location and effects on lifespan and growth hormone levels. J Hered 71:187–190

Ellsworth BS, Butts DL, Camper SA (2008) Mechanisms underlying pituitary hypoplasia and failed cell specification in Lhx3-deficient mice. Dev Biol 313:118–129

Enwright JF 3rd, Kawecki-Crook MA, Voss TC, Schaufele F, Day RN (2003) A PIT-1 homeodomain mutant blocks the intranuclear recruitment of the CCAAT/enhancer binding protein alpha required for prolactin gene transcription. Mol Endocrinol 17:209–222

Ericson J, Norlin S, Jessell TM, Edlund T (1998) Integrated FGF and BMP signaling controls the progression of progenitor cell differentiation and the emergence of pattern in the embryonic anterior pituitary. Development 125:1005–1015

Fang Q, George AS, Brinkmeier ML, Mortensen AH, Gergics P, Cheung LY, Daly AZ, Ajmal A, Perez Millan MI, Ozel AB et al (2016) Genetics of combined pituitary hormone deficiency: roadmap into the genome era. Endocr Rev 37:636–675

Fauquier T, Guerineau NC, McKinney RA, Bauer K, Mollard P (2001) Folliculostellate cell network: a route for long-distance communication in the anterior pituitary. Proc Natl Acad Sci USA 98:8891–8896

Fauquier T, Rizzoti K, Dattani M, Lovell-Badge R, Robinson IC (2008) SOX2-expressing progenitor cells generate all of the major cell types in the adult mouse pituitary gland. Proc Natl Acad Sci USA 105:2907–2912

Fekete C, Lechan RM (2014) Central regulation of hypothalamic-pituitary-thyroid axis under physiological and pathophysiological conditions. Endocr Rev 35:159–194

Filges I, Bischof-Renner A, Rothlisberger B, Potthoff C, Glanzmann R, Gunthard J, Schneider J, Huber AR, Zumsteg U, Miny P et al (2012) Panhypopituitarism presenting as life-threatening heart failure caused by an inherited microdeletion in 1q25 including LHX4. Pediatrics 129: e529–e534

Fofanova OV, Takamura N, Kinoshita E, Yoshimoto M, Tsuji Y, Peterkova VA, Evgrafov OV, Dedov II, Goncharov NP, Yamashita S (1998) Rarity of PIT1 involvement in children from Russia with combined pituitary hormone deficiency. Am J Med Genet 77:360–365

Frawley LS, Boockfor FR, Hoeffler JP (1985) Identification by plaque assays of a pituitary cell type that secretes both growth hormone and prolactin. Endocrinology 116:734–737

Gage PJ, Camper SA (1997) Pituitary homeobox 2, a novel member of the bicoid-related family of homeobox genes, is a potential regulator of anterior structure formation. Hum Mol Genet 6:457–464

Gage PJ, Lossie AC, Scarlett LM, Lloyd RV, Camper SA (1995) Ames dwarf mice exhibit somatotrope commitment but lack growth hormone-releasing factor response. Endocrinology 136:1161–1167

Gage PJ, Brinkmeier ML, Scarlett LM, Knapp LT, Camper SA, Mahon KA (1996) The Ames dwarf gene, df, is required early in pituitary ontogeny for the extinction of Rpx transcription and initiation of lineage-specific cell proliferation. Mol Endocrinol 10:1570–1581

Gage PJ, Suh H, Camper SA (1999) Dosage requirement of Pitx2 for development of multiple organs. Development 126:4643–4651

Gaston-Massuet C, Andoniadou CL, Signore M, Jayakody SA, Charolidi N, Kyeyune R, Vernay B, Jacques TS, Taketo MM, Le Tissier P et al (2011) Increased wingless (Wnt) signaling in pituitary progenitor/stem cells gives rise to pituitary tumors in mice and humans. Proc Natl Acad Sci USA 108:11482–11487

Gergics P, Brinkmeier ML, Camper SA (2015) Lhx4 deficiency: increased cyclin-dependent kinase inhibitor expression and pituitary hypoplasia. Mol Endocrinol 29:597–612

Giacomini D, Paez-Pereda M, Theodoropoulou M, Gerez J, Nagashima AC, Chervin A, Berner S, Labeur M, Refojo D, Renner U et al (2006) Bone morphogenetic protein-4 control of pituitary pathophysiology. Front Horm Res 35:22–31

Goldberg LB, Aujla PK, Raetzman LT (2011) Persistent expression of activated Notch inhibits corticotrope and melanotrope differentiation and results in dysfunction of the HPA axis. Dev Biol 358:23–32

Gordon DF, Lewis SR, Haugen BR, James RA, McDermott MT, Wood WM, Ridgway EC (1997) Pit-1 and GATA-2 interact and functionally cooperate to activate the thyrotropin beta-subunit promoter. J Biol Chem 272:24339–24347

Gordon DF, Tucker EA, Tundwal K, Hall H, Wood WM, Ridgway EC (2006) MED220/thyroid receptor-associated protein 220 functions as a transcriptional coactivator with Pit-1 and GATA-2 on the thyrotropin-beta promoter in thyrotropes. Mol Endocrinol 20:1073–1089

Gregory LC, Humayun KN, Turton JP, McCabe MJ, Rhodes SJ, Dattani MT (2015) Novel lethal form of congenital hypopituitarism associated with the first recessive LHX4 mutation. J Clin Endocrinol Metab 100:2158–2164

Gubbay J, Collignon J, Koopman P, Capel B, Economou A, Munsterberg A, Vivian N, Goodfellow P, Lovell-Badge R (1990) A gene mapping to the sex-determining region of the mouse Y chromosome is a member of a novel family of embryonically expressed genes. Nature 346:245–250

Guner B, Ozacar AT, Thomas JE, Karlstrom RO (2008) Graded hedgehog and fibroblast growth factor signaling independently regulate pituitary cell fates and help establish the pars distalis and pars intermedia of the zebrafish adenohypophysis. Endocrinology 149:4435–4451

Guran T, Buonocore F, Saka N, Ozbek MN, Aycan Z, Bereket A, Bas F, Darcan S, Bideci A, Guven A et al (2016) Rare causes of primary adrenal insufficiency: genetic and clinical characterization of a large Nationwide cohort. J Clin Endocrinol Metab 101:284–292

Gurnett CA, Alaee F, Kruse LM, Desruisseau DM, Hecht JT, Wise CA, Bowcock AM, Dobbs MB (2008) Asymmetric lower-limb malformations in individuals with homeobox PITX1 gene mutation. Am J Hum Genet 83:616–622

Halvorson LM, Kaiser UB, Chin WW (1996) Stimulation of luteinizing hormone beta gene promoter activity by the orphan nuclear receptor, steroidogenic factor-1. J Biol Chem 271:6645–6650

Hasegawa T, Fukami M, Sato N, Katsumata N, Sasaki G, Fukutani K, Morohashi K, Ogata T (2004) Testicular dysgenesis without adrenal insufficiency in a 46,XY patient with a heterozygous inactive mutation of steroidogenic factor-1. J Clin Endocrinol Metab 89:5930–5935

Hermesz E, Mackem S, Mahon KA (1996) Rpx: a novel anterior-restricted homeobox gene progressively activated in the prechordal plate, anterior neural plate and Rathke's pouch of the mouse embryo. Development 122:41–52

Higham CE, Johannsson G, Shalet SM (2016) Hypopituitarism. Lancet 388:2403–2415

Himes AD, Raetzman LT (2009) Premature differentiation and aberrant movement of pituitary cells lacking both Hes1 and Prop1. Dev Biol 325:151–161

Ho Y, Tadevosyan A, Liebhaber SA, Cooke NE (2008) The juxtaposition of a promoter with a locus control region transcriptional domain activates gene expression. EMBO Rep 9:891–898

Ho Y, Shewchuk BM, Liebhaber SA, Cooke NE (2013) Distinct chromatin configurations regulate the initiation and the maintenance of hGH gene expression. Mol Cell Biol 33:1723–1734

Ho Y, Cooke NE, Liebhaber SA (2015) An autoregulatory pathway establishes the definitive chromatin conformation at the pit-1 locus. Mol Cell Biol 35:1523–1532

Honda S, Morohashi K, Omura T (1990) Novel cAMP regulatory elements in the promoter region of bovine P-450(11 beta) gene. J Biochem 108:1042–1049

Honda S, Morohashi K, Nomura M, Takeya H, Kitajima M, Omura T (1993) Ad4BP regulating steroidogenic P-450 gene is a member of steroid hormone receptor superfamily. J Biol Chem 268:7494–7502

Horn F, Windle JJ, Barnhart KM, Mellon PL (1992) Tissue-specific gene expression in the pituitary: the glycoprotein hormone alpha-subunit gene is regulated by a gonadotrope-specific protein. Mol Cell Biol 12:2143–2153

Horvath E, Lloyd RV, Kovacs K (1990) Propylthiouracyl-induced hypothyroidism results in reversible transdifferentiation of somatotrophs into thyroidectomy cells. A morphologic study of the rat pituitary including immunoelectron microscopy. Lab Investig 63:511–520

Hosoyama T, Nishijo K, Garcia MM, Schaffer BS, Ohshima-Hosoyama S, Prajapati SI, Davis MD, Grant WF, Scheithauer BW, Marks DL et al (2010) A postnatal Pax7 progenitor gives rise to pituitary adenomas. Genes Cancer 1:388–402

Igarashi M, Takasawa K, Hakoda A, Kanno J, Takada S, Miyado M, Baba T, Morohashi KI, Tajima T, Hata K et al (2017) Identical NR5A1 missense mutations in two unrelated 46,XX individuals with testicular tissues. Hum Mutat 38:39–42

Ikeda Y, Shen WH, Ingraham HA, Parker KL (1994) Developmental expression of mouse steroidogenic factor-1, an essential regulator of the steroid hydroxylases. Mol Endocrinol 8:654–662

Ikeda Y, Luo X, Abbud R, Nilson JH, Parker KL (1995) The nuclear receptor steroidogenic factor 1 is essential for the formation of the ventromedial hypothalamic nucleus. Mol Endocrinol 9:478–486

Imam A, Iqbal J, Blair HC, Davies TF, Huang CL, Zallone A, Zaidi M, Sun L (2009) Role of the pituitary-bone axis in skeletal pathophysiology. Curr Opin Endocrinol Diabetes Obes 16:423–429

Ingraham HA, Chen RP, Mangalam HJ, Elsholtz HP, Flynn SE, Lin CR, Simmons DM, Swanson L, Rosenfeld MG (1988) A tissue-specific transcription factor containing a homeodomain specifies a pituitary phenotype. Cell 55:519–529

Ingraham HA, Lala DS, Ikeda Y, Luo X, Shen WH, Nachtigal MW, Abbud R, Nilson JH, Parker KL (1994) The nuclear receptor steroidogenic factor 1 acts at multiple levels of the reproductive axis. Genes Dev 8:2302–2312

Jacobs SB, Coss D, McGillivray SM, Mellon PL (2003) Nuclear factor Y and steroidogenic factor 1 physically and functionally interact to contribute to cell-specific expression of the mouse follicle-stimulating hormone-beta gene. Mol Endocrinol 17:1470–1483

Jayakody SA, Andoniadou CL, Gaston-Massuet C, Signore M, Cariboni A, Bouloux PM, Le Tissier P, Pevny LH, Dattani MT, Martinez-Barbera JP (2012) SOX2 regulates the hypothalamic-pituitary axis at multiple levels. J Clin Invest 122:3635–3646

Jin JM, Yang WX (2014) Molecular regulation of hypothalamus-pituitary-gonads axis in males. Gene 551:15–25

Jones BK, Monks BR, Liebhaber SA, Cooke NE (1995) The human growth hormone gene is regulated by a multicomponent locus control region. Mol Cell Biol 15:7010–7021

Jostes B, Walther C, Gruss P (1990) The murine paired box gene, Pax7, is expressed specifically during the development of the nervous and muscular system. Mech Dev 33:27–37

Jullien N, Romanet P, Philippon M, Quentien MH, Beck-Peccoz P, Bergada I, Odent S, Reynaud R, Barlier A, Saveanu A et al (2019) Heterozygous LHX3 mutations may lead to a mild phenotype of combined pituitary hormone deficiency. Eur J Hum Genet 27:216–225

Kaiser UB, Halvorson LM, Chen MT (2000) Sp1, steroidogenic factor 1 (SF-1), and early growth response protein 1 (egr-1) binding sites form a tripartite gonadotropin-releasing hormone response element in the rat luteinizing hormone-beta gene promoter: an integral role for SF-1. Mol Endocrinol 14:1235–1245

Karpova T, Presley J, Manimaran RR, Scherrer SP, Tejada L, Peterson KR, Heckert LL (2005) A FTZ-F1-containing yeast artificial chromosome recapitulates expression of steroidogenic factor 1 in vivo. Mol Endocrinol 19:2549–2563

Kassar-Duchossoy L, Giacone E, Gayraud-Morel B, Jory A, Gomes D, Tajbakhsh S (2005) Pax3/Pax7 mark a novel population of primitive myogenic cells during development. Genes Dev 19:1426–1431

Kelberman D, Rizzoti K, Avilion A, Bitner-Glindzicz M, Cianfarani S, Collins J, Chong WK, Kirk JM, Achermann JC, Ross R et al (2006) Mutations within Sox2/SOX2 are associated with abnormalities in the hypothalamo-pituitary-gonadal axis in mice and humans. J Clin Invest 116:2442–2455

Kelberman D, Rizzoti K, Lovell-Badge R, Robinson IC, Dattani MT (2009) Genetic regulation of pituitary gland development in human and mouse. Endocr Rev 30:790–829

Keller-Wood M (2015) Hypothalamic-pituitary—adrenal axis-feedback control. Compr Physiol 5:1161–1182

Keri RA, Nilson JH (1996) A steroidogenic factor-1 binding site is required for activity of the luteinizing hormone beta subunit promoter in gonadotropes of transgenic mice. J Biol Chem 271:10782–10785

Keri RA, Wolfe MW, Saunders TL, Anderson I, Kendall SK, Wagner T, Yeung J, Gorski J, Nett TM, Camper SA et al (1994) The proximal promoter of the bovine luteinizing hormone beta-subunit gene confers gonadotrope-specific expression and regulation by gonadotropin-releasing hormone, testosterone, and 17 beta-estradiol in transgenic mice. Mol Endocrinol 8:1807–1816

Kioussi C, Briata P, Baek SH, Rose DW, Hamblet NS, Herman T, Ohgi KA, Lin C, Gleiberman A, Wang J et al (2002) Identification of a Wnt/Dvl/beta-catenin→Pitx2 pathway mediating cell-type-specific proliferation during development. Cell 111:673–685

Kitamura K, Miura H, Miyagawa-Tomita S, Yanazawa M, Katoh-Fukui Y, Suzuki R, Ohuchi H, Suehiro A, Motegi Y, Nakahara Y et al (1999) Mouse Pitx2 deficiency leads to anomalies of the ventral body wall, heart, extra- and periocular mesoderm and right pulmonary isomerism. Development 126:5749–5758

Klopocki E, Kahler C, Foulds N, Shah H, Joseph B, Vogel H, Luttgen S, Bald R, Besoke R, Held K et al (2012) Deletions in PITX1 cause a spectrum of lower-limb malformations including mirror-image polydactyly. Eur J Hum Genet 20:705–708

Kragesteen BK, Brancati F, Digilio MC, Mundlos S, Spielmann M (2019) H2AFY promoter deletion causes PITX1 endoactivation and Liebenberg syndrome. J Med Genet 56:246–251

Kristrom B, Zdunek AM, Rydh A, Jonsson H, Sehlin P, Escher SA (2009) A novel mutation in the LIM homeobox 3 gene is responsible for combined pituitary hormone deficiency, hearing impairment, and vertebral malformations. J Clin Endocrinol Metab 94:1154–1161

Kulig E, Camper SA, Kuecker S, Jin L, Lloyd RV (1998) Remodeling of hyperplastic pituitaries in hypothyroid us-subunit knockout mice after thyroxine and 1713-estradiol treatment: role of apoptosis. Endocr Pathol 9:261–274

Lafont C, Desarmenien MG, Cassou M, Molino F, Lecoq J, Hodson D, Lacampagne A, Mennessier G, El Yandouzi T, Carmignac D et al (2010) Cellular in vivo imaging reveals coordinated regulation of pituitary microcirculation and GH cell network function. Proc Natl Acad Sci USA 107:4465–4470

Lala DS, Rice DA, Parker KL (1992) Steroidogenic factor I, a key regulator of steroidogenic enzyme expression, is the mouse homolog of fushi tarazu-factor I. Mol Endocrinol 6:1249–1258

Lamolet B, Pulichino AM, Lamonerie T, Gauthier Y, Brue T, Enjalbert A, Drouin J (2001) A pituitary cell-restricted T box factor, Tpit, activates POMC transcription in cooperation with Pitx homeoproteins. Cell 104:849–859

Lamonerie T, Tremblay JJ, Lanctot C, Therrien M, Gauthier Y, Drouin J (1996) Ptx1, a bicoid-related homeo box transcription factor involved in transcription of the pro-opiomelanocortin gene. Genes Dev 10:1284–1295

Lanctot C, Gauthier Y, Drouin J (1999) Pituitary homeobox 1 (Ptx1) is differentially expressed during pituitary development. Endocrinology 140:1416–1422

Langlais D, Couture C, Balsalobre A, Drouin J (2012) The Stat3/GR interaction code: predictive value of direct/indirect DNA recruitment for transcription outcome. Mol Cell 47:38–49
Lebl J, Vosahlo J, Pfaeffle RW, Stobbe H, Cerna J, Novotna D, Zapletalova J, Kalvachova B, Hana V, Weiss V et al (2005) Auxological and endocrine phenotype in a population-based cohort of patients with PROP1 gene defects. Eur J Endocrinol 153:389–396
Lee SL, Sadovsky Y, Swirnoff AH, Polish JA, Goda P, Gavrilina G, Milbrandt J (1996) Luteinizing hormone deficiency and female infertility in mice lacking the transcription factor NGFI-A (Egr-1). Science 273:1219–1221
Lemos MC, Gomes L, Bastos M, Leite V, Limbert E, Carvalho D, Bacelar C, Monteiro M, Fonseca F, Agapito A et al (2006) PROP1 gene analysis in Portuguese patients with combined pituitary hormone deficiency. Clin Endocrinol 65:479–485
Li S, Crenshaw EB 3rd, Rawson EJ, Simmons DM, Swanson LW, Rosenfeld MG (1990) Dwarf locus mutants lacking three pituitary cell types result from mutations in the POU-domain gene pit-1. Nature 347:528–533
Li H, Witte DP, Branford WW, Aronow BJ, Weinstein M, Kaur S, Wert S, Singh G, Schreiner CM, Whitsett JA et al (1994) Gsh-4 encodes a LIM-type homeodomain, is expressed in the developing central nervous system and is required for early postnatal survival. EMBO J 13:2876–2885
Li M, Pevny L, Lovell-Badge R, Smith A (1998) Generation of purified neural precursors from embryonic stem cells by lineage selection. Curr Biol 8:971–974
Lin CR, Kioussi C, O'Connell S, Briata P, Szeto D, Liu F, Izpisua-Belmonte JC, Rosenfeld MG (1999) Pitx2 regulates lung asymmetry, cardiac positioning and pituitary and tooth morphogenesis. Nature 401:279–282
Lin L, Philibert P, Ferraz-de-Souza B, Kelberman D, Homfray T, Albanese A, Molini V, Sebire NJ, Einaudi S, Conway GS et al (2007) Heterozygous missense mutations in steroidogenic factor 1 (SF1/Ad4BP, NR5A1) are associated with 46,XY disorders of sex development with normal adrenal function. J Clin Endocrinol Metab 92:991–999
Lodge EJ, Santambrogio A, Russell JP, Xekouki P, Jacques TS, Johnson RL, Thavaraj S, Bornstein SR, Andoniadou CL (2019) Homeostatic and tumourigenic activity of SOX2+ pituitary stem cells is controlled by the LATS/YAP/TAZ cascade. Elife 8:e43996
Lourenco D, Brauner R, Lin L, De Perdigo A, Weryha G, Muresan M, Boudjenah R, Guerra-Junior G, Maciel-Guerra AT, Achermann JC et al (2009) Mutations in NR5A1 associated with ovarian insufficiency. N Engl J Med 360:1200–1210
Lu MF, Pressman C, Dyer R, Johnson RL, Martin JF (1999) Function of Rieger syndrome gene in left-right asymmetry and craniofacial development. Nature 401:276–278
Luo X, Ikeda Y, Parker KL (1994) A cell-specific nuclear receptor is essential for adrenal and gonadal development and sexual differentiation. Cell 77:481–490
Mallet D, Bretones P, Michel-Calemard L, Dijoud F, David M, Morel Y (2004) Gonadal dysgenesis without adrenal insufficiency in a 46, XY patient heterozygous for the nonsense C16X mutation: a case of SF1 haploinsufficiency. J Clin Endocrinol Metab 89:4829–4832
Mansouri A, Gruss P (1998) Pax3 and Pax7 are expressed in commissural neurons and restrict ventral neuronal identity in the spinal cord. Mech Dev 78:171–178
Mansouri A, Stoykova A, Torres M, Gruss P (1996) Dysgenesis of cephalic neural crest derivatives in Pax7-/- mutant mice. Development 122:831–838
Matsumoto A, Ishii S (1992) Atlas of endocrine organs: vertebrates and invertebrates. Springer, Berlin
Mayran A, Khetchoumian K, Hariri F, Pastinen T, Gauthier Y, Balsalobre A, Drouin J (2018) Pioneer factor Pax7 deploys a stable enhancer repertoire for specification of cell fate. Nat Genet 50:259–269
McCormick A, Wu D, Castrillo JL, Dana S, Strobl J, Thompson EB, Karin M (1988) Extinction of growth hormone expression in somatic cell hybrids involves repression of the specific trans-activator GHF-1. Cell 55:379–389

Melamed P, Zhu Y, Tan SH, Xie M, Koh M (2006) Gonadotropin-releasing hormone activation of c-jun, but not early growth response factor-1, stimulates transcription of a luteinizing hormone beta-subunit gene. Endocrinology 147:3598–3605

Melmed S (2011) Pathogenesis of pituitary tumors. Nat Rev Endocrinol 7:257–266

Metherell LA, Savage MO, Dattani M, Walker J, Clayton PE, Farooqi IS, Clark AJ (2004) TPIT mutations are associated with early-onset, but not late-onset isolated ACTH deficiency. Eur J Endocrinol 151:463–465

Mollard P, Hodson DJ, Lafont C, Rizzoti K, Drouin J (2012) A tridimensional view of pituitary development and function. Trends Endocrinol Metab 23:261–269

Morohashi K, Honda S, Inomata Y, Handa H, Omura T (1992) A common trans-acting factor, Ad4-binding protein, to the promoters of steroidogenic P-450s. J Biol Chem 267:17913–17919

Mortensen AH, MacDonald JW, Ghosh D, Camper SA (2011) Candidate genes for panhypopituitarism identified by gene expression profiling. Physiol Genomics 43:1105–1116

Nakano T, Ando S, Takata N, Kawada M, Muguruma K, Sekiguchi K, Saito K, Yonemura S, Eiraku M, Sasai Y (2012) Self-formation of optic cups and storable stratified neural retina from human ESCs. Cell Stem Cell 10:771–785

Nasonkin IO, Ward RD, Raetzman LT, Seasholtz AF, Saunders TL, Gillespie PJ, Camper SA (2004) Pituitary hypoplasia and respiratory distress syndrome in Prop1 knockout mice. Hum Mol Genet 13:2727–2735

Navardauskaite R, Dusatkova P, Obermannova B, Pfaeffle RW, Blum WF, Adukauskiene D, Smetanina N, Cinek O, Verkauskiene R, Lebl J (2014) High prevalence of PROP1 defects in Lithuania: phenotypic findings in an ethnically homogenous cohort of patients with multiple pituitary hormone deficiency. J Clin Endocrinol Metab 99:299–306

Navin N, Kendall J, Troge J, Andrews P, Rodgers L, McIndoo J, Cook K, Stepansky A, Levy D, Esposito D et al (2011) Tumour evolution inferred by single-cell sequencing. Nature 472:90–94

Nelson C, Crenshaw EB 3rd, Franco R, Lira SA, Albert VR, Evans RM, Rosenfeld MG (1986) Discrete cis-active genomic sequences dictate the pituitary cell type-specific expression of rat prolactin and growth hormone genes. Nature 322:557–562

Netchine I, Sobrier ML, Krude H, Schnabel D, Maghnie M, Marcos E, Duriez B, Cacheux V, Moers A, Goossens M et al (2000) Mutations in LHX3 result in a new syndrome revealed by combined pituitary hormone deficiency. Nat Genet 25:182–186

Ohta K, Nobukuni Y, Mitsubuchi H, Fujimoto S, Matsuo N, Inagaki H, Endo F, Matsuda I (1992) Mutations in the Pit-1 gene in children with combined pituitary hormone deficiency. Biochem Biophys Res Commun 189:851–855

Ohuchi H, Hori Y, Yamasaki M, Harada H, Sekine K, Kato S, Itoh N (2000) FGF10 acts as a major ligand for FGF receptor 2 IIIb in mouse multi-organ development. Biochem Biophys Res Commun 277:643–649

Okamoto N, Wada Y, Ida S, Koga R, Ozono K, Chiyo H, Hayashi A, Tatsumi K (1994) Monoallelic expression of normal mRNA in the PIT1 mutation heterozygotes with normal phenotype and biallelic expression in the abnormal phenotype. Hum Mol Genet 3:1565–1568

Olson LE, Tollkuhn J, Scafoglio C, Krones A, Zhang J, Ohgi KA, Wu W, Taketo MM, Kemler R, Grosschedl R et al (2006) Homeodomain-mediated beta-catenin-dependent switching events dictate cell-lineage determination. Cell 125:593–605

Oustanina S, Hause G, Braun T (2004) Pax7 directs postnatal renewal and propagation of myogenic satellite cells but not their specification. EMBO J 23:3430–3439

Ozone C, Suga H, Eiraku M, Kadoshima T, Yonemura S, Takata N, Oiso Y, Tsuji T, Sasai Y (2016) Functional anterior pituitary generated in self-organizing culture of human embryonic stem cells. Nat Commun 7:10351

Peel MT, Ho Y, Liebhaber SA (2018) Transcriptome analyses of female Somatotropes and Lactotropes reveal novel regulators of cell identity in the pituitary. Endocrinology 159:3965–3980

Perez Millan MI, Brinkmeier ML, Mortensen AH, Camper SA (2016) PROP1 triggers epithelial-mesenchymal transition-like process in pituitary stem cells. Elife 5:e14470

Perez C, Dastot-Le Moal F, Collot N, Legendre M, Abadie I, Bertrand AM, Amselem S, Sobrier ML (2012) Screening of LHX2 in patients presenting growth retardation with posterior pituitary and ocular abnormalities. Eur J Endocrinol 167:85–91

Pfaeffle RW, Savage JJ, Hunter CS, Palme C, Ahlmann M, Kumar P, Bellone J, Schoenau E, Korsch E, Bramswig JH et al (2007) Four novel mutations of the LHX3 gene cause combined pituitary hormone deficiencies with or without limited neck rotation. J Clin Endocrinol Metab 92:1909–1919

Pfaeffle RW, Hunter CS, Savage JJ, Duran-Prado M, Mullen RD, Neeb ZP, Eiholzer U, Hesse V, Haddad NG, Stobbe HM et al (2008) Three novel missense mutations within the LHX4 gene are associated with variable pituitary hormone deficiencies. J Clin Endocrinol Metab 93:1062–1071

Phifer RF, Midgley AR, Spicer SS (1973) Immunohistologic and histologic evidence that follicle-stimulating hormone and luteinizing hormone are present in the same cell type in the human pars distalis. J Clin Endocrinol Metab 36:125–141

Porter FD, Drago J, Xu Y, Cheema SS, Wassif C, Huang SP, Lee E, Grinberg A, Massalas JS, Bodine D et al (1997) Lhx2, a LIM homeobox gene, is required for eye, forebrain, and definitive erythrocyte development. Development 124:2935–2944

Poulin G, Turgeon B, Drouin J (1997) NeuroD1/beta2 contributes to cell-specific transcription of the proopiomelanocortin gene. Mol Cell Biol 17:6673–6682

Poulin G, Lebel M, Chamberland M, Paradis FW, Drouin J (2000) Specific protein-protein interaction between basic helix-loop-helix transcription factors and homeoproteins of the Pitx family. Mol Cell Biol 20:4826–4837

Pow DV, Perry VH, Morris JF, Gordon S (1989) Microglia in the neurohypophysis associate with and endocytose terminal portions of neurosecretory neurons. Neuroscience 33:567–578

Pulichino AM, Vallette-Kasic S, Couture C, Gauthier Y, Brue T, David M, Malpuech G, Deal C, Van Vliet G, De Vroede M et al (2003a) Human and mouse TPIT gene mutations cause early onset pituitary ACTH deficiency. Genes Dev 17:711–716

Pulichino AM, Vallette-Kasic S, Tsai JP, Couture C, Gauthier Y, Drouin J (2003b) Tpit determines alternate fates during pituitary cell differentiation. Genes Dev 17:738–747

Qiao S, Nordstrom K, Muijs L, Gasparoni G, Tierling S, Krause E, Walter J, Boehm U (2016) Molecular plasticity of male and female murine gonadotropes revealed by mRNA sequencing. Endocrinology 157:1082–1093

Radovick S, Nations M, Du Y, Berg LA, Weintraub BD, Wondisford FE (1992) A mutation in the POU-homeodomain of Pit-1 responsible for combined pituitary hormone deficiency. Science 257:1115–1118

Raetzman LT, Ward R, Camper SA (2002) Lhx4 and Prop1 are required for cell survival and expansion of the pituitary primordia. Development 129:4229–4239

Raetzman LT, Ross SA, Cook S, Dunwoodie SL, Camper SA, Thomas PQ (2004) Developmental regulation of Notch signaling genes in the embryonic pituitary: Prop1 deficiency affects Notch2 expression. Dev Biol 265:329–340

Ragge NK, Lorenz B, Schneider A, Bushby K, de Sanctis L, de Sanctis U, Salt A, Collin JR, Vivian AJ, Free SL et al (2005) SOX2 anophthalmia syndrome. Am J Med Genet A 135:1–7. discussion 8

Rajab A, Kelberman D, de Castro SC, Biebermann H, Shaikh H, Pearce K, Hall CM, Shaikh G, Gerrelli D, Grueters A et al (2008) Novel mutations in LHX3 are associated with hypopituitarism and sensorineural hearing loss. Hum Mol Genet 17:2150–2159

Ramzan K, Bin-Abbas B, Al-Jomaa L, Allam R, Al-Owain M, Imtiaz F (2017) Two novel LHX3 mutations in patients with combined pituitary hormone deficiency including cervical rigidity and sensorineural hearing loss. BMC Endocr Disord 17:17

Rappaport R, Amselem S (2001) Hypothalamic-pituitary development: genetic and clinical aspects. Karger, Basel

Regal M, Paramo C, Sierra SM, Garcia-Mayor RV (2001) Prevalence and incidence of hypopituitarism in an adult Caucasian population in northwestern Spain. Clin Endocrinol 55:735–740

Relaix F, Rocancourt D, Mansouri A, Buckingham M (2005) A Pax3/Pax7-dependent population of skeletal muscle progenitor cells. Nature 435:948–953

Relaix F, Montarras D, Zaffran S, Gayraud-Morel B, Rocancourt D, Tajbakhsh S, Mansouri A, Cumano A, Buckingham M (2006) Pax3 and Pax7 have distinct and overlapping functions in adult muscle progenitor cells. J Cell Biol 172:91–102

Rhodes SJ, Chen R, DiMattia GE, Scully KM, Kalla KA, Lin SC, Yu VC, Rosenfeld MG (1993) A tissue-specific enhancer confers Pit-1-dependent morphogen inducibility and autoregulation on the pit-1 gene. Genes Dev 7:913–932

Rice DA, Mouw AR, Bogerd AM, Parker KL (1991) A shared promoter element regulates the expression of three steroidogenic enzymes. Mol Endocrinol 5:1552–1561

Riepe FG, Partsch CJ, Blankenstein O, Monig H, Pfaffle RW, Sippell WG (2001) Longitudinal imaging reveals pituitary enlargement preceding hypoplasia in two brothers with combined pituitary hormone deficiency attributable to PROP1 mutation. J Clin Endocrinol Metab 86:4353–4357

Rinehart JF, Farquhar MG (1953) Electron microscopic studies of the anterior pituitary gland. J Histochem Cytochem 1:93–113

Rizzoti K (2015) Genetic regulation of murine pituitary development. J Mol Endocrinol 54:R55–R73

Rizzoti K, Lovell-Badge R (2005) Early development of the pituitary gland: induction and shaping of Rathke's pouch. Rev Endocr Metab Disord 6:161–172

Rizzoti K, Akiyama H, Lovell-Badge R (2013) Mobilized adult pituitary stem cells contribute to endocrine regeneration in response to physiological demand. Cell Stem Cell 13:419–432

Robyn C, Leleux P, Vanhaelst L, Golstein J, Herlant M, Pasteels JL (1973) Immunohistochemical study of the human pituitary with anti-luteinizing hormone, anti-follicle stimulating hormone and anti-thyrotrophin sera. Acta Endocrinol 72:625–642

Rochette C, Jullien N, Saveanu A, Caldagues E, Bergada I, Braslavsky D, Pfeifer M, Reynaud R, Herman JP, Barlier A et al (2015) Identifying the deleterious effect of rare LHX4 allelic variants, a challenging issue. PLoS One 10:e0126648

Roh M, Paterson AJ, Asa SL, Chin E, Kudlow JE (2001) Stage-sensitive blockade of pituitary somatomammotrope development by targeted expression of a dominant negative epidermal growth factor receptor in transgenic mice. Mol Endocrinol 15:600–613

Rosso L, Mienville JM (2009) Pituicyte modulation of neurohormone output. Glia 57:235–243

Ruf-Zamojski F, Fribourg M, Ge Y, Nair V, Pincas H, Zaslavsky E, Nudelman G, Tuminello SJ, Watanabe H, Turgeon JL et al (2018) Regulatory architecture of the LbetaT2 Gonadotrope cell underlying the response to gonadotropin-releasing hormone. Front Endocrinol 9:34

Sadeghi-Nejad A, Senior B (1974) Autosomal dominant transmission of isolated growth hormone deficiency in iris-dental dysplasia (Rieger's syndrome). J Pediatr 85:644–648

Sajedi E, Gaston-Massuet C, Signore M, Andoniadou CL, Kelberman D, Castro S, Etchevers HC, Gerrelli D, Dattani MT, Martinez-Barbera JP (2008) Analysis of mouse models carrying the I26T and R160C substitutions in the transcriptional repressor HESX1 as models for septo-optic dysplasia and hypopituitarism. Dis Model Mech 1:241–254

Salisbury TB, Binder AK, Grammer JC, Nilson JH (2009) GnRH-regulated expression of Jun and JUN target genes in gonadotropes requires a functional interaction between TCF/LEF family members and beta-catenin. Mol Endocrinol 23:402–411

Sarro R, Kocher AA, Emera D, Uebbing S, Dutrow EV, Weatherbee SD, Nottoli T, Noonan JP (2018) Disrupting the three-dimensional regulatory topology of the Pitx1 locus results in overtly normal development. Development 145:dev158550

Sato T, Vries RG, Snippert HJ, van de Wetering M, Barker N, Stange DE, van Es JH, Abo A, Kujala P, Peters PJ et al (2009) Single Lgr5 stem cells build crypt-villus structures in vitro without a mesenchymal niche. Nature 459:262–265

Schaeffer M, Hodson DJ, Lafont C, Mollard P (2010) Functional importance of blood flow dynamics and partial oxygen pressure in the anterior pituitary. Eur J Neurosci 32:2087–2095

Schaible R, Gowen JW (1961) A new dwarf mouse. Genetics 46:896

Seale P, Sabourin LA, Girgis-Gabardo A, Mansouri A, Gruss P, Rudnicki MA (2000) Pax7 is required for the specification of myogenic satellite cells. Cell 102:777–786

Sekido R, Lovell-Badge R (2008) Sex determination involves synergistic action of SRY and SF1 on a specific Sox9 enhancer. Nature 453:930–934

Semina EV, Reiter R, Leysens NJ, Alward WL, Small KW, Datson NA, Siegel-Bartelt J, Bierke-Nelson D, Bitoun P, Zabel BU et al (1996) Cloning and characterization of a novel bicoid-related homeobox transcription factor gene, RIEG, involved in Rieger syndrome. Nat Genet 14:392–399

Shapiro E, Biezuner T, Linnarsson S (2013) Single-cell sequencing-based technologies will revolutionize whole-organism science. Nat Rev Genet 14:618–630

Sheng HZ, Zhadanov AB, Mosinger B Jr, Fujii T, Bertuzzi S, Grinberg A, Lee EJ, Huang SP, Mahon KA, Westphal H (1996) Specification of pituitary cell lineages by the LIM homeobox gene Lhx3. Science 272:1004–1007

Sheng HZ, Moriyama K, Yamashita T, Li H, Potter SS, Mahon KA, Westphal H (1997) Multistep control of pituitary organogenesis. Science 278:1809–1812

Shima Y, Zubair M, Ishihara S, Shinohara Y, Oka S, Kimura S, Okamoto S, Minokoshi Y, Suita S, Morohashi K (2005) Ventromedial hypothalamic nucleus-specific enhancer of Ad4BP/SF-1 gene. Mol Endocrinol 19:2812–2823

Shima Y, Zubair M, Komatsu T, Oka S, Yokoyama C, Tachibana T, Hjalt TA, Drouin J, Morohashi K (2008) Pituitary homeobox 2 regulates adrenal4 binding protein/steroidogenic factor-1 gene transcription in the pituitary gonadotrope through interaction with the intronic enhancer. Mol Endocrinol 22:1633–1646

Simon FR, Fortune J, Iwahashi M, Sutherland E (2002) Sexual dimorphic expression of ADH in rat liver: importance of the hypothalamic-pituitary-liver axis. Am J Physiol Gastrointest Liver Physiol 283:G646–G655

Singh G, Kaur S, Stock JL, Jenkins NA, Gilbert DJ, Copeland NG, Potter SS (1991) Identification of 10 murine homeobox genes. Proc Natl Acad Sci USA 88:10706–10710

Skowronska-Krawczyk D, Ma Q, Schwartz M, Scully K, Li W, Liu Z, Taylor H, Tollkuhn J, Ohgi KA, Notani D et al (2014) Required enhancer-matrin-3 network interactions for a homeodomain transcription program. Nature 514:257–261

Smith PE, MacDowell EC (1930) An hereditary anterior-pituitary deficiency in the mouse. Anat Rec 46:249–257

Snell GD (1929) Dwarf, a new Mendelian recessive character of the house mouse. Proc Natl Acad Sci USA 15:733–734

Sobrier ML, Brachet C, Vie-Luton MP, Perez C, Copin B, Legendre M, Heinrichs C, Amselem S (2012) Symptomatic heterozygotes and prenatal diagnoses in a nonconsanguineous family with syndromic combined pituitary hormone deficiency resulting from two novel LHX3 mutations. J Clin Endocrinol Metab 97:E503–E509

Sobrier ML, Tsai YC, Perez C, Leheup B, Bouceba T, Duquesnoy P, Copin B, Sizova D, Penzo A, Stanger BZ et al (2016) Functional characterization of a human POU1F1 mutation associated with isolated growth hormone deficiency: a novel etiology for IGHD. Hum Mol Genet 25:472–483

Soji T, Mabuchi Y, Kurono C, Herbert DC (1997) Folliculo-stellate cells and intercellular communication within the rat anterior pituitary gland. Microsc Res Tech 39:138–149

Sornson MW, Wu W, Dasen JS, Flynn SE, Norman DJ, O'Connell SM, Gukovsky I, Carriere C, Ryan AK, Miller AP et al (1996) Pituitary lineage determination by the prophet of Pit-1 homeodomain factor defective in Ames dwarfism. Nature 384:327–333

Spielmann M, Brancati F, Krawitz PM, Robinson PN, Ibrahim DM, Franke M, Hecht J, Lohan S, Dathe K, Nardone AM et al (2012) Homeotic arm-to-leg transformation associated with genomic rearrangements at the PITX1 locus. Am J Hum Genet 91:629–635

Stallings NR, Hanley NA, Majdic G, Zhao L, Bakke M, Parker KL (2002) Development of a transgenic green fluorescent protein lineage marker for steroidogenic factor 1. Endocr Res 28:497–504

Suga H, Kadoshima T, Minaguchi M, Ohgushi M, Soen M, Nakano T, Takata N, Wataya T, Muguruma K, Miyoshi H et al (2011) Self-formation of functional adenohypophysis in three-dimensional culture. Nature 480:57–62

Suh H, Gage PJ, Drouin J, Camper SA (2002) Pitx2 is required at multiple stages of pituitary organogenesis: pituitary primordium formation and cell specification. Development 129:329–337

Swartz JM, Ciarlo R, Guo MH, Abrha A, Diamond DA, Chan YM, Hirschhorn JN (2017a) Two unrelated undervirilized 46,XY males with inherited NR5A1 variants identified by whole-exome sequencing. Horm Res Paediatr 87:264–270

Swartz JM, Ciarlo R, Guo MH, Abrha A, Weaver B, Diamond DA, Chan YM, Hirschhorn JN (2017b) A 46,XX ovotesticular disorder of sex development likely caused by a steroidogenic factor-1 (NR5A1) variant. Horm Res Paediatr 87:191–195

Szeto DP, Ryan AK, O'Connell SM, Rosenfeld MG (1996) P-OTX: a PIT-1-interacting homeodomain factor expressed during anterior pituitary gland development. Proc Natl Acad Sci USA 93:7706–7710

Szeto DP, Rodriguez-Esteban C, Ryan AK, O'Connell SM, Liu F, Kioussi C, Gleiberman AS, Izpisua-Belmonte JC, Rosenfeld MG (1999) Role of the bicoid-related homeodomain factor Pitx1 in specifying hindlimb morphogenesis and pituitary development. Genes Dev 13:484–494

Taira M, Jamrich M, Good PJ, Dawid IB (1992) The LIM domain-containing homeo box gene Xlim-1 is expressed specifically in the organizer region of Xenopus gastrula embryos. Genes Dev 6:356–366

Tajima T, Hattori T, Nakajima T, Okuhara K, Tsubaki J, Fujieda K (2007) A novel missense mutation (P366T) of the LHX4 gene causes severe combined pituitary hormone deficiency with pituitary hypoplasia, ectopic posterior lobe and a poorly developed sella turcica. Endocr J 54:637–641

Tajima T, Yorifuji T, Ishizu K, Fujieda K (2010) A novel mutation (V101A) of the LHX4 gene in a Japanese patient with combined pituitary hormone deficiency. Exp Clin Endocrinol Diabetes 118:405–409

Takahashi K, Yamanaka S (2006) Induction of pluripotent stem cells from mouse embryonic and adult fibroblast cultures by defined factors. Cell 126:663–676

Takahashi K, Tanabe K, Ohnuki M, Narita M, Ichisaka T, Tomoda K, Yamanaka S (2007) Induction of pluripotent stem cells from adult human fibroblasts by defined factors. Cell 131:861–872

Tatsumi K, Miyai K, Notomi T, Kaibe K, Amino N, Mizuno Y, Kohno H (1992) Cretinism with combined hormone deficiency caused by a mutation in the PIT1 gene. Nat Genet 1:56–58

Theill LE, Castrillo JL, Wu D, Karin M (1989) Dissection of functional domains of the pituitary-specific transcription factor GHF-1. Nature 342:945–948

Therrien M, Drouin J (1991) Pituitary pro-opiomelanocortin gene expression requires synergistic interactions of several regulatory elements. Mol Cell Biol 11:3492–3503

Therrien M, Drouin J (1993) Cell-specific helix-loop-helix factor required for pituitary expression of the pro-opiomelanocortin gene. Mol Cell Biol 13:2342–2353

Thomas PQ, Rathjen PD (1992) HES-1, a novel homeobox gene expressed by murine embryonic stem cells, identifies a new class of homeobox genes. Nucleic Acids Res 20:5840

Thomas PQ, Johnson BV, Rathjen J, Rathjen PD (1995) Sequence, genomic organization, and expression of the novel homeobox gene Hesx1. J Biol Chem 270:3869–3875

Thomas PQ, Dattani MT, Brickman JM, McNay D, Warne G, Zacharin M, Cameron F, Hurst J, Woods K, Dunger D et al (2001) Heterozygous HESX1 mutations associated with isolated congenital pituitary hypoplasia and septo-optic dysplasia. Hum Mol Genet 10:39–45

Topilko P, Schneider-Maunoury S, Levi G, Trembleau A, Gourdji D, Driancourt MA, Rao CV, Charnay P (1998) Multiple pituitary and ovarian defects in Krox-24 (NGFI-A, Egr-1)-targeted mice. Mol Endocrinol 12:107–122

Trapnell C, Cacchiarelli D, Grimsby J, Pokharel P, Li S, Morse M, Lennon NJ, Livak KJ, Mikkelsen TS, Rinn JL (2014) The dynamics and regulators of cell fate decisions are revealed by pseudotemporal ordering of single cells. Nat Biotechnol 32:381–386

Treier M, Gleiberman AS, O'Connell SM, Szeto DP, McMahon JA, McMahon AP, Rosenfeld MG (1998) Multistep signaling requirements for pituitary organogenesis in vivo. Genes Dev 12:1691–1704

Treier M, O'Connell S, Gleiberman A, Price J, Szeto DP, Burgess R, Chuang PT, McMahon AP, Rosenfeld MG (2001) Hedgehog signaling is required for pituitary gland development. Development 128:377–386

Tremblay JJ, Drouin J (1999) Egr-1 is a downstream effector of GnRH and synergizes by direct interaction with Ptx1 and SF-1 to enhance luteinizing hormone beta gene transcription. Mol Cell Biol 19:2567–2576

Tremblay JJ, Lanctot C, Drouin J (1998) The pan-pituitary activator of transcription, Ptx1 (pituitary homeobox 1), acts in synergy with SF-1 and Pit1 and is an upstream regulator of the Lim-homeodomain gene Lim3/Lhx3. Mol Endocrinol 12:428–441

Tremblay JJ, Marcil A, Gauthier Y, Drouin J (1999) Ptx1 regulates SF-1 activity by an interaction that mimics the role of the ligand-binding domain. EMBO J 18:3431–3441

Tsai YC, Cooke NE, Liebhaber SA (2016) Long-range looping of a locus control region drives tissue-specific chromatin packing within a multigene cluster. Nucleic Acids Res 44:4651–4664

Vallette-Kasic S, Brue T, Pulichino AM, Gueydan M, Barlier A, David M, Nicolino M, Malpuech G, Dechelotte P, Deal C et al (2005) Congenital isolated adrenocorticotropin deficiency: an underestimated cause of neonatal death, explained by TPIT gene mutations. J Clin Endocrinol Metab 90:1323–1331

von Möllendorff W, Bargmann W (1940) Handbuch der mikroskopischen Anatomie des Menschen: Hypofyse. In: Blutgefäss- und Lymphgefässapparat; Innersekretorische Drüsen. Springer, Berlin

Wang J, Scully K, Zhu X, Cai L, Zhang J, Prefontaine GG, Krones A, Ohgi KA, Zhu P, Garcia-Bassets I et al (2007) Opposing LSD1 complexes function in developmental gene activation and repression programmes. Nature 446:882–887

Ward RD, Raetzman LT, Suh H, Stone BM, Nasonkin IO, Camper SA (2005) Role of PROP1 in pituitary gland growth. Mol Endocrinol 19:698–710

Ward RD, Davis SW, Cho M, Esposito C, Lyons RH, Cheng JF, Rubin EM, Rhodes SJ, Raetzman LT, Smith TP et al (2007) Comparative genomics reveals functional transcriptional control sequences in the Prop1 gene. Mamm Genome 18:521–537

Wen S, Ai W, Alim Z, Boehm U (2010) Embryonic gonadotropin-releasing hormone signaling is necessary for maturation of the male reproductive axis. Proc Natl Acad Sci USA 107:16372–16377

Wu W, Cogan JD, Pfaffle RW, Dasen JS, Frisch H, O'Connell SM, Flynn SE, Brown MR, Mullis PE, Parks JS et al (1998) Mutations in PROP1 cause familial combined pituitary hormone deficiency. Nat Genet 18:147–149

Xia Y, Nivet E, Sancho-Martinez I, Gallegos T, Suzuki K, Okamura D, Wu MZ, Dubova I, Esteban CR, Montserrat N et al (2013) Directed differentiation of human pluripotent cells to ureteric bud kidney progenitor-like cells. Nat Cell Biol 15:1507–1515

Xu Y, Baldassare M, Fisher P, Rathbun G, Oltz EM, Yancopoulos GD, Jessell TM, Alt FW (1993) LH-2: a LIM/homeodomain gene expressed in developing lymphocytes and neural cells. Proc Natl Acad Sci USA 90:227–231

Xue L, Cai JY, Ma J, Huang Z, Guo MX, Fu LZ, Shi YB, Li WX (2013) Global expression profiling reveals genetic programs underlying the developmental divergence between mouse and human embryogenesis. BMC Genomics 14:568

Young MD, Willson TA, Wakefield MJ, Trounson E, Hilton DJ, Blewitt ME, Oshlack A, Majewski IJ (2011) ChIP-seq analysis reveals distinct H3K27me3 profiles that correlate with transcriptional activity. Nucleic Acids Res 39:7415–7427

Zappone MV, Galli R, Catena R, Meani N, De Biasi S, Mattei E, Tiveron C, Vescovi AL, Lovell-Badge R, Ottolenghi S et al (2000) Sox2 regulatory sequences direct expression of a (beta)-geo transgene to telencephalic neural stem cells and precursors of the mouse embryo, revealing regionalization of gene expression in CNS stem cells. Development 127:2367–2382

Zhao L, Bakke M, Krimkevich Y, Cushman LJ, Parlow AF, Camper SA, Parker KL (2001) Steroidogenic factor 1 (SF1) is essential for pituitary gonadotrope function. Development 128:147–154

Zhao Y, Mailloux CM, Hermesz E, Palkovits M, Westphal H (2010) A role of the LIM-homeobox gene Lhx2 in the regulation of pituitary development. Dev Biol 337:313–323

Zheng GX, Terry JM, Belgrader P, Ryvkin P, Bent ZW, Wilson R, Ziraldo SB, Wheeler TD, McDermott GP, Zhu J et al (2017) Massively parallel digital transcriptional profiling of single cells. Nat Commun 8:14049

Zhu X, Zhang J, Tollkuhn J, Ohsawa R, Bresnick EH, Guillemot F, Kageyama R, Rosenfeld MG (2006) Sustained Notch signaling in progenitors is required for sequential emergence of distinct cell lineages during organogenesis. Genes Dev 20:2739–2753

Zhu X, Gleiberman AS, Rosenfeld MG (2007) Molecular physiology of pituitary development: signaling and transcriptional networks. Physiol Rev 87:933–963

Ziegenhain C, Vieth B, Parekh S, Reinius B, Guillaumet-Adkins A, Smets M, Leonhardt H, Heyn H, Hellmann I, Enard W (2017) Comparative analysis of single-cell RNA sequencing methods. Mol Cell 65(631–643):e634

Part II

Developmental Modulators and Epigenetic Factors

Hypothalamic Development: Role of GABA 7

M. Stratton and S. Tobet

Abstract

A number of cellular developmental processes occur in an orchestrated fashion to mediate the transition from a simple neural tube to the complexity of the adult brain. A combination of internal programming and external cues provides the molecular specificity needed to direct cells to the correct location, initiate correct gene expression, and connect to the appropriate circuits. The heterogeneous cells of the hypothalamus affect all aspects of physiology through the regulation of the autonomic nervous system, motivated behavior, and endocrine balance. Small disruptions (genetic or environmental) to hypothalamic development can impact adult physiology and contribute to pathology. GABAergic signaling has drawn the attention of developmental neuroendocrinologists as a potential effector system of hypothalamic development. This chapter discusses the role of GABA in directing the development of the hypothalamus, specifically related to the migration of neurons that release gonadotropin-releasing hormone (GnRH), and in the formation of the ventromedial (VMH) and paraventricular (PVN) nuclei of the hypothalamus (Fig. 7.1).

Keywords

Neuron migration · Hypothalamic development · Gamma-aminobutyric acid (GABA) · Gonadotropin-releasing hormone (GnRH) · Paraventricular nucleus of the hypothalamus (PVN) · Ventromedial nucleus of the hypothalamus (VMH) · Developmental patterning

M. Stratton (✉)
Department of Physiology and Cell Biology, Ohio State University, Columbus, OH, USA
e-mail: Mattthew.Stratton@osumc.edu

S. Tobet (✉)
Department of Biomedical Sciences, School of Biomedical Engineering, Colorado State University, Fort Collins, CO, USA
e-mail: stuart.tobet@colostate.edu

S. Wray, S. Blackshaw (eds.), *Developmental Neuroendocrinology*, Masterclass in Neuroendocrinology 9, https://doi.org/10.1007/978-3-030-40002-6_7

Abbreviations

AOB	Accessory olfactory bulb
GABA	Gamma aminobutyric acid
GAD	Glutamic acid decarboxylase
GnRH	Gonadotropin releasing hormone
OB	Olfactory bulb
OE	Olfactory epithelium
PVN	Paraventricular nucleus of the hypothalamus
VMH	Ventromedial nucleus of the hypothalamus
VNO	Vomeronasal organ

7.1 Hypothalamic Organization

The hypothalamus is a master regulator of physiological homeostasis. Cells in the hypothalamus survey information from the periphery through either systemic circulation (blood/cerebrospinal fluid/lymph) or peripheral sensory inputs and integrate information from other brain regions to elicit coordinated responses through neuroendocrine and autonomic outlets. The complexity of this master regulator is astounding given the sheer diversity of inputs and outputs. The oldest parcellations of cell grouping in the brain were based on a simple dye-based method known as Nissl staining, which visualizes neuronal cell bodies according to the accumulation of rough endoplasmic reticulum. For more than 100 years, people have subdivided the hypothalamus into different numbers of subdivisions. In general, there are about a dozen nuclear groups and several named areas delineated based on Nissl stains and functional circuits in adults. From another perspective (Puelles and Rubenstein 2015), you might consider 33 molecularly discrete hypothalamic progenitor areas that produce as many as 150 derived "nuclei" (distinct cell populations).

Early chapters in this series deal with the molecular specification of hypothalamic cells (e.g., transcription factors) and some deal with expressed genes that directly signal developmental processes (e.g., sonic hedgehog). The chapter presented here discusses a secreted factor that is the product of an enzymatic reaction (e.g., the decarboxylation of glutamate) that is a well-known neurotransmitter in the central nervous system; gamma-aminobutyric acid (GABA) (Fig. 7.1).

Beginning with studies in the telencephalon and extending to studies of the diencephalon, evidence suggests that GABA release in early development is capable of impacting the cytoarchitectural landscape of the brain and for this chapter, in particular, the hypothalamus. The effects of GABA on cell positioning and cell phenotype within the hypothalamus during early development (and interactions with perinatal stress and sex) are antecedent to hypothalamic regulation of adult homeostasis. This chapter focuses on the formation of the ventromedial (VMH) and paraventricular (PVN) hypothalamic nuclei as well as neurons that secrete

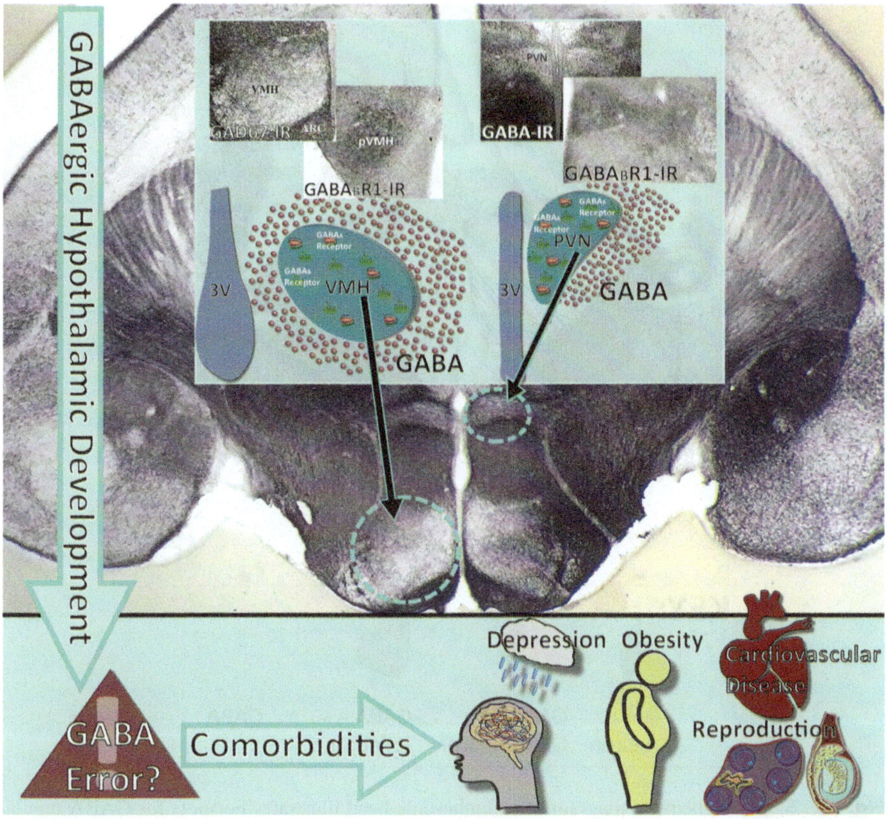

Fig. 7.1 Introduction to the role of GABA in hypothalamic development and potential implications for adult physiology and pathophysiology. There is a striking absence of GABA and GABA-producing enzymes inside select developing regions of the hypothalamus (e.g., VMH–GAD67 depicted and PVN–GABA depicted) which is contrasted by enrichment of GABA receptors within these regions ($GABA_BR1$ depicted). Subsequent text and figures summarize work done to determine the role of GABA in directing the development of these regions and for the GnRH system as well as potential consequences of altered GABA signaling during hypothalamic development. The digital image in the background was taken from archival material in the Tobet laboratory and shows immunoreactive GAD67 in adult mouse brain; dashed circles highlight the VMH and PVN nuclei. Inset VMH images were adapted from Davis et al. (2002) and inset PVN images were adapted from McClellan et al. (2010)

gonadotropin-releasing hormone (GnRH), which are scattered through the basal forebrain including in the hypothalamus (Fig. 7.2). Functions associated with these hypothalamic regions/cells include reproduction, feeding, drinking, stress responses, cardiac output, blood pressure, growth signaling, and arousal.

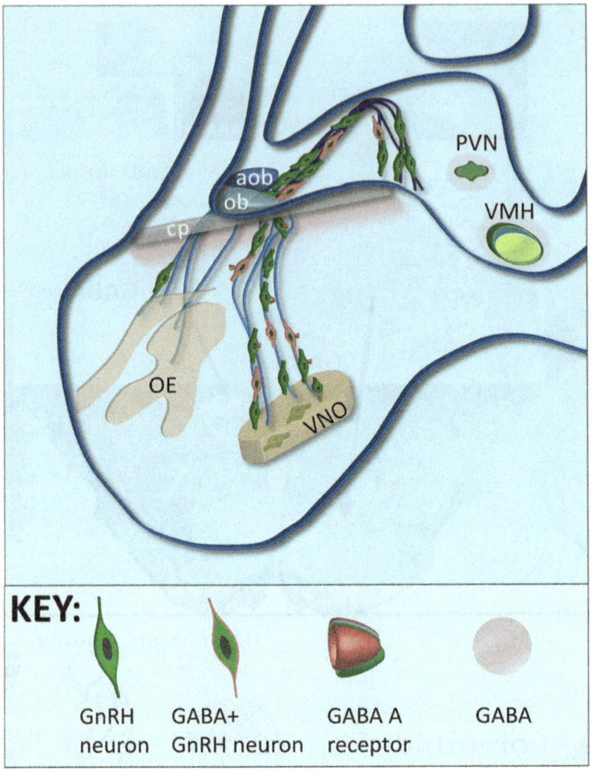

Fig. 7.2 Sagittal schematic diagram of an embryonic head illustrates hotspots for GABA regulation of neuron migration in the developing hypothalamus. GABA has been shown to act generally as a "brake" for hypothalamic cells as they exit nasal cavity and enter the brain (GnRH neurons), in the developing ventromedial nucleus of the hypothalamus (VMH) and in the developing paraventricular nucleus of the hypothalamus (PVN). GnRH neurons mostly track fibers of the vomeronasal nerve from the vomeronasal organ (VNO) across the cortical plate (CP) toward the accessory olfactory bulb (AOB). When a subset of fibers bifurcates and turn toward the basal hypothalamus, GnRH neurons start making directional choices that are likely influenced by GABA from neighboring GnRH neurons and from other neurons in the region. The other major source of fibers is from main olfactory neuroepithelial cells (OE) that travel to the main olfactory bulb (OB). For the PVN and VMH, cells originate medially and enter the region of the nuclei before they are impacted by GABA at the boundaries of the nuclei as they coalesce

7.1.1 Patterning Processes

From a mechanistic perspective, there are two major organizational schemes in brain development that individual cells must navigate; joining layers (e.g., cerebral cortex) or groups (e.g., diencephalic nuclei). GnRH neurons indicate a less common third case that does not do either (in most species) as these cells are not distributed in either layer or confined to nuclear group. Due to the inability to categorize cells solely based on distance from a proliferative zone, developmental studies of the hypothalamus may be intrinsically more complex than in areas like the cortex. As in any tissue, the processes of proliferation, growth, migration, differentiation, and death mediate the development of the hypothalamus. Process extension (axon and dendrite) and circuit integration are additional crucial steps in brain development. During development, each of these steps impacts the other and altered regulation of one process can lead to profound effects in downstream processes. For instance, the microenvironment and epigenetic program of the progenitor pool directs gene expression which influences migration. The microenvironment through which a cell migrates, further influences its epigenetic and gene expression program, which influences its growth, differentiation, process extension, and circuit integration.

Regardless of whether you categorize functional entities of brain development into larger units like prosomeres or split things into smaller units based on molecular genetic characteristics (early chapters, Puelles and Rubenstein 2015), the business end of development is operated by the molecules that provide signals from one cell to another (Fig. 7.3). They can be recognition, adhesion or receptive entities on a cell surface, or secreted, diffusive molecules that signal attraction or repulsion. The current chapter addresses the secreted molecule GABA as a contributor to cell movement and ultimately anatomical position. For secreted molecules to provide patterning information they must have a way to address the issue of spatial distribution relative to nuclear boundaries that characterize the hypothalamus. Whereas a radially migrating cell in the cerebral cortex can get 2-dimensional data to signal stop or go relative to a layer, radially migrating cells in the hypothalamus must sense curvilinear boundaries.

In regions outside of the hypothalamus, GABA had been suggested to elicit multiple effects in neurodevelopment. These include effects on proliferation, migration, DNA synthesis, differentiation, growth, and morphology (reviewed in Wu and Sun 2015). These developmental effects of GABA are thought to be mediated by non-synaptic release/secretion of GABA and typically occur prior to the emergence of structural synapses. Likely mechanisms for GABA release through non-synaptic mechanisms include GABA transporters or through non-vesicular channel-mediated release.

7.1.2 The Birth and Organization of Hypothalamic Neurons

For hypothalamic cells, the first cells to leave the proliferative regions are usually the ones that migrate the farthest. This is in contrast to structures such as the cerebral cortex, where cells that populate the farthest layers, are the last ones to leave the proliferative zone. Derived from the ventral and inferior lobes of the diencephalic neuroepithelium, Altman and Bayer (1986) describe three waves of neurogenesis in development of the rat hypothalamus. The first wave is between embryonic day (E) 13 and E15 (E0 = first day post timed breeding; for mouse E11–E13) and these neurons form lateral hypothalamic structures. The second wave is between E15 and E17 with these neurons becoming part of the medial hypothalamus. The final wave occurs between E17 and E19. These neurons make up the periventricular

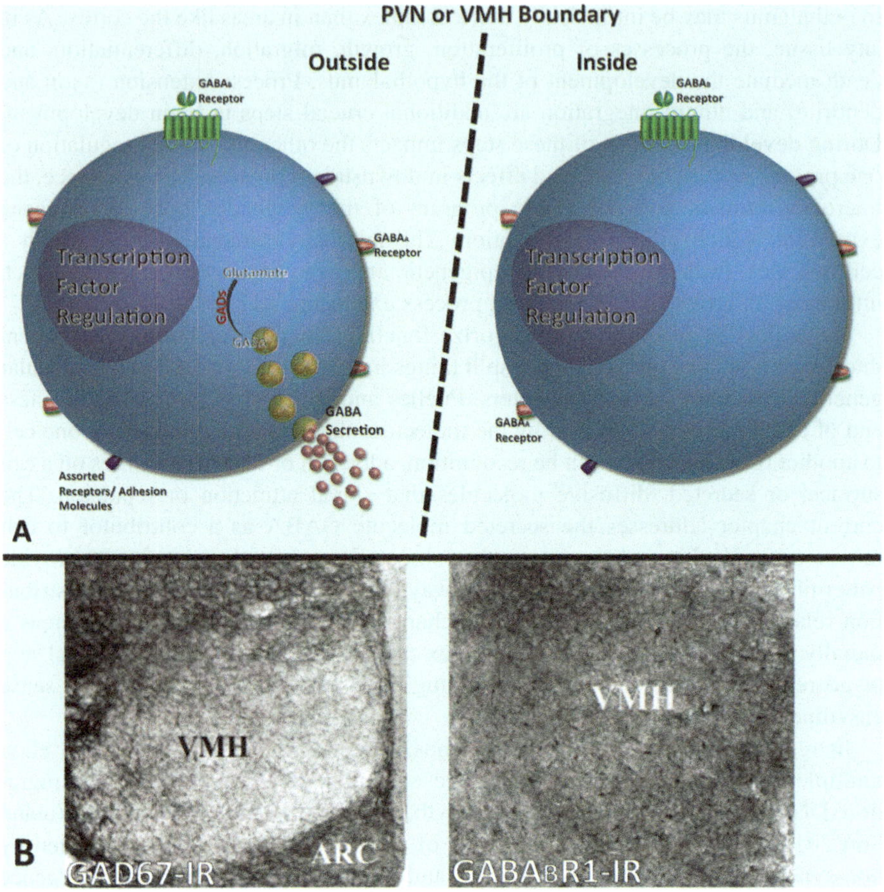

Fig. 7.3 Schematic diagram illustrates a cellular perspective of the developmental boundaries of the PVN and VMH that are defined by the presence/absence of GABA synthesizing cells. The cell on the left in (**a**) suggests a cell found outside of the boundary that synthesizes and secretes GABA following glutamic acid decarboxylase (GAD) acting on glutamate precursor. The cell on the right in (**a**) suggests a cell found inside of the boundary that receives GABA through $GABA_A$ (forming a chloride channel) or $GABA_B$ (G-protein coupled) receptors. Transcription factor programs initiated by classic hypothalamic differentiation genes dictate cells that secrete and/or respond to GABA signaling. Digital images in (**b**) depict the developmental boundaries of the VMH. Images demonstrate GAD67 (left) and $GABA_BR1$ (right) immunoreactivities in the developing VMH, and were adapted from Davis et al. (2002)

hypothalamic populations with notable exception of the PVN, for which most neurons complete terminal mitosis prior to E15.

The field of birthdate labeling with tritiated thymidine to follow cells was pioneered by investigators such as Sidman, Rakic, Altman, and Bayer. After 1988 (Miller and Nowakowski 1988), the field was gradually commandeered by studies using bromodeoxyuridine (BrdU) because results could be generated more quickly by immunohistochemistry instead of waiting for the development of radioautographic images. Interestingly, the technique of choice is shifting again, this time to 5-ethynyl-2′-deoxyuridine (EdU) with the publication of a simpler method that makes it even easier to process tissue (Salic and Mitchison 2008). Tritiated thymidine remains the gold standard by virtue of the least modification to the nucleic acid being introduced to the DNA for tracing purposes.

Neurons that synthesize an essential peptide for the regulation of reproduction, gonadotropin-releasing hormone (GnRH), traverse significant distances in almost all vertebrates that have been so characterized. The realization that the GnRH neurons originate outside of the CNS in the olfactory periphery was a major neuroendocrine "aha" moment for 1989 (Schwanzel-Fukuda and Pfaff 1989; Wray et al. 1989). Both publications indicated that GnRH neurons were born in a narrow temporal window in mice between embryonic days 10 and 13 based on tritiated thymidine autoradiography. This time frame fits well, albeit a little early, with other hypothalamic neurons that do not migrate nearly as far. GnRH neurons originating in the nasal compartment migrate over the nasal septum, cross the cribriform plate, and enter the basal forebrain where they must turn toward the hypothalamus. While their fibers are directed to two major sites (organum vasculosum of the lamina terminalis and median eminence), the cell bodies scatter across a much larger expanse. The signals that direct this unusual migration include soluble factors that can work by diffusing across a distance as well as factors that regulate GnRH neuron adhesion to guiding fibers (reviewed Wierman et al. 2011), and it is likely that none of the selective signals drive more than 30–40% of the GnRH neuronal population. This redundancy of signaling is probably a major factor in ensuring reproductive success leading to the preservation of species.

Medial-to-lateral migration in the hypothalamus, like in the cerebral cortex, is aided by the presence of radial glia, which arises at the stage of the neural tube and stretches from the ventricle to the pia surface. This creates a series of railroad tracks like fibers that immature neurons can crawl along during the course of migration (Rakic 1972). While the rules and roles of radial glia are well studied in the cortex, fewer studies exist in the diencephalon. Nonetheless, there is ample evidence of radial glial fibers throughout the developing diencephalon in a number of species (Tobet et al. 1995).

Combining birthdate labeling with radial glia labeling in the same experiment allowed for the observation that radial glia may provide paths for cells to cross putative boundaries into diencephalic regions (Tobet et al. 1995) as well as paths across forming cell groups such as the VMH (Dellovade et al. 2001). GABA regulation of cell adhesion to guidance fibers (radial glial or neuronal) may be one mechanism of migratory control. Importantly for this chapter, there are patterns of GABAergic patterning elements present throughout hypothalamic cell migratory periods that may influence cell movements (Bless et al. 2000; Davis et al. 2004).

Several hypothalamic cell groups are defined by neurochemical phenotypes of cells in regions (e.g., GABA, see Fig. 7.1) before they become recognizable using Nissl stains. This is readily apparent in mice for the VMH (Tobet et al. 1999) and the PVN (McClellan et al. 2010). Neither VMH or PVN is discernible as a cell group using Nissl stains in mice until E17. However, many of the neurons that make up these nuclear groups are present in groups earlier and express immunoreactive peptides before E17. For example, immunoreactive corticotropin-releasing hormone (CRH) and arginine vasopressin (AVP) can be found in the PVN by E15. It is not yet known what event occurs between E15 and E17 to allow for Nissl stain detection, although possible mechanisms include the proliferation of glia and/or tighter cell packing mediated by migration. Since there is a GABAergic surround to both VMH and PVN (see below), it is possible that GABA contributes to the fine-tuning of cell positions that results in nuclear appearance.

7.2 Reconstructing GABA and GABA Receptor Distributions Reveals Unique Patterns

The first requirement for connecting a signaling factor with hypothalamic development is the demonstration of presence. Within the developing hypothalamus, multiple immature cell groupings, including those destined to form PVN and VMH, are enriched with GABA receptors while being bordered by areas with many cells containing immunoreactive GABA or expressing GABA synthesizing enzymes (Figs. 7.1 and 7.2). This striking pattern, first observed in the 1990s, provided the first suggestions of an important role for GABA in directing hypothalamic development.

7.2.1 GABAergic Elements

GABA Sources: The patterning of GABA and its synthesizing enzymes can be visualized efficiently through immunohistochemistry. Notably, GABA is found in many cells throughout the hypothalamus where its synthetic enzymes glutamic acid decarboxylases 1 and 2 (GAD aka GAD67 and GAD65) are found in cell bodies that are excluded from locations within the developing VMH and PVN during development (Tobet et al. 1999; McClellan et al. 2008; e.g., Fig. 7.1) and in adulthood (by in situ hybridization for GAD; Okamura et al. 1990). Thus, the embryonic PVN and VMH are ringed by regions of cells that contain immunoreactive GABA and GAD while there is an absence of GABA and GAD within cells of the developing nuclear groups. This pattern of expression also holds for GABA transporters GAT1 and GAT4 (Jursky and Nelson 1996).

This void of GABAergic components in developing nuclei is apparent as early as E12–13 in mice when the various developing nuclei are first detectable by immunohistochemistry (but not yet Nissl). As GABAergic axons grow into these hypothalamic nuclei, the pattern is lost (earlier for PVN than VMH) and the entire hypothalamus becomes highly immunoreactive for GABAergic *protein* elements soon after birth (GAD mRNA is still restricted to cells outside of the VMH and PVN; Okamura et al. 1990). It is unclear if all of the sources of GABA are neuronal or if some might be glial in nature. Thoughts of GAD as strictly neuronal might be only partially true as there is evidence of glial expression of GAD enzymes. Glia can also synthesize GABA through the putrescine degradation pathway and is dependent on the activity of monoamine oxidase B (Yoon and Lee 2014). More recent work also points toward endothelial cells associated with developing blood vessels as being sources of GABA in other brain regions (Li et al. 2018).

The case is less clear for patterning of GABAergic elements along the developmental path of GnRH neurons. Consistent with this notion, a subset of GnRH neurons (~30%) contain GABA during development (Tobet et al. 1996a) while exposure to GABA varies across the long path of migration from the nasal compartment to destinations in the hypothalamus and other brain regions. It is likely that GABA from within and outside the population is one of the soluble secreted factors that contribute to their migration.

7.2.2 The A's, B's, and C's of GABA Receptors

GABA binds to three classes of cell surface receptors, $GABA_A$, $GABA_A$-rho (previously $GABA_C$), and $GABA_B$ receptors. Given the paucity of literature investigating $GABA_A$-rho in the development of the hypothalamus, the $GABA_A$-rho receptor, a ligand-gated Cl^- channel composed of 5 rho subunits predominantly studied in the retina, will not be discussed further in this chapter. Notably, $GABA_A$ and $GABA_B$ receptor subunits are enriched in the developing PVN (McClellan et al. 2010) and VMH (Dellovade et al. 2001; Davis et al. 2002) at least as early as E15. $GABA_A$ receptors have been found in GnRH neurons from the earliest ages examined (E12.5; Temple and Wray 2005) and virtually all GnRH neurons contain $GABA_A$ receptors, but with heterogeneous complements of subunit composition (Simonian et al. 2000).

$GABA_A$ receptors are a family of ligand-gated Cl^- channels. Five subunits (e.g., 2 α subunits, 2 β subunits, and 1 γ subunit) comprise a functional $GABA_A$ channel. There are at least 21 subunits that can be used to build the channel (Lujan et al. 2005). Depending on the intracellular versus extracellular concentration gradient of Cl^-, the activation of $GABA_A$ receptors can exert a depolarizing or hyperpolarizing influence on membrane potential. In the embryo, Cl^- concentrations are higher inside the cell than outside, and thus $GABA_A$ receptor activation is excitatory in the embryo. The Cl^- gradient usually reverses at some point in development with the expression of the K^+/Cl^- cotransporter KCC2, which shunts Cl^- ions to the outside of the cell. There have been several reports of this embryonic Cl^- gradient being present in specialized locations within the adult brain, for example, the progenitor pool of the adult hippocampus and some magnocellular neurons in the PVN, which coincide with a lack of KCC2 expression.

$GABA_B$ receptors, among the first G protein-coupled receptors to be identified, were cloned after searching for molecules that could explain responses to GABA that were insensitive to chloride and bicuculline ($GABA_A$ antagonist). $GABA_B$ receptors have been extensively studied and excellent reviews are available (e.g., Frangaj and Fan 2018). The functional G protein-coupled receptor is a heterodimer of R1 and R2 subunits, for which multiple splice variants exist. $GABA_B$ receptor activation has been shown to affect the adenylate cyclase system, decrease Ca^{2+} conductance, and increase K^+ conductance. Originally thought to be mediated by pertussis toxin-sensitive $G_{i\alpha}$ and $G_{o\alpha}$ G proteins, some pertussis toxin insensitive baclofen (agonist) effects have been found in magnocellular neurons of PVN and supraoptic nucleus (SON), in presynaptic receptors and whole-brain co-immunoprecipitation mass spectrometry studies point toward $G_{\beta1/\beta2}$, $G_{\gamma2}$, and $G_{\alpha ?}$ as being predominant effector molecules (Schwenk et al. 2016).

7.3 Organizing Principles

The role of homeotic genes in the formation of the hypothalamus is complicated and all the more so by a lack of understanding downstream molecular consequences. When molecular genetics have been used to disrupt the expression of any of a number of transcription factors, hypothalamic patterning has been dramatically altered (see preceding chapters in this series). Whether the mechanisms are through proliferation, migration, and terminal differentiation is often an open question. Determining the impact on the molecules that mediate cell–cell communication and induction of cellular processes is still less understood and infrequently attempted. GABAergic signaling falls in this latter category.

Relative to understanding a role for GABA, a major homeodomain protein of interest is the vertebrate homolog of the fly gene distal-less (DLX 1 and 2). DLX protein exerts a direct regulatory impact on the transcription of glutamic acid decarboxylases (GAD)65 and 67 (Le et al. 2017) and establishes the distribution of GABA sources in multiple brain regions including the developing hypothalamus. When a number of different transcription factors have been disrupted, a common phenotype is aberrant *Dlx* expression and, presumably GABA. One homeobox-containing gene, Orthopedia (Otp) is expressed in neurons that come to reside in the PVN, SON, anterior periventricular, and arcuate nuclei. In mice in which *Otp* was replaced by a lacZ reporter gene, cells migrated abnormally and were apparently displaced by cells expressing *Dlx1*, in a manner suggesting that *Dlx1*-positive cells replaced PVN-committed cells (positive for lacZ) that migrated abnormally (Acampora et al. 1999). Interestingly, these types of knockout mice usually, when they can be raised to adulthood, display altered physiology and behavior associated with the hypothalamus including, response to dehydration, feeding, sexual function, and stress responses. A key question is what downstream molecular changes follow the transcription factor action(s) to alter cell positions? Are they secreted protein-coding genes like sonic hedgehog or wnt's or might they be GAD 65 or 67 leading to changes in GABA that drive the movement of other cells?

The transcription factor HES1, a Notch effecter gene, is also expressed in the developing pituitary and hypothalamus. In mice lacking HES1, distribution of GAD 67 immunoreactivity was shifted from surrounding the PVN to within the PVN. In this case, AVP neurons were found to be located outside of the PVN boundaries in the HES1 null (Aujla et al. 2011). The SON also showed ectopic GAD expression

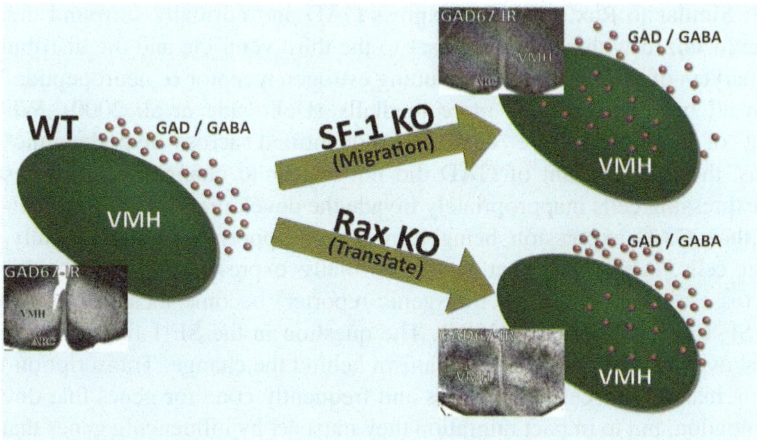

Fig. 7.4 Schematic diagram illustrates genetic manipulations that alter GABAergic cell populations relative to boundaries in the developing ventromedial nucleus of the hypothalamus (VMH) by different mechanisms. Loss of SF-1 function most likely causes cellular rearrangements that result in GABAergic cells moving into the location of the VMH (Dellovade et al. 2000; Davis et al. 2004). Loss of Rax function (knockout—KO) most likely causes cells in the location of the VMH to change fate (transfate) to express glutamic acid decarboxylases (GAD) and make GABA (Lu et al. 2013). In both cases, the phenotypes of cells in the VMH are altered and importantly, GABA synthesizing cells become inappropriately located within the nucleus. Digital images from sections taken from wild-type (WT) and SF1-KO mice were adapted from Dellovade et al. (2000). The image of a section from a Rax KO mouse was adapted from one provided by Dr. D. Kurrasch related to Lu et al. (2013)

in the HES1 null, which corresponded with the dispersion of typical AVP expressing neurons in the SON. Based on the hypothesis that an inappropriately located cell is exposed to alternate signaling milieus that have further functional consequences, axons from these misplaced AVP neurons were unable to appropriately target the posterior pituitary.

The gene *Rax* codes for a homeodomain protein that was found to be critical to the formation of the VMH based on changes in the fates or molecular identities of multiple cells (Lu et al. 2013). Mice lacking functional *Rax* also exhibited a change in the pattern of GAD immunoreactivity. Cells containing GAD become localized inside the VMH in contrast to its normal pattern of expression in cells surrounding the nucleus. The authors pose two possible models to explain the *Rax* phenotype. First, that mutant VMH cells "transfate" into a hypothalamic non-VMH identity, or second that hypothalamic non-VMH cells (e.g., GAD containing) could migrate into the mutant VMH region (Fig. 7.4). However, the former change in fate hypothesis was considered more likely, based on the lack of change in the distribution of cells expressing another transcription factor, *Nkx2.1*. Nonetheless, the new area of GAD immunoreactivity within the VMH in the *Rax* null excludes several VMH cell types that typically reside there.

When the transcription factor steroidogenic factor 1 (SF-1) was disrupted, the positions of cells in and around the VMH change, including those expressing

Nkx2.1. Similar to Rax, cells that express GAD and normally surround the VMH become localized within regions closer to the third ventricle and the distribution of other markers of cell phenotype, including estrogen receptor-α, neuropeptide Y, and galanin all become localized more medially (Dellovade et al. 2000). When the amount of immunoreactive GAD was quantified across hypothalamic tissue sections, the total amount of GAD did not appear to change. This suggests that GAD expressing cells inappropriately invade the developing VMH in the SF-1 null, rather than GAD expression being turned on inappropriately in typically VMH resident cells. Also, cells that would normally express SF-1 and reside in the VMH (assessed by SF-1-GFP transgenic reporter) become located more laterally in the SF-1 null (Davis et al. 2004). The question in the SF-1 findings, like those noted above, is identifying the mechanism behind the change. Transcription factors are great markers for cell phenotypes and frequently code for genes that direct cell differentiation, but to impact migration they must act by influencing genes that affect cell movements. In this case, if GABA is an active agent influencing movement, and if GABA acts as a stop signal for normal movement of SF-1 positive cells (more discussion below), then when SF-1 is disrupted, GABA must no longer provide that stop signal (Davis et al. 2004).

7.4 Finding Useful Model Animals

As alluded to above, a major tool for determining the impact of gene expression on hypothalamic nuclear formation has been derived from animal models with selective gene disruptions. It is particularly useful when the expression of a key regulatory gene is somewhat selective for a given cell group. This is true for *Sf-1* being relatively restricted to the VMH and to some extent true for *Sim-1*, which is somewhat restricted to the PVN (but does extend laterally and ventrally more so than is usually discussed). Many of the other transcription factors discussed above (e.g., *Dlx*) are expressed in many regions, but with patterns that still allow localized interpretations when there are genetic disruptions. Unfortunately for GnRH neurons, the only reliable marker is GnRH itself. Most of the genetic disruptions of transcription factors that have been characterized as impacting GnRH neuron development, do so indirectly, acting on elements other than the neurons themselves. However, an advantage in this system is that most GnRH neurons found in immunohistochemical studies can now be linked to one site of origin. Thus, genetic disruptions that influence GnRH neuron movement can be characterized by changes in the distribution of GnRH immunoreactive cells at different developmental stages. The same cannot be said of most other neurons in the brain. For example, neurons containing CRH can arise from several different progenitor populations.

Much of the work done to investigate GABA's role in directing neuron migration in the hypothalamus has used a mouse line generated with YFP expressed under the control of a modified neuronal Thy-1 promoter. In efforts to find a good reporter to visualize synaptogenesis in motor neurons and retinal ganglion cells, 25 fluorescent reporter lines were generated. Sixteen of these lines included YFP. Though the same genetic element was used to drive fluorescent reporter expression, the lines have remarkably divergent expression patterns. One of these lines showed subpopulation-specific fluorescence in the hypothalamus and has been used as a tool to query mechanisms of hypothalamic development (reviewed in Tobet et al. 2003). For studies of GnRH migration, transgenic fluorescent mice (Suter et al. 2000) again played an important role (Bless et al. 2005), although significant work was done before GnRH promoter driving GFP mice were available.

7.5 Choosing Ex Vivo Approaches

A key question posed in Sect. 7.3 is whether the molecular changes following transcription factor actions involve secreted protein-coding genes like *sonic hedgehog* or *wnt*'s or might they be acting through GAD 65/67 leading to changes in GABA that drive the movement of other cells? This is not a question easily answered in vivo. To get a better look at the movement of cells in real-time and in response to specific factors that influence motion, in vitro methods are needed. There are two usual methods to evaluate influences on motion. In one case, dispersed cell cultures have been used. This is perhaps most appropriate when there is a relatively homogenous population to work with, such as immortalized cell lines that represent GnRH neurons (e.g., NLT, GT-1). Transwell migration assays and wound-healing migration assays were both able to demonstrate roles for secretory products impacting NLT cells in vitro (reviewed Wierman et al. 2011). The problem for GnRH neurons in vitro and for many other regions in the hypothalamus is the cellular diversity and the dependence of cell behaviors on the surrounding cellular environment. For GABA, the transwell migration assays have been done with cerebral cortical cells (e.g., Behar et al. 1998), but were also complemented by later studies using organotypic slice cultures (e.g., Behar et al. 2000).

Organotypic cultures have an advantage when cell diversity and local environment is important for understanding influences. It is important to note that in the hypothalamus, the plane of sectioning is critical as the tissue slice should contain the source of the migrating cells, the destination, and any cues used by the cell to direct migration. These are different planes of sectioning for radial hypothalamic migration (coronal) versus migration of GnRH neurons, which originate in the nasal compartment (sagittal). Tissue explants offer relatively intact systems to test hypotheses of cellular behavior but suffer limitations of inclusion of all necessary cues, imaging hurdles for depth, as well as tissue health deep in tissue that has been removed from nutrient and oxygen vascular perfusion. Two approaches have been used for creating in vitro models of GnRH neuron migration. The first was an explant model (Fueshko and Wray 1994) and then a head slice model (Tobet et al. 1996b). Both models have been used to test the role of GABA; first in explants and then in head slices (see Wierman et al. 2011). Live views of GABA influences on GnRH neurons were discerned in slices (Bless et al. 2005) and then in explants (Casoni et al. 2012). The slice model maintained the arrangements of axonal fiber guides as they existed in vivo, while the explant model allows free extension of those axons away from their nasal origins. Each technique has its place, and all have been successfully used to shed light on many aspects of neurodevelopment.

7.6 Manipulating GABA Signaling Alters Hypothalamic Cell Behavior

Complementary Pharmacology and Molecular Genetics Results from organotypic brain slice experiments suggest that GABA acts via $GABA_A$ and $GABA_B$ receptors to restrict cell movements in the hypothalamus. This is apparent as receptor agonists slow cell movement speeds while receptor antagonists increase cell movement speeds, indicating that basal migration may be dampened by endogenous GABA gradients emanating from developmental boundaries. This appears true for GnRH neurons, as well as for cells in the VMH and PVN that are discussed separately below. The availability of selective models of genetic disruption makes it possible to consider complementary studies of in vivo function relative to secreted factors (gene product, e.g., WNTs or BDNF, or enzymes that produce ligands of interest, e.g., GAD for GABA) and the receptors for any secreted factor. Gene disruptions are not simple in interpretation. Multi-subunit structures like the $GABA_A$ receptor present the challenge of choosing which of 21 subunits to disrupt (not all of which impact function and there are no more than 5 subunits per receptor). Since there are several routes to GABA production, it is also unclear what GABA-producing enzymes would be most beneficial to disrupt to assay GABA effects in particular aspects of development in particular regions. Nonetheless, the complementary data of pharmacology and genetics helps draw more generalizable conclusions.

GnRH Neurons: Pharmacology First tested via in utero exposure then organotypic head slices exposed to the $GABA_A$ receptor agonist muscimol or antagonist bicuculline, GABA was shown to regulate the transition of migrating GnRH neurons exiting the nasal compartment and entering the brain at the cribriform plate (Fig. 7.5). Muscimol decreased the number of GnRH neurons in the

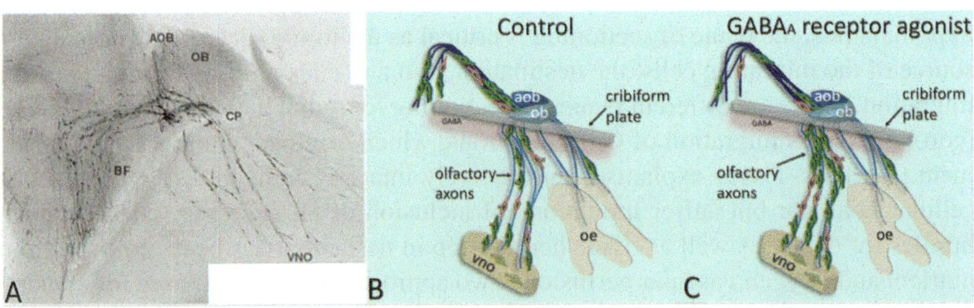

Fig. 7.5 Schematic diagram illustrates $GABA_A$ receptor activation decreasing GnRH neuron exit from the nasal compartment. (**a**) Montage shows immunochemically labeled GnRH neurons in migration in a mouse sagittal brain section at E15 (adapted from Tobet et al. 2001). (**b**) GnRH neurons originate in the vomeronasal organ (VNO) and migrate toward the brain along olfactory axons, crossing the cribriform plate. (**c**) When slice cultures are treated with a $GABA_A$ receptor agonist, fewer GnRH neurons cross the cribriform plate. Additional molecular complexity is present in this system: GABA-producing GnRH neurons are shown in red and $GABA_A$ receptor containing cells marked with red tags. *BF* Basal forebrain, *OE* olfactory epithelia, *AOB* accessory olfactory bulb, *OB* olfactory bulb

olfactory bulb and increased their number in the nasal compartment. Meanwhile, bicuculline treatment resulted in GnRH neurons more scattered in the brain and olfactory bulb (Bless et al. 2000). When fluorescence video microscopy was used to measure acute effects of GABA signaling on GnRH cell movement, $GABA_A$ receptor blockade resulted in increased migration and decreased turning (Bless et al. 2005). These results were confirmed in explants where picrotoxin, a $GABA_A$ receptor antagonist increased total distance traveled for GnRH neurons while muscimol again restricted movement (Casoni et al. 2012). Interestingly, the work addressing GABA's role in GnRH cell movements was preceded by work showing that GABA regulates GnRH electrophysiology in explant cultures started from mouse embryos at E11.5 (Kusano et al. 1995) when GnRH neuron birth and migration is just beginning. Interpretations of GABA effects on GnRH migration are complicated by the fact that ~30% of migrating GnRH neurons are themselves GABA positive and thus are not simply receivers of GABA cues (see genetic manipulation below).

GnRH Neurons: Genetics Given the complications of multiple sources of GABA along the migratory path of GnRH neurons, a gain-of-function approach was used in which the synthetic enzyme for GABA—glutamic acid decarboxylase—was driven by the GnRH promoter. This approach was attractive as a minority subset of GnRH neurons already express GAD. Transgenic mice with GAD67 selectively overexpressed in GnRH neurons increased the number impacted directly by GABA, but still only influenced a percentage of the total population (reviewed in Wierman et al. 2011). In contrast, GAD67 null mice displayed increased numbers of GnRH neurons out of the nasal placode at E14.5 and E17.5, although the changes were not balanced along the migratory route, making assessment of cell migration difficult to evaluate (Lee et al. 2008). More recent data using GAD67 and GAD65 knockout mice showed differential effects of knocking out the two forms of GAD and although the results were slightly different from the earlier study, the bottom line was a confirmation for a role of GABA in GnRH neuronal development (Vastagh et al. 2015). The role of $GABA_A$ receptors was probed using mice in which the $\gamma 2$ subunit was disrupted (Simonian et al. 2000), but this had no apparent effect on the distribution of GnRH neurons. There was a preliminary examination of mice with $GABA_A$, $\beta 3$ subunit knockdown using cohorts of mice used for studies of VMH (Dellovade et al. 2001); however, the results were inconclusive.

VMH: Pharmacology Manipulation of $GABA_A$ receptors affected cell movements and the distribution of identified cells in the region of the VMH when examined ex vivo (Dellovade et al. 2001). When treated with the antagonist bicuculline, the number of migrating cells increased (Dellovade et al. 2001) and the speed of movement increased (McClellan et al. 2008). Since bicuculline treatment also decreased the movement of neurons along the orientation of radial glial fibers in the central VMH, a possible mechanism of GABA action may be the regulation cell associations with radial glial fibers. Interestingly, at the extreme medial and lateral regions of the VMH, where GABA concentrations should be highest, cell movements are typically perpendicular to the radial fibers. When treated with bicuculline, movement patterns in these areas switched to being more parallel to the orientation of radial fibers. This data, in agreement with work done in the cortex (Behar et al. 1998), indicates that the functional result of GABA signaling can change based on location, GABA concentration, or developmental stage. GABA signaling through the $GABA_B$ receptor also has a strong influence on VMH development. Activation of $GABA_B$ receptors with baclofen decreased cell movements in the VMH (Davis et al. 2002) while inhibition with saclofen increased cell movement speed (McClellan et al. 2008).

VMH: Genetics Many $GABA_A$ receptor subunits are expressed in the developing hypothalamus, some of which are expressed in subregions. As the β3 subunit is highly expressed in the VMH and its disruption was known to reduce $GABA_A$ receptor binding by 50% across the brain, β3 subunit-disrupted mice were a logical place to start dissecting the role of $GABA_A$ receptors from a genetic perspective. Mice lacking $GABA_A$ β3 subunits had larger VMH areas and cells containing ERα were more widely dispersed than in controls (Dellovade et al. 2001). In mice lacking functional $GABA_B$ receptors, the size of the VMH also was larger and there was increased dorsal-to-ventral spread of cells containing ERα in the ventrolateral quadrant (McClellan et al. 2008). Both of these findings are consistent with the hypothesis that GABA produced at the lateral edges of the VMH provides a slowing or stop signal (Fig. 7.4).

PVN: Pharmacology The influence of $GABA_A$ receptor signaling on cell movements in this region has not been robustly characterized, however, when treated in utero with bicuculline, the PVN and lateral hypothalamic areas contain significantly fewer cells that express ERα (McClellan et al. 2010), pointing to potential roles in proliferation, survival, or gene expression regulation. More of the work investigating GABA influences in the development of the PVN has centered on actions through the $GABA_B$ receptor. When the $GABA_B$ receptor was chemically inhibited in utero, a lateral shift in the placement of cells containing ERα was noted in neonates, which was preserved through adulthood. This corresponded with increased cell movement speeds in tissue slices taken from the embryonic brain when the $GABA_B$ receptor was inhibited with antagonists, either saclofen or CGP 55845, suggesting that GABA acts to slow or stop migrating cells at predefined boundaries of the immature PVN (Stratton et al. 2014).

PVN: Genetics It was not surprising that the disruption of the β3 subunit of the $GABA_A$ receptor did not impact the PVN as β3 subunit expression in the PVN was found to be quite low (McClellan et al. 2010). On the other hand, in mice lacking functional $GABA_B$ receptors, the cytoarchitecture of the PVN was altered. Specifically, relative to littermate controls, cells containing nNOS were less tightly packed, cells containing ERα are found more laterally, and there was decreased expression of BDNF in the PVN (McClellan et al. 2010) while CRH expression was increased (Stratton et al. 2011). There was an unexpected finding of reduced blood vessel density in these mice (Frahm et al. 2012) that provided a new target and new mechanisms hypothalamic differentiation and disruption based on glucocorticoid actions.

GABA Receptor Synergy? As alluded to above, GABA may influence cell positioning by affecting the guidance of cells by neuronal (as per GnRH; Bless et al. 2000) or glial (as per VMH; Dellovade et al. 2001) fibers. Manipulation of $GABA_A$ receptor signaling altered the orientation of cell movements (Dellovade et al. 2001), whereas $GABA_B$ receptor signaling decreased the rate of cell movement (Davis et al. 2002). Since the two receptors may be differentially sensitive to GABA levels, the interaction of the two mechanisms may be spatially regulated by the locations of different concentrations of GABA (e.g., high at the nuclear boundary). It is still unknown whether $GABA_A$ and $GABA_B$ receptors act synergistically, antagonistically, or independently to influence movements in the developing hypothalamus.

7.7 Important Discoveries Along the Way Functional Significance

Taken together, the breadth of studies on the developing hypothalamus indicate significant effects of GABA on cell placement and/or gene/protein expression in the developing hypothalamus. In this section, we consider the potential functional consequences of such early actions. When using genetic approaches, it is difficult to link early critical period GABA impacts to adult function because GABA is a potent neurotransmitter throughout the life span. Given that the hypothalamus impacts many aspects of organismal homeostasis and how individuals respond to their environment, there are many potential functions to test. While it would be best if long-term functional consequences of GABA signaling were able to be interpreted from in vivo experiments where the time and location of manipulation were tightly controlled, that data does not yet exist. We will discuss below the consequences of manipulating GABA signaling globally but for discreet developmentally confined periods using chemicals and consequences of manipulating GABA signaling throughout the life span using genetics. A handful of data sets allow evaluation of the impact of GABA signaling where genetic manipulation was accomplished selectively in GnRH cell populations throughout the life span.

By virtue of GnRH being a hypothalamic peptide that directly regulates pituitary function, it is easy to consider what to examine for GnRH neuron failures. Nevertheless, GnRH neurons can still be difficult to evaluate for functional impact if the endpoint is a reproductive failure. Assuming that GABA manipulations affect all of the GnRH neurons that might be sensitive to GABA effects on movement that might only be 30% of the GnRH neuronal population (even though virtually all GnRH neurons contain a form of GABA receptor). Is 30% enough to impact fertility? One study indicated that fertility in males is preserved with only 12% of the GnRH neuronal population, while in females between 12 and 34% are necessary (Herbison et al. 2008). Treatments that influence GnRH neurons in development also influence reproductive functions in adults. Corresponding with developmental effects noted above, female mice that overexpressed GAD67 in a subset of GnRH neurons displayed altered estrus cyclicity and rates of pregnancy, although it did not disrupt the onset of sexual maturation. GABA may also play a role in normal reproductive function in the adult through other receptors since the examination of $GABA_BR1$ subunit null mice demonstrated again a normal onset of sexual maturation, but abnormal estrus cyclicity and impaired fertility (reviewed in Wierman et al. 2011).

Functionally, exposure to a $GABA_B$ receptor antagonist affected stress-related physiology and behaviors associated with the PVN. Adult mice who were embryonically exposed to the $GABA_B$ receptor antagonist, CGP55845 behaved differently in elevated plus maze, open field, tail suspension, and sucrose preference tests. These same animals displayed altered HPA regulation as indicated by altered post-restraint corticosterone levels and FOS activation in the PVN. Many of these same behavioral changes are present in the $GABA_B$ R1 knockout mice, which displayed increased anxiety-like behavior, reduced depression-like behaviors, and were also hyperactive (reviewed in Tiao and Bettler 2007).

It is also important to highlight the interaction between sex of the animal and susceptibility of the hypothalamus to GABA manipulation. Specifically, in the $GABA_B$ receptor manipulation studies, there are quite dramatic sex-selective effects. In the receptor knockout studies, CRH expression in the PVN was only affected in females, while lateral spread of ERα-containing cells was more predominant in females. With pharmacological manipulation of the $GABA_B$ receptor, cells in ex vivo slices from females preferentially showed altered cell movement speeds, and increased lateral spreads for in vivo cell placement. Finally, females exposed in utero to $GABA_B$ receptor inhibition displayed increased anxiety-like behaviors as adults while littermate males displayed a hyperactivity phenotype.

7.8 Perspectives

GABA is intimately tied to hypothalamic development, whether it be in the richness of GABAergic cells within particular areas (e.g., preoptic area) or nuclei (e.g., arcuate) or in the more striking pattern of cells surrounding particular nuclei (e.g., VMH and PVN). In the PVN and VMH, there is evidence of GABAergic role(s) in affecting cell placements or differentiation. GnRH neurons are different by nature with a unique origin outside the CNS yet also falls into the category of being subject to GABAergic influences on cell movements/migration. Understanding molecular mechanisms by which GABA receptor activation impact these changes has been more challenging, largely due to the heterogeneity of the hypothalamus. While some insight can be gained from the investigation of mechanisms underlying GABA effects in the cortex, these have not all been tested in the hypothalamus. Furthermore, directly demonstrating cause/effect relationships of physiological and behavioral consequences to altered GABA signaling during hypothalamic development is extremely complex and circumstantial evidence has not been conclusive. Even when one can tie a particular cellular phenotype to a particular molecule in a particular location using genetic targeting, it is difficult to tie that cellular phenotype to a particular functional endpoint at a later time point in the animal's life. Although it may be easier to tie a phenotype to a particular time point with chemical targeting, if the drug is given systemically, it is still difficult to prove the functional phenotype is brought about via chemical activity at that particular location of interest.

Despite these limitations, there is sufficient evidence to suspect phenotypes/disorders that are related to hypothalamic function (and other brain regions) in humans subjected to in utero exposure to drugs that manipulate GABA signaling. Exposures to barbiturates, benzodiazepines, and other drugs (e.g., valproate) that impact GABAergic systems can occur when the mother is being treated for epilepsy, anxiety, or depression disorders. This is particularly true for pregnancies prior to the mid-1980s when evidence for potential effects on the fetus began to mount. While these drugs are at times, still used during pregnancy, it is after an informed risk–benefit analysis. Most studies assessing effects in the fetus are limited in time of follow-up and typically do not extend into the post-pubescent period (or even middle age) when subtle alterations in hypothalamic function may manifest. In one rare study that followed the offspring of mothers treated with barbiturates, potential effects were seen in sexual preference and gender identity (Dessens et al. 1999).

What other phenotypes might be seen? For cells in hypothalamic cell groups such as the VMH, PVN, and ARC that are all part of the classic feeding/obesity circuitry, it is easy to consider that as a functional endpoint. The PVN also regulates systemic glucocorticoids and autonomic tone in the cardiovascular system. Thus, developmental GABA signaling in the hypothalamus represents a potential mechanism that might link the highly comorbid conditions of depression, cardiovascular disease (CVD), and obesity (Tobet et al. 2013). Comorbidity may point toward interactions in adult physiology, for which obvious mechanisms exist between obesity and CVD, and mechanisms linking depression to CVD or obesity are being developed (Fig. 7.6). Nonetheless, a common developmental root to all three may lie in common hypothalamic or even GABAergic etiology. Strikingly, Prader–Willi Syndrome (PWS) includes in its spectrum of symptoms, depression, obesity, CVD, and reproductive phenotypes that could be due to effects in the GnRH system, and additional behavioral phenotypes potentially rooted in the VMH. Dysregulation of multiple hypothalamic circuits/functions is well documented in PWS patients. Though the exact molecular roots of PWS are complex and still under investigation, the spectrum of symptoms may well represent the potential phenotypes, albeit severe, to be expected in humans subjected to conditions that impact the function of GABAergic systems in the hypothalamus during early development.

Box 7.1 Multiple Roads Lead to GABA

Historically, decarboxylation of glutamic acid was thought to be the predominant if not only source of physiologically relevant gamma-aminobutyric acid (GABA). Decarboxylase enzymes GAD1 and GAD2 efficiently remove terminal COOH from glutamic acid ($C_5H_{10}NO_4$) to form GABA ($C_4H_9NO_2$) (Roberts and Frankel 1951). More recent work has demonstrated that a number

(continued)

Box 7.1 (continued)

of enzymes have the capacity to convert pyrroline (C_4H_7N) and other four carbon amine containing molecules to GABA (e.g. Laschet et al. 1992). These precursor molecules are typically sourced from amino acid metabolism. Mechanisms that utilize non-glutamic acid precursors are critical for GABA production in several contexts including in glia (Laschet et al. 1992) and in dopamine neurons of the mid-brain (Kim et al. 2015). GABA can also be converted to succinic semialdehyde ($C_4H_6O_3$) and subsequently succinate ($C_4H_6O_4$) for use in the tricarboxylic acid (TCA) cycle (e.g. Bessman et al. 1953). The TCA cycle generates electromotive force in the mitochondria for ATP production. Precursors of glutamic acid include glutamine ($C_5H_{10}N_2O_3$) and α-ketoglutarate ($C_5H_6O_5$) (Peng et al. 1991), another TCA cycle intermediate. Thus, GABA production and the myriad important functions of GABA can be intimately linked with two ubiquitous metabolic processes, the TCA cycle and amino acid metabolism (Fig. 7.7).

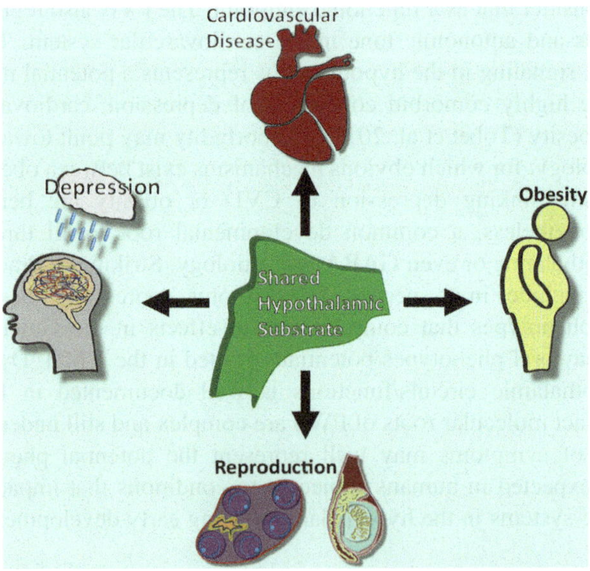

Fig. 7.6 Schematic diagram illustrates potential comorbid consequences of altered GABAergic signaling. Hypothalamic dysfunction may provide a shared anatomical substrate for clinical comorbidity. There are high rates of comorbidity among depression, cardiovascular disease, obesity, and reproductive dysfunction. Altered hypothalamic development through genetic or environmental insult could predispose individuals to one or more of the above conditions. Once present, pathophysiology in one system is likely to disrupt physiology of the other systems and cause additional disease. Brain circuits that regulate mood, cardiovascular function, metabolism, and sexual function are all connected within the hypothalamus

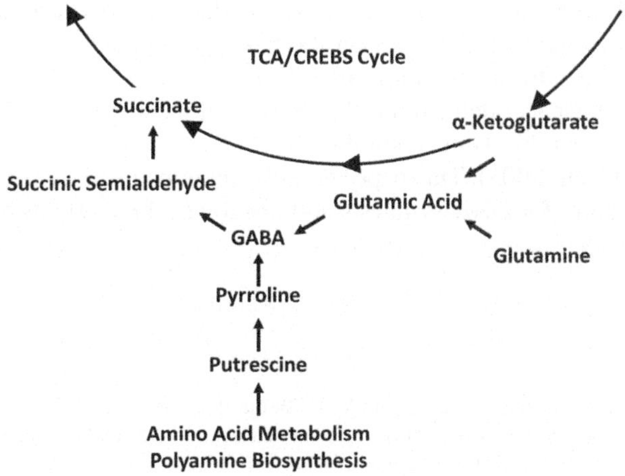

Fig. 7.7 The "GABA Shunt" uses α-Ketoglutarate, an intermediate of the TCA cycle, to produce glutamic acid, which is decarboxylated by GAD enzymes to GABA. The 4-carbon chain of GABA can re-enter the TCA cycle following conversion to succinic semialdehyde and ultimately succinate. GABA is also produced in putrescine degradation or again from glutamic acid but derived from glutamine. Many of these reactions are also reversible (e.g. glutamic acid can be converted to α-ketoglutarate)

Acknowledgments We would like to thank Dr. Kristy McClellan for helpful discussion, Mr. Luke Schwerdtfeger for expert assistance in preparing figures, and Dr. Deborah Kurrasch for providing an original digital image for use in Fig. 7.4. MS was supported by 1K01AG056848.

Key References

Altman and Bayer (1986). This is a classic description of the birth of cells in the diencephalon.

Frahm et al. (2012). First paper to indicate a role for regulated development of hypothalamic vasculature.

Herbison et al. (2008). This paper is notable for being the first to directly address the redundancy in the GnRH population needed for fertility.

Miller and Nowakowski (1988). This paper introduced the BrdU method to the neuroscience community and kicked off a plethora of studies of cell proliferation in the nervous system due to the relative ease compared to the use of tritiated thymidine.

Okamura et al. (1990). This paper was the first to note the special relationship between cell bodies for GABA synthesis based on the localization of glutamic acid decarboxylases and how they lie outside of major hypothalamic cell groups.

Puelles and Rubenstein (2015). This paper updates history and approaches to parcellation of the hypothalamus on the basis of large anatomical assumptions to smaller units informed by molecular data.

Rakic (1972). This paper defined the classic view of radial glia providing migratory guidance for neuroblasts in cerebral cortex.

Salic and Mitchison (2008). This paper provides an impetus to switch from BrdU to EdU as a marker for DNA synthesis and thereby (indirectly) proliferation.

References

Acampora D, Postiglione MP, Avantaggiato V, Di Bonito M, Vaccarino FM, Michaud J, Simeone A (1999) Progressive impairment of developing neuroendocrine cell lineages in the hypothalamus of mice lacking the Orthopedia gene. Genes Dev 13(21):2787–2800

Altman J, Bayer SA (1986) The development of the rat hypothalamus. Advances in anatomy, embryology and cell biology. Springer, Berlin. Monograph

Aujla PK, Bora A, Monahan P, Sweedler JV, Raetzman LT (2011) The Notch effector gene Hes1 regulates migration of hypothalamic neurons, neuropeptide content and axon targeting to the pituitary. Dev Biol 353(1):61–71

Behar TN, Schaffner AE, Scott CA, O'Connell C, Barker JL (1998) Differential response of cortical plate and ventricular zone cells to GABA as a migration stimulus. J Neurosci 18(16):6378–6387

Behar TN, Schaffner AE, Scott CA, Greene CL, Barker JL (2000) GABA receptor antagonists modulate postmitotic cell migration in slice cultures of embryonic rat cortex. Cereb Cortex 10(9):899–909

Bessman SP, Rosen J, Layne EC (1953) Gamma-aminobutyric acid-glutamic acid transamination in brain. J Biol Chem 201(1):385–391

Bless EP, Westaway WA, Schwarting GA, Tobet SA (2000) Effects of gamma-aminobutyric acid (A) receptor manipulation on migrating gonadotropin-releasing hormone neurons through the entire migratory route in vivo and in vitro. Endocrinology 141(3):1254–1262

Bless EP, Walker HJ, Yu KW, Knoll JG, Moenter SM, Schwarting GA, Tobet SA (2005) Live view of gonadotropin-releasing hormone containing neuron migration. Endocrinology 146(1):463–468

Casoni F, Hutchins BI, Donohue D, Fornaro M, Condie BG, Wray S (2012) SDF and GABA interact to regulate axophilic migration of GnRH neurons. J Cell Sci 125(Pt 21):5015–5025

Davis AM, Henion TR, Tobet SA (2002) Gamma-aminobutyric acid B receptors and the development of the ventromedial nucleus of the hypothalamus. J Comp Neurol 449(3):270–280

Davis AM, Seney ML, Stallings NR, Zhao L, Parker KL, Tobet SA (2004) Loss of steroidogenic factor 1 alters cellular topography in the mouse ventromedial nucleus of the hypothalamus. J Neurobiol 60(4):424–436

Dellovade TL, Young M, Ross EP, Henderson R, Caron K, Parker K, Tobet SA (2000) Disruption of the gene encoding SF-1 alters the distribution of hypothalamic neuronal phenotypes. J Comp Neurol 423(4):579–589

Dellovade TL, Davis AM, Ferguson C, Sieghart W, Homanics GE, Tobet SA (2001) GABA influences the development of the ventromedial nucleus of the hypothalamus. J Neurobiol 49(4):264–276

Dessens AB, Cohen-Kettenis PT, Mellenbergh GJ, vd Poll N, Koppe JG, Boer K (1999) Prenatal exposure to anticonvulsants and psychosexual development. Arch Sex Behav 28(1):31–44

Frahm KA, Schow MJ, Tobet SA (2012) The vasculature within the paraventricular nucleus of the hypothalamus in mice varies as a function of development, subnuclear location, and GABA signaling. Horm Metab Res 44(8):619–624

Frangaj A, Fan QR (2018) Structural biology of GABAB receptor. Neuropharmacology 136(Pt A):68–79. https://doi.org/10.1016/j.neuropharm.2017.10.011

Fueshko S, Wray S (1994) LHRH cells migrate on peripherin fibers in embryonic olfactory explant cultures: an in vitro model for neurophilic neuronal migration. Dev Biol 166(1):331–348

Herbison AE, Porteous R, Pape JR, Mora JM, Hurst PR (2008) Gonadotropin-releasing hormone neuron requirements for puberty, ovulation, and fertility. Endocrinology 149(2):597–604

Jursky F, Nelson N (1996) Developmental expression of GABA transporters GAT1 and GAT4 suggests involvement in brain maturation. J Neurochem 67(2):857–867

Kim K, Ganesan S, Luo SX, Wu Y, Park E, Huang EJ, Chen L, Ding JB (2015) Aldehyde dehydrogenase 1a1 mediates a GABA synthesis pathway in midbrain dopaminergic neurons. Science 350(6256):102–106

Kusano K, Fueshko S, Gainer H, Wray S (1995) Electrical and synaptic properties of embryonic luteinizing hormone-releasing hormone neurons in explant cultures. Proc Natl Acad Sci U S A 92(9):3918–3922

Laschet J, Grisar T, Bureau M, Guillaume D (1992) Characteristics of putrescine uptake and subsequent GABA formation in primary cultured astrocytes from normal C57BL/6J and epileptic DBA/2J mouse brain cortices. Neuroscience 48:151–157

Le TN, Zhou QP, Cobos I, Zhang S, Zagozewski J, Japoni S, Vriend J, Parkinson T, Du G, Rubenstein JL, Eisenstat DD (2017) GABAergic interneuron differentiation in the basal forebrain is mediated through direct regulation of glutamic acid decarboxylase isoforms by Dlx homeobox transcription factors. J Neurosci 37(36):8816–8829

Lee JM, Tiong J, Maddox DM, Condie BG, Wray S (2008) Temporal migration of gonadotrophin-releasing hormone-1 neurones is modified in GAD67 knockout mice. J Neuroendocrinol 20(1):93–103

Li S, Kumar TP, Joshee S, Kirschstein T, Subburaju S, Khalili JS, Kloepper J, Du C, Elkhal A, Szabó G, Jain RK, Köhling R, Vasudevan A (2018) Endothelial cell-derived GABA signaling modulates neuronal migration and postnatal behavior. Cell Res 28(2):221–248

Lu F, Kar D, Gruenig N, Zhang ZW, Cousins N, Rodgers HM, Swindell EC, Jamrich M, Schuurmans C, Mathers PH, Kurrasch DM (2013) Rax is a selector gene for mediobasal hypothalamic cell types. J Neurosci 33(1):259–272

Luján R, Shigemoto R, López-Bendito G (2005) Glutamate and GABA receptor signalling in the developing brain. Neuroscience 130(3):567–580

McClellan KM, Calver AR, Tobet SA (2008) GABAB receptors role in cell migration and positioning within the ventromedial nucleus of the hypothalamus. Neuroscience 151(4):1119–1131

McClellan KM, Stratton MS, Tobet SA (2010) Roles for gamma-aminobutyric acid in the development of the paraventricular nucleus of the hypothalamus. J Comp Neurol 518(14):2710–2728

Miller MW, Nowakowski RS (1988) Use of bromodeoxyuridine-immunohistochemistry to examine the proliferation, migration and time of origin of cells in the central nervous system. Brain Res 457(1):44–52

Okamura H, Abitbol M, Julien JF, Dumas S, Bérod A, Geffard M, Kitahama K, Bobillier P, Mallet J, Wiklund L (1990) Neurons containing messenger RNA encoding glutamate decarboxylase in rat hypothalamus demonstrated by in situ hybridization, with special emphasis on cell groups in medial preoptic area, anterior hypothalamic area and dorsomedial hypothalamic nucleus. Neuroscience 39(3):675–699

Peng L, Schousboe A, Hertz L (1991) Utilization of alpha-ketoglutarate as a precursor for transmitter glutamate in cultured cerebellar granule cells. Neurochem Res 16:29

Puelles L, Rubenstein JL (2015) A new scenario of hypothalamic organization: rationale of new hypotheses introduced in the updated prosomeric model. Front Neuroanat 9:27. https://doi.org/10.3389/fnana.2015.00027

Rakic P (1972) Mode of cell migration to the superficial layers of fetal monkey neocortex. J Comp Neurol 145(1):61–83

Roberts E, Frankel S (1951) Glutamic acid decarboxylase in brain. J Biol Chem 188:789–795

Salic A, Mitchison TJ (2008) A chemical method for fast and sensitive detection of DNA synthesis in vivo. Proc Natl Acad Sci U S A 105(7):2415–2420

Schwanzel-Fukuda M, Pfaff DW (1989) Origin of luteinizing hormone-releasing hormone neurons. Nature 338(6211):161–164

Schwenk J, Pérez-Garci E, Schneider A, Kollewe A, Gauthier-Kemper A, Fritzius T, Raveh A, Dinamarca MC, Hanuschkin A, Bildl W, Klingauf J, Gassmann M, Schulte U, Bettler B, Fakler B (2016) Modular composition and dynamics of native GABAB receptors identified by high-resolution proteomics. Nat Neurosci 19(2):233–242

Simonian SX, Skynner MJ, Sieghart W, Essrich C, Luscher B, Herbison AE (2000) Role of the GABA(A) receptor gamma2 subunit in the development of gonadotropin-releasing hormone neurons in vivo. Eur J Neurosci 12(10):3488–3496

Stratton MS, Searcy BT, Tobet SA (2011) GABA regulates corticotropin releasing hormone levels in the paraventricular nucleus of the hypothalamus in newborn mice. Physiol Behav 104(2):327–333

Stratton MS, Staros M, Budefeld T, Searcy BT, Nash C, Eitel C, Carbone D, Handa RJ, Majdic G, Tobet SA (2014) Embryonic GABA(B) receptor blockade alters cell migration, adult hypothalamic structure, and anxiety- and depression-like behaviors sex specifically in mice. PLoS One 9(8):e106015

Suter KJ, Wuarin JP, Smith BN, Dudek FE, Moenter SM (2000) Whole-cell recordings from preoptic/hypothalamic slices reveal burst firing in gonadotropin-releasing hormone neurons identified with green fluorescent protein in transgenic mice. Endocrinology 141(10):3731–3736

Temple JL, Wray S (2005) Developmental changes in GABA receptor subunit composition within the gonadotrophin-releasing hormone-1 neuronal system. J Neuroendocrinol 17(9):591–599

Tiao J, Bettler B (2007) Characteristics of GABAB receptor mutant mice. In: The GABA receptors, 3rd edn. Springer, Berlin

Tobet SA, Paredes RG, Chickering TW, Baum MJ (1995) Telencephalic and diencephalic origin of radial glial processes in the developing preoptic area/anterior hypothalamus. J Neurobiol 26(1):75–86

Tobet SA, Chickering TW, King JC, Stopa EG, Kim K, Kuo-Leblank V, Schwarting GA (1996a) Expression of gamma-aminobutyric acid and gonadotropin-releasing hormone during neuronal migration through the olfactory system. Endocrinology 137(12):5415–5420

Tobet SA, Hanna IK, Schwarting GA (1996b) Migration of neurons containing gonadotropin releasing hormone (GnRH) in slices from embryonic nasal compartment and forebrain. Brain Res Dev Brain Res 97(2):287–292

Tobet SA, Henderson RG, Whiting PJ, Sieghart W (1999) Special relationship of gamma-aminobutyric acid to the ventromedial nucleus of the hypothalamus during embryonic development. J Comp Neurol 405(1):88–98

Tobet SA, Bless EP, Schwarting GA (2001) Developmental aspect of the gonadotropin-releasing hormone system. Mol Cell Endocrinol 185(1–2):173–184

Tobet SA, Walker HJ, Seney ML, Yu KW (2003) Viewing cell movements in the developing neuroendocrine brain. Integr Comp Biol 43(6):794–801

Tobet SA, Handa RJ, Goldstein JM (2013) Sex-dependent pathophysiology as predictors of comorbidity of major depressive disorder and cardiovascular disease. Pflugers Arch 465(5):585–594

Vastagh C, Schwirtlich M, Kwakowsky A, Erdélyi F, Margolis FL, Yanagawa Y, Katarova Z, Szabó G (2015) The spatiotemporal segregation of GAD forms defines distinct GABA signaling functions in the developing mouse olfactory system and provides novel insights into the origin and migration of GnRH neurons. Dev Neurobiol 75(3):249–270

Wierman ME, Kiseljak-Vassiliades K, Tobet S (2011) Gonadotropin-releasing hormone (GnRH) neuron migration: initiation, maintenance and cessation as critical steps to ensure normal reproductive function. Front Neuroendocrinol 32(1):43–52

Wray S, Nieburgs A, Elkabes S (1989) Spatiotemporal cell expression of luteinizing hormone-releasing hormone in the prenatal mouse: evidence for an embryonic origin in the olfactory placode. Brain Res Dev Brain Res 46(2):309–318

Wu C, Sun D (2015) GABA receptors in brain development, function, and injury. Metab Brain Dis 30(2):367–379

Yoon BE, Lee CJ (2014) GABA as a rising gliotransmitter. Front Neural Circuits 8:141

Epigenetic and Transcriptional Regulation of the Reproductive Hypothalamus

8

Carlos Francisco Aylwin and Alejandro Lomniczi

Abstract

During the last few years, new genes required for the advent of puberty have been identified thanks to the use of new deep sequencing associated technologies. Evidence has emerged suggesting that these genes, organized into functional networks, are regulated by a mechanism of epigenetic control. While the Polycomb group (PcG) of transcriptional repressors is in charge of the repressive arm of this mechanism, the Trithorax Group (TrxG) of epigenetic activators counteract the PcG-mediated silencing during the infantile to juvenile stage where increased *Kiss1* expression is needed for activation of gonadotropin-releasing hormone secretion from the hypothalamus. Here we will discuss the recent advances and tools used for large-scale genomic and epigenomic studies of the developing hypothalamus.

Keywords

Epigenetic regulators · Female puberty gene networks · GnRH neurons · KNDy neurons · Reproductive development · Transcriptional activators · Transcriptional repressors

8.1 Introduction

Puberty begins when pulsatile gonadotropin-releasing hormone (GnRH) release from neurosecretory neurons of the hypothalamus increases in a diurnal manner for a prolonged period of time, inducing the pituitary gland to increase its pulsatile release of luteinizing hormone (LH). Changes in *trans*-synaptic and glial inputs to

C. F. Aylwin · A. Lomniczi (✉)
Division of Neuroscience, Oregon National Primate Research Center, Oregon Health and Science University, Beaverton, OR, USA
e-mail: lomniczi@ohsu.edu

S. Wray, S. Blackshaw (eds.), *Developmental Neuroendocrinology*, Masterclass in Neuroendocrinology 9, https://doi.org/10.1007/978-3-030-40002-6_8

GnRH neurons are responsible for the increased GnRH secretion. These changes involve a reduction in inhibitory *trans*-synaptic inputs combined with increased *trans*-synaptic and glial excitatory inputs to the GnRH neuronal network (Ojeda and Skinner 2006).

The time of puberty has a strong genetic component, but very recently epigenetic mechanisms have been implicated as part of the regulatory control of the developmental process by which GnRH release is kept in check during infancy, and also during the increase in GnRH secretion that brings about the pubertal process. Agreeing with this emergent concept, a central target of epigenetic control is the transcriptional machinery of neurons in charge of stimulating GnRH release. In this chapter, we review key findings as well as research methodologies used to support the existence of an epigenetically controlled gene regulatory network residing at the core of the neuroendocrine process that controls pubertal timing.

8.2 Neuroendocrine Regulatory Mechanisms Involved in the Hormonal Changes of Puberty

An increase in pulsatile LH release is the first endocrine indication of the initiation of puberty (Boyar et al. 1972). In peripubertal girls, before any somatic manifestation of pubertal development is evident, the amplitude of LH pulses detected in the systemic circulation increases shortly after the initiation of sleep in the absence of changes in circulating ovarian steroid levels (Boyar et al. 1972). Increased LH pulsatility is also observed in rodents and occurs in the afternoon hours at the end of the juvenile period, at postnatal days (PND) 28–30 (Urbanski and Ojeda 1985, 1987). As a consequence of this, gonadal output of sex steroids rises, leading to the development of secondary sexual characteristics. In females, another event needs to occur for puberty to be accomplished; the first preovulatory surge of gonadotropins, which occurs upon maturation of the central mechanism of estrogen positive feedback. It is clear that the pubertal alteration in LH output is determined by an increase in pulsatile GnRH release from the hypothalamus [reviewed in Ojeda and Terasawa (2002)]. GnRH neurons are able to synthesize and release GnRH long before puberty, thus the ultimate responsibility for the initiation of puberty rests on the neuronal and glial networks that control GnRH neuron excitability (Ojeda and Terasawa 2002).

It is considered that the primary *trans*-synaptic mechanism underlying pulsatile GnRH release to be the synchronized action of a subclass of arcuate nucleus (ARC) neurons termed KNDy neurons (Lehman et al. 2010; Navarro et al. 2011) because they produce *K*isspeptin, *N*KB (neurokinin B), and *D*ynorphin (Navarro et al. 2011; Wakabayashi et al. 2010). NKB released from KNDy neurons acts on other KNDy neurons via tachykinin receptors to stimulate kisspeptin release (Navarro et al. 2011; Wakabayashi et al. 2010). NKB and kisspeptin are released in a rhythmic fashion that appears to be determined by an inhibitory, periodic effect of dynorphin on NKB release (Meijs-Roelofs et al. 1985; Navarro et al. 2011). Direct evidence for a role of KNDy neurons in the genesis of pulsatile LH release was recently provided (Beale et al. 2014). A second population of kisspeptin neurons, which do not contribute to the control of pulsatile GnRH release, is located in the anteroventral periventricular nucleus (AVPV) of rodents [reviewed in Pinilla et al. (2012)]. These neurons are involved in the preovulatory surge of gonadotropins (Pinilla et al. 2012) and they do not seem to be involved in the initiation of female puberty, because the gonadotropin surge occurs only after the pubertal process is well underway.

During the prepubertal period (or infancy), the secretory activity of GnRH neurons is under predominant *trans*-synaptic inhibitory control. At puberty, this inhibition is lifted while there is a concomitant increase of excitatory inputs to the GnRH network (Ojeda 1991; Ojeda and Terasawa 2002). This concept is strongly supported by the evidence that activation of kisspeptin neurons, which provide a substantial portion of the stimulatory inputs controlling GnRH neurons [reviewed in Pinilla et al. (2012)], is essential for puberty to occur (de Roux et al. 2003; Seminara et al. 2003).

Recent evidence suggests that this *trans*-synaptic regulatory mechanism is controlled by a molecular switch that regulates the timing of puberty by epigenetically coordinating the transcriptional activity of neurons implicated in stimulating GnRH release (Lomniczi et al. 2013a, 2015; Toro et al. 2018).

8.2.1 Gene Networks Involved in the Neuroendocrine Control of Puberty

With the advent of new high-throughput genomic technologies, coupled with a novel and developing systems biology tools, great advances have been made in determining new pathways involved in the neuroendocrine control of puberty. This novel approach is based on the concept that a diversity of genes that affect the time of puberty are functionally organized into networks capable of generating a coordinated output of biological signals. About 10 years ago, using expression data from female rats and monkeys, we described the first gene network operating in the peripubertal hypothalamus. This gene network is organized into a structure containing "central" nodes that reside at the core of the network, and a number of peripherally located subordinate genes that are transcriptionally controlled by the central nodes. The central nodes of this network were earlier identified as being implicated in tumor suppression/tumor formation (Roth et al. 2007), now referred to as Tumor-Related Genes (TRGs). The TRG network has five central nodes or "hubs": *CDP/CUTL1/CUX1, MAF, p53, YY1, and USF2*. These hubs are strongly connected to each other and also to additional upper-echelon genes (*OCT2, TTF1, and EAP1*) previously found to be involved in the transcriptional regulation of the pubertal process (Heger et al. 2007; Mastronardi et al. 2006; Ojeda et al. 1999), and TRGs themselves (Lomniczi et al. 2013b).

We (Heger et al. 2007; Lomniczi et al. 2012; Mueller et al. 2011, 2012) and others (Cukier et al. 2013) have experimentally verified the role of *EAP1* as a central transcriptional regulator of subordinate genes like *KISS1* and *GnRH.* Interestingly, *KISS1* was formerly known as a Metastasis Suppressor or Metastin (Steeg et al. 2003), while *EAP1* was described as part of a transcriptional repressive complex that regulates apoptosis in breast cancer (Yeung et al. 2011). Most recently, we found a series of transcriptional regulators with epigenetic capabilities to be involved in the transcriptional control of the *Kiss1* gene. We validated the function of EED, a member of the Polycomb family of transcriptional repressors (Lomniczi et al. 2013a) as well as MLL1 and MLL3 (Toro et al. 2018), two members of the Trithorax complex of transcriptional activators. These two families, also involved in cancer biology, are in charge of epigenetically controlling the expression pattern of *Kiss1* and *Tac3* during the infantile–pubertal transition (Lomniczi et al. 2013a; Toro et al. 2018). Finally, *SynCAM1* (also known as a tumor suppressor of lung cancer, *TSLC1*) plays a central role in the developmental control of glia to GnRH neuron adhesive communications. This adhesion molecule is part of the mechanism by which astrocytic erbB4 receptors facilitate GnRH release and modulate female reproductive function (Sandau et al. 2011a, b).

In addition to the TRG network, there is a posttranscriptional repressor structure that may contribute to pubertal development. *LIN28b* encodes an RNA-binding protein that prevents the maturation of let7 miRNAs (Hagan et al. 2009; Heo et al. 2009; Lehrbach et al. 2009), a family of microRNAs with tumor suppressor activity (Chang et al. 2009). Genome-wide association studies found a single nucleotide polymorphism near the human *LIN28B* gene associated with earlier puberty and shorter stature in girls (Ong et al. 2009; Perry et al. 2009; Sulem et al. 2009). In the hypothalamus of male and female rats, the expression of *Lin28b* decreases during pubertal development, while two well-known LIN28b targets, *let7a* and *let7b,* increase (Sangiao-Alvarellos et al. 2013). As the repressive effect of LIN28b on *let7a* and *let7b* maturation diminishes, degradation of *let7* target genes increases due to increased availability of these miRNAs. It is predicted that the Lin28b-let7 posttranscriptional regulatory pathway should predominantly control inhibitory genes in the pubertal process. Still today, there are no formal targets for this mode of action in the hypothalamic control of pubertal development.

8.3 Genome-Wide Technologies

With the advent of relatively inexpensive whole-genome sequencing technologies (see Box 8.1), we can today interrogate the mammalian hypothalamus for RNA expression changes using RNA-seq (Wang et al. 2009), Global Run-On sequencing (GRO-seq) (Garcia-Martinez et al. 2004), or Precision nuclear Run-On sequencing (PRO-seq) (Kwak et al. 2013). Using the same sequencing technologies, we can study gene regulation by Chromatin Immunoprecipitation (CHIP) followed by deep sequencing (CHIP-seq), Chromatin conformation analysis like 3C, 4C, Hi-C, and Chia-Pet (Barutcu et al. 2016) or chromatin accessibility assays to determine nucleosome-free areas of the genome like Assay for Transposase-Accessible Chromatin sequencing (ATAC-seq) (Buenrostro et al. 2015), Formaldehyde-Assisted Isolation of Regulatory Elements sequencing (FAIRE-seq) (Giresi et al. 2007) or DNase I hypersensitive sites sequencing (DNase-seq) (Crawford et al. 2006). In this section, we highlight the methods used to study whole-genome changes during pubertal development.

Box 8.1 Next-Generation Sequencing Technologies

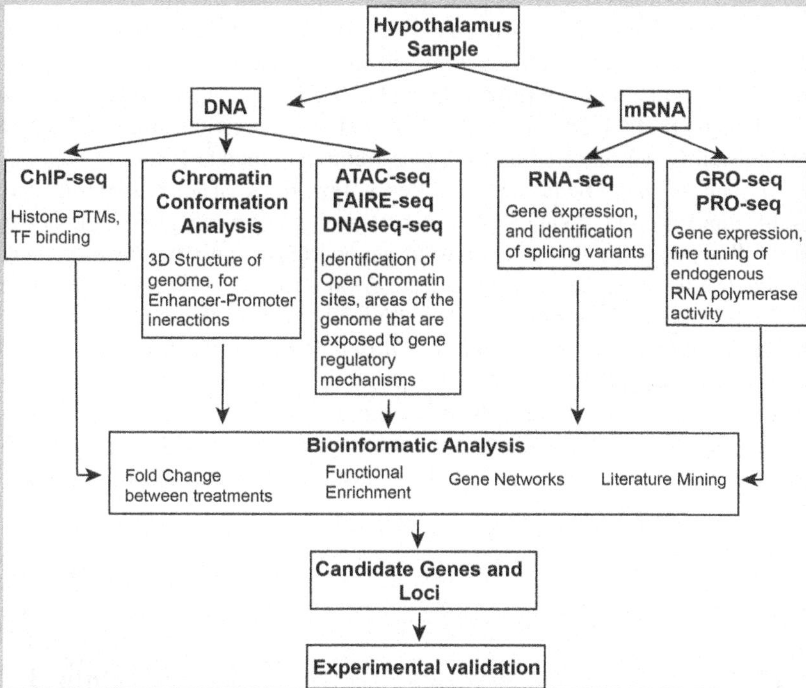

Next-generation sequencing (NGS) allows for massively parallel sequencing reactions that enable us to analyze up to hundreds of millions of sequencing products at the same time, supporting a broad range of applications such as genome (DNA)- and transcriptome (mRNA)-based methods. Methods differ in the sample isolation and library preparation steps and can be processed by Illumina platforms. DNA technologies can be used to detect direct genome-wide protein to DNA interactions (CHIP-seq), 3D DNA conformation to identify promoter/enhancer interactions (3C, 4C, etc.) or regions of open chromatin that allows for transcription factor binding (ATAC-seq, FAIRE-seq, etc.). RNA technologies are mainly used to identify new transcripts, splicing variants or to quantify mRNA synthesis globally (RNA-seq) or nascent transcripts associated with endogenous RNA polymerase activity (GRO and PRO-seq). Data analysis requires bioinformatics tools in order to filter output data by passing quality controls, whole-genome alignment, quantitation, and statistical analysis. Experimental validation by qPCR is required for accurate results.

8.3.1 Phenotypic Characterization and Tissue Dissection

The first step into analyzing RNA expression or chromatin modifications throughout critical developmental periods is to characterize the phenotype of interest, link it to a specific day of pre- or postnatal development and to a tissue or brain region. In the rat model, the Hypothalamus–Pituitary–Gonadal (HPG) axis remains quiescent until postnatal day 28 (PND28) when the first endocrine manifestations of increased GnRH pulsatility are detected as LH pulse frequency increases during the evening hours (Urbanski and Ojeda 1985). After a few days of increased diurnal LH pulsatility, and around PND32–34, the first preovulatory surge of gonadotropins occurs, followed by the first ovulation. In general terms, pubertal development can be divided into three stages: juvenile period (Juv = PND21), late juvenile (LJ = PND28), and late proestrus (LP = PND32–34) (Fig. 8.1).

In the rhesus macaque, we define four different stages of reproductive development: neonatal, infantile, juvenile, and pubertal. As in humans, during the first months to a year of life, high levels of LH pulsatility are detected in a period called "mini-puberty" This is followed by the infantile period in which there is a marked quiescence of the gonadotropin axis. During the early pubertal period, there is a

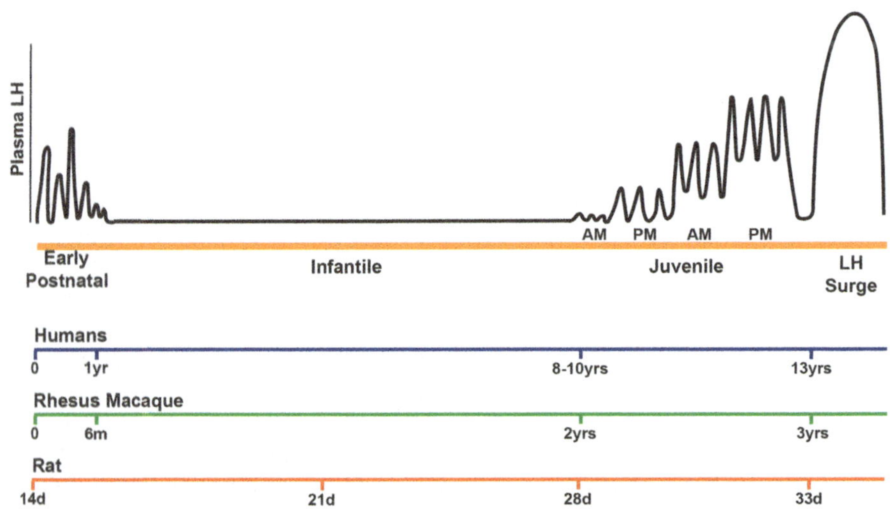

Fig. 8.1 Plasma LH concentrations during postnatal development in humans, rhesus macaques, and rats. In humans and monkeys, during the early postnatal period, high LH pulsatility is detected in a period called "mini-puberty." This is followed by the infantile period where a "repressed" central drive generates low LH levels. During the juvenile stage, 8–9 years of age in humans and approximately 2 years of age for the rhesus macaques, there is a reawakening of the GnRH network evidenced by increased LH pulsatility during the night hours. In the rodent, this same phenomenon is first detected in the evening of postnatal days 28–29. As puberty approaches in females, ovarian estrogen secretion increases, reaching a sustained magnitude capable of activating the positive feedback mechanism that results in the preovulatory surge of gonadotropins. Days (d), months (m), and years (yrs)

reawakening of the GnRH pulse generator and therefore increased LH pulsatility generating a clear a.m./p.m. difference in LH pulse frequency. Finally, during puberty, the rhesus macaque female is capable of sustaining ovulatory cycles with the presence of a dominant ovulatory follicle and a sustained luteal phase followed by menses, as is typical of primates (Fig. 8.1).

The rodent model is ideal to detect genome changes during peripubertal development because of the anatomical nature of the GnRH network. The AVPV Kiss1–GnRH neurons involved in the preovulatory surge of gonadotropins are distinctively separated from the ARC KNDy neurons to GnRH terminals implicated in GnRH pulsatility. ARC and AVPV tissue dissection can be achieved simply by rapid microdissection using surgical Iris Scissors under binocular magnifying glasses, or by micro-punching the area of interest from freshly prepared 400 μm vibratome sections. The advantage of direct microdissection is that the tissue is more rapidly recovered and flash frozen in liquid nitrogen, while sectioning and micro-punching a more accurate region is recovered but with longer processing times (Fig. 8.2a, b).

8.3.2 RNA-Seq: mRNA Isolation and Library Preparation

A typical medial basal hypothalamic fragment weighs between 3 and 5 mg and provides in average 5 μg of total RNA. Total RNA can be extracted using TRIzol reagent (Invitrogen cat# 15596026) or any of the ion exchange-based methods like the RNeasy mini kit (Qiagen cat# 74104). High-quality total RNA samples must have an RNA Integrity Number (RIN) greater or equal to 8 and a ratio 28S/18S rRNA of 2:1 (Fig. 8.3a). For mRNA library preparation, total RNA must be depleted

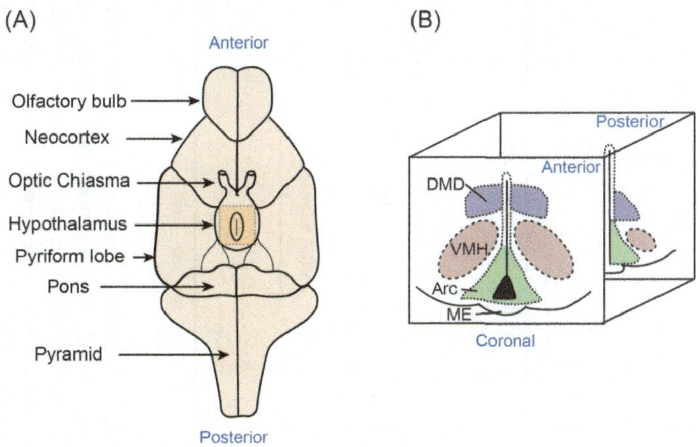

Fig. 8.2 Schematic representation of the hypothalamic region of the rat. (**a**) Ventral view of the rat brain. Major regions (olfactory bulb, neocortex, optic chiasma, hypothalamus) are labeled. The hypothalamic region, shaded in orange, is used for micro-punching or coronal slicing. (**b**) Coronal view of the hypothalamic region depicting the arcuate nucleus (Arc), median eminence (ME), ventromedial hypothalamus (VMH), and dorsomedial hypothalamus (DMD)

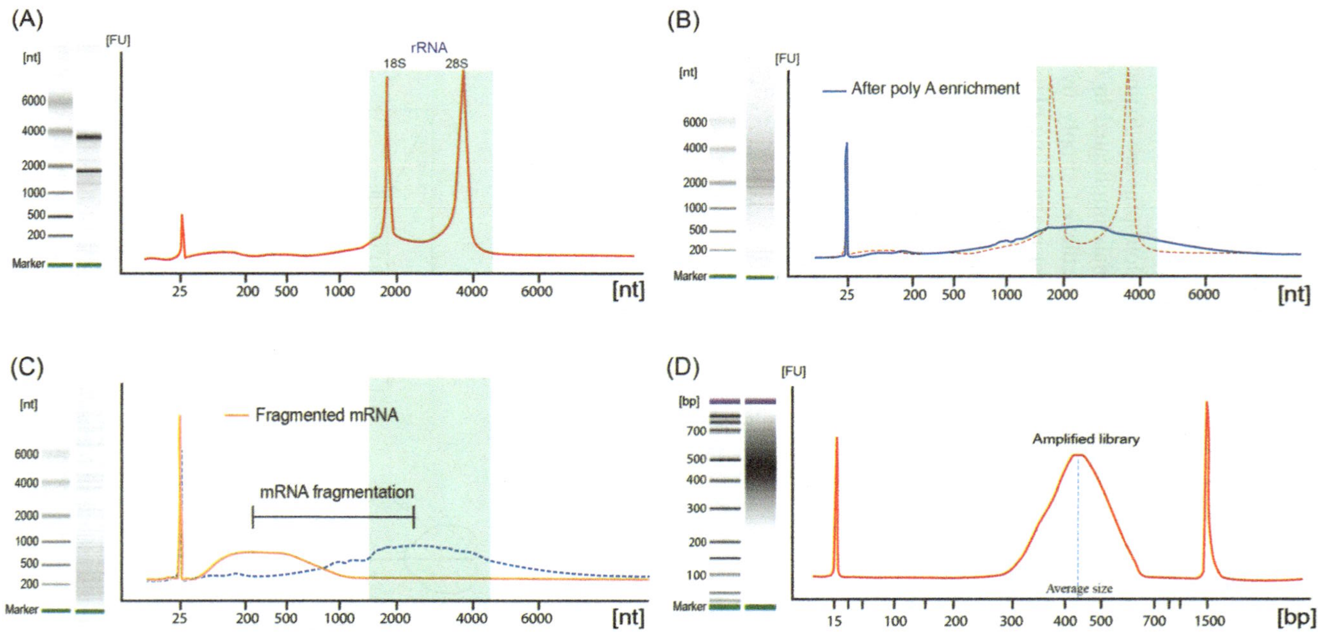

Fig. 8.3 RNA-Seq library preparation process from rat MBH tissue. (**a**) Agilent Bioanalyzer electrophoretic trace of a high-quality total RNA sample. Shaded green area shows the 18S and 28S rRNA bands. (**b**) Electrophoretic trace after rRNA depletion using poly-A magnetic beads. The green shaded area highlights the absence of 18S and 28S signals depicted by dotted lines. The blue line corresponds to the purified mRNA 1000–5000 nt. (**c**) Electrophoretic trace after chemical fragmentation of mRNA. Fragmented mRNA can be observed by a shift of the enriched area toward the lower molecular weight, 100–800 nt (orange line) as opposed to intact mRNA (dotted blue line). (**d**) Electrophoretic trace of a finalized sequencing library after amplification using primers P5 and P7 followed by magnetic beads size selection to discard both adapter and primer dimers. An enrichment in 400–600 bp long amplicons is detected

of ribosomal RNA using either active ribosome binding technology like Ribo-Zero rRNA removal kit (Illumina cat# MRZH116) or by further isolating the poly-A mRNA from the total RNA fraction by means of poly-(A) paramagnetic beads (Bioo Scientific Cat# NOVA-51979). These magnetic beads are coupled to oligo (dT) probes capable of isolating poly-A mRNA starting from 10 ng to 10 µg of total RNA (See Box 8.2). Poly-A beads are more price effective than any of the ribosomal depletion methods, with the caveat of generating libraries that only represent poly-A containing mRNA. The first step is to test the best condition for mRNA isolation. One efficient way to do this is to use equal volumes of poly-(A) beads with increasing amounts of total RNA. Efficient removal of ribosomal RNA can be identified by the absence of 18S and 28S fractions commonly seen in total RNA samples in Bioanalyzer RNA nano- or pico chip runs (Agilent cat# 5067-1511 and 5067-1514) (Fig. 8.3b).

Box 8.2 RNA-Seq Library Preparation Pipeline

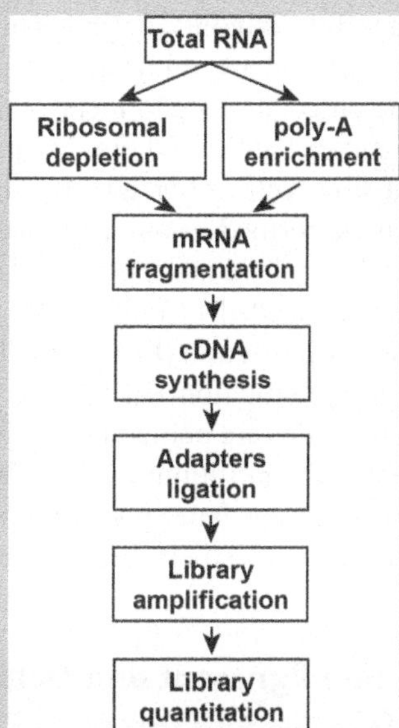

RNA transcripts can be isolated from total RNA by two different methods: (a) Ribosomal depletion, a method where all ribosomal RNAs are captured and discarded from the sample, allows for isolation of both mRNA and lncRNA and (b) Poly-A-enrichment, specifically captures and purifies poly-A-containing mRNAs using poly-T-containing magnetic beads. After isolation, samples are subjected to fragmentation and cDNA synthesis. Adapter ligation allows for the attachment of Illumina compatible adapter sequences and barcodes for downstream sequencing and sample identification. Before sequencing, library quantification by qPCR is necessary for accuracy.

Several Illumina compatible kits for RNA-seq library preparation are available on the market. In general, the basic steps to prepare a next-generation sequencing (NGS) library include (a) fragmentation of the RNA sample; (b) converting RNA into double-stranded DNA; (c) attaching oligonucleotide adapters to each end of the target fragments; and (d) quantitating the library by qPCR. The first step of this process is RNA fragmentation and can be achieved by (a) physical fragmentation like acoustic or hydrodynamic shearing; (b) enzymatic shearing like RNase III treatment; and (c) chemical fragmentation by incubating the mRNA sample at 95 °C with divalent cations. The optimal size fragmentation for Illumina compatible libraries ranges from 100 to 400 nt, this can also be detected by Bioanalyzer RNA nano- or pico chip (Fig. 8.3c). After fragmentation, first strand synthesis is performed with random primers, if it is desired to retain the strand orientation of the transcripts, dUTP is added into the second strand synthesis, which prevents further amplification with the PCR polymerase that is unable to recognize uracil in the template strand (NEXTflex Rapid Directional RNA-seq kit, Bioo Scientific cat# NOVA-5138-07). Bead cleanup using Agencourt AMPure XP Magnetic Beads is performed in order to remove free oligonucleotides and dNTPs from the reaction mix. This is followed by 3′ end adenylation and adapter ligation. The Illumina compatible adapters carry the barcodes used to identify the sample after sequencing as well as the primer recognition sites used for library amplification as well as sequencing on the Illumina flow cell. After adapter ligation, limited cycle PCR amplification of the library is performed in order to reach the minimum amount of DNA needed for sequencing. Size selection is performed on the library with AMPure XP Magnetic beads, this step is needed to remove the primers from the mix that could interfere with the sequencing reaction (Fig. 8.3d). The library is quantitated by qPCR using the Kapa Library Quantitation Kit (Roche cat# 07960140001). Six to eight samples with different barcodes are pooled into a single tube at a concentration of 3 nM and send for sequencing in an Illumina HiSeq 3000 or 4000 sequencer that delivers around 300 million reads per lane, providing about 50 million reads per sample when using pools of six samples per sequencing lane.

8.3.3 RNA-Seq: Genome Alignment and Data Analysis

The Illumina platform delivers raw reads in fasqc formatted files. The first step is to run quality controls on the sequencing procedure, followed by alignment of the sequencing reads to a reference genome and quantitation of reads per gene. All the computational tools needed to analyze NGS datasets are freely available at the Bioconductor webpage (http://www.bioconductor.org/), and are coded on R language (https://www.r-project.org/). Another platform widely used for high-throughput data analysis is Galaxy (https://usegalaxy.org/), which is especially helpful for users with less experience in programming on R. Finally, DNASTAR (https://www.dnastar.com/) is a commercial package that fulfills all the steps from quality control, data alignment, transcript quantitation as well as data visualization and gene ontology analysis.

FastQC (http://www.bioinformatics.babraham.ac.uk/projects/fastqc/) is a standard tool used for quality control of the raw reads that include GC content, adaptor identification, duplicated reads, and overrepresented *k*-mers. These parameters help identify sequencing errors and PCR artifacts. As sequencing quality decreases toward the 3′ end of the reads, the Trimmomatic (Bolger et al. 2014) tool is run to trim the ends of the sequence and discard low-quality reads and trim adaptor sequences. Alignment of the remaining reads to a reference genome or transcriptome is performed by Bowtie (Langmead et al. 2009) or STAR (Dobin et al. 2013) aligners. When transcriptome annotation is comprehensive, Bowtie performs non-gapped alignment in a very fast and efficient manner. In the case of less studied genomes, genome alignment helps to identify novel genes or transcripts, which is performed with gapped or spliced mappers capable of reading the splice junctions correctly. The most important parameters to consider when aligning transcripts are the strandness of the library, the number of mismatches to accept and the length of the sequenced fragments.

To estimate gene expression, featureCounts (Liao et al. 2014) quantitates the aggregate raw counts of the previously exonic mapped reads. This measure is affected by factors like transcript length, total number of reads, and library size. In order to compare expression levels among samples, normalization methods are used to correct for the aforementioned biases. RPKM (Mortazavi et al. 2008) (reads per kilobase of exon per million reads), a within sample normalization method, will eliminate the feature length and library size effects. This normalization method performs well only when transcript distribution is relatively homogeneous. When highly and differentially expressed features skew the count distribution, TPM (Li et al. 2008) (transcripts per million reads) and TMM (Robinson and Smyth 2007) (trimmed mean of *M*-values) normalization methods are preferred because they disregard highly variable and highly expressed features. Since RNA-seq quantitation is based on read counts that are probabilistically assigned to different transcripts, discrete probabilistic distributions, like Poisson or the negative binomial distribution are used to compute differential expression (Anders and Huber 2010; Robinson et al. 2010 #7592). Several R packages are available to perform multiple comparison analyses including edgeR (Robinson et al. 2010), DESeq (Anders and Huber 2010), and limma-voom (Law et al. 2014). We recommend that RNA-seq experiments be run in tri- or quadruplicates in order for the differential expression methods to generate reproducible results in the ample spectrum of mRNA expression levels in the brain.

8.3.4 CHIP-Seq: DNA Isolation, Fragmentation, and Library Preparation

CHIP-seq analysis allows for the determination of transcription factor binding or positioning of histone PTMs modification throughout the genome. We have successfully performed several CHIP-qPCR and CHIP-seq studies from rat MBH (Lomniczi et al. 2013a, 2015; Toro et al. 2018). The basic principle of this technique is immobilization of the protein–DNA interactions by crosslinking the tissue with 4% paraformaldehyde (PF) followed by chromatin extraction and fragmentation by sonication. Finally, a fraction of the chromatin is immunoprecipitated with the antibody of choice using protein-A or -G beads. The immune-isolated DNA fragments are purified and used for Illumina NGS library preparation followed by deep sequencing.

A single 5 mg MBH produces about 2–3 μg of DNA, enough material to perform about 10–12 independent CHIP assays. First, the frozen tissue is crosslinked with 4% PF for 10 min at room temperature. This is followed by sonication in SDS lysis buffer. Sonication needs to generate fragments of DNA with an average size of 150–500 bp. In our laboratory, we generate these fragments with about 45 s of sonication time using a microtip sonicator. It is extremely important that the investigator sets up the fragmentation protocol by testing different sonication times and checking for DNA fragmentation in agarose gel (Fig. 8.4a) or Bioanalyzer DNA nano chip (Agilent # 5067-4626). Fragmented chromatin is diluted, aliquoted, and stored at −80 °C. In order to determine the linear range of detection, it is necessary to establish the best chromatin-to-antibody ratio. For that, we perform CHIP reactions with a fixed amount of antibody and increasing volumes of chromatin followed by qPCR of promoter regions corresponding to high and low expressed genes (Fig. 8.4b). The selection of antibodies is crucial, several commercial antibodies have been validated for CHIP, especially for histone modifications (see: https://www.diagenode.com/en/categories/antibodies and https://genome.ucsc.edu/encode/antibodies.html). If antibodies have not been validated, it is safe to assume that an antibody that works for immunocytochemistry could work in CHIP assays. Negative controls are usually species-specific IgGs or antibodies directed against proteins that are not present in the sample. We have successfully used antibodies anti-Green Fluorescent Protein or anti-Beta Galactosidase as negative controls for CHIPs from rat hypothalamus. The antibody–protein immune complexes are isolated using protein A or G agarose beads or paramagnetic Dynabeads (ThermoFisher cat# 10001D and 10007D). The use of paramagnetic beads has particularly improved this technique because they have extremely low background DNA and protein binding. DNA is recovered from the paramagnetic beads using a high pH SDS buffer at high temperatures and purified using commercially available columns (Zymo research cat# D5205). Qubit (ThermoFisher cat# Q33226) fluorometric quantitation is used to determine the DNA concentration that is usually in the range 0.2–2 ng/μl.

Several Illumina compatible kits are available for library preparation from sub-nanogram samples. NEBNext Ultra Library Prep Kit (New England cat# NEB E7370) or DNA SMART CHIP-Seq Kit (Takara cat# 634865) are well suited to generate Illumina compatible libraries starting with around 1 ng of total DNA. Library preparation is similar to the description in Sect. 8.3.3. Every biological sample will be sequenced two times, one will correspond to the immunoprecipitated chromatin while the other corresponds to the "Input DNA" or non-precipitated chromatin that serves as total loading in the qPCR and as peak detection after NGS. It is highly recommended to perform double size selection with AMPure XP

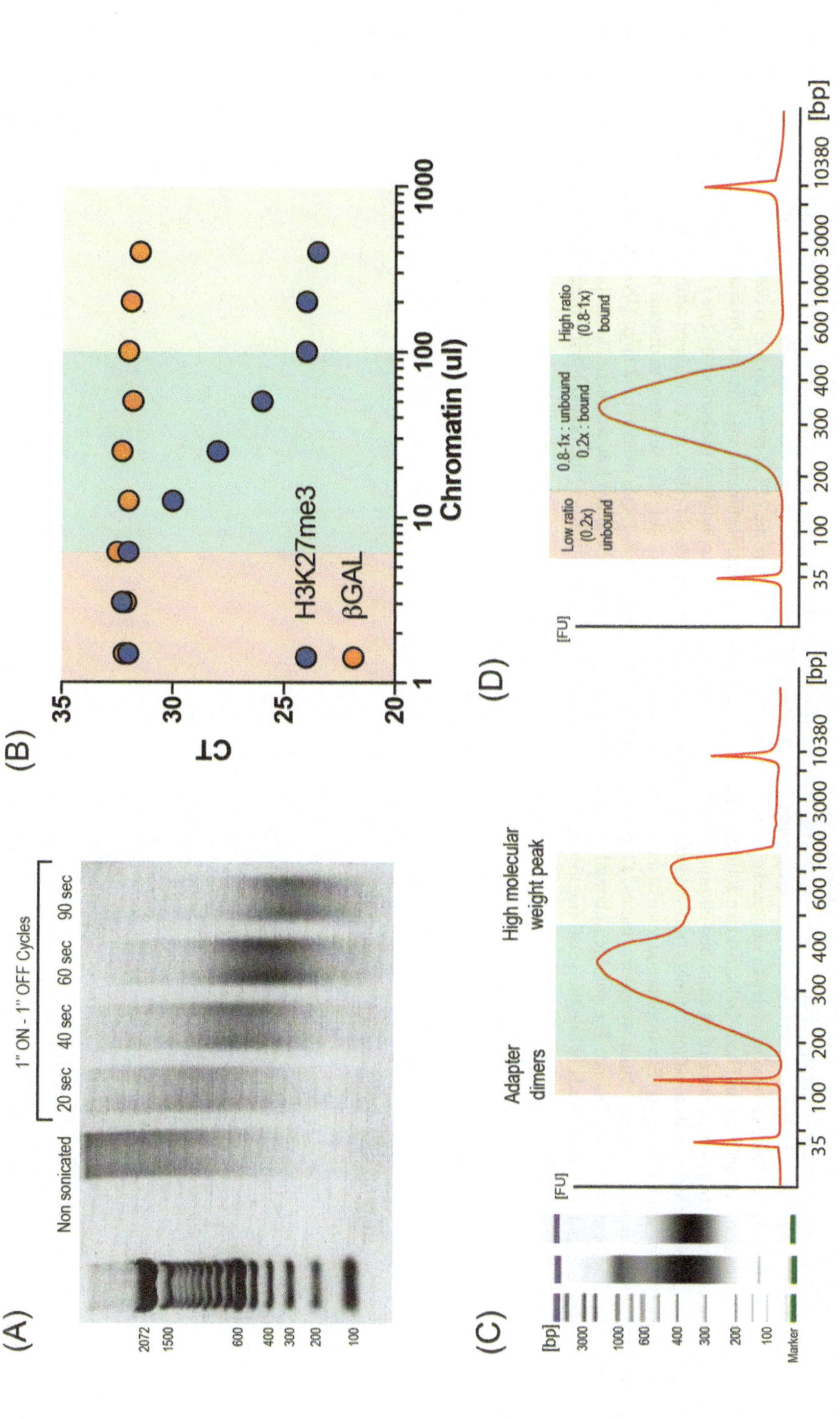

Fig. 8.4 CHIP-Seq library preparation process from rat MBH tissue. **(a)** In order to determine the sonication time to generate optimum chromatin fragmentation, the tissue is treated with increasing number of pulses followed by de-crosslinking, DNA purification, and gel electrophoresis. Increasing the

Fig. 8.4 (continued) number of pulses reduces the average size of the fragments. Large fragments (>500 bp) reduce the accuracy of the CHIP procedure. The optimum fragment size is between 150 and 500 bp. (**b**) Linearity of chromatin capture is determined by using a constant amount of Antibody (2–5 ug) and increasing amounts of chromatin. After the CHIP assay is completed, cycle threshold (ct) values amplification of a specific loci by qPCR must decrease linearly according to the amount of chromatin used (blue circles corresponding to CHIPed H3K27me3). The green shaded area shows the linear range of chromatin/antibody interaction. The red and yellow areas depict saturated chromatin/antibody conditions where the qPCR is not capable to detect changes in DNA capture. The orange circles correspond to an anti beta-Galactosidase antibody (βGAL) used as negative control showing no efficient chromatin/antibody interactions. (**c**) Agilent Bioanalyzer electrophoretic trace of an amplified chromatin library generates a signal corresponding to the barcoded library (green shaded area). The low molecular weight peak corresponds to adapter dimers (red shaded area), while the high molecular weight peaks are considered artifacts due to overamplification of the library (yellow shaded area). (**d**) Electrophoretic trace of an amplified chromatin library after double size selection to discard both low- and high-molecular peaks. First, and to cleanup large molecular weight fragments, a high magnetic bead:library ratio (0.8–1×) is used to bind large DNA fragments while the library and the small fragments remain unbound. The soluble unbound DNA is transferred into a new tube and a low magnetic bead:library ratio (0.2×) is used, the desired library is captured by the beads, while the small fragments remain unbound. After several washes, the bound library fragments are eluted

Magnetic beads before PCR amplification of the library, to remove free unligated adapters and high molecular weight DNA that has been selected during immunoprecipitation (Fig. 8.4c, d). To correct for technical variability during CHIP/Library preparation, it is suggested to add a spike-in control to each CHIP sample (Egan et al. 2016). We usually add 1% of chromatin from *Drosophila* and antibodies to immunoprecipitate histone H2Av, only present in this species. Since this step will pull down the same amount of *Drosophila* histone 2Av between samples, it allows for normalization of the CHIP-seq values from the mammalian genome to the total number of reads obtained from the *Drosophila* genome.

8.3.5 CHIP-Seq: Genome Alignment and Data Analysis

Platforms used for CHIP-seq analysis are the same as for RNA-seq. It is estimated that a reading depth of about 20 million reads per sample is adequate to detect transcription factor binding in the mammalian genome. Proteins with more binding sites as well as histone marks require up to 60 million reads. After assessment of the reads quality metrics, a mapping tool like Bowtie (Langmead et al. 2009) or MAQ (Li and Durbin 2009) is used to align the short reads to the genome. The analysis is followed by peak calling algorithms like MACS (Zhang et al. 2008) that predict the regions of the genome where the immunoprecipitated protein is bound. Peak callers will provide a p-value and FDR as enrichment metrics for each peak. For enzyme or transcription factor binding, peaks are readily detected, but in the case of histone PTMs where marks are more broadly positioned, it is more accurate to define "chromatin domains." Quantitative analysis is based on the total read count per region of interest. DBCHIP (Liang and Keles 2012) and MAnorm (Shao et al. 2012) compute read density over peak regions and give a precise statistical assessment of differentially affected regions across biological conditions. Peaks or domains identified by CHIP-seq are coded as BED or GFF files that can be uploaded to a genome browser for further identification of promoter and enhancer regions as well as gene transcription start sites.

8.4 Important Discoveries Made with NGS Technologies

The first evidence of epigenetic mechanisms involved in pubertal development was published by our laboratory (Lomniczi et al. 2013a, 2015). In these papers, we looked for transcriptional repressors highly expressed in the MBH during the juvenile period and lower expression as the animal reaches puberty. First, using the rat model, we demonstrated that the Polycomb Group (PcG) of transcriptional repressors prevents the initiation of puberty by repressing ARC *Kiss1* expression (Lomniczi et al. 2013a). During the prepubertal period, two central members of the PcG, *Eed,* and *Cbx7,* are highly expressed within ARC kisspeptin neurons, and their protein products are coupled to the *Kiss1* promoter (Lomniczi et al. 2013a). As the animal reaches puberty, DNA methylation of *Eed* and *Cbx7* promoter regions increases, lowering their expression, inducing the loss of PcG components from the *Kiss1* promoter. This is accompanied by increased levels of the epigenetic marks associated with gene activation like H3K9ac, H3K14ac, and H3K4me3, ultimately inducing increased *Kiss1* mRNA expression in the ARC (Lomniczi et al. 2013a). In a second paper, and using *rhesus macaques* as animal model, we provided evidence that members of the zinc finger (ZNF) family of transcriptional repressors play a role in keeping the GnRH pulse generator in check during prepubertal development (Lomniczi et al. 2015). Using MBH of both ovary-intact females and agonadal male rhesus monkeys during the juvenile–pubertal transition we demonstrated that while *KISS1* and *TAC3* mRNA expression increases, *GATAD1* expression decreases. GATAD1 recruits the histone eraser KDM1A, inducing a decrease in the activating H3K4me3/2 marks at the Kiss1 and TAC3 genes 5′-regulatory regions (Lomniczi et al. 2015; Maisonpierre et al. 1991; Shi et al. 2004). Stereotaxic delivery of lentiviral particles overexpressing *GATAD1* in the ARC of immature rats, delays puberty, and disrupts estrous cyclicity (Lomniczi et al. 2015). The decrease in GATAD1 and KDM1A association to the *KISS1* and *TAC3* promoters and the concomitant increase in H3K4me2 abundance observed in the monkey MBH at the time of puberty, strongly support the notion that during juvenile development there is a mechanism of epigenetic repression, that is lifted during the reawakening of the GnRH pulse generator (Lomniczi et al. 2015).

The finding that the chromatin landscape at the *Kiss1* promoter switches from a primarily repressed state, brought about by the PcG and GATAD1/KDM1A complex, to an activating state characterized by trimethylation and acetylation of H3 at lysines 4 and 9, respectively, during the peri-pubertal period indicates the action of an activating transcriptional complex occurring simultaneously to the loss of inhibition. This idea was studied by RNA-seq analysis of the rat MBH during peri-pubertal development. In this study, we found that members of the Trithorax Group (TrxG) of transcriptional activators, well known for antagonizing PcG activity (Shilatifard 2012; Simon and Kingston 2009), increase their expression in the MBH as puberty progresses (Toro et al. 2018). MLL1 and MLL3, two members of the TrxG complex, act on the *Kiss1* promoter and enhancer regions, respectively, to provide *trans*-activational activity at the time when the inhibitory influence of the PcG complex is declining (Toro et al. 2018). To confirm TrxG function, siRNA-mediated knockdown of *Mll1* in the ARC, inhibited *Kiss1* expression and delayed puberty (Toro et al. 2018).

Since the TrxG member MLL3 has a preference for binding at enhancer sites, and active enhancers display high levels of H3K4me1 and H3K27ac (Kim and Shiekhattar 2015), we performed CHIP-seq analysis of these two histone modifications in the MBH of prepubertal rats to pinpoint potential enhancers of the ARC *Kiss1* gene. In a 12-kb segment of the *Kiss1* gene 5′ upstream region, we detected two areas showing high levels of H3K4me1 and H3K27ac. CHIP-qPCR analysis of Site 1 and Site 2 in the rat genome revealed that only Site 1 (about −3000 bp from TSS) displayed increased association of p300/CBP and H3K27ac during prepubertal development, a hallmark of active enhancers (Kim et al. 2010; Kim and Shiekhattar 2015). In agreement with this concept, MLL3 recruitment to

Site 1 (but not Site 2) increased during pubertal development, while the EED-mediated H3K27m3 mark weans down (Toro et al. 2018). In order to determine the function of this new *Kiss1* enhancer region, we used a CRISPR-CAS9-KRAB epigenetic remodeling approach that inhibits the action of Mll3 by site-specific deposition of the repressive H3K9m3 histone mark. Animals stereotaxically injected with an AAV9 virus carrying this construct into the ARC experienced a downregulation of *Kiss1* expression, delayed puberty and disrupted estrous cyclicity (Toro et al. 2018). In conclusion, the TrxG of epigenetic activators is capable of controlling *Kiss1* expression in the ARC, by implementing histone PTM changes at promoter and enhancer regions around the time of puberty.

Box 8.3 The Epigenetic Yin-Yang of Puberty

During the last few years, we learned that there is a basic mechanism of inhibition–activation that operates at a transcriptional level of ARC kisspeptin neurons. This mechanism, of epigenetic nature, ultimately controls pulsatile GnRH secretion. In 2013, we demonstrated that *Kiss1* expression in the ARC of prepubertal female rats is subjected to an epigenetic repressive state exerted by the polycomb group of transcriptional repressors. At puberty, the association of the polycomb proteins to the *Kiss1* promoter declines, coinciding with increased *Kiss1* expression and changes in the chromatin structure of the *Kiss1* promoter. Increased levels of two histone marks associated with gene activation are detected, while a gradual loss was found in repressive histone modifications catalyzed by the polycomb group complex. In 2017, we determined that the increase in activating histone marks at puberty is mediated by the action of two members of the Trithorax Group of epigenetic activators. Using deep-sequencing techniques, we demonstrated that the Trithorax binds to the *Kiss1* promoter and enhancer regions during the pubertal period to facilitate open chromatin states, ultimately enhancing *Kiss1* expression and hence, increased GnRH release.

8.5 Limitations and Caveats of NGS Technologies

NGS technologies, used by researchers and clinicians, have provided invaluable information into genome variation, stability, as well as gene regulation. With the widespread use of this technology several limitations have arisen. First, for efficient library preparation and sequencing, DNA samples need to be above the ng range, limiting the size of the starting material. In our hands, we start with 200–300 ng of chromatin from rat MBH (about 1/10th of the total chromatin obtained from a single MBH), and at the end of the CHIP protocol we end up with 500 pg to 5 ng total DNA available for library preparation depending on the antibody used for immunoprecipitation. This approach, while not ideal because it generates data from a complex mixture of different cell types, is still today the only method available to overcome the critical mass needed for library preparation and sequencing. While single-cell mRNA sequencing technologies have been developed (Eberwine et al. 2014), allowing for picogram amounts of starting material to be used, heavy amplification is needed for sample and library preparation, resulting in uneven coverage, high noise, and inaccurate quantitation. Needless to say, CHIP-seq technologies have improved tremendously, but efficiency and sequence quality get compromised when starting material lies below 10^4 cells (Gilfillan et al. 2012).

The second limitation is related to the massive amount of information generated and the wide range of platforms and algorithms that can be used to analyze and visualize the data. Although the ENCODE Consortium (https://www.encodeproject.org/) has set the standard for the analysis of most of these technologies, researchers have to rely on external bioinformatics cores, recruit informaticians in their laboratories or even learn coding and statistics for high-throughput data analysis. This can be especially challenging for small laboratories that do not have easy access to core facilities and other informatics resources.

8.6 Gene Network Visualization

With the widespread use of NGS technologies, scientists struggle to find ways to identify the relevant genes or genome signatures from long gene lists. For instance, genes are scored by their differential expression or chromatin signatures between different biological states. Using an arbitrary expression threshold (we usually use 1.5-fold expression change), gene lists are screened for the top most variable genes. Searching for enrichment of predefined functionally related genes can be easily done with stand-alone software like Ingenuity Pathway Analysis (IPA) from Qiagen (https://www.qiagenbioinformatics.com/products/ingenuity-pathway-analysis/), web-based program like DAVID Bioinformatic Database (https://david.ncifcrf.gov/home.jsp) (Dennis et al. 2003) and GeneMANIA (https://genemania.org/) (Mostafavi et al. 2008) or open-source options like several plugins for Cytoscape (http://www.cytoscape.org/) (Cline et al. 2007).

8.6.1 Enrichment Analysis

Enrichment analysis is a statistically powerful method to analyze and interpret large datasets using a priori knowledge. Most of the enrichment algorithms use Gene Ontology (GO) annotations to identify gene sets that participate or belong to a certain category (e.g., participation in a signaling pathway, biological process, protein domain, chromosomal location, specific disease, and genetic perturbation). DAVID uses fuzzy gene clusters to identify enriched gene sets, classifies them into partially overlapped groups and display results in a tabular form (Dennis et al. 2003). For fast and easy visualization, the plugins BiNGO (Maere et al. 2005) and ClueGO (Bindea et al. 2009) for Cytoscape display the hierarchical structure of a network of enriched GO terms. For annotated gene function like co-expression, co-localization,

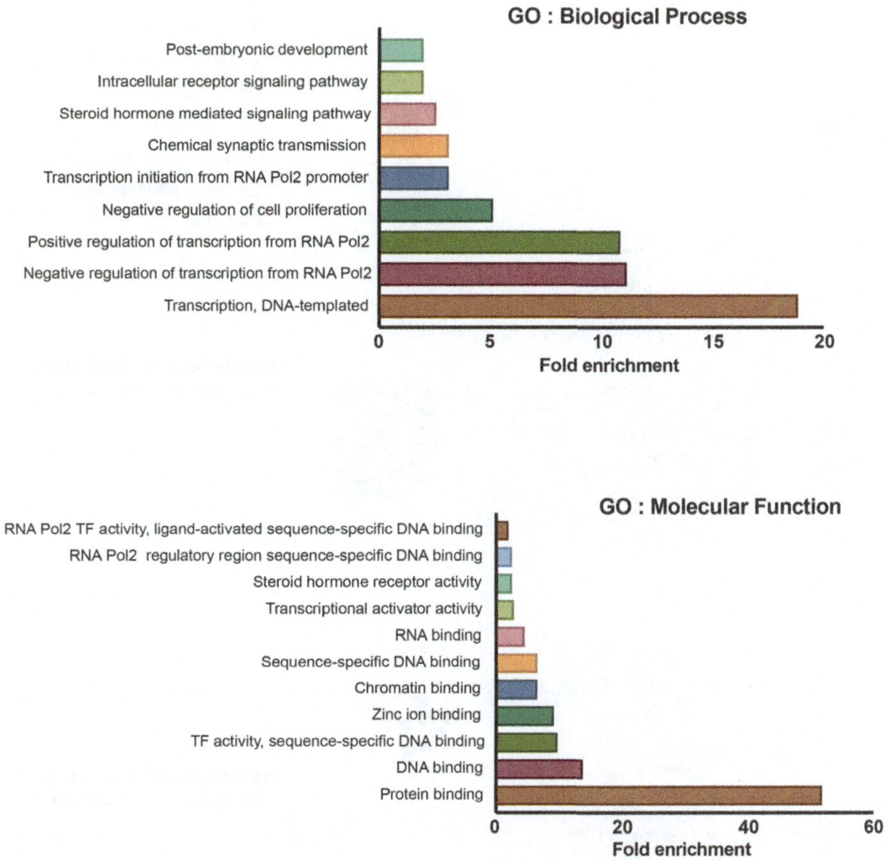

Fig. 8.5 Gene ontology (GO) analysis of the genes with SNPs associated with age at menarche. DAVID (https://david.ncifcrf.gov/home.jsp) GO categories: Biological Process and Molecular Function of the 339 genes with SNPs associated with the age at menarche show a high enrichment in, e.g., gene transcription, chemical synapses, steroid signaling, and others

genetic interactions, physical interactions, and pathway classification, web-based GeneMANIA, as well as the plugin, for Cytoscape is a great tool for network visualization for validated gene interactions.

As an example, we used DAVID to analyze the GO function of the genes located nearest to the 389 SNPs reported as associated with age at menarche (Day et al. 2017). Surprisingly, this gene set is significantly enriched in functions related to transcriptional regulation like Negative/Positive regulation of transcription, chromatin binding, and DNA binding in addition to steroid hormone signaling and chemical synaptic transmission (Fig. 8.5). To determine if this group of genes is co-regulated in any way to the three families of epigenetic modifiers we described in our laboratory, we ran GeneMANIA on the same data set and added the Polycomb group of epigenetic writers (EED, EZH1, and EZH2), the epigenetic reader/eraser

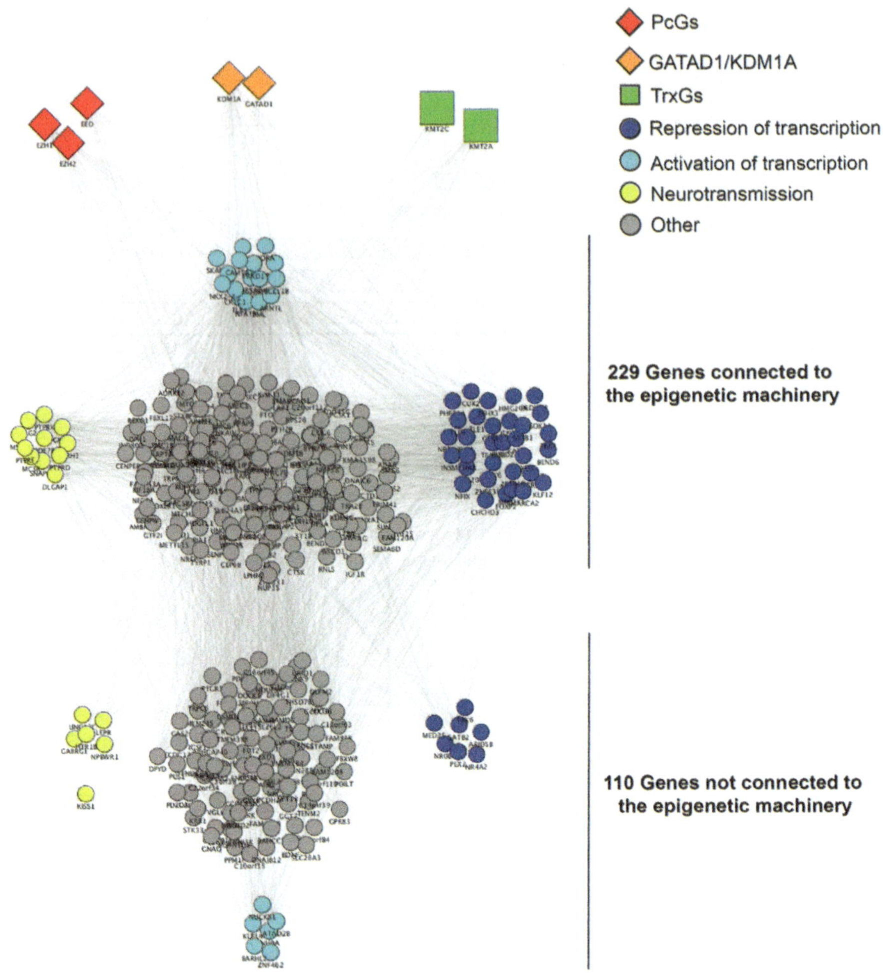

Fig. 8.6 Gene network representation of the genes with SNPs associated with age at menarche and their GeneMANIA connections with the epigenetic machinery. The round nodes of the network are the 339 genes associated with age at menarche, red diamonds are genes of the Polycomb Group (PcG) of transcriptional repressors, orange diamonds represent Gatad1 and Kdm1a and green squares are the Trithorax Group (TrxG) of transcriptional activators. The epigenetic machinery (depicted on the first layer of the network) is connected to 229 out of 339 of the genes associated with age at menarche (second layer of nodes). The third layer depicts the 110 genes that are not directly connected to the epigenetic machinery. Yellow nodes are genes involved in neurotransmission, blue nodes are transcriptional repressors and cyan nodes are transcriptional activators. Gray lines, or edges, are connections established by the GeneMANIA database representing: Genetic interactions, Pathways, Physical interactions, Predicted, and Shared protein domains

complex (GATAD1/LSD1) and the recently described Trithorax members (MLL1 and MLL3) (Fig. 8.6). Interestingly, most of the human dataset (229 out of 339 genes) seems to be co-regulated or controlled by at least one of the epigenetic complexes mentioned before, raising the tantalizing possibility that the majority of the SNP's described by Day et al. (2017) as associated with the age at menarche, are in some way altering the epigenetic milieu of their downstream target genes.

8.6.2 Gene Co-expression Analysis

Another way to infer gene function is by co-expression network analysis, a method that creates networks of genes whose expression across samples is highly correlated and thus, potentially co-regulated. Gene co-expression networks only specify which genes are active simultaneously, often indicating that they are active in the same biological process (Amar et al. 2013; Xue et al. 2013; Pierson et al. 2015). This approach serves in the identification of transcription factors or methylation patterns that can affect the expression of expression modules (Glass et al. 2013). The first step in the analysis is to express the individual relationship between genes as a correlation measure between each pair of genes. This correlation measure describes the similarity between expression patterns of each gene across all samples (Steuer et al. 2002; Margolin et al. 2006). Typically, a Pearson's or Spearman's correlation, as well as a Bayesian approach, can be used to create a co-expression network (Friedman et al. 2000). In the second step of the analysis, the associations between genes are used to construct gene networks where the nodes represent each gene and the edges represent the strength of the co-expression correlation. The final step is to use the k-means clustering and hierarchical clustering to identify modules of genes or "clusters" to group genes with similar patterns of expression (D'Haeseleer 2005). These modules can finally be studied for functional enrichment to identify overrepresented functional categories using the enrichment analysis tools mentioned above.

An example of this approach is our study of the function of GATAD1 on pubertal development. To this aim, we analyzed the network configuration and connectivity of 147 selected transcripts important for pubertal development. We used a compressive-sensing approach that deduces gene networks by identifying robust co-expression links between genes (Lomniczi et al. 2015). We showed that the top 20 genes gaining cumulative connectivity are sparsely connected to themselves, while the top 20 losing connectivity is strongly connected to themselves in the MBH at PND14 (Fig. 8.7a). Later on, during the juvenile-to-puberty transition (at PND28) connectivity shifts and the 20 genes gaining connectivity become more interconnected. When overexpressing GATAD1 in the ARC, the pattern of connectivity resembles more that of an immature (PND14) than of a pubertal (PND28) animal (Lomniczi et al. 2015) (Fig. 8.7a). Gene Ontology (GO) enrichment analysis showed that for the epigenetic repressors *Kdm1a, Kdm5b, Ehmt1, Pcgf2*, and *Rnf2* their loss of connectivity during pubertal maturation is reversed in the presence of GATAD1 (Fig. 8.7b). Another set of epigenetic regulators, *Kdm1b, Ezh1*, and *Hdac1*, that gain connectivity during normal puberty, fail to do so when overexpressing GATAD1, keeping gene connectivity in an "undeveloped" or infantile stage (Fig. 8.7b). It is known that GATAD1 recruits lysine demethylase 1A (KDM1A/LSD1) to promoter regions to induce the loss of the activating histone 3 trimethylated at lysine 4 (H3K4me3). These analyses assisted us in the

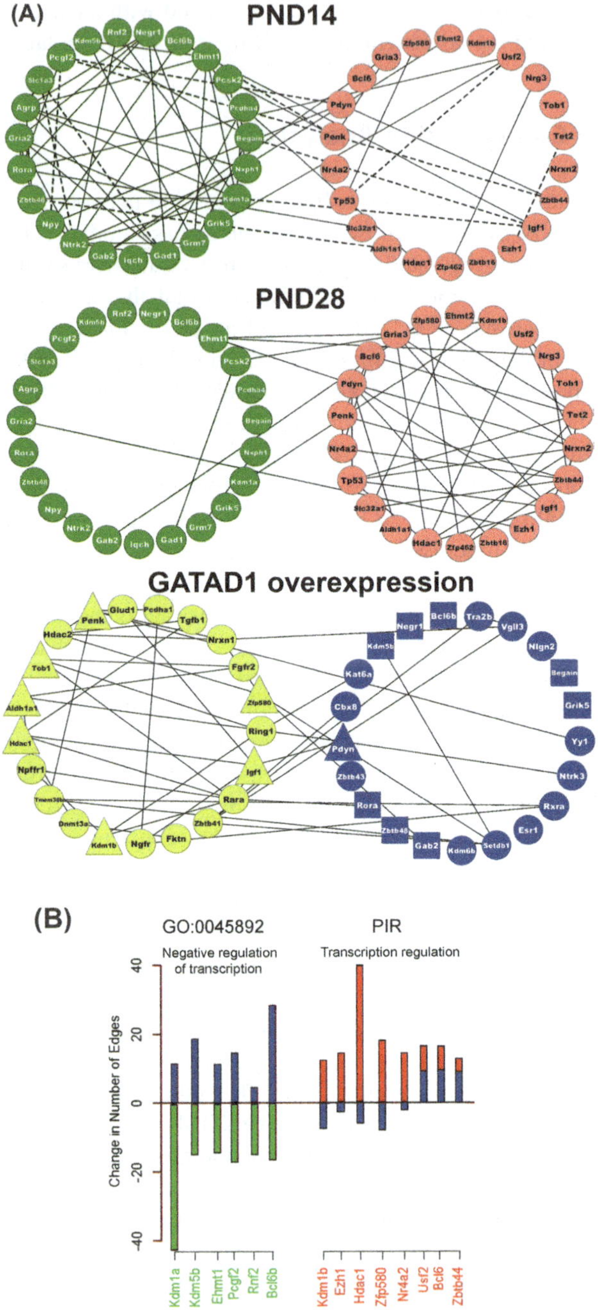

Fig. 8.7 Effect of GATAD1 overexpression on gene co-expression networks in the hypothalamus of rhesus macaques. (**a**) Inferred co-expression network between the top 20 genes losing edges (green) and the top 20 genes gaining edges/connections (red) during prepubertal development, based on the sum of changes in connectivity between PND14–PND21 and PND21–PND28. Network at PND 28 with the top 20 genes losing edges (yellow) and gaining edges (blue) under

demonstration that during the infantile period, when *Kiss1* expression is low, the GATAD1/LSD1 complex is associated to the Kiss1 promoter keeping low levels of H3K4me3. As puberty approaches and *Gatad1* expression decreases, the GATAD1/LSD1 complex is evicted from the promoter, H3K4me2/3 rises, and *Kiss1* expression increases (Lomniczi et al. 2015).

8.6.3 Limitations and Caveats of Network Visualization

One of the biggest limitations is related to the animal species of choice. Most of the interactomics databases is limited to human and/or mouse datasets. During the last few years, efforts have been made to widen this repertoire to include rats, *Drosophila*, *C. elegans,* and others. With that in mind, researchers many times opt to use human or mouse databases when investigating rat genomic interactions due to the higher characterization of gene function and interactions in these species. Another issue that comes out frequently is that not all genes are yet classified on the Gene Ontology database, producing bias toward a subclass of previously well-studied genes present in the dataset.

Network analysis and visualization are powerful tools for whole-genome studies, but it is crucial for investigators to take all the precautions for the correct statistical analysis and subsequent validation of the identified pathways or gene interactions.

8.7 Perspectives

High-throughput sequencing technologies have transformed basic biology as well as clinical research. As sequencing costs keep falling and informatics tools become more easily available, these technologies will be embraced by a higher number of laboratories involved in neuroendocrine research. With this in mind, it is key that data become openly available for the discovery of common pathways across different models.

We just started uncovering the first epigenetic complexes involved in normal pubertal progression, but the technologies discussed in this chapter will be crucial in

Fig. 8.7 (continued) GATAD1 overexpression as compared with the network inferred at PND28. In this network, genes that are among the top 20 genes gaining edges during normal prepubertal development are shown as triangles, while genes that are among the top 20 losers of connectivity during development are shown as squares. For all networks, positive correlation edges are represented by solid black lines, negative correlation edges are shown as dashed lines. (**b**) Changes in the connectivity of genes belonging to selected transcription-related functional categories that lose (Gene Ontology category GO:0045892: negative regulation of transcription, DNA templated) or gain (Protein Information Resource, http://pir.georgetown.edu/; Keyword "transcription regulation") connectivity (green and red bars, respectively) from PND14 to PND28. For each gene, the alteration in connectivity resulting from GATAD1 overexpression as compared with PND28 is shown as blue bars. Modified from Lomniczi et al. (2015)

studying alterations of reproductive development induced by external factors like metabolism, endocrine disruptors, or changes in circadian rhythms.

Acknowledgments We thank every past member of our laboratory who has contributed to the understanding of the genomic events involved in the maturation of the reproductive hypothalamus. The authors are supported by the National Institute of Health (1R01HD084542 and P51OD011092).

Key References

Boyar et al. (1972). This study was the first demonstrating that LH pulsatility increases before sleep in pubertal patients.

Lomniczi et al. (2013). This study describes how the Polycomb family of epigenetic repressors control the epigenetic landscape at the Kiss1 promoter region, keeping ARC Kiss1 transcription low during the infantile period.

Lomniczi et al. (2015). This study describes how the GATAD1/LSD1 controls the epigenetic landscape at the Kiss1 promoter region in rhesus macaques, keeping ARC Kiss1 transcription low during the infantile period.

Roth et al. (2007). This study was the first proposing a tumor-related gene network that controls mammalian pubertal development.

Toro et al. (2018). This study describes how the Trithorax family of epigenetic activators control the epigenetic landscape at the Kiss1 promoter/enhancer regions during the pubertal activation on Kiss1 transcription in the ARC.

References

Amar D, Safer H, Shamir R (2013) Dissection of regulatory networks that are altered in disease via differential co-expression. PLoS Comput Biol 9:E1002955

Anders S, Huber W (2010) Differential expression analysis for sequence count data. Genome Biol 11:R106

Barutcu AR, Fritz AJ, Zaidi SK, Van Wijnen AJ, Lian JB, Stein JL, Nickerson JA, Imbalzano AN, Stein GS (2016) C-Ing the genome: a compendium of chromosome conformation capture methods to study higher-order chromatin organization. J Cell Physiol 231:31–35

Beale KE, Kinsey-Jones JS, Gardiner JV, Harrison EK, Thompson EL, Hu MH, Sleeth ML, Sam AH, Greenwood HC, McGavigan AK, Dhillo WS, Mora JM, Li XF, Franks S, Bloom SR, O'Byrne KT, Murphy KG (2014) The physiological role of arcuate kisspeptin neurons in the control of reproductive function in female rats. Endocrinology 155(3):1091–1098

Bindea G, Mlecnik B, Hackl H, Charoentong P, Tosolini M, Kirilovsky A, Fridman WH, Pages F, Trajanoski Z, Galon J (2009) Cluego: a cytoscape plug-in to decipher functionally grouped gene ontology and pathway annotation networks. Bioinformatics 25:1091–1093

Bolger AM, Lohse M, Usadel B (2014) Trimmomatic: a flexible trimmer for illumina sequence data. Bioinformatics 30:2114–2120

Boyar R, Finkelstein J, Roffwarg H, Kapen S, Weitzman E, Hellman L (1972) Synchronization of augmented luteinizing hormone secretion with sleep during puberty. N Engl J Med 287:582–

586. [This study was the first demonstrating that LH pulsatility increases before sleep in pubertal patients]

Buenrostro JD, Wu B, Chang HY, Greenleaf WJ (2015) Atac-seq: a method for assaying chromatin accessibility genome-wide. Curr Protoc Mol Biol 109:21.29.1–21.29.9

Chang TC, Zeitels LR, Hwang HW, Chivukula RR, Wentzel EA, Dews M, Jung J, Gao P, Dang CV, Beer MA, Thomas-Tikhonenko A, Mendell JT (2009) Lin-28B transactivation is necessary for Myc-mediated let-7 repression and proliferation. Proc Natl Acad Sci U S A 106(9):3384–3389

Cline MS, Smoot M, Cerami E, Kuchinsky A, Landys N, Workman C, Christmas R, Avila-Campilo I, Creech M, Gross B, Hanspers K, Isserlin R, Kelley R, Killcoyne S, Lotia S, Maere S, Morris J, Ono K, Pavlovic V, Pico AR, Vailaya A, Wang PL, Adler A, Conklin BR, Hood L, Kuiper M, Sander C, Schmulevich I, Schwikowski B, Warner GJ, Ideker T, Bader GD (2007) Integration of biological networks and gene expression data using cytoscape. Nat Protoc 2:2366–2382

Crawford GE, Holt IE, Whittle J, Webb BD, Tai D, Davis S, Margulies EH, Chen Y, Bernat JA, Ginsburg D, Zhou D, Luo S, Vasicek TJ, Daly MJ, Wolfsberg TG, Collins FS (2006) Genome-wide mapping of dnase hypersensitive sites using massively parallel signature sequencing (Mpss). Genome Res 16:123–131

Cukier P, Wright H, Rulfs T, Silveira LF, Teles MG, Mendonca BB, Arnhold IJ, Heger S, Latronico AC, Ojeda SR, Brito VN (2013) Molecular and gene network analysis of thyroid transcription factor 1 (TTF1) and enhanced at puberty (EAP1) genes in patients with GnRH-dependent pubertal disorders. Horm Res Paediatr 80(4):257–266

D'haeseleer P (2005) How does gene expression clustering work? Nat Biotechnol 23:1499–1501

Day FR, Thompson DJ, Helgason H, Chasman DI, Finucane H, Sulem P, Ruth KS, Whalen S, Sarkar AK, Albrecht E, Altmaier E, Amini M, Barbieri CM, Boutin T, Campbell A, Demerath E, Giri A, He C, Hottenga JJ, Karlsson R, Kolcic I, Loh PR, Lunetta KL, Mangino M, Marco B, Mcmahon G, Medland SE, Nolte IM, Noordam R, Nutile T, Paternoster L, Perjakova N, Porcu E, Rose LM, Schraut KE, Segre AV, Smith AV, Stolk L, Teumer A, Andrulis IL, Bandinelli S, Beckmann MW, Benitez J, Bergmann S, Bochud M, Boerwinkle E, Bojesen SE, Bolla MK, Brand JS, Brauch H, Brenner H, Broer L, Bruning T, Buring JE, Campbell H, Catamo E, Chanock S, Chenevix-Trench G, Corre T, Couch FJ, Cousminer DL, Cox A, Crisponi L, Czene K, Davey Smith G, De Geus E, De Mutsert R, De Vivo I, Dennis J, Devilee P, Dos-Santos-Silva I, Dunning AM, Eriksson JG, Fasching PA, Fernandez-Rhodes L, Ferrucci L, Flesch-Janys D, Franke L, Gabrielson M, Gandin I, Giles GG, Grallert H, Gudbjartsson DF, Guenel P, Hall P, Hallberg E, Hamann U, Harris TB, Hartman CA, Heiss G, Hooning MJ, Hopper JL, Hu F, Hunter DJ, Ikram MA, Im HK, Jarvelin MR, Joshi PK, Karasik D, Kellis M et al (2017) Genomic analyses identify hundreds of variants associated with age at menarche and support a role for puberty timing in cancer risk. Nat Genet 49:834–841

de Roux N, Genin E, Carel J-C, Matsuda F, Chaussain J-L, Milgrom E (2003) Hypogonadotropic hypogonadism due to loss of function of the KiSS1-derived peptide receptor GPR54. Proc Natl Acad Sci U S A 100:10972–10976

Dennis G Jr, Sherman BT, Hosack DA, Yang J, Gao W, Lane HC, Lempicki RA (2003) David: database for annotation, visualization, and integrated discovery. Genome Biol 4:P3

Dobin A, Davis CA, Schlesinger F, Drenkow J, Zaleski C, Jha S, Batut P, Chaisson M, Gingeras TR (2013) Star: ultrafast universal RNA-seq aligner. Bioinformatics 29:15–21

Eberwine J, Sul JY, Bartfai T, Kim J (2014) The promise of single-cell sequencing. Nat Methods 11:25–27

Egan B, Yuan CC, Craske ML, Labhart P, Guler GD, Arnott D, Maile TM, Busby J, Henry C, Kelly TK, Tindell CA, Jhunjhunwala S, Zhao F, Hatton C, Bryant BM, Classon M, Trojer P (2016) An alternative approach To ChIP-seq normalization enables detection of genome-wide changes in histone h3 lysine 27 trimethylation upon EZH2 inhibition. PLoS One 11:E0166438

Friedman N, Linial M, Nachman I, Pe'er D (2000) Using Bayesian networks to analyze expression data. J Comput Biol 7:601–620

Garcia-Martinez J, Aranda A, Perez-Ortin JE (2004) Genomic run-on evaluates transcription rates for all yeast genes and identifies gene regulatory mechanisms. Mol Cell 15:303–313

Gilfillan GD, Hughes T, Sheng Y, Hjorthaug HS, Straub T, Gervin K, Harris JR, Undlien DE, Lyle R (2012) Limitations and possibilities of low cell number ChIP-seq. BMC Genomics 13:645

Giresi PG, Kim J, Mcdaniell RM, Iyer VR, Lieb JD (2007) Faire (formaldehyde-assisted isolation of regulatory elements) isolates active regulatory elements from human chromatin. Genome Res 17:877–885

Glass K, Huttenhower C, Quackenbush J, Yuan GC (2013) Passing messages between biological networks to refine predicted interactions. PLoS One 8:E64832

Hagan JP, Piskounova E, Gregory RI (2009) Lin 28 recruits the TUTase Zcchc11 to inhibit let-7 maturation in mouse embryonic stem cells. Nat Struct Mol Biol 16(10):1021–1025

Heger S, Mastronardi C, Dissen GA, Lomniczi A, Cabrera R, Roth CL, Jung H, Galimi F, Sippell W, Ojeda SR (2007) Enhanced at puberty 1 (EAP1) is a new transcriptional regulator of the female neuroendocrine reproductive axis. J Clin Invest 117(8):2145–2154

Heo I, Joo C, Kim YK, Ha M, Yoon MJ, Cho J, Yeom KH, Han J, Kim VN (2009) TUT4 in concert with Lin28 suppresses microRNA biogenesis through pre-microRNA uridylation. Cell 138 (4):696–708

Kim TK, Shiekhattar R (2015) Architectural and functional commonalities between enhancers and promoters. Cell 162(5):948–959

Kim TK, Hemberg M, Gray JM, Costa AM, Bear DM, Wu J, Harmin DA, Laptewicz M, Barbara-Haley K, Kuersten S, Markenscoff-Papadimitriou E, Kuhl D, Bito H, Worley PF, Kreiman G, Greenberg ME (2010) Widespread transcription at neuronal activity-regulated enhancers. Nature 465(7295):182–187

Kwak H, Fuda NJ, Core LJ, Lis JT (2013) Precise maps of RNA polymerase reveal how promoters direct initiation and pausing. Science 339:950–953

Langmead B, Trapnell C, Pop M, Salzberg SL (2009) Ultrafast and memory-efficient alignment of short DNA sequences to the human genome. Genome Biol 10:R25

Law CW, Chen Y, Shi W, Smyth GK (2014) Voom: precision weights unlock linear model analysis tools for RNA-seq read counts. Genome Biol 15:R29

Lehman MN, Coolen LM, Goodman RL (2010) Minireview: kisspeptin/neurokinin B/dynorphin (KNDy) cells of the arcuate nucleus: a central node in the control of gonadotropin-releasing hormone secretion. Endocrinology 151(8):3479–3489

Lehrbach NJ, Armisen J, Lightfoot HL, Murfitt KJ, Bugaut A, Balasubramanian S, Miska EA (2009) LIN-28 and the poly(U) polymerase PUP-2 regulate let-7 microRNA processing in Caenorhabditis elegans. Nat Struct Mol Biol 16(10):1016–1020

Li H, Durbin R (2009) Fast and accurate short read alignment with Burrows-Wheeler transform. Bioinformatics 25:1754–1760

Li H, Ruan J, Durbin R (2008) Mapping short DNA sequencing reads and calling variants using mapping quality scores. Genome Res 18:1851–1858

Liang K, Keles S (2012) Detecting differential binding of transcription factors with ChIP-seq. Bioinformatics 28:121–122

Liao Y, Smyth GK, Shi W (2014) Featurecounts: an efficient general purpose program for assigning sequence reads to genomic features. Bioinformatics 30:923–930

Lomniczi A, Garcia-Rudaz C, Ramakrishnan R, Wilmont B, Khouangsathiene S, Ferguson B, Dissen GA, Ojeda SR (2012) A single nucleotide polymorphism in the EAP1 gene is associated with amenorrhea/oligomenorrhea in nonhuman primates. Endocrinology 153(1):339–349

Lomniczi A, Loche A, Castellano JM, Ronnekleiv OK, Bosh M, Kaidar G, Knoll JG, Wright H, Pfeifer GP, Ojeda SR (2013a) Epigenetic control of female puberty. Nat Neurosci 16:281–289. [This study describes how the Polycomb family of epigenetic repressors control the epigenetic landscape at the Kiss1 promoter region, keeping ARC Kiss1 transcription low during the infantile period]

Lomniczi A, Wright H, Castellano JM, Sonmez K, Ojeda SR (2013b) A system biology approach to identify regulatory pathways underlying the neuroendocrine control of female puberty in rats and nonhuman primates. Horm Behav 64(2):175–186

Lomniczi A, Wright H, Castellano JM, Matagne V, Toro CA, Ramaswamy S, Plant TM, Ojeda SR (2015) Epigenetic regulation of puberty via Zinc finger protein-mediated transcriptional repression. Nat Commun 6:10195. [This study describes how the GATAD1/LSD1 controls the epigenetic landscape at the Kiss1 promoter region in rhesus macaques, keeping ARC Kiss1 transcription low during the infantile period]

Maere S, Heymans K, Kuiper M (2005) Bingo: a cytoscape plugin to assess overrepresentation of gene ontology categories in biological networks. Bioinformatics 21:3448–3449

Maisonpierre PC, Le Beau MM, Espinosa R III, Ip NY, Belluscio L, de la Monte SM, Squinto S, Furth ME, Yancopoulos GD (1991) Human and rat brain-derived neurotrophic factor and neurotrophin-3: gene structures, distributions, and chromosomal localizations. Genomics 10:558–568

Margolin AA, Nemenman I, Basso K, Wiggins C, Stolovitzky G, Dalla Favera R, Califano A (2006) Aracne: an algorithm for the reconstruction of gene regulatory networks in a mammalian cellular context. BMC Bioinformatics 7(Suppl 1):S7

Mastronardi C, Smiley GG, Raber J, Kusakabe T, Kawaguchi A, Matagne V, Dietzel A, Heger S, Mungenast AE, Cabrera R, Kimura S, Ojeda SR (2006) Deletion of the Ttf1 gene in differentiated neurons disrupts female reproduction without impairing basal ganglia function. J Neurosci 26:13167–13179

Meijs-Roelofs HMA, Kramer P, Cappellen WAV, Woutersen PJA (1985) Changes in serum concentration and ovarian content of 5a-androstane-3a-17b-diol in the female rat approaching first ovulation. Biol Reprod 32:301–308

Mortazavi A, Williams BA, McCue K, Schaeffer L, Wold B (2008) Mapping and quantifying mammalian transcriptomes by RNA-Seq. Nat Methods 5:621–628

Mostafavi S, Ray D, Warde-Farley D, Grouios C, Morris Q (2008) Genemania: a real-time multiple association network integration algorithm for predicting gene function. Genome Biol 9(Suppl 1):S4

Mueller JK, Dietzel A, Lomniczi A, Loche A, Tefs K, Kiess W, Danne T, Ojeda SR, Heger S (2011) Transcriptional regulation of the human KiSS1 gene. Mol Cell Endocrinol 342(1–2):8–19

Mueller JK, Koch I, Lomniczi A, Loche A, Rulfs T, Castellano JM, Kiess W, Ojeda S, Heger S (2012) Transcription of the human EAP1 gene is regulated by upstream components of a puberty-controlling tumor suppressor gene network. Mol Cell Endocrinol 351(2):184–198

Navarro VM, Gottsch ML, Wu M, Garcia-Galiano D, Hobbs SJ, Bosch MA, Pinilla L, Clifton DK, Dearth A, Ronnekleiv OK, Braun RE, Palmiter RD, Tena-Sempere M, Alreja M, Steiner RA (2011) Regulation of NKB pathways and their roles in the control of Kiss1 neurons in the arcuate nucleus of the male mouse. Endocrinology 152(11):4265–4275

Ojeda SR (1991) The mystery of mammalian puberty: how much more do we know? Perspect Biol Med 34:365–383

Ojeda SR, Hill J, Hill DF, Costa ME, Tapia V, Cornea A, Ma YJ (1999) The Oct-2 POU-domain gene in the neuroendocrine brain: a transcriptional regulator of mammalian puberty. Endocrinology 140:3774–3789

Ojeda SR, Terasawa E (2002) Neuroendocrine regulation of puberty. In: Pfaff D et al (eds) Hormones, brain and behavior, vol 4. Elsevier, New York, pp 589–659

Ojeda SR, Skinner MK (2006) Puberty in the rat. In: Neill JD (ed) The physiology of reproduction, 3rd edn. Academic Press/Elsevier, San Diego, pp 2061–2126

Ong KK, Elks CE, Li S, Zhao JH, Luan J, Andersen LB, Bingham SA, Brage S, Smith GD, Ekelund U, Gillson CJ, Glaser B, Golding J, Hardy R, Khaw KT, Kuh D, Luben R, Marcus M, McGeehin MA, Ness AR, Northstone K, Ring SM, Rubin C, Sims MA, Song K, Strachan DP, Vollenweider P, Waeber G, Waterworth DM, Wong A, Deloukas P, Barroso I, Mooser V, Loos RJ, Wareham NJ (2009) Genetic variation in LIN28B is associated with the timing of puberty. Nat Genet 41(6):729–733

Perry JR, Stolk L, Franceschini N, Lunetta KL, Zhai G, McArdle PF, Smith AV, Aspelund T, Bandinelli S, Boerwinkle E, Cherkas L, Eiriksdottir G, Estrada K, Ferrucci L, Folsom AR, Garcia M, Gudnason V, Hofman A, Karasik D, Kiel DP, Launer LJ, van MJ, Nalls MA,

Rivadeneira F, Shuldiner AR, Singleton A, Soranzo N, Tanaka T, Visser JA, Weedon MN, Wilson SG, Zhuang V, Streeten EA, Harris TB, Murray A, Spector TD, Demerath EW, Uitterlinden AG, Murabito JM (2009) Meta-analysis of genome-wide association data identifies two loci influencing age at menarche. Nat Genet 41(6):648–650

Pierson E, GTEx Consortium, Koller D, Battle A, Mostafavi S, Ardlie KG, Getz G, Wright FA, Kellis M, Volpi S, Dermitzakis ET (2015) Sharing and specificity of co-expression networks across 35 human tissues. PLoS Comput Biol 11:E1004220

Pinilla L, Aguilar E, Dieguez C, Millar RP, Tena-Sempere M (2012) Kisspeptins and reproduction: physiological roles and regulatory mechanisms. Physiol Rev 92(3):1235–1316

Robinson MD, Smyth GK (2007) Moderated statistical tests for assessing differences in tag abundance. Bioinformatics 23:2881–2887

Robinson MD, Mccarthy DJ, Smyth GK (2010) Edger: a bioconductor package for differential expression analysis of digital gene expression data. Bioinformatics 26:139–140

Roth CL, Mastronardi C, Lomniczi A, Wright H, Cabrera R, Mungenast AE, Heger S, Jung H, Dubay C, Ojeda SR (2007) Expression of a tumor-related gene network increases in the mammalian hypothalamus at the time of female puberty. Endocrinology 148(11):5147–5161. [This study was the first proposing a tumor-related gene network that controls mammalian pubertal development]

Sandau US, Mungenast AE, Alderman Z, Sardi SP, Fogel AI, Taylor B, Parent AS, Biederer T, Corfas G, Ojeda SR (2011a) SynCAM1, a synaptic adhesion molecule, is expressed in astrocytes and contributes to erbB4 receptor-mediated control of female sexual development. Endocrinology 152(6):2364–2376

Sandau US, Mungenast AE, McCarthy J, Biederer T, Corfas G, Ojeda SR (2011b) The synaptic cell adhesion molecule, SynCAM1, mediates astrocyte-to-astrocyte and astrocyte-to-GnRH neuron adhesiveness in the mouse hypothalamus. Endocrinology 152(6):2353–2363

Sangiao-Alvarellos S, Manfredi-Lozano M, Ruiz-Pino F, Navarro V, Sanchez-Garrido MA, Leon S, Dieguez C, Cordido F, Matagne V, Dissen GA, Ojeda SR, Pinilla L, Tena-Sempere M (2013) Changes in hypothalamic expression of the Lin28/let-7 system and related microRNAs during postnatal maturation and after experimental manipulations of puberty. Endocrinology 154:942–955

Seminara SB, Messager S, Chatzidaki EE, Thresher RR, Acierno JS Jr, Shagoury JK, Bo-Abbas Y, Kuohung W, Schwinof KM, Hendrick AG, Zahn D, Dixon J, Kaiser UB, Slaugenhaupt SA, Gusella JF, O'Rahilly S, Carlton MB, Crowley WF Jr, Aparicio SA, Colledge WH (2003) The GPR54 gene as a regulator of puberty. N Engl J Med 349:1614–1627

Shao Z, Zhang Y, Yuan GC, Orkin SH, Waxman DJ (2012) Manorm: a robust model for quantitative comparison of ChIP-seq data sets. Genome Biol 13:R16

Shi Y, Lan F, Matson C, Mulligan P, Whetstine JR, Cole PA, Casero RA, Shi Y (2004) Histone demethylation mediated by the nuclear amine oxidase homolog LSD1. Cell 119(7):941–953

Shilatifard A (2012) The COMPASS family of histone H3K4 methylases: mechanisms of regulation in development and disease pathogenesis. Annu Rev Biochem 81:65–95

Simon JA, Kingston RE (2009) Mechanisms of polycomb gene silencing: knowns and unknowns. Nat Rev Mol Cell Biol 10(10):697–708

Steeg PS, Ouatas T, Halverson D, Palmieri D, Salerno M (2003) Metastasis suppressor genes: basic biology and potential clinical use. Clin Breast Cancer 4:51–62

Steuer R, Kurths J, Daub CO, Weise J, Selbig J (2002) The mutual information: detecting and evaluating dependencies between variables. Bioinformatics 18(Suppl 2):S231–S240

Sulem P, Gudbjartsson DF, Rafnar T, Holm H, Olafsdottir EJ, Olafsdottir GH, Jonsson T, Alexandersen P, Feenstra B, Boyd HA, Aben KK, Verbeek AL, Roeleveld N, Jonasdottir A, Styrkarsdottir U, Steinthorsdottir V, Karason A, Stacey SN, Gudmundsson J, Jakobsdottir M, Thorleifsson G, Hardarson G, Gulcher J, Kong A, Kiemeney LA, Melbye M, Christiansen C, Tryggvadottir L, Thorsteinsdottir U, Stefansson K (2009) Genome-wide association study identifies sequence variants on 6q21 associated with age at menarche. Nat Genet 41(6):734–738

Toro CA, Wright H, Aylwin CF, Ojeda SR, Lomniczi A (2018) Trithorax dependent changes in chromatin landscape at enhancer and promoter regions drive female puberty. Nat Commun 9 (1):57. [This study describes how the Trithorax family of epigenetic activators control the epigenetic landscape at the Kiss1 promoter/enhancer regions during the pubertal actiation on Kiss1 transcription in the ARC]

Urbanski HF, Ojeda SR (1985) The juvenile-peripubertal transition period in the female rat: establishment of a diurnal pattern of pulsatile luteinizing hormone secretion. Endocrinology 117:644–649

Urbanski HF, Ojeda SR (1987) Gonadal-independent activation of enhanced afternoon luteinizing hormone release during pubertal development in the female rat. Endocrinology 121:907–913

Wakabayashi Y, Nakada T, Murata K, Ohkura S, Mogi K, Navarro VM, Clifton DK, Mori Y, Tsukamura H, Maeda K, Steiner RA, Okamura H (2010) Neurokinin B and dynorphin A in kisspeptin neurons of the arcuate nucleus participate in generation of periodic oscillation of neural activity driving pulsatile gonadotropin-releasing hormone secretion in the goat. J Neurosci 30(8):3124–3132

Wang Z, Gerstein M, Snyder M (2009) RNA-seq: a revolutionary tool for transcriptomics. Nat Rev Genet 10:57–63

Xue Z, Huang K, Cai C, Cai L, Jiang CY, Feng Y, Liu Z, Zeng Q, Cheng L, Sun YE, Liu JY, Horvath S, Fan G (2013) Genetic programs in human and mouse early embryos revealed by single-cell RNA sequencing. Nature 500:593–597

Yeung KT, Das S, Zhang J, Lomniczi A, Ojeda S, Xu CF, Neubert TA, Samuels HH (2011) A novel transcription complex that selectively modulates apoptosis of breast cancer cells through regulation of FASTKD2. Mol Cell Biol 31(11):2287–2298

Zhang Y, Liu T, Meyer CA, Eeckhoute J, Johnson DS, Bernstein BE, Nusbaum C, Myers RM, Brown M, Li W, Liu XS (2008) Model-based analysis of ChIP-seq (Macs). Genome Biol 9: R137

Epigenetic Regulation of the *GnRH* and *Kiss1* Genes: Developmental Perspectives

9

Joseph R. Kurian and Ei Terasawa

Abstract

Gonadotropin-releasing hormone (GnRH) and kisspeptin neurons are indispensable for reproductive function. This chapter focuses on the epigenetic regulation of the *GnRH* and kisspeptin (*Kiss1*) genes in the context of neuronal development, puberty onset, and maintenance of adult reproductive function. Proper function of GnRH and kisspeptin neurons in the hypothalamus requires coordinated embryonic and postnatal maturation. Recent studies indicate that diverse epigenetic phenomena including the formation of chromatin loops, activation of bivalent domains, maintenance of "stable" histone modifications, and DNA methylation, active demethylation, and hydroxymethylation are all involved in epigenetic regulation of these neurons. This chapter focuses on how these epigenetic regulation components interact with each other and which enzymes or binding factors are involved in specific stages of neuronal development or aging. By emphasizing similar epigenetic mechanisms operating within these two peptidergic neuronal populations, we highlight likely targets for future investigations of environmental influences over puberty timing and progression.

Keywords

Gonadotropin-releasing hormone (GnRH) · Kisspeptin · Puberty · Bivalent chromatin domain · DNA hydroxymethylation · Tet enzymes

J. R. Kurian
Departments of Obstetrics/Gynecology and Internal Medicine, Southern Illinois University School of Medicine, Springfield, IL, USA

St. Johns Hospital, Carol Jo Vecchie Women and Children's Center, Springfield, IL, USA

E. Terasawa (✉)
Department of Pediatrics, University of Wisconsin-Madison, Madison, WI, USA

Wisconsin National Primate Research Center, University of Wisconsin-Madison, Madison, WI, USA
e-mail: terasawa@primate.wisc.edu

S. Wray, S. Blackshaw (eds.), *Developmental Neuroendocrinology*, Masterclass in Neuroendocrinology 9, https://doi.org/10.1007/978-3-030-40002-6_9

9.1 GnRH Neurons

Gonadotropin-releasing hormone (GnRH) neurons are critical relays in the axis controlling reproductive function. These neurons reside in the preoptic area and basal hypothalamus and project to the median eminence, where they release the decapeptide GnRH into the portal circulation. GnRH subsequently stimulates anterior pituitary release of the gonadotropins luteinizing hormone (LH) and follicle-stimulating hormone (FSH), which promote gametogenesis and gonadal steroidogenesis in both sexes as well as ovulation in females. In order to properly stimulate gonadotropin release, GnRH must be supplied to the portal circulation in sufficient, episodic pulses. In fact, the onset of puberty and maintenance of reproductive function depends on the elevated and finely coordinated pulsatility of GnRH release. While upstream mechanisms fine-tune GnRH neuronal activity, GnRH neurons also have an intrinsic capacity to release GnRH peptide in a pulsatile manner (Terasawa 2001).

GnRH-expressing neurons reside in a complex milieu of other neurons and glia; this presents a challenge for studying the cellular and molecular mechanisms of neuronal differentiation and function. Fortunately, their unique ontogeny provides an opportunity to isolate GnRH-expressing neurons for in vitro studies. In the rhesus monkey, these neurons differentiate from progenitor cells in the nasal placode between embryonic days (E) 32–34 (Ronnekleiv and Resko 1990; Quanbeck et al. 1997). GnRH neurons subsequently begin migrating into the brain at about E36 and settle in the hypothalamus by E55 (Terasawa et al. 2001). Isolating the nasal placode after E34 but prior to migration provides a neuronal population enriched in GnRH neurons. We found that placode tissue isolated on E36 developed typical patterns of mature activity (e.g., GnRH peptide release; Terasawa et al. 1993, 1999; Kurian et al. 2010a) after about 2–2.5 weeks in vitro. Wray and colleagues, who developed a similar murine in vitro culture model (Fueshko and Wray 1994), have also reported a period of gradual maturation after isolation from the nasal placode (Constantin et al. 2009). In addition, they report that development of GnRH peptide release patterns is paralleled by increasing GnRH gene expression and peptide biosynthesis (Maurer and Wray 1997; Moore and Wray 2000). A question arises: what mechanism triggers increasing levels of gene expression during maturation of GnRH-expressing neurons?

9.2 DNA Methylation in GnRH Neurons

The genetic control of GnRH gene expression is well characterized and depends on several *cis*-regulatory sequences in the 5′ region of the gene (Box 9.1). The rat gene contains a neuron-specific enhancer region between −1863 and −1571 (Kepa et al. 1992, 1996b; Clark and Mellon 1995; Whyte et al. 1995). This spans a major region of homology between rat (−1786 to −1559) and human (−2766 to −2539) genes. Interestingly, however, this portion of the human gene does not appear to activate gene expression. In fact, based on serial truncations of the 5′ human GnRH gene in luciferase assay constructs, it appears that this region represses gene expression

(Kepa et al. 1996a). Importantly, this area has sequence similarity to a 5′ portion of the rhesus monkey *GnRH* gene. We noticed that this distal 5′ region of the rhesus monkey gene contains a 243-bp segment (−2126 to −1863) that has 60% GC content and a CpG (cytosine–guanine) dinucleotide observed to expected ratio of 0.65 (14 CpG sites). These characteristics define the region as a CpG island (CGI; Gardiner-Garden and Frommer 1987). CGIs, when associated with gene promoters, are related to the epigenetic regulation (DNA methylation) of gene expression (Deaton and Bird 2011).

Box 9.1 Genetic and Epigenetic Regulation of the *GnRH* Gene During Development

Understanding transcriptional regulation of the *GnRH* gene is important, as deficiency of this process leads to abnormal timing of puberty and infertility.

The mammalian form of GnRH (GnRH1) encoded by the *GnRH* gene, consists of a ten amino acid-peptide (pyro-Glu-His-Trp-Ser-Tyr-Gly-Leu-Arg-Pro-Gly-NH2), cleaved from the 92-amino acid GnRH precursor protein (prepro-GnRH peptide). The *GnRH* gene contains four exons and three introns. The 5′ *GnRH* gene-regulatory region comprises a proximal promoter and three enhancer regions. Transcription factors bind to conserved enhancer regions and the proximal promoter region.

Genetic Regulation: Since the 1990s several groups reported using in vitro GnRH cell lines and in vivo transgenic mouse approaches to identify various transcription factors that bind the 5′ *GnRH* gene-regulatory region, and activate or repress *GnRH* gene transcription. Among the activators are the homeodomain transcription factors OCT1, OTX2, Six3/6, MSX1 and VAX1, neuron-specific elements, GATA4 (Radovick et al. 1994; Clark and Mellon 1995; Lawson et al. 1998; Kelley et al. 2000; Kim et al. 2007; Larder and Mellon. 2009; Gan et al. 2012; Xie et al. 2015; Larder et al. 2013; Hoffmann et al. 2018), an estrogen response element (human/primates only, Radovick et al. 1991b, 1994), and a kisspeptin response element (Novaira et al. 2012, 2016). Repressors are C/EBPβ (Belsham and Mellon 2000; Messina et al. 2016), MSX1/2 (Givens et al. 2005, Rave-Harel et al. 2005), NKX2.1/TTF-1 (Provenzano et al. 2010), and ZEB1 (Messina et al. 2016). See further discussion in Hoffmann and Mellon (2018).

Epigenetic Regulation: More recently, *GnRH* gene expression was also found to be altered by environmental factors through mechanisms such as DNA methylation and histone modifications without changes in nucleotide sequences. Recent findings are summarized in this chapter.

DNA methylation is the covalent addition of a methyl (–CH3) group to nucleotides. Mammalian DNA-methyltransferase (DNMT) enzymes catalyze this reaction. Mammalian DNA methylation occurs primarily at the 5′ carbon of cytosines in CpG dinucleotides and to a lesser extent at CpH (H=A, T, C) dinucleotides (Ramsahoye et al. 2000; Guo et al. 2014). There are three primary

DNMT enzymes (DNMT1, 3a, and 3b), each critical to development as demonstrated by embryonic or early postnatal lethality in monogenic null mouse models (Li et al. 1992; Okano et al. 1999). DNMT1 is responsible for faithful maintenance of DNA methylation after replication through cell division (Bestor and Ingram 1983; Bestor et al. 1988; Hermann et al. 2004). DNMTs 3a and 3b are de novo methyltransferases responsible for newly acquired methylation, such as during the initial establishment of methylation patterns during early embryonic development (Okano et al. 1998, 1999) and throughout neuronal maturation and aging (Guo et al. 2014). Once established, DNA methylation can have several impacts on gene transcription. Methylated DNA can directly alter transcription factor binding to cis-regulatory sequences or attract methyl-binding proteins, which in turn block transcription factor binding. In addition, methyl-binding proteins (e.g., MeCP2, Mbd2, and Kaiso) interact with histone-modifying factors to alter chromatin structure.

Until our recent studies (Kurian et al. 2010a), there were no reports on the epigenetic aspects of GnRH neuron maturation or function. The CGI in the 5′ region of the rhesus monkey *GnRH* gene suggested to us that DNA methylation had some role in neuronal function (Fig. 9.1). We hypothesized that increasing peptide release during in vitro maturation of GnRH neurons resulted from increased *GnRH* transcription, and changing DNA methylation patterns across the rhesus monkey *GnRH* gene, particularly within the 5′CGI. As suspected, we found that GnRH mRNA levels were low at day 0 but rose dramatically by day 20 of in vitro culture. This increase was paralleled by a dramatic decrease in CpG methylation status in the 5′ CGI. We also found that a comparable phenomenon occurs across puberty. Similar to the report by Plant and colleagues which showed that GnRH mRNA levels increase between juvenile and pubertal stages in the medial basal hypothalamus (MBH) of orchidectomized male rhesus monkeys (El Majdoubi et al. 2000), we observed that GnRH mRNA levels increase across puberty in the male monkey MBH. Furthermore, methylation status of the 5′ CGI of the *GnRH* gene was lower in adults compared to prepubertal male rhesus monkeys (Kurian et al. 2011; Kurian and Terasawa 2013). This suggests that the developmental rise in GnRH gene expression is, at least partly, the result of DNA demethylation across a CGI in the GnRH gene.

When evaluating these findings, it is important to consider the distinct pattern of GnRH neuronal activity across development. In primates, GnRH release is elevated, as indicated by peripheral luteinizing hormone levels, during a brief perinatal period (this is commonly known as “mini-puberty”), but then decreases during juvenile development before gradually increasing again through puberty. Our findings suggest that each period of elevated GnRH release is linked to sequential waves of demethylation of the *GnRH* gene CGI. Specifically, demethylation observed during in vitro maturation mirrors that seen during embryonic development, and leads to elevated GnRH expression during the postnatal period. In contrast, measurements in MBH tissue reflect developmental changes across pubertal maturation. The decrease in DNA methylation seen during puberty is indicative of two potential scenarios. One possibility is that the process of DNA demethylation initiated in embryonic development might stall during the juvenile period, and subsequently resume at

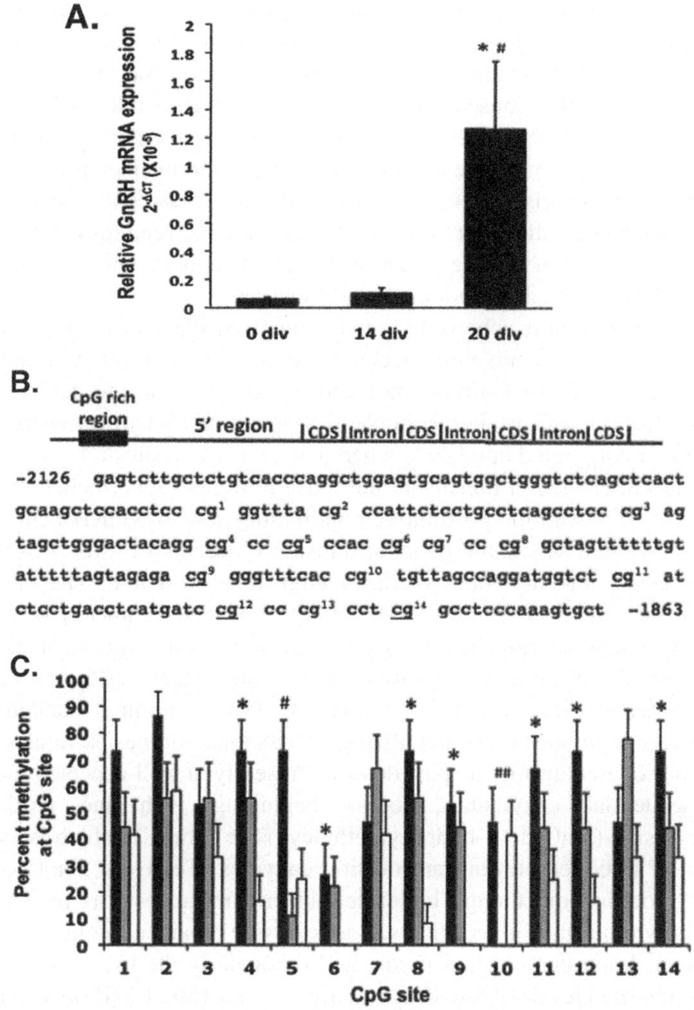

Fig. 9.1 Embryonic development of primate GnRH neurons is marked by a dramatic shift in *Gnrh* gene expression and active *de*methylation of a 5′ CpG island. (**a**) Total RNA was extracted from nasal placode cultures containing GnRH neurons at 0, 14, and 20 days *(div)*. *Gnrh* mRNA levels, measured by quantitative PCR, started to increase after 14 div, reaching the highest level at 20 div. ($*P <$ 0.05 versus 0 div; $\#P =$ 0.05 versus 14 div). (**b**) A schematic representation of the rhesus monkey *GnRH* gene depicting the location of the 5′ CpG island (CGI) and nucleotide sequence of this region. This region is the only classifiable CGI within 2500 bases upstream of the *Gnrh* transcription start site. CpG sites are indicated with superscript numbers corresponding to the CpG sites in the bottom graph. (**c**) Nasal placode regions, containing GnRH neurons, were dissected from two rhesus monkey embryos at E36 and E37, plated and then harvested on 0, 14, or 20 div. DNA extracted from pooled samples (four cultures) at each time point was bisulfite sequenced. Percent changes in methylation at each CpG site on 0 div (black bars), 14 div (gray bars), and 20 div (white bars) are shown. CpG methylation status was significantly higher at 0 div compared with 20 div at sites 4, 5, 6, 8, 9, 11, 12, and 14 ($*P \leq 0.01$). CpG methylation status was significantly higher at 0 div compared with 14 div at site 5 ($\#P < 0.05$) and at 0 div compared with 14 div but not 20 div at site 10 ($\#\#P <$ 0.05). Modified from Kurian et al. Endocrinology 151: 5359–5368 (2010a)

puberty onset. Alternatively, the CGI may become re-methylated during juvenile development, and again demethylated at puberty onset. The latter scenario would suggest that lower CpG methylation status must be maintained to enable elevated GnRH gene expression. Consequently, a mechanism responsible for DNA demethylation and perhaps another mechanism responsible for maintaining hypomethylated DNA might be necessary for the transition to puberty and maintenance of reproductive function. A comparison of CpG methylation status between perinatal and early pubertal MBH tissue will be necessary to differentiate between these two scenarios. Nonetheless, our current findings suggest that DNA demethylation is an important aspect of GnRH neuronal development and function.

DNA demethylation is achieved via two general mechanisms, passive or active. Passive demethylation occurs through cell divisions and interruption of maintenance methyltransferase activity. Current pharmacological approaches for DNA demethylation target this mechanism. For example, 5-azacytidine (5-aza) is metabolized and subsequently incorporated into DNA, where it then acts as a substrate that covalently traps DNMTs after methyl transfer to the 5-aza nitrogen (Schermelleh et al. 2005; Svedruzic 2008). Given the postmitotic/nondividing state of GnRH neurons in our in vitro cultures and across the pubertal transition measured in MBH samples, the process of DNA demethylation we discovered must be active. The mechanisms responsible for active DNA demethylation are not well characterized, and until recently skepticism has remained over the existence of this process in mammalian systems. Several mechanisms are now proposed, and intense efforts to validate or better characterize these mechanisms continue. For a thorough background, we suggest a recent review (Wu and Zhang 2010) that outlines several promising pathways discovered during the past decade. Presently, a well-accepted mechanism is the sequential enzymatic process beginning with the oxidation of 5-methylcytosine (5mC) to 5-hydroxymethylcytosine (5hmC), which is carried out by anyone of three ten-eleven-translocation enzymes (Tet 1–3). 5hmC is a stable epigenetic modification, though under certain circumstances it is recognized and excised by thymine DNA glycosylases. Base excision repair subsequently completes the transition. Importantly, brain tissue, and particularly the hypothalamus, shows the highest reported levels of 5-hydroxymethylcytosine (5hmC) (Branco et al. 2012).

Emerging physiological evidence also supports a role for Tet enzymes in neuroendocrine development, and particularly the control of reproductive function. For example, *Tet1* knockout mice exhibits deficits in fecundity. This effect is primarily a female-specific abnormality, as typical litter sizes result from crossing wild-type females with Tet1 knockout males, whereas wild-type males mated to *Tet1* knockout females produce significantly fewer offspring per litter (Dawlaty et al. 2011). A subsequent report suggests this defect in fecundity might be the consequence of abnormal progression of female germ cell development through the second meiotic division prior to ovulation (Yamaguchi et al. 2012). Interestingly, stimulation of the second meiotic division and ovulation requires a GnRH-driven LH surge (Mehlmann 2005). Consequently, given the apparent active DNA demethylation associated with GnRH neuronal development, the effect of Tet depletion on abnormal reproductive function might be the consequence of altered Tet-mediated differentiation of the neuroendocrine hypothalamus. While hypothalamic expression patterns of Tet enzymes are not yet reported, we recently found that basal hypothalamic *TET2* expression increases across puberty in female rhesus monkeys (Kurian and Terasawa; unpublished observation). That the deficit in fecundity was even more pronounced in *Tet1/Tet2* double knockout mice (Dawlaty et al. 2013) gives further credence to the suggestion that Tet enzymes act in the hypothalamus to promote differentiation of neurons controlling stage-specific reproductive functions.

9.3 Histone Modifications in GnRH Neurons

Histones are an integral component of nucleosomes, the primary units for genome organization or compaction. Consequently, through their interaction with DNA, these proteins have a critical role in determining gene expression patterns. There are four primary classes of histones, 1 through 4. Histones 2 through 4 are components of the core octamer, which DNA circumnavigates in about 146 base pairs to form a single nucleosome. Histone 1 is a scaffold protein, which tightly packages nucleosomes when present. Several variants of this histone are distinguishable by sensitivity to hormones (Banks et al. 2001). Histone 2 also has several variants including A, AX, and B. Histone 2AX is particularly intriguing in the context of neuronal maturation (Lee et al. 2010) and function based on its association with activity-dependent DNA double-strand breaks in neurons (Suberbielle et al. 2013). Histones 3 and 4 complete the nucleosome octamer with histone 3 the most heavily investigated in the context of neuroendocrine function. Histone 3 has two variants, H3.3A and H3.3B. These variants are integral to DNA replication-independent histone switching, which would be presumed an important mechanism in the regulation of post-mitotic cell activity. To our knowledge, nothing is reported regarding the relationship between histone switching and hypothalamic neuronal maturation or function.

Histone proteins package DNA largely due to their predominant positive charge attracting negatively charged DNA. Posttranslational modifications (PTMs) alter the strength of that attraction, and recruit or repel transcriptional machinery and histone-modifying enzymes. Consequently, these PTMs alter gene accessibility and rates of transcription. Several known modifications include acetylation, phosphorylation, methylation, ubiquitination, sumoylation, and GlcNAcylation. To date, measurements of histone PTMs in neuroendocrine systems have focused on acetylation and methylation. Acetylation leaves a more negative charge on histones and consequently promotes transcription. Methylation is neutral, and depending on the location and degree (mono, di, or tri-methylation), it can either promote or inhibit transcription.

Mellon and colleagues were the first to report a pattern of permissive histone modifications enabling elevated or mature *Gnrh* gene transcription (Iyer et al. 2011). Their studies capitalized on the distinct stages of development between two GnRH neuronal cell lines. GN11 cells, originally isolated from a tumor in the mouse nasal placode (Radovick et al. 1991a), express *Gnrh* at very low levels, whereas GT1 cells, which were isolated from an analogous tumor in the mouse hypothalamus (Mellon et al. 1990), are characterized by mature activity patterns including elevated *Gnrh* expression. In essence, comparisons between these two cell lines are similar to our evaluations of embryonic nasal placode derived neurons from days 0 and 20 in vitro described above. They found that the GnRH gene promoter and enhancer regions in immature GN cells were more heavily associated with a repressive histone modification: histone 3 (H3) lysine 9 (K9) di-methylation (me2). On the other hand, the same genomic regions in GT1 cells were associated with the permissive H3K9 acetylation and H3K4me3 PTMs. For comparison, they also evaluated these histone PTM patterns in a nonneuronal (NIH3T3) cell line. The repressive PTMs were high in NIH3T3 cells, intermediate in GN cells and low in GT1 cells. The presence of permissive PTMs was low and similar between NIH3T3 and GN cells but significantly higher in mature GT1 cells.

The intermediate repressive chromatin state in GN cells is indicative of a bivalent promoter, where both repressive and permissive histone PTMs maintain genes in a repressed albeit primed position. These chromatin domains were first described in embryonic stem (ES) cells and associated with developmentally regulated transcription factors (Azuara et al. 2006; Bernstein et al. 2006; Pan et al. 2007). Several recent reports, taken together, indicate that establishment of bivalent domains through polycomb repressive complex 2 (PRC2) and COMPASS families is critical for neural lineage differentiation from ES cells (Yu et al. 1995; Yagi et al. 1998; Glaser et al. 2006; Pasini et al. 2007; Shen et al. 2008). The PRC2 component, Ezh2, is responsible for H3K27 methylation, which is subsequently bound by the PRC1 complex to maintain tri-methylation at H3K27. COMPASS complexes methylate H3K4. Together, these complexes are proposed to establish a bivalent promoter with the heavily repressive H3K27me3 mark in close proximity to H3K4me3. Interestingly, this bivalent domain also appears related to the juvenile repression and subsequent activation of *Kiss1* (kisspeptin) gene expression during the pubertal transition in female rats (Lomniczi et al. 2013). As a major stimulant of GnRH release, this epigenetic regulation of *Kiss1* expression (discussed in detail below) may be instrumental in pubertal development. Together, this also suggests that bivalent domain activation may be a consistent mechanism across reproductive neuroendocrine systems.

9.4 Tet Enzyme Activity in GnRH Neurons and Reproductive Function

A recent discovery points to Tet enzymes as critical mediators of activation at bivalent promoters during neuronal differentiation. Specifically, while Ezh2 (the H3K27 methyltransferase component of PRC2) is critical for progression of neuronal precursor cells toward a neuronal fate, Tet2 or Tet3 appear to complete the process of differentiation (Hahn et al. 2013). In fact, these studies also found that accumulation of intragenic 5-hydroxymethylcytosine is associated with loss of H3K27me3 near regions with the most significant gene activation during neuronal differentiation. In addition, Tet2 was recently reported to promote H3K4me3 through association with the Set1/COMPASS complex (Deplus et al. 2013). Based on this background, our recent studies have focused on Tet enzyme activity in the context of GnRH neuron development and function. Initially we utilized the developmentally distinct GnRH neuronal cell lines. We discovered that *Tet2* expression is low in immature GN11 cells, but substantially higher in mature GT1–7 cells. Subsequently we found that overexpression of Tet2 in immature cells led to significant elevation of *Gnrh* gene expression. This increase was potentiated when cultured cells were supplemented with glutamine, which enhances Tet2 activity by increasing intracellular concentrations of the cofactor alpha-ketoglutarate (Yang et al. 2014; Carey et al. 2015). Using chromatin immunoprecipitation, we evaluated Tet2 binding as well as H3K4me3 and H3K27me3 abundance across the mouse *GnRH* gene 5′ region (Kurian 2016; Fig. 9.2). Chromatin conditions were evaluated in four conditions: (1) GN11, (2) GT1–7, (3) GN11 cells overexpressing Tet2, and (4) GT1–7 cells with Tet2 eliminated by CRISPR/Cas9-mediated gene disruption. Tet2 binding was low in GN11 cells across the entire 5′ region, but significantly elevated in GT1–7 cells at the promoter and neuron-specific enhancer region. The activating H3K4me3 modification largely mirrored Tet2 binding patterns, with significantly higher abundance in GT1–7 compared to GN11 cells. Overexpression of Tet2 in GN11 cells significantly increased Tet2 binding to the *Gnrh* promoter and, to a lesser extent, the neuron-specific enhancer; H3K4me3 levels were also significantly higher at the promoter compared to non-transfected GN11 cells. Disruption of Tet2 in GT1–7 cells decreased Tet2 binding across the entire region; remarkably, H3K4me3 levels significantly dropped at the neuron-specific enhancer after *Tet2* disruption in GT1–7 cells. H3K27 levels were largely consistent between all conditions at each site evaluated. Altogether, these results show that Tet2 influences bivalent promoter-specific histone modification patterns near the *Gnrh* gene and contributes to the elevation and maintenance of elevated *Gnrh* transcription (Fig. 9.3). Our subsequent studies evaluating the impact of Tet2 activity on puberty progression and reproductive function support this conclusion, as described below.

Using Cre-lox technology, we recently generated mice with *Tet2* loss of function in GnRH neurons (gTKO animals). We expected that this would disrupt GnRH neuron epigenetic maturation and consequently the acquisition of reproductive competence (i.e., puberty onset and progression). Surprisingly, we found no evidence of pubertal disruption though, remarkably, gTKO males exhibited an age-dependent decrease of circulating LH and impairment in fecundity not seen in wild type animals (Kurian et al. 2014). Considered together with the loss of H3K4me3 abundance at the *Gnrh* promoter after *Tet2* disruption in a cell line derived from mature GnRH neurons, these physiological findings suggest that epigenetic patterns must be actively maintained, as opposed to established during development.

In fact, Messina and colleagues (2016), focusing on the micro-RNAs (miR-200 and miR-155), have demonstrated the reliance on active maintenance of *Gnrh* gene expression. Using Cre-lox technology in mice, they specifically disrupted the miRNA processing enzyme *Dicer* in GnRH neurons. Compared to control animals,

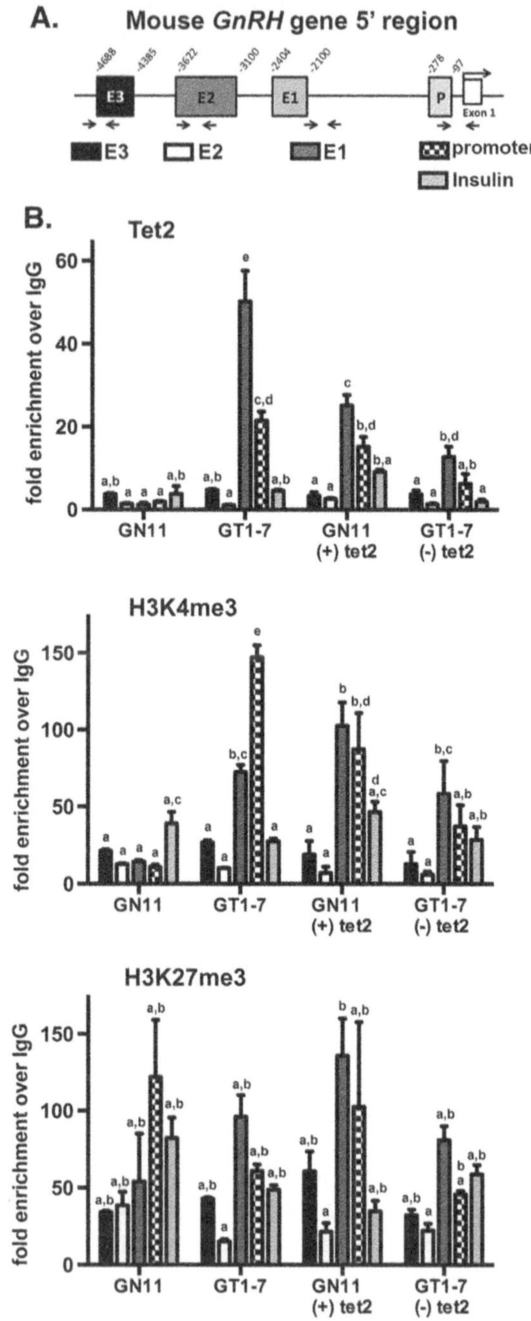

Fig. 9.2 Tet2 binds the mouse *Gnrh* gene and promotes accumulation and maintenance of an activating histone modification. (**a**) Schematic diagram of the mouse GnRH gene showing relative locations of 3 enhancer regions (E1–3) and the gene promoter (P) [adapted from Iyer et al. (2010)]. (**b**) Graphs represent relative abundance of Tet2, H3K4me3 and H3K27me3 detected at each *GnRH* gene region using Chromatin Immunoprecipitation from immature GnRH neurons (GN11), differentiated GnRH neurons (GT1–7), GN11 cells transfected with Tet2 (GN11 (+) tet2) and GT1–7 cells with Tet2 knockdown (GT1–7 (−) tet2). Tet2 binds the promoter and neuron-specific enhancer (E1) in GT1–7 and GN11 (+) tet2 cells. H3K4me3 is significantly more abundant at the promoter and E1 region in GT1–7 and GN11 (+) tet2 compared to GN11 cells. H3K27me3 abundance was similar across all cell lines at all regions evaluated. Comparisons were made within each chromatin-associated factor. Levels assigned different letters are significantly different ($P < 0.05$) from each other by Tukey post hoc after Two-Way ANOVA across all cell lines and genomic regions. From Kurian et al. Endocrinology 157: 3588–3603 (2016)

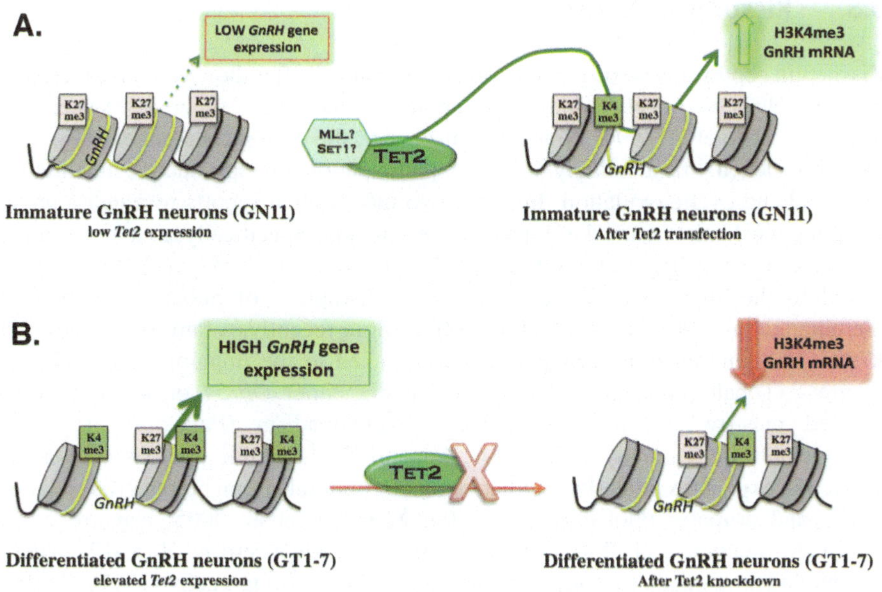

Fig. 9.3 Tet2 influences accumulation and maintenance of the activating H3K4 trimethylation in immature and differentiated GnRH neurons, respectively. (**a**) Immature GnRH neuronal cell lines (GN11) express *Gnrh* at very low levels and less *Tet2* than differentiated GnRH neuronal cell lines (GT1–7). Increasing Tet2 expression in GN11 cells by transient transfection leads to an elevation of *Gnrh* mRNA and H3K4me3 at the *Gnrh* gene neuron-specific enhancer and promoter. The histone modifications are suspected to be generated by the Tet2 interacting histone-modifying enzyme (MLL: mixed lineage leukemia) or complex (SET1: a histone methyltransferase complex) (**b**) Knockdown of *Tet2* in differentiated GnRH neuronal cultures (GT1–7 cells) leads to a loss of Tet2 binding near the *Gnrh* gene promoter, loss of H3K4me3 abundance in the same region and reduction in *Gnrh* mRNA levels. Modified from Kurian et al. Endocrinology 157: 3588–3603 (2016)

these mice exhibited a gradual decline in GnRH neurons, from typical numbers and distribution at P0 to no detectable *Gnrh* expression in adult animals. These animals also developed hypogonadotropic hypogonadism and had significantly lower circulating LH and FSH. While mRNA and miRNA expression patterns have not been profiled in mice lacking *Dicer* in GnRH neurons, it is clear that miR-200 and miR-155, which target the *GnRH* gene repressors Zeb-1 and Cebpp, respectively, are critical for GnRH neuron maintenance in rodents (Messina et al. 2016). Moreover, the work by Messina et al. (2016) supports the concept that a set of microRNAs regulates *Gnrh* gene expression during sexual development. Whether additional miRNAs are involved in rodents, whether a similar mechanism can be seen in primates, and whether the microRNA-dependent epigenetic regulation of GnRH release is involved in the pathophysiology of disease states, such as idiopathic hypogoanadotropic hypogonadism in human patients remain open questions.

9.5 Kisspeptin Neurons

Kisspeptin peptides were originally coined "metastins" after they were discovered in the mid-1990s as novel ligands that suppressed metastasis of breast cancer and melanoma (Lee et al. 1996). Since that discovery, the biological roles of kisspeptin have broadened, most notably for their participation in reproductive maturation (puberty) and cyclic ovulation. In 2003, two independent reports revealed a novel signaling mechanism crucial to reproductive function. Specifically, mutations in the human gene encoding G-protein coupled receptor 54 (*GPR54* or *KISS1R*) were linked to the absence of puberty or severe disruption of pubertal progression (Seminara et al. 2003; de Roux et al. 2003). More recently, a familial mutation in *KISS1* (the gene encoding kisspeptins) was linked to the impairment of pubertal progress (Topaloglu et al. 2012). Mice lacking either *Gpr54* or *Kiss1* exhibits delayed puberty and hypogonadotropic hypogonadism (Funes et al. 2003; d'Anglemont de Tassigny et al. 2007; Lapatto et al. 2007), whereas exogenous kisspeptin exposures potently stimulate gonadotropin release in prepubertal rodents, sheep, and primates (Gottsch et al. 2004; Messager et al. 2005; Navarro et al. 2005a, b; Shahab et al. 2005; Dhillo et al. 2007; Guerriero et al. 2012). These kisspeptin stimulations of gonadotropin release occur through activation of GPR54 (KISS1R) located on GnRH neurons (Hrabovszky et al. 2010; Kirilov et al. 2013; Yeo et al. 2014), suggesting GnRH neurons are capable of adult levels of activity prior to puberty onset, and that developmental activation of kisspeptin systems may play a role in the timing of puberty. Consequently, based on the proposed environmental influence over timing of puberty, significant efforts are currently directed toward characterizing the epigenetic development and regulation of kisspeptin neurons.

Kisspeptin cell bodies are concentrated in the arcuate nucleus (ARC) and the anteroventral periventricular nucleus (AVPV)/periventricular nucleus (PEN) (Gottsch et al. 2004; Smith et al. 2005; Clarkson and Herbison 2006; Kauffman et al. 2007). In rodents, the presence of kisspeptin cell bodies in the medial amygdala is also described (Kim et al. 2011). While *KISS1* expression in the ARC is consistent across species (Smith et al. 2005, 2010; Ramaswamy et al. 2008; Shahab et al. 2005; Smith 2008; Alcin et al. 2013), there are some species differences in the distribution pattern of kisspeptin neurons in the anterior population. In rodents, the AVPV population is very clear but in primates, including humans, a similar but less conspicuous population is present (Rometo et al. 2007; Hrabovszky et al. 2012; Smith et al. 2010; Watanabe et al. 2014; Vargas Trujillo et al. 2017). In sheep and goats, in contrast, kisspeptin neurons are diffusely distributed in the preoptic area (POA), although within the POA a surge activated population is restricted in the AVPV area (Franceschini et al. 2006; Smith et al. 2009; Hoffman et al. 2011; Matsuda et al. 2015). A clear sexual dimorphism of the AVPV population, i.e., a larger number of kisspeptin neurons in female, exists in mice and rats (Clarkson and Herbison 2006; Kauffman et al. 2007; Ohkura et al. 2009, Semaan and Kauffman 2010), and this reflects the significance of AVPV kisspeptin neurons in the preovulatory surge in rodents. However, there is little sex difference in the number of the anterior kisspeptin neurons in sheep and monkeys and anterior kisspeptin neurons respond to the estrogen challenge in castrated male monkeys, but not in the castrated ram (Cheng et al. 2010; Watanabe et al. 2014). Therefore, despite the well-established concept that negative feedback effects of estradiol are mediated through estrogen receptor alpha (ERα) located in kisspeptin neurons in the ARC, whereas the positive feedback effects of estradiol are mediated through ERα in kisspeptin neurons in the rodent AVPV (Smith 2013; Herbison 2016), the role of kisspeptin neurons in regulation of GnRH release in primates and ruminants requires further investigation.

In rodents, kisspeptin neuronal systems develop in characteristic regional and sex-specific patterns. *Kiss1* is detectable in the rat and mouse ARC by E13–14 (Desroziers et al. 2012a; Kumar et al. 2014) and persists stably throughout postnatal and pubertal development (Poling and Kauffman 2013; Semaan and Kauffman 2015). In contrast, AVPV *Kiss1* expression becomes apparent at P7–10 (Semaan et al. 2010; Desroziers et al. 2012b; Clarkson et al. 2009) with only a small amount of kisspeptin peptide detectable by P15 (Clarkson et al. 2009) but levels substantially elevate prior to the onset of puberty (Clarkson et al. 2009; Semaan and Kauffman 2015). In addition, the appearance of *Kiss1* AVPV briefly lags behind in males and is significantly lower in pubertal and adult males compared to females (Kauffman et al. 2007; Semaan et al. 2010; Clarkson and Herbison 2006; Homma et al. 2009). The sex- and region-specific differences in kisspeptin expression are critical to sex-specific reproductive function and have therefore been a focus of early investigations of rodent kisspeptin neuron epigenetic regulation. The origin and prenatal development of prenatal nonhuman primate kisspeptin neurons have not been studied. Nevertheless, the kisspeptin neurons during the neonatal mini-puberty are active at levels consistent with adult neurons, and subsequently enter a dormant state during the prepubertal period until the onset of puberty (Ramaswamy et al. 2013; Shahab et al. 2018).

9.6 Epigenetic Aspects of Sexual Differentiation of Kisspeptin Neurons

Sex differences in the brain are largely created during a prenatal or perinatal "critical period" by a sex-specific steroid environment (Phoenix et al. 1959). During this critical period, a male-specific acute surge of testicular testosterone secretion coordinates the masculinization/defeminization of the brain and associated physiology and behaviors (Simerly 1998, 2002). Importantly, prior to acute sex steroid exposure, the brain exhibits the potential to develop both male- and female-specific neural circuitry. While in rodents testosterone aromatized into estradiol is responsible for this organizational action (Arnold and Gorski 1984; McCarthy 1994), in ruminants and primates, androgens themselves are responsible (Wallen 2005; Puttabyatappa and Padmanabhan 2017). As evidence, male-like brain development can be initiated by perinatal estradiol treatment of female rats and mice, whereas androgen treatment of early to mid-embryonic sheep and monkey embryos directs male-like brain development (Phoenix et al. 1959; Arai and Gorski 1968; see Wallen

2009). Conversely, castration of newborn male rats and mice leads to a female-like brain in adulthood (Döhler et al. 1984). The critical period in rodents extends through approximately P10, after which sex steroid manipulations no longer alter sexually dimorphic brain structures (Gorski 1985). Some brain traits, however, appear to be influenced by steroid exposures during the pubertal stage (Schulz and Sisk 2006; Schulz et al. 2009).

Importantly, in addition to sex steroids, sexually dimorphic experiences during the perinatal critical period also appear to masculinize the brain. For example, male rodents experience a higher degree of maternal grooming than female littermates during this period. Subjecting female rats to a sex steroid surge or simulated maternal grooming in the first week of life is reported to masculinize the ERα gene CpG methylation pattern and gene expression profile in the MBH (Kurian et al. 2010b). That study was the first to report an equivalent impact of steroids and environmental influence over sexual differentiation of epigenetic patterns in the brain. This suggested that epigenetic mechanisms likely had a significant role in the sexual differentiation of the brain in general. In fact, several "critical periods" for sex steroid exposures were soon after related to epigenetic factors, including bed nucleus of the stria terminalus (BNST) size and vasopressin fiber projections (Murray et al. 2011), male sexual behavior in rats and olfactory behavior in mice (Matsuda et al. 2011), and steroid receptor and DNA methyltransferase 3a expression (Schwarz et al. 2010; Kolodkin and Auger 2011).

To date, only one study has investigated whether sexual differences of the two kisspeptin systems are driven by epigenetic factors. Specifically, Semaan, Kauffman, and colleagues evaluated whether histone deacetylation and DNA methylation contributed to sexually dimorphic regulation of AVPV *Kiss1* gene expression and cell number (Semaan et al. 2012). They pharmacologically blocked histone deacetylation during the postnatal critical period with valproic acid, and evaluated male and female AVPV *Kiss1* expression in adulthood. While the authors observed the expected disruption of sex differences in the BNST similar to that reported by Murray et al. (2011), *Kiss1* gene expression in the AVPV was elevated in both sexes compared to same sex controls and there was no sex-specific effect per se. Whether, the elevation of *Kiss1* expression in males had a functional impact (i.e., feminization of estradiol-induced LH surge potential) was not evaluated and remains an important question. Additionally, potential off-target effects of valproic acid are a significant concern in interpreting these results, as valproic acid is also a voltage-gated sodium channel blocker and elevates GABA levels in the brain, which could indirectly alter GnRH neuronal activity.

In the same study, Semaan et al. (2012) further evaluated DNA methylation over 70 unique sites of the *Kiss1* gene in microdissected AVPV and ARC tissues. CpG methylation was significantly higher in females, primarily at the putative *Kiss1* promoter in the AVPV but not in the ARC. This finding was unexpected, as methylation is typically associated with reduced gene expression, though AVPV Kiss1 expression is substantially higher in females than males. Consequently, it is currently unclear how methylation alters Kiss 1 gene expression, though it is likely involved in chromatin conformation, which is critical to hormone-dependent Kiss

1 expression in the AVPV (discussed in the following section). Whether the perinatal hormone surge drives epigenetic changes in the *Kiss1* gene in the AVPV also remains an open critical question.

9.7 Epigenetic Aspects of Estrogen Signaling on *Kiss1* Gene Expression

An essential aspect of female reproductive function is the ability to generate LH surges in response to estradiol. LH surges are critical to gamete maturation and ovulation and require ERα signaling in AVPV kisspeptin neurons (d'Anglemont de Tassigny and Colledge 2010; Dubois et al. 2015). As stated above, acute estradiol (E2) exposure in adult females induces suppression of *Kiss1* gene expression in the ARC, whereas it activates *Kiss1* expression in the AVPV (Smith et al. 2005). This phenomenon was recently utilized to interrogate the epigenetic effects of E2 action on *Kiss1* gene expression in both regions (Fig. 9.4).

First, using only adult female mice, Tomikawa and colleagues reported that E2 exposure increased ERα binding at the *Kiss1* promoter in the AVPV, but not in the ARC (Tomikawa et al. 2012). In addition, total histone 3 (H3) acetylation at the *Kiss1* promoter increased in the AVPV and simultaneously dropped in the ARC during E2 exposure. Interestingly, H3 acetylation in only the ARC also changed in a 5′ region upstream of the *Kiss1* promoter (increased acetylation) and a 3′ intergenic region (decreased acetylation) located downstream of the last exon of the *Kiss1* gene. These later differences in H3 acetylation are probably the consequence of brain region specific chromatin loop conformations. Specifically, using chromatin conformation capture assays, Tomikawa and colleagues exquisitely found that in the ARC, the *Kiss1* gene is held in a tight loop, with interactions between a 5′ region and a proximal 3′ region centered at the promoter. In contrast, no chromatin loop is apparent in the AVPV; however, during E2 exposure, a loop forms between the promoter and a distal intergenic 3′ region, later shown to act as an enhancer in the AVPV. That same distal 3′ region also associates with the promoter in the ARC during E2 exposure. Whereas this region is necessary for enhancing E2-driven AVPV *Kiss1* expression, it is also necessary for E2 inhibition of ARC *Kiss1* expression. We consequently refer to this region as a "transcriptional regulatory region." Importantly, while histone acetylation is typically associated with increased gene expression, the ARC 5′ region exhibits increased H3 acetylation even while *Kiss1* expression drops significantly. Examination of the DNA loop structures generated by E2 exposure in the ARC (Fig. 9.4) sheds light on this apparent discordance. It is possible that the tight looping of DNA results in histone displacement without any loss of histone acetylation, thereby shifting the regional associations between the *Kiss1* gene and histone core. In essence, this would mean no net gain of histone acetylation, but rather that a regional shift in H3 acetylation patterns. To clarify the relationship between histone acetylation and *Kiss 1* gene expression in the ARC and AVPV, closer examination of the histone

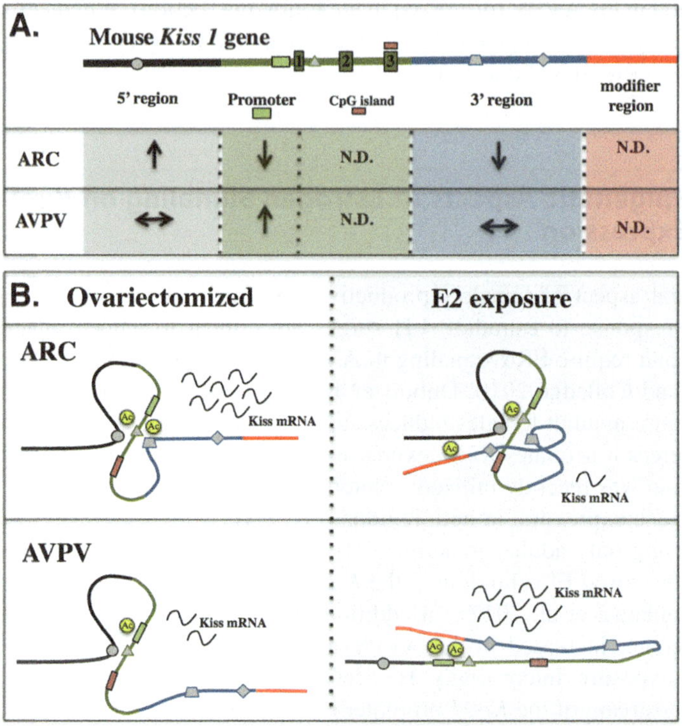

Fig. 9.4 Estradiol (E2) exposure has brain region-specific influence over Kiss1 gene expression, chromatin conformation and histone 3 (H3) acetylation (Tomikawa et al. 2012). (**a**) A schematic representation of the mouse *Kiss1* gene depicts several critical regions as defined by color and shapes including a 5′ region (black), a coding region (green) that includes a promoter (lime green box) 3 exons (numbered boxes) and a CpG island (red box, within the third exon), a 3′ region (blue) and a transcriptional modifier region (red). Small gray shapes indicate sites utilized for analysis by Tomikawa et al. (2012). The table indicates the direction of H3 acetylation abundance detected in each gene region in the ARC and AVPV during E2 exposure. H3 acetylation significantly increased (↑), decreased (↓), or remained the same (↔); *N.D.* not determined. (**b**) Representations of DNA looping at the *Kiss1* gene in the ARC and AVPV. Yellow circles represent H3 acetylation abundance. Location of yellow circles enables visualization of *Kiss1* regional interactions with the acetylated histone. Prior to E2 exposure, the *Kiss1* gene structure is similar in the ACR and AVPV, though a second loop is apparent in only the ARC. E2 exposure causes formation of a complex DNA structure and decrease in *Kiss1* expression in the ARC. On the other hand, the AVPV 5′ loop is lost in favor of a long range 3′ loop and increased *Kiss1* expression during E2 exposure. Created, based on Tomikawa et al., Proc Natl Acad Sci USA 109:E1294-1301 (2012), Uenoyama et al., Neuroendocrinology 103:640–649 (2016), and Kurian et al., Endocrinology 157: 3588–3603 (2016)

acetyltransferase and histone deacetylase activities in kisspeptin neurons during E2 exposure is necessary.

Based on these intriguing discoveries, critical questions arise. First, how do these different populations of kisspeptin neurons evolve to create distinct orientations of the kisspeptin gene? Are there sex-specific differences in *Kiss1* gene conformation?

Also, how might DNA methylation status affect chromatin conformation? Tomikawa and colleagues found no substantial differences in CpG methylation status in the basal or hormone exposed states, but they only assessed six CpG sites near the transcription start site. Importantly, the most pronounced CpG island (CGI) of the *Kiss1* gene is located in the third exon. There are presently no reports of differences in methylation status across this region between ARC and AVPV tissue. This could be critical, as the *Kiss1* gene contains dense CCCTC-binding factor (CTCF) sites surrounding the last exon as well as between the promoter and first exon. As CTCF binding and consequential modification of chromatin loop structure is profoundly influenced by DNA methylation status (Ito et al. 2013; Ong and Corces 2014), it is tempting to speculate that DNA methylation could factor heavily into brain region-specific *Kiss1* gene conformations. Nevertheless, at this time, there are no reports of CTCF binding within the *Kiss1* gene or differences in methylation status across the third exon CGI.

9.8 Epigenetic Regulation of the *Kiss1* Gene and Kisspeptin Cells in the Context of Puberty

Kisspeptin signaling is essential for puberty onset and progression. Because puberty onset is heavily influenced by environmental conditions including climate, stress, body weight, and environmental estrogens, interest in the epigenetic regulation of kisspeptin cell development and puberty-related *Kiss1* gene expression has grown substantially. As mentioned earlier, similar to the findings of Mellon and colleagues comparing *Gnrh* gene promoter structure of immature and mature cell lines, Lomniczi et al. (2013) reported that the female rat pubertal transition is accompanied by increased prevalence of activating histone PTMs at the *Kiss1* promoter. Specifically, prior to puberty onset, the kisspeptin promoter was associated with a bivalent domain (i.e., H3K27me3 and H3K4me3) and occupied by a component of a polycomb repressive complex, EED. The transition to puberty from prepubertal period was accompanied by decreased EED occupancy of the kisspeptin promoter, a gradual loss of the repressive H3K27me3 PTM, and increased levels of the permissive H3K9,14 acetylation and H3K4me3 PTMs. The authors suggest this process is the consequence of increased DNA methylation of the EED promoter, leading to lower EED expression and consequential decreased occupancy of the kisspeptin promoter by the repressive PRC2 (EED, SUZ12, Ezh2) complex. However, these conclusions were largely based on observations of dramatically delayed puberty in female mice when DNA methylation was inhibited by peripheral administration of the pharmacological DNA methyltransferase inhibitor 5-azacytidine (5-aza). While the findings are consistent with the expected results, this 5-aza activity may not be specific to inhibition of postmitotic neuronal DNA methylation. Because the well-characterized mechanism of 5-aza requires nucleoside incorporation into DNA (Schermelleh et al. 2005; Stresemann and Lyko 2008; Svedruzic 2008) the mechanism by which this compound inhibits DNMT in postmitotic cells remains unclear. In addition, the significant reduction in growth rate and elevated levels of circulating corticosterone that occurs following initiation of drug treatment (Lomniczi et al. 2013) is indicative of toxicity that likely contributes to delays in sexual maturation. Because of these concerns, more direct approaches (e.g., cell-specific genetic or enzyme expression manipulations) are necessary to clarify the relationship between DNA methylation, chromatin modifications, and puberty onset. Nonetheless, these studies and a recent report that isolated another vital repressor gene (GATAD1) through an extensive gene array analysis in orchiectomized male monkey hypothalamus (Lomniczi et al. 2015), further establish the concept that structural modification of bivalent promoters in the neuroendocrine hypothalamus is an integral step toward puberty onset and reproductive function in primates and rodents.

Lomniczi and colleagues' focused approach toward measuring histone modification status during development at one gene has tremendous value for clarifying the temporal progression of bivalent promoter transactivation. Importantly, they report that permissive histone PTMs (H3K4me3 and H3K9,14 acetylation) accumulate near the kisspeptin promoter during the transition from juvenile to early pubertal stages. This preceded the loss of the repressive H3K27me3 PTM, suggesting that the recruitment of activating complexes is imperative for transactivation and increased gene expression. Based on the temporal relationship between H3K4me3 accumulation and loss of H3K27me3, as well as the association between Tet2 and H3K4me3, we recently evaluated the impact of kisspeptin neuron-specific ablation of *Tet2* (kTKO animals) on puberty onset and progression in mice (Kurian et al. 2014). Remarkably, *Tet2* disruption delayed puberty onset and progression in both male and female kTKO animals. Whether this impact on puberty is the consequence of Tet2-mediated DNA demethylation, altered DNA hydroxymethylation patterns, H3K4me3 accumulation or another mechanism remain open and important questions. More recent evidence indicates that histone-modifying Tet2 interactors (i.e., MLL1 and MLL3) are also critical mediators of increased rodent *kiss1* expression and pubertal progression; suggesting factors that drive Tet2/MLL interactions could represent signals that are critical for initiating puberty (Toro, et al. 2018). Which genomic regions in kisspeptin neurons, in addition to *Kiss1*, are altered following *Tet2* disruption also remain unknown.

The similar genomic targets as integral positions of epigenetic regulation in kisspeptin cells are predicated on the recent discoveries of additional peptides that are co-expressed in ARC kisspeptin neurons, namely neurokinin B (Tac2) and dynorphin. Neurons that express all three peptides are known as KNDy neurons (Goodman et al. 2014). In fact, *Tac2* expression in the ARC in mice is an early marker of puberty onset (Gill et al. 2012a), and is likely regulated by a bivalent promoter. As evidence, recent preliminary studies (Gill et al. 2012b) found that female mice heterozygous for a loss-of-function mutation in *Lsd1*, a histone lysine demethylase of di- and mono-methylated H3K4, exhibit precocious vaginal opening and first ovulation (3–4 days prior to wild-type littermates), with early elevations of circulating gonadotropins and hypothalamic expression of *Tac2* (Gill et al. 2012b). Importantly, these findings parallel those of Lomniczi et al. (2013), in that lower expression of an epigenetic repressive enzyme is related to the elevated hypothalamic expression of a puberty-related gene. However, it is not known whether Lsd1 expression or activity changes across development. In addition, Lsd1 also stimulates hormone (ligand associated androgen receptor) mediated gene activation through demethylation of H3K9 (Metzger et al. 2005). Consequently, while preliminary evidence suggests the intriguing possibility that Lsd1 directly alters hypothalamic development (including kisspeptin neurons) during the pubertal transition, models for cell or region-specific genetic manipulations will be necessary to verify this interpretation. These models will be critical for determining the primary activity of Lsd1 (i.e., H3K4 or H3K9 demethylation) as it relates to reproductive maturation. Nonetheless, it is increasingly clear that *Tac2* is an additional genomic loci within kisspeptin neurons that is the target of repressive and activating epigenetic complexes related to the transition to puberty (Toro et al. 2018).

The concept that KNDy neurons in the arcuate nucleus are critical to the regulation of GnRH release has been proposed, based on the finding that kisspeptin, neurokinin B and dynorphin are all co-localized in the same neuron in rodents and ruminants (Goodman et al. 2014). In primates, including humans, however, the rate of co-localization of these 3 peptides is not as high as reported in other mammalian species (Ramaswamy et al. 2010; Hrabovszky 2014). Nevertheless, studies in this lab indicate that kisspeptin, neurokinin B and dynorphin have considerable interactions in the median eminence of the hypothalamus as each independent entry (Garcia et al. 2017, 2018). Moreover, the degree of interactions between these 3 peptidergic neurons varies during development, and different hormonal and physiological conditions (Terasawa et al. 2018). Whether and how epigenetic mechanisms are involved in the interactions between 3 peptidergic neurons remain as a great challenge.

9.9 Conclusion

Taken together, the findings detailed in this chapter highlights exciting connections between epigenetic regulations in different cells of the reproductive neuroendocrine hypothalamus. Nevertheless, our understanding of epigenetic regulation of GnRH and kisspeptin neurons and reproductive neuroendocrine function, in general, is still in its infancy. In fact, the findings described in this chapter are all based on changes in known epigenetic markers in a given physiological condition and the question of whether actual modifications of environmental conditions result in similar epigenetic modifications in GnRH and kisspeptin neurons, along with altered physiology, has not been examined to date. For example, it is known that exposure of prepubertal animals to stress results in delayed puberty, whereas exposure of prepubertal animals to a high-calorie diet induces precocious puberty. Do these environmental changes cause consistent opposite epigenetic modifications to *GnRH*? The field is progressing rapidly. We are inching closer to the ultimate promise of neuroendocrine epigenetic research: to define how the environment influences our physiology.

Acknowledgments The authors express their sincere appreciation for many current and past postdoctoral research fellows, graduate students, and research specialists including Kim L. Keen, who has contributed to research in the Terasawa laboratory as well as Somaja Louis and Jennifer Li, who contributed to research in the Kurian laboratory. This work was supported by the NIH grants, HD011355, HD015344, and HD077447 for ET, ES020878 for JRK, and OD01106 and RR0061 for the WNPRC.

Key References

Clark and Mellon (1995). This is the first paper identifying that the homeodomain transcription factor Oct 1 regulates the GnRH gene transcription.

Iyer et al. (2011). This is the first paper demonstrating chromatin modifications of GnRH gene expression during neuronal differentiation.

Kepa et al. (1996a). This is one of the first works to analyze the structure of GnRH gene promoter.

Kurian et al. (2010a). This is the first paper to demonstrate that epigenetic changes occur during GnRH neuronal maturation.

Lomniczi et al. (2015). This is a seminal work demonstrating that Zinc finger protein is involved in prepubertal GnRH suppression in non-human primates.

Messina et al. (2016). This paper identified that microRNAs regulate GnRH gene expression, which correlates with the timing of puberty.

References

Alcin E, Sahu A, Ramaswamy S, Hutz ED, Keen KL, Terasawa E, Bethea CL, Plant TM (2013) Ovarian regulation of kisspeptin neurones in the arcuate nucleus of the rhesus monkey (*Macaca mulatta*). J Neuroendocrinol 5:488–496

Arai Y, Gorski RA (1968) Critical exposure time for androgenization of the developing hypothalamus in the female rat. Endocrinology 82:1010–1014

Arnold AP, Gorski RA (1984) Gonadal steroid induction of structural sex differences in the central nervous system. Annu Rev Neurosci 7:413–442

Azuara V, Perry P, Sauer S, Spivakov M, Jørgensen HF, John RM, Gouti M, Casanova M, Warnes G, Merkenschlager M, Fisher AG (2006) Chromatin signatures of pluripotent cell lines. Nat Cell Biol 8:532–538

Banks GC, Deterding LJ, Tomer KB, Archer TK (2001) Hormone-mediated dephosphorylation of specific histone H1 isoforms. J Biol Chem 276:36467–36473

Belsham DD, Mellon PL (2000) Transcription factors Oct-1 and C/EBPbeta (CCAAT/enhancer-binding protein-beta) are involved in the glutamate/nitric oxide/cyclic-guanosine 5'-monophosphate-mediated repression of mediated repression of gonadotropin-releasing hormone gene expression. Mol Endocrinol 14:212–228

Bernstein BE, Mikkelsen TS, Xie X, Kamal M, Huebert DJ, Cuff J, Fry B, Meissner A, Wernig M, Plath K, Jaenisch R, Wagschal A, Feil R, Schreiber SL, Lander ES (2006) A bivalent chromatin structure marks key developmental genes in embryonic stem cells. Cell 125:315–326

Bestor TH, Ingram VM (1983) Two DNA methyltransferases from murine erythroleukemia cells: purification, sequence specificity, and mode of interaction with DNA. Proc Natl Acad Sci U S A 80:5559–5563

Bestor T, Laudano A, Mattaliano R, Ingram V (1988) Cloning and sequencing of a cDNA encoding DNA methyltransferase of mouse cells. The carboxyl-terminal domain of the mammalian enzymes is related to bacterial restriction methyltransferases. J Mol Biol 203:971–983

Branco MR, Ficz G, Reik W (2012) Uncovering the role of 5-hydroxymethylcytosine in the epigenome. Nat Rev Genet 13:7–13

Carey BW, Finley LW, Cross JR, Allis CD, Thompson CB (2015) Intracellular α-ketoglutarate maintains the pluripotency of embryonic stem cells. Nature 518(7539):413–416

Cheng G, Coolen LM, Padmanabhan V, Goodman RL, Lehman MN (2010) The kisspeptin/neurokinin B/dynorphin (KNDy) cell population of the arcuate nucleus: sex differences and effects of prenatal testosterone in sheep. Endocrinology 151:301–311

Clark ME, Mellon PL (1995) The POU homeodomain transcription factor Oct-1 is essential for activity of the gonadotropin-releasing hormone neuron-specific enhancer. Mol Cell Biol 15:6169–6177

Clarkson J, Herbison AE (2006) Postnatal development of kisspeptin neurons in mouse hypothalamus; sexual dimorphism and projections to gonadotropin-releasing hormone neurons. Endocrinology 147:5817–5825

Clarkson J, Boon WC, Simpson ER, Herbison AE (2009) Postnatal development of an estradiol-kisspeptin positive feedback mechanism implicated in puberty onset. Endocrinology 150:3214–3220

Constantin S, Caraty A, Wray S, Duittoz AH (2009) Development of gonadotropin-releasing hormone-1 secretion in mouse nasal explants. Endocrinology 150:3221–3227

d'Anglemont de Tassigny X, Colledge WH (2010) The role of kisspeptin signaling in reproduction. Physiology 25:207–217

d'Anglemont de Tassigny X, Fagg LA, Dixon JP, Day K, Leitch HG, Hendrick AG, Zahn D, Franceschini I, Caraty A, Carlton MB, Aparicio SA, Colledge WH (2007) Hypogonadotropic hypogonadism in mice lacking a functional Kiss1 gene. Proc Natl Acad Sci U S A 104:10714–10719

Dawlaty MM, Ganz K, Powell BE, Hu YC, Markoulaki S, Cheng AW, Gao Q, Kim J, Choi SW, Page DC, Jaenisch R (2011) Tet1 is dispensable for maintaining pluripotency and its loss is compatible with embryonic and postnatal development. Cell Stem Cell 9:166–175

Dawlaty MM, Breiling A, Le T, Raddatz G, Barrasa MI, Cheng AW, Gao Q, Powell BE, Li Z, Xu M, Faull KF, Lyko F, Jaenisch R (2013) Combined deficiency of Tet1 and Tet2 causes epigenetic abnormalities but is compatible with postnatal development. Dev Cell 24:310–323

de Roux N, Genin E, Carel JC, Matsuda F, Chaussain JL, Milgrom E (2003) Hypogonadotropic hypogonadism due to loss of function of the KiSS1-derived peptide receptor GPR54. Proc Natl Acad Sci U S A 100:10972–10976

Deaton AM, Bird A (2011) CpG islands and the regulation of transcription. Genes Dev 25:1010–1022

Deplus R, Delatte B, Schwinn MK, Defrance M, Méndez J, Murphy N, Dawson MA, Volkmar M, Putmans P, Calonne E, Shih AH, Levine RL, Bernard O, Mercher T, Solary E, Urh M, Daniels DL, Fuks F (2013) TET2 and TET3 regulate GlcNAcylation and H3K4 methylation through OGT and SET1/COMPASS. EMBO J 32:645–655

Desroziers E, Droguerre M, Bentsen AH, Robert V, Mikkelsen JD, Caraty A, Tillet Y, Duittoz A, Franceschini I (2012a) Embryonic development of kisspeptin neurones in rat. J Neuroendocrinol 24:1284–1295

Desroziers E, Mikkelsen JD, Duittoz A, Franceschini I (2012b) Kisspeptin-immunoreactivity changes in a sex- and hypothalamic-region-specific manner across rat postnatal development. J Neuroendocrinol 24:1154–1165

Dhillo WS, Chaudhri OB, Thompson EL, Murphy KG, Patterson M, Ramachandran R, Nijher GK, Amber V, Kokkinos A, Donaldson M, Ghatei MA, Bloom SR (2007) Kisspeptin-54 stimulates gonadotropin release most potently during the preovulatory phase of the menstrual cycle in women. J Clin Endocrinol Metab 92:3958–3966

Döhler KD, Hancke JL, Srivastava SS, Hofmann C, Shryne JE, Gorski RA (1984) Participation of estrogens in female sexual differentiation of the brain; neuroanatomical, neuroendocrine and behavioral evidence. Prog Brain Res 61:99–117

Dubois SL, Acosta-Martínez M, DeJoseph MR, Wolfe A, Radovick S, Boehm U, Urban JH, Levine JE (2015) Positive, but not negative feedback actions of estradiol in adult female mice require estrogen receptor α in kisspeptin neurons. Endocrinology 156:1111–1120

El Majdoubi M, Sahu A, Ramaswamy S, Plant TM (2000) Neuropeptide Y: a hypothalamic brake restraining the onset of puberty in primates. Proc Natl Acad Sci U S A 97:6179–6184

Franceschini I, Lomet D, Cateau M, Delsol G, Tillet Y, Caraty A (2006) Kisspeptin immunoreactive cells of the ovine preoptic area and arcuate nucleus co-express estrogen receptor alpha. Neurosci Lett 401:225–230

Fueshko S, Wray S (1994) LHRH cells migrate on peripherin fibers in embryonic olfactory explant cultures: an in vitro model for neurophilic neuronal migration. Dev Biol 166:331–348

Funes S, Hedrick JA, Vassileva G, Markowitz L, Abbondanzo S, Golovko A, Yang S, Monsma FJ, Gustafson EL (2003) The KiSS-1 receptor GPR54 is essential for the development of the murine reproductive system. Biochem Biophys Res Commun 312:1357–1363

Gan L, Ni PY, Ge Y, Xiao YF, Sun CY, Deng L, Zhang W, Wu SS, Liu Y, Jiang W, Xin HB (2012) Histone deacetylases regulate gonadotropin-releasing hormone I gene expression via modulating Otx2-driven transcriptional activity. PLoS One 7:e39770

Garcia JP, Guerriero KA, Keen KL, Kenealy BP, Seminara SB, Terasawa E (2017) Kisspeptin and neurokinin B signaling network underlies the pubertal increase in GnRH release in female rhesus monkeys. Endocrinology 158:3269–3280

Garcia JP, Keen KL, Kenealy BP, Seminara SB, Terasawa E (2018) Role of kisspeptin and neurokinin B signaling in male rhesus monkey puberty. Endocrinology 159:3048–3060

Gardiner-Garden M, Frommer M (1987) CpG islands in vertebrate genomes. J Mol Biol 196:261–282

Gill JC, Navarro VM, Kwong C, Noel SD, Martin C, Xu S, Clifton DK, Carroll RS, Steiner RA, Kaiser UB (2012a) Increased neurokinin B (Tac2) expression in the mouse arcuate nucleus is an early marker of pubertal onset with differential sensitivity to sex steroid-negative feedback than Kiss1. Endocrinology 153:4883–4893

Gill J, Kwong C, Clark E, Carroll RS, Shi YG, Kaiser UB (2012b) A role for the histone demethylase LSD1 in controlling the timing of pubertal onset. In: Abstracts for Endocrine Society Annual Meeting, Huston, TX, No. OR12-3

Givens ML, Rave-Harel N, Goonewardena VD, Kurotani R, Berdy SE, Swan CH, Rubenstein JL, Robert B, Mellon PL (2005) Developmental regulation of gonadotropin-releasing hormone gene expression by the MSX and DLX homeodomain protein families. J Biol Chem 280:19156–19165

Glaser S, Schaft J, Lubitz S, Vintersten K, van der Hoeven F, Tufteland KR, Aasland R, Anastassiadis K, Ang SL, Stewart AF (2006) Multiple epigenetic maintenance factors implicated by the loss of Mll2 in mouse development. Development 133:1423–1432

Goodman RL, Coolen LM, Lehman MN (2014) A role for neurokinin B in pulsatile GnRH secretion in the ewe. Neuroendocrinology 99:18–32

Gorski RA (1985) Sexual dimorphisms of the brain. J Anim Sci 61(Suppl 3):38–61

Gottsch ML, Cunningham MJ, Smith JT, Popa SM, Acohido BV, Crowley WF, Seminara S, Clifton DK, Steiner RA (2004) A role for kisspeptins in the regulation of gonadotropin secretion in the mouse. Endocrinology 145:4073–4077

Guerriero KA, Keen KL, Millar RP, Terasawa E (2012) Developmental changes in GnRH release in response to kisspeptin agonist and antagonist in female Rhesus monkeys (*Macaca mulatta*): implication for the mechanism of puberty. Endocrinology 153:825–836

Guo JU, Su Y, Shin JH, Shin J, Li H, Xie B, Zhong C, Hu S, Le T, Fan G, Zhu H, Chang Q, Gao Y, Ming GL, Song H (2014) Distribution, recognition and regulation of non-CpG methylation in the adult mammalian brain. Nat Neurosci 17:215–222

Hahn MA, Qiu R, Wu X, Li AX, Zhang H, Wang J, Jui J, Jin SG, Jiang Y, Pfeifer GP, Lu Q (2013) Dynamics of 5-hydroxymethylcytosine and chromatin marks in mammalian neurogenesis. Cell Rep 3:291–300

Herbison AE (2016) Control of puberty onset and fertility by gonadotropin-releasing hormone neurons. Nat Rev Endocrinol 12:452–466

Hermann A, Goyal R, Jeltsch A (2004) The Dnmtl DNA-(cytosine-C5)-methyltransferase methylates DNA processively with high preference for hemimethylated target sites. J Biol Chem 279:48350–48359

Hoffmann HM, Mellon PL (2018) Regulation of GnRH gene expression. In: Herbison AE, Plant TM (eds) Master class in neuroendocrinology series.The GnRH neuron and its control. Wiley, Hoboken, NJ, pp 95–119

Hoffmann GE, Le WW, Franceschini I, Caraty A, Advis JP (2011) Expression of fos and in vivo median eminence release of LHRH identifies an active role for preoptic area kisspeptin neurons in synchronized surges of LH and LHRH in the ewe. Endocrinology 152:214–222

Hoffmann HM, Gong P, Tamrazian A, Mellon PL (2018) Transcriptional interaction between cFOS and the homeodomain-binding transcription factor VAX1 on the GnRH promoter controls Gnrh1 expression levels in a GnRH neuron maturation specific manner. Mol Cell Endocrinol 461:143–154

Homma T, Sakakibara M, Yamada S, Kinoshita M, Iwata K, Tomikawa J, Kanazawa T, Matsui H, Takatsu Y, Ohtaki T, Matsumoto H, Uenoyama Y, Maeda K, Tsukamura H (2009) Significance of neonatal testicular sex steroids to defeminize anteroventral periventricular kisspeptin neurons and the GnRH/LH surge system in male rats. Biol Reprod 81:1216–1225

Hrabovszky E (2014) Neuroanatomy of the human hypothalamic kisspeptin system. Neuroendocrinology 99:33–48

Hrabovszky E, Ciofi P, Vida B, Horvath MC, Keller E, Caraty A, Bloom SR, Ghatei MA, Dhillo WS, Liposits Z, Kallo I (2010) The kisspeptin system of the human hypothalamus: sexual dimorphism and relationship with gonadotropin-releasing hormone and neurokinin B neurons. Eur J Neurosci 31:1984–1998

Hrabovszky E, Sipos MT, Molnár CS, Ciofi P, Borsay BÁ, Gergely P, Herczeg L, Bloom SR, Ghatei MA, Dhillo WS, Liposits Z (2012) Low degree of overlap between kisspeptin, neurokinin B, and dynorphin immunoreactivities in the infundibular nucleus of young male human subjects challenges the KNDy neuron concept. Endocrinology 153:4978–4989

Ito Y, Nativio R, Murrell A (2013) Induced DNA demethylation can reshape chromatin topology at the IGF2-H19 locus. Nucleic Acids Res 41:5290–5302

Iyer AK, Miller NL, Yip K, Tran BH, Mellon PL (2010) Enhancers of GnRH transcription embedded in an upstream gene use homeodomain proteins to specify hypothalamic expression. Mol Endocrinol 24:1949–1964

Iyer AK, Brayman MJ, Mellon PL (2011) Dynamic chromatin modifications control GnRH gene expression during neuronal differentiation and protein kinase C signal transduction. Mol Endocrinol 25:460–473

Kauffman AS, Gottsch ML, Roa J (2007) Sexual differentiation of Kiss1 gene expression in the brain of the rat. Endocrinology 148:1774–1783

Kelley CG, Lavorgna G, Clark ME, Boncinelli E, Mellon PL (2000) The Otx2 homeoprotein regulates expression from the gonadotropin-releasing hormone proximal promoter. Mol Endocrinol 14:1246–1256

Kepa JK, Wang C, Neeley CI, Raynolds MV, Gordon DF, Wood WM, Wierman ME (1992) Structure of the rat gonadotropin releasing hormone (rGnRH) gene promoter and functional analysis in hypothalamic cells. Nucleic Acids Res 20:1393–1399

Kepa JK, Spaulding AJ, Jacobsen BM, Fang Z, Xiong X, Radovick S, Wierman ME (1996a) Structure of the distal human gonadotropin releasing hormone (hGnRH) gene promoter and functional analysis in GT1-7 neuronal cells. Nucleic Acids Res 24:3614–3620

Kepa JK, Jacobsen BM, Boen EA, Prendergast P, Edwards DP, Takimoto G, Wierman ME (1996b) Direct binding of progesterone receptor to nonconsensus DNA sequences represses rat GnRH. Mol Cell Endocrinol 117:3927–3910

Kim HH, Wolfe A, Cohen RN, Eames SC, Johnson AL, Wieland CN, Radovick S (2007) In vivo identification of a 107-base pair promoter element mediating neuron-specific expression of mouse gonadotropin-releasing hormone. Mol Endocrinol 21:457–471

Kim J, Semaan SJ, Clifton DK, Steiner RA, Dhamija S, Kauffman AS (2011) Regulation of Kiss1 expression by sex steroids in the amygdala of the rat and mouse. Endocrinology 152:2020–2030

Kirilov M, Clarkson J, Liu X, Roa J, Campos P, Porteous R, Schütz G, Herbison AE (2013) Dependence of fertility on kisspeptin-Gpr54 signaling at the GnRH neuron. Nat Commun 4:2492

Kolodkin MH, Auger AP (2011) Sex difference in the expression of DNA methyltransferase 3a in the rat amygdala during development. J Neuroendocrinol 23:577–583

Kumar D, Freese M, Drexler D, Hermans-Borgmeyer I, Marquardt A, Boehm U (2014) Murine arcuate nucleus kisspeptin neurons communicate with GnRH neurons in utero. J Neurosci 34:3756–3766

Kurian JR (2016) Epigenetic regulation of the GnRH and Kiss1 genes. In: Binder E (ed) Epigenetics and neuroendocrinology, vol 1. Springer Nature, Switzerland, pp 243–264

Kurian JR, Terasawa E (2013) Epigenetic control of gonadotropin releasing hormone neurons. Front Endocrinol (Lausanne) 4:61

Kurian JR, Keen KL, Terasawa E (2010a) Epigenetic changes coincide with in vitro primate GnRH neuronal maturation. Endocrinology 151:5359–5368

Kurian JR, Olesen KM, Auger AP (2010b) Sex differences in epigenetic regulation of the estrogen receptor-alpha promoter within the developing preoptic area. Endocrinology 151:2297–2305

Kurian JR, Kapke JA, Terasawa E (2011) Epigenetic contribution to GnRH neuronal development in male pubertal development. In: Abstracts for the Society for Neuroscience Annual Meeting, Washington, DC, No. 500.03

Kurian JR, Terasawa E, Levine JE (2014) Ten eleven translocase 2 (Tet2) drives GnRH gene expression and enables the typical progression through puberty. In: Abstracts for the Endocrine Society Annual Meeting, Chicago, IL, No.OR03-3

Lapatto R, Pallais JC, Zhang D, Chan YM, Mahan A, Cerrato F, Le WW, Hoffman GE, Seminara SB (2007) Kiss1−/− mice exhibit more variable hypogonadism than Gpr54−/− mice. Endocrinology 148:4927–4936

Larder R, Mellon PL (2009) Otx2 induction of the gonadotropin-releasing hormone promoter is modulated by direct interactions with Grg co-repressors. J Biol Chem 284:16966–16978

Larder R, Kimura I, Meadows J, Clark DD, Mayo S, Mellon PL (2013) Gene dosage of Otx2 is important for fertility in male mice. Mol Cell Endocrinol 377:16–22

Lawson MA, Buhain AR, Jovenal JC, Mellon PL (1998) Multiple factors interacting at the GATA sites of the gonadotropin-releasing hormone neuron-specific enhancer regulate gene expression. Mol Endocrinol 12:364–377

Lee JH, Miele ME, Hicks DJ, Phillips KK, Trent JM, Weissman BE, Welch DR (1996) KiSS-1, a novel human malignant melanoma metastasis-suppressor gene. J Natl Cancer Inst 88:1731–1737

Lee SY, Lau AT, Jeong CH, Shim JH, Kim HG, Kim J, Bode AM, Dong Z (2010) Histone XH2AX is required for Xenopus anterior neural development: critical role of threonine 16 phosphorylation. J Biol Chem 285:29525–29534

Li E, Bestor TH, Jaenisch R (1992) Targeted mutation of the DNA methyltransferase gene results in embryonic lethality. Cell 69:915–926

Lomniczi A, Loche A, Castellano JM, Ronnekleiv OK, Bosch M, Kaidar G, Knoll JG, Wright H, Pfeifer GP, Ojeda SR (2013) Epigenetic control of female puberty. Nat Neurosci 16:281–289

Lomniczi A, Wright H, Castellano JM, Matagne V, Toro CA, Ramaswamy S, Plant TM, Ojeda SR (2015) Epigenetic regulation of puberty via zinc finger protein-mediated transcriptional repression. Nat Commun 6:10195

Matsuda KI, Mori H, Nugent BM, Pfaff DW, McCarthy MM, Kawata M (2011) Histone deacetylation during brain development is essential for permanent masculinization of sexual behavior. Endocrinology 152:2760–2767

Matsuda F, Nakatsukasa K, Suetomi Y, Naniwa Y, Ito D, Inoue N, Wakabayashi Y, Okamura H, Maeda KI, Uenoyama Y, Tsukamura H, Ohkura S (2015) The luteinising hormone surge-generating system is functional in male goats as in females: involvement of kisspeptin neurones in the medial preoptic area. J Neuroendocrinol 27:57–65

Maurer JA, Wray S (1997) Luteinizing hormone-releasing hormone (LHRH) neurons maintained in hypothalamic slice explant cultures exhibit a rapid LHRH mRNA turnover rate. J Neurosci 17:9481–9491

McCarthy MM (1994) Molecular aspects of sexual differentiation of the rodent brain. Psychoneuroendocrinology 19:415–427

Mehlmann LM (2005) Stops and starts in mammalian oocytes: recent advances in understanding the regulation of meiotic arrest and oocyte maturation. Reproduction 30:791–799

Mellon PL, Windle JJ, Goldsmith PC, Padula CA, Roberts JL, Weiner RI (1990) Immortalization of hypothalamic GnRH neurons by genetically targeted tumorigenesis. Neuron 5:1–10

Messager S, Chatzidaki EE, Ma D, Hendrick AG, Zahn D, Dixon J, Thresher RR, Malinge I, Lomet D, Carlton MB, Colledge WH, Caraty A, Aparicio SA (2005) Kisspeptin directly stimulates gonadotropin-releasing hormone release via G protein-coupled receptor 54. Proc Natl Acad Sci U S A 102:1761–1766

Messina A, Langlet F, Chachlaki K, Roa J, Rasika S, Jouy N, Gallet S, Gaytan F, Parkash J, Tena-Sempere M, Giacobini P, Prevot V (2016) A microRNA switch regulates the rise in hypothalamic GnRH production before puberty. Nat Neurosci 19:835–844

Metzger E, Wissmann M, Yin N, Müller JM, Schneider R, Peters AH, Günther T, Buettner R, Schüle R (2005) LSD1 demethylates repressive histone marks to promote androgen-receptor-dependent transcription. Nature 437:436–439

Moore JP Jr, Wray S (2000) Luteinizing hormone-releasing hormone biosynthesis and secretion in embryonic LHRH neurons. Endocrinology 141:4486–4495

Murray EK, Varnum MM, Fernandez JL, de Vries GJ, Forger NG (2011) Effects of neonatal treatment with valproic acid on vasopressin immunoreactivity and olfactory behaviour in mice. J Neuroendocrinol 10:906–914

Navarro VM, Castellano JM, Fernández-Fernández R, Tovar S, Roa J, Mayen A, Barreiro ML, Casanueva FF, Aguilar E, Dieguez C, Pinilla L, Tena-Sempere M (2005a) Effects of KiSS-1 peptide, the natural ligand of GPR54, on follicle-stimulating hormone secretion in the rat. Endocrinology 146:1689–1697

Navarro VM, Castellano JM, Fernández-Fernández R, Tovar S, Roa J, Mayen A, Nogueiras R, Vazquez MJ, Barreiro ML, Magni P, Aguilar E, Dieguez C, Pinilla L, Tena-Sempere M (2005b) Characterization of the potent luteinizing hormone-releasing activity of KiSS-1 peptide, the natural ligand of GPR54. Endocrinology 146:156–163

Novaira HJ, Fadoju D, Diaczok D, Radovick S (2012) Genetic mechanisms mediating kisspeptin regulation of GnRH gene expression. J Neurosci 32:17391–17400

Novaira HJ, Sonko ML, Radovick S (2016) Kisspeptin induces dynamic chromatin modifications to control GnRH gene expression. Mol Neurobiol 53:3315–3325

Ohkura S, Takase K, Matsuyama S, Mogi K, Ichimaru T, Wakabayashi Y, Uenoyama Y, Mori Y, Steiner RA, Tsukamura H, Maeda KI, Okamura H (2009) Gonadotrophin-releasing hormone pulse generator activity in the hypothalamus of the goat. J Neuroendocrinol 21:813–821

Okano M, Xie S, Li E (1998) Cloning and characterization of a family of novel mammalian DNA (cytosine-5) methyltransferases. Nat Genet 19:219–220

Okano M, Bell DW, Haber DA, Li E (1999) DNA methyltransferases Dnmt3a and Dnmt3b are essential for de novo methylation and mammalian development. Cell 99:247–257

Ong CT, Corces VG (2014) CTCF: an architectural protein bridging genome topology and function. Nat Rev Genet 15:234–246

Pan G, Tian S, Nie J, Yang C, Ruotti V, Wei H, Jonsdottir GA, Stewart R, Thomson JA (2007) Whole-genome analysis of histone H3 lysine 4 and lysine 27 methylation in human embryonic stem cells. Cell Stem Cell 1:299–312

Pasini D, Bracken AP, Hansen JB, Capillo M, Helin K (2007) The polycomb group protein Suz12 is required for embryonic stem cell differentiation. Mol Cell Biol 27:3769–3779

Phoenix CH, Goy RW, Gerall AA, Young WC (1959) Organizing action of prenatally administered testosterone propionate on the tissues mediating mating behavior in the female guinea pig. Endocrinology 65:369–382

Poling MC, Kauffman AS (2013) Organizational and activational effects of sex steroids on kisspeptin neuron development. Front Neuroendocrinol 34:3–17

Provenzano C, Pascucci B, Lupari E, Civitareale D (2010) Large scale analysis of transcription factor TTF-1/NKX2.1 target genes in GnRH secreting cell line GT1-7. Mol Cell Endocrinol 323:215–223

Puttabyatappa M, Padmanabhan V (2017) Prenatal testosterone programming of insulin resistance in the female sheep. Adv Exp Med Biol 1043:575–596

Quanbeck C, Sherwood NM, Millar RP, Terasawa E (1997) Two populations of luteinizing hormone-releasing hormone neurons in the forebrain of the rhesus macaque during embryonic development. J Comp Neurol 380:293–309

Radovick S, Wray S, Lee E, Nicols DK, Nakayama Y, Weintraub BD, Westphal H, Cutler GB Jr, Wondisford FE (1991a) Migratory arrest of gonadotropin-releasing hormone neurons in transgenic mice. Proc Natl Acad Sci U S A 88:3402–3406

Radovick S, Ticknor CM, Nakayama Y, Notides AC, Rahman A, Weintraub BD, Cutler GB Jr, Wondisford FE (1991b) Evidence for direct estrogen regulation of the human gonadotropin-releasing hormone gene. J Clin Invest 88:1649–1655

Radovick S, Wray S, Muglia L, Westphal H, Olsen B, Smith E, Patriquin E, Wondisford FE (1994) Steroid hormone regulation and tissue-specific expression of the human GnRH gene in cell culture and transgenic animals. Horm Behav 28:520–529

Ramaswamy S, Guerriero KA, Gibbs RB, Plant TM (2008) Structural interactions between kisspeptin and GnRH neurons in the mediobasal hypothalamus of the male rhesus monkey (Macaca mulatta) as revealed by double immunofluorescence and confocal microscopy. Endocrinology 149:4387–4395

Ramaswamy S, Seminara SB, Ali B, Ciofi P, Amin NA, Plant TM (2010) Neurokinin B stimulates GnRH release in the male monkey (*Macaca mulatta*) and is colocalized with kisspeptin in the arcuate nucleus. Endocrinology 151:4494–4503

Ramaswamy S, Dwarki K, Ali B, Gibbs RB, Plant TM (2013) The decline in pulsatile GnRH release, as reflected by circulating LH concentrations, during the infant-juvenile transition in the agonadal male rhesus monkey (*Macaca mulatta*) is associated with a reduction in kisspeptin content of KNDy neurons of the arcuate nucleus in the hypothalamus. Endocrinology 154:1845–1853

Ramsahoye BH, Biniszkiewicz D, Lyko F, Clark V, Bird AP, Jaenisch R (2000) Non-CpG methylation is prevalent in embryonic stem cells and may be mediated by DNA methyltransferase 3a. Proc Natl Acad Sci U S A 97:5237–5242

Rave-Harel N, Miller NL, Givens ML, Mellon PL (2005) The Groucho-related gene family regulates the gonadotropin-releasing hormone gene through interaction with the homeodomain proteins MSX1 and OCT1. J Biol Chem 280:30975–30983

Rometo AM, Krajewski SJ, Voytko ML, Rance NE (2007) Hypertrophy and increased kisspeptin gene expression in the hypothalamic infundibular nucleus of postmenopausal women and ovariectomized monkeys. J Clin Endocrinol Metab 92:2744–2750

Ronnekleiv OK, Resko JA (1990) Ontogeny of gonadotropin-releasing hormone-containing neurons in early fetal development of rhesus macaques. Endocrinology 126:498–511

Schermelleh L, Spada F, Easwaran HP, Zolghadr K, Margot JB, Cardoso MC, Leonhardt H (2005) Trapped in action: direct visualization of DNA methyltransferase activity in living cells. Nat Methods 2:751–756

Schulz KM, Sisk CL (2006) Pubertal hormones, the adolescent brain, and the maturation of social behaviors: lessons from the Syrian hamster. Mol Cell Endocrinol 254–255:120–126

Schulz KM, Molenda-Figueira HA, Sisk CL (2009) Back to the future: the organizational-activational hypothesis adapted to puberty and adolescence. Horm Behav 55:597–604

Schwarz JM, Nugent BM, McCarthy MM (2010) Developmental and hormone-induced epigenetic changes to estrogen and progesterone receptor genes in brain are dynamic across the life span. Endocrinology 151:4871–4881

Semaan SJ, Kauffman AS (2010) Sexual differentiation and development of forebrain reproductive circuits. Curr Opin Neurobiol 20:424–431

Semaan SJ, Kauffman AS (2015) Daily successive changes in reproductive gene expression and neuronal activation in the brains of pubertal female mice. Mol Cell Endocrinol 401:84–97

Semaan SJ, Murray EK, Poling MC, Dhamija S, Forger NG, Kauffman AS (2010) BAX-dependent and BAX-independent regulation of Kiss1 neuron development in mice. Endocrinology 151:5807–5817

Semaan SJ, Dhamija S, Kim J, Ku EC, Kauffman AS (2012) Assessment of epigenetic contributions to sexually-dimorphic Kiss1 expression in the anteroventral periventricular nucleus of mice. Endocrinology 153:1875–1886

Seminara SB, Messager S, Chatzidaki EE, Thresher RR, Acierno JS Jr, Shagoury JK, Bo-Abbas Y, Kuohung W, Schwinof KM, Hendrick AG, Zahn D, Dixon J, Kaiser UB, Slaugenhaupt SA, Gusella JF, O'Rahilly S, Carlton MB, Crowley WF Jr, Aparicio SA, Colledge WH (2003) The GPR54 gene as a regulator of puberty. N Engl J Med 349:1614–1627

Shahab M, Mastronardi C, Seminara SB, Crowley WF, Ojeda SR, Plant TM (2005) Increased hypothalamic GPR54 signaling: a potential mechanism for initiation of puberty in primates. Proc Natl Acad Sci U S A 102:2129–2134

Shahab M, Lippincott M, Chan YM, Davies A, Merino PM, Plummer L, Mericq V, Seminara S (2018) Discordance in the dependence on kisspeptin signaling in mini puberty vs adolescent puberty: human genetic evidence. J Clin Endocrinol Metab 103:1273–1276

Shen X, Liu Y, Hsu YJ, Fujiwara Y, Kim J, Mao X, Yuan GC, Orkin SH (2008) EZH1 mediates methylation on histone H3 lysine 27 and complements EZH2 in maintaining stem cell identity and executing pluripotency. Mol Cell 32:491–502

Simerly RB (1998) Organization and regulation of sexually dimorphic neuroendocrine pathways. Behav Brain Res 92:195–203

Simerly RB (2002) Wired for reproduction: organization and development of sexually dimorphic circuits in the mammalian forebrain. Annu Rev Neurosci 25:507–536

Smith JT (2008) Kisspeptin signalling in the brain: steroid regulation in the rodent and ewe. Brain Res Rev 57:288–298

Smith JT (2013) Sex steroid regulation of kisspeptin circuits. Adv Exp Med Biol 784:275–295

Smith JT, Cunningham MJ, Rissman EF, Clifton DK, Steiner RA (2005) Regulation of Kiss1 gene expression in the brain of the female mouse. Endocrinology 146:3686–3692

Smith JT, Li Q, Pereira A, Clarke IJ (2009) Kisspeptin neurons in the ovine arcuate nucleus and preoptic area are involved in the preovulatory luteinizing hormone surge. Endocrinology 150:5530–5538

Smith JT, Shahab M, Pereira A, Pau KY, Clarke IJ (2010) Hypothalamic expression of KISS1 and gonadotropin inhibitory hormone genes during the menstrual cycle of a non-human primate. Biol Reprod 83:568–577

Stresemann C, Lyko F (2008) Modes of action of the DNA methyltransferase inhibitors azacytidine and decitabine. Int J Cancer 123:8–13

Suberbielle E, Sanchez PE, Kravitz AV, Wang X, Ho K, Eilertson K, Devidze N, Kreitzer AC, Mucke L (2013) Physiologic brain activity causes DNA double-strand breaks in neurons, with exacerbation by amyloid-β. Nat Neurosci 16:613–621

Svedruzic ZM (2008) Mammalian cytosine DNA methyltransferase Dnmt1: enzymatic mechanism, novel mechanism-based inhibitors, and RNA-directed DNA methylation. Curr Med Chem 15:92–106

Terasawa E (2001) Luteinizing hormone-releasing hormone (LHRH) neurons: mechanism of pulsatile LHRH release. Vitam Horm 63:91–129

Terasawa E, Quanbeck CD, Schulz CA, Burich AJ, Luchansky LL, Claude P (1993) A primary cell culture system of luteinizing hormone releasing hormone (LHRH) neurons derived from fetal olfactory placode in the rhesus monkey. Endocrinology 133:2379–2390

Terasawa E, Keen KL, Mogi K, Claude P (1999) Pulsatile release of luteinizing hormone-releasing hormone (LHRH) in cultured LHRH neurons derived from the embryonic olfactory placode of the rhesus monkey. Endocrinology 140:1432–1441

Terasawa E, Busser BW, Luchansky LL, Sherwood NM, Jennes L, Millar RP, Glucksman MJ, Roberts JL (2001) Presence of luteinizing hormone-releasing hormone fragments in the rhesus monkey forebrain. J Comp Neurol 439:491–504

Terasawa E, Garcia JP, Seminara SB, Keen KL (2018) Role of kisspeptin and neurokinin B in puberty in female non-humanprimates. Front Endocrinol (Lausanne) 9:148

Tomikawa J, Uenoyama Y, Ozawa M, Fukanuma T, Takase K, Goto T, Abe H, Ieda N, Minabe S, Deura C, Inoue N, Sanbo M, Tomita K, Hirabayashi M, Tanaka S, Imamura T, Okamura H, Maeda K, Tsukamura H (2012) Epigenetic regulation of Kiss1 gene expression mediating estrogen-positive feedback action in the mouse brain. Proc Natl Acad Sci U S A 109:1294–1301

Topaloglu AK, Tello JA, Kotan LD, Ozbek MN, Yilmaz MB, Erdogan S, Gurbuz F, Temiz F, Millar RP, Yuksel B (2012) Inactivating KISS1 mutation and hypogonadotropic hypogonadism. N Engl J Med 366:629–635

Toro CA, Wright H, Aylwin CF, Ojeda SR, Lomniczi A (2018) Trithorax dependent changes in chromatin landscape at enhancer and promoter regions drive female puberty. Nat Commun 9:57

Uenoyama Y, Tomikawa J, Inoue N, Goto T, Minabe S, Ieda N, Nakamura S, Watanabe Y, Ikegami K, Matsuda F, Ohkura S, Maeda K, Tsukamura H (2016) Molecular and epigenetic mechanism regulating hypothalamic Kiss1 gene expression in mammals. Neuroendocrinology 103:640–649

Vargas Trujillo M, Kalil B, Ramaswamy S, Plant TM (2017) Estradiol upregulates kisspeptin expression in the preoptic area of both the male and female Rhesus monkey (Macaca mulatta): implications for the hypothalamic control of ovulation in highly evolved primates. Neuroendocrinology 105:77–89

Wallen K (2005) Hormonal influences on sexually differentiated behavior in nonhuman primates. Front Neuroendocrinol 26:7–26

Wallen K (2009) The organizational hypothesis: reflections on the 50th anniversary of the publication of Phoenix, Goy, Gerall, and Young (1959). Horm Behav 55:561–565

Watanabe Y, Uenoyama Y, Suzuki J, Takase K, Suetomi Y, Ohkura S, Inoue N, Maeda KI, Tsukamura H (2014) Oestrogen-induced activation of preoptic kisspeptin neurones may be involved in the luteinising hormone surge in male and female Japanese monkeys. J Neuroendocrinol 26:909–917

Whyte DB, Lawson MA, Belsham DD, Eraly SA, Bond CT, Adelman JP, Mellon PL (1995) A neuron-specific enhancer targets expression of the gonadotropin-releasing hormone gene to hypothalamic neurosecretory neurons. Mol Endocrinol 9:467–477

Wu SC, Zhang Y (2010) Active DNA demethylation: many roads lead to Rome. Nat Rev Mol Cell Biol 11:607–620

Xie H, Hoffmann HM, Meadows JD, Mayo SL, Trang C, Leming SS, Maruggi C, Davis SW, Larder R, Mellon PL (2015) Homeodomain proteins six3 and six6 regulate gonadotrope-specific genes during pituitary development. Mol Endocrinol 29:842–855

Yagi H, Deguchi K, Aono A, Tani Y, Kishimoto T, Komori T (1998) Growth disturbance in fetal liver hematopoiesis of Mll-mutant mice. Blood 92:108–117

Yamaguchi S, Hong K, Liu R, Shen L, Inoue A, Diep D, Zhang K, Zhang Y (2012) Tet1 controls meiosis by regulating meiotic gene expression. Nature 492:443–447

Yang H, Lin H, Xu H, Zhang L, Cheng L, Wen B, Shou J, Guan K, Xiong Y, Ye D (2014) TET-catalyzed 5-methylcytosine hydroxylation is dynamically regulated by metabolites. Cell Res 24:1017–1020

Yeo SH, Clarkson J, Herbison AE (2014) Kisspeptin-gpr54 signaling at the GnRH neuron is necessary for negative feedback regulation of luteinizing hormone secretion in female mice. Neuroendocrinology 100:191–197

Yu BD, Hess JL, Horning SE, Brown GA, Korsmeyer SJ (1995) Altered Hox expression and segmental identity in Mll-mutant mice. Nature 378:505–508

Imprinted Genes and Hypothalamic Function

10

Michela Pulix and Antonius Plagge

Abstract

Genomic imprinting, a specific type of inherited epigenetic regulation, controls a number of genes with important functions in the hypothalamus and its associated neuroendocrinology. Imprinted genes are expressed monoallelically, dependent on the parental origin of the allele, whereby maternally or paternally inherited DNA methylation marks determine gene activity. In this chapter, we provide an overview of imprinted genes with roles in the hypothalamic regulation of whole-body energy homeostasis, neuroendocrine hormone functions, and circadian rhythmicity. As an example for imprinted genes that impact on the central regulation of energy balance, we describe the *Gnas* locus in more detail, since it generates parental allele-specific gene products with antagonistic roles in energy expenditure and sympathetic nervous system activity. Several human neuroendocrine disorders are caused by defects in imprinted genes as described here for Prader–Willi Syndrome and Angelman Syndrome. Furthermore, we outline methodological approaches for the investigation of DNA methylation marks and allelic gene expression, including bisulfite conversion of DNA and pyrosequencing. Recent technological advances to resolve methylation and allelic expression at the single-cell level are introduced.

M. Pulix
Cellular and Molecular Physiology, Institute of Translational Medicine, University of Liverpool, Liverpool, UK

Department of Neuroscience, Sheffield Institute for Translational Neuroscience, University of Sheffield, Sheffield, UK

A. Plagge (✉)
Cellular and Molecular Physiology, Institute of Translational Medicine, University of Liverpool, Liverpool, UK
e-mail: plagge@liverpool.ac.uk

S. Wray, S. Blackshaw (eds.), *Developmental Neuroendocrinology*, Masterclass in Neuroendocrinology 9, https://doi.org/10.1007/978-3-030-40002-6_10

Keywords
Genomic imprinting · Hypothalamus · Gnas · Energy homeostasis · Prader–Willi syndrome · Angelman syndrome · DNA methylation · Bisulfite conversion of DNA · Pyrosequencing

10.1 Introduction

The term genomic imprinting refers to the expression of certain genes from only one of the two parental alleles. This is dependent on the parental origin, i.e., some genes are monoallelically expressed from the maternal, others from the paternal allele, while the opposite alleles are silenced. The allelic silencing is already determined in the parental germ cells through epigenetic "marks" and maintained in the somatic tissues of the offspring (Fig. 10.1). To maintain this parent-of-origin-specific gene expression, the epigenetic "marks" are erased in the primordial germline cells of the developing offspring and reset according to the sex of the offspring. This process ensures that the maternal or paternal allelic expression, respectively, of an imprinted gene is inherited from generation to generation. To date, circa 150 imprinted genes have been confirmed in humans and mice, but imprinted gene expression also exists in other mammals (http://www.mousebook.org/imprinting-gene-list; http://igc.otago.ac.nz/home.html). Most imprinted genes occur in conserved clusters, but the imprinting of some genes is species specific.

Many imprinted genes have essential roles in embryonic development and placental functions (Plasschaert and Bartolomei 2014). However, they also play important roles in the brain, including regulation of neurogenesis, synaptic transmission, and behavior (Perez et al. 2015; Peters 2014). Within the brain, the hypothalamus has proven to be the main site of expression of imprinted genes, and from studies of knockout mouse models as well as human inherited disorders, it is evident that they influence neuroendocrine pathways, energy homeostasis, circadian rhythm, and social interactions (Ivanova and Kelsey 2011; Perez et al. 2015). In this chapter, we provide an overview of the hypothalamic functions of imprinted genes and highlight a few disease-relevant examples. In addition, we discuss techniques used in the identification and epigenetic analysis of imprinted genes and indicate currently open questions and trends in the field.

10.2 Main Features of Imprinted Gene Clusters

The importance of imprinting in mammals became evident from embryological studies, in which pro-nuclei of newly fertilized mouse eggs were transplanted in order to generate a genetically diploid bi-maternal (parthenogenetic) or a bi-paternal (androgenetic) conceptus. The embryos that developed from these concepti died in utero and showed growth aberration in embryonic and extraembryonic tissues (Ferguson-Smith 2011). These observations demonstrated that a diploid genome as

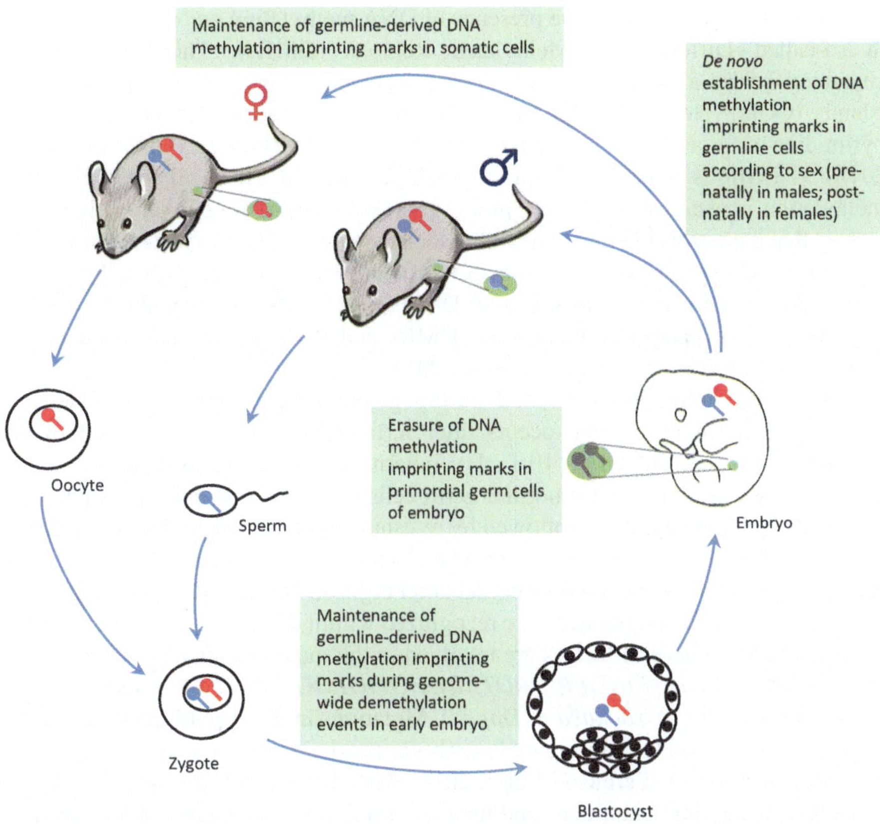

Fig. 10.1 Inheritance cycle of DNA methylation imprinting marks through the germline and maintenance in somatic tissues. The genomic DNA methylation imprints are established in the germline cells in a sex-specific manner and transmitted to the zygote, where they resist reprogramming events and global changes in DNA methylation, which take place after fertilization. While the genomic imprinting marks are maintained in somatic cells throughout life, they are erased in the primordial germ cells of the developing embryo and re-established de novo according to the sex of the organism, to be transmitted to the next generation. Re-establishment of imprinting marks in gametogenesis occurs prenatally in males and at postnatal stages in females [Adapted from Bartolomei and Ferguson-Smith (2011)]

such is not sufficient for normal mammalian development, that the two parental genomes make different functional contributions to embryonic development and that there must be some kind of "mark," which mediates the distinct functions of the maternal and paternal genomes, respectively. Thus, genomic imprinting became a paradigm that drove research into "epigenetic" mechanisms (Ferguson-Smith 2011). We know today that those epigenetic "marks" represent methylation of genomic DNA, histone modifications, and noncoding RNA expression.

In the majority of cases, imprinted genes are typically found in clusters of 3–12 genes spread over 20 Kb–3.7 Mb of DNA (Ferguson-Smith 2011; Peters, 2014) and they are usually regulated by the presence of DNA methylation on one parental allele in so-called Differentially Methylated Regions (DMRs). These regions are characterized by a high density of CpG dinucleotides (above 55%), termed CpG islands (CGIs). Here, methylation of cytosines occurs at the C5 position of the pyrimidine ring and is catalyzed by DNA methyltransferases (DNMTs) (Lyko 2018). A DMR is termed an Imprinting Control Region (ICR), if the differential methylation is established already in the germ cells and if functional studies have proven that it controls the imprinting of associated genes within a cluster. Thus, ICRs regulate gene expression and chromatin structure over long distances (Ferguson-Smith 2011). Maternally methylated DMRs and ICRs are usually located at promoters while paternally methylated DMRs and ICRs are located in intergenic regions (Bartolomei and Ferguson-Smith 2011).

DNA methylation is established during gametogenesis, resists the extensive genomic reprogramming that occurs after fertilization and is maintained in the offspring throughout life (Fig. 10.1). For inheritance of imprinted gene expression to the next generation, the primordial germ cells of an embryo undergo a phase of erasure of DNA methylation, followed by re-establishment of methylation at DMRs/ICRs according to the sex of the individual and mediated by DNMT3A and the accessory protein DNMT3L (Bartolomei and Ferguson-Smith 2011; Plasschaert and Bartolomei 2014). How the ICRs are recognized within the genome is still unknown, but it is clear that several factors are involved in the maintenance of DNA methylation, including *ZINC FINGER PROTEIN HOMOLOG 57* (*ZFP57*) and *Developmental pluripotency-associated 3* (*Dppa3*). Mutations in these genes result in loss of imprinting at several loci during early embryogenesis, which in turn causes a range of disease symptoms (Ferguson-Smith 2011; Plasschaert and Bartolomei 2014).

Before discussing techniques and methods used in the analysis of allelic expression and epigenetic regulation of imprinted genes, we first provide an overview about the functional roles of imprinted genes in the hypothalamus and, where applicable, their involvement in human neurodevelopmental disorders. Then, in the context of the Prader–Willi syndrome (PWS)/Angelman syndrome (AS) imprinting cluster and the *Gnas* locus, we discuss the structure and organization of regulatory elements of imprinted genes.

10.3 Functions of Imprinted Genes in the Hypothalamus

When the first imprinted genes (*Igf2*, *Igf2r,* and *H19*) were identified, they were found to affect mainly embryonic growth, placental function, and nutrient supply and demand at the maternal–fetal interface (Ferguson-Smith 2011; Peters 2014; Plasschaert and Bartolomei 2014). However, it soon became apparent that a number of imprinted genes were expressed in defined brain regions, and many of them showed high expression levels specifically in the hypothalamus (Ivanova and Kelsey 2011). Phenotype analyses of knockout mouse models and clinical case studies of human imprinting disorders revealed that many imprinted genes continue to impact growth and whole-body energy homeostasis beyond embryonic development and into adulthood via influencing neuroendocrine pathways and the autonomic nervous system. Other core functions of the hypothalamus, including circadian rhythms, reproductive functions, and maternal-infant care behavior are also deregulated upon the deficiency of some imprinted genes. We will first describe the hypothalamic roles of the *Gnas* and the PWS-AS loci since they have been analyzed in some detail, before summarizing findings related to other imprinted genes.

10.3.1 Hypothalamic Regulation of Energy Homeostasis by *Gnas* and *Gnasxl*

The *Gnas* imprinted locus on mouse chromosome 2 and human chromosome 20 comprises a complex arrangement of protein-coding and noncoding transcripts, alternative promoters and epigenetically controlled regulatory elements (Fig. 10.2a) (Chen et al. 2011; Peters 2014). The two protein-coding transcripts *Gnas* and *Gnasxl* derive from different first exons, but share exons 2–12. As the open reading frames are conserved, the two proteins Gαs and XLαs only differ at their NH_2-termini, thus forming alternative versions of the G-protein α-stimulatory subunit. Both can mediate signal transduction from seven transmembrane receptors to adenylate cyclase and cAMP formation, although XLαs can also stimulate the inositol 1,4,5-trisphosphate (IP3)/protein kinase C (PKC) second messenger pathway in some cell types (He et al. 2015). While *Gnasxl* is exclusively expressed from the paternal allele, *Gnas* is biallelically expressed in most tissues, but monoallelically transcribed from the maternal allele in specific cell types of the brain, renal proximal tubules, thyroid gland, pituitary somatotroph cells and the ovary (Chen et al. 2011). The third protein-coding transcript, maternal allele-derived *Nesp*, generates a neuroendocrine secretory protein (Nesp55) from an open reading frame that is confined to its upstream exon, although the RNA is spliced onto downstream exons 2–12 of *Gnas*. The regions of differential DNA methylation (ICRs at *Nespas-Gnasxl* and *exon-1a*; DMR at *Nesp*), the noncoding transcripts *Nespas* and *exon-1a*, as well as the *Nesp* transcript have a role in the establishment and regulation of genomic imprinting at the cluster (Plagge 2012). Both ICRs become methylated in the maternal germ cells, and the methylation on the maternal allele is maintained in somatic tissues of the offspring. The ICR at *Nespas-Gnasxl* controls the imprinting of all transcripts of the locus, while the ICR at *exon-1a* only regulates the monoallelic expression of *Gnas* in those specific tissues mentioned above. The DNA methylation on the maternal allele inhibits the transcription of *Gnasxl*, *Nespas*, and *exon-1a*. On the paternal allele, the unmethylated ICRs serve as promoters for the noncoding RNAs *Nespas* and *exon-1a*, which silence *in cis* the expression of *Nesp* and *Gnas* via unclear mechanisms that involve histone modifications (Mehta et al. 2015).

Hypothalamic functions have been shown for *Gnas* and *Gnasxl*, while behavioral phenotypes of *Nesp* knockout mice could not be specifically attributed to a role in this brain region, although *Nesp* is expressed in the hypothalamus (Plagge et al. 2005). In humans, *GNAS* and *GNASXL* mutations or epigenetic deregulation cause

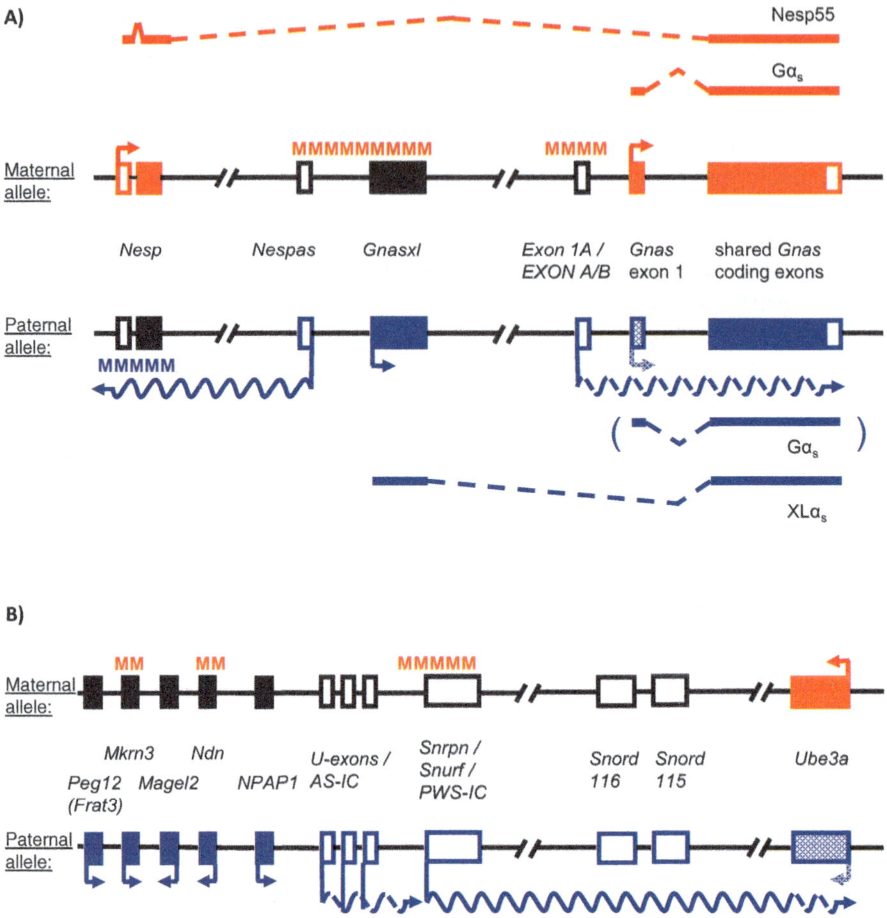

Fig. 10.2 Simplified schemes of the *Gnas* and the PWS-AS (*Snrpn-Ube3a*) imprinted loci. The features of the maternal and paternal alleles are indicated in red and blue, respectively. Genes or transcripts are named in the central part. Arrows mark transcription start sites and undulating lines noncoding RNAs. Open and filled boxes represent noncoding and coding exons, respectively, while black boxes indicate silenced genes. Differentially methylated regions of DNA (DMRs) are marked by MMM. (**a**) For the *Gnas* locus, the alternatively spliced coding transcripts and proteins are named above and below the alleles. DMRs at *Nespas* and *Exon 1A* (*EXON A/B* in human) are established in the maternal germline, while methylation at *Nesp* occurs during early embryonic development. *Gnas* is expressed biallelically, but is silenced on the paternal allele in some tissues (hatched box). *Nesp* represents a coding transcript (ORF limited to *Nesp* exon 2). *Nespas* is expressed from the unmethylated imprinting control region (ICR) of the locus. (**b**) The PWS-AS-associated *Snrpn-Ube3a* imprinting cluster contains several genes (represented by single-exon boxes) that show monoallelic expression in the brain. *Ube3a* is biallelically expressed in most tissues, but silenced on the paternal allele in neurons (hatched box). The *Snrpn* long noncoding RNA occurs in multiple, variably processed forms, including brain-specific variants that overlap with *Ube3a* (interrupted undulating lines). *Snord115* and *Snord116* represent clusters of C/D box snoRNAs, which are generated from the *Snrpn* transcript. The bipartite ICR of the human locus is indicated as AS-IC and PWS-IC, the latter being conserved in mouse around the main *Snrpn* start site. *NPAP1* is not present in the murine locus, but *Peg12* is imprinted in mice. Somatic DMRs at *Ndn* and *Mkrn3* are established during development

complex neuroendocrine disorders termed Albright's Hereditary Osteodystrophy (AHO) and Pseudohypoparathyroidism (PHP), the symptoms of which are caused by hormone resistance in multiple tissues, including disruption of parathyroid hormone signaling in the kidney (Table 10.1) (Mantovani et al. 2018; Turan and Bastepe 2013). Hypothalamus-associated symptoms are related to postnatal growth, feeding, and regulation of energy balance at the adult stage. Homozygous mutations of *Gnas* in mice are embryonic lethal and heterozygous mutations show different phenotypes, dependent on whether they were maternally or paternally inherited, whereby the mouse phenotypes largely reproduce the human disease symptoms. Due to genomic imprinting, maternally inherited mutations in shared exons only affect the *Gnas* transcript, but in those tissues, in which it is imprinted, Gαs function will be completely lost, while a 50% reduction occurs in cells with biallelic expression. On the other hand, paternally inherited mutations in shared exons will result in loss of XLαs and a 50% reduction of Gαs in cells with biallelic expression.

Gnas and *Gnasxl* transcript- and tissue-specific knockout mouse models have shed some light onto the functions of the two proteins in the hypothalamus and revealed opposite or antagonistic impacts of Gαs and XLαs on energy balance and sympathetic nervous system (SNS) activity (Weinstein 2014). Conditional gene targeting of the *Gnas*-specific exon 1 and its brain-specific deletion (via Nestin-Cre strain) on the maternally inherited chromosome causes severe obesity due to decreased energy expenditure and O_2 consumption, lower locomotor activity, reduced SNS outflow, body temperature, and heart rate (Table 10.1). Glucose intolerance and insulin resistance were also observed. These symptoms are, however, not found in the paternally inherited, brain-specific knockout, which show normal energy homeostasis. These findings indicate genomic imprinting in the brain. The response to stimulation of the melanocortin receptors (MC3R, MC4R), which mediate the anorexigenic response downstream of leptin signaling, was found to be blunted in these mice. However, the maternally inherited *Gnas* mutation specifically disrupts the energy expenditure-stimulating functions of melanocortins, while their food intake-reducing effects remained intact (Chen et al. 2009). The mechanism (s) underlying these responses are unclear. The decreased energy expenditure in the mutants has been linked to reduced SNS outflow as levels of norepinephrine and uncoupling protein 1 (UCP1) in brown adipose tissue (BAT), as well as heart rate and blood pressure are reduced.

Monoallelic maternal expression of *Gnas* is found in the paraventricular nucleus (PVN) of the hypothalamus (Chen et al. 2009). A specific maternal allele knockout of *Gnas* in the PVN, using a Sim1-Cre strain, reproduced a much milder obesity phenotype than observed with the Nestin-Cre line described above suggesting that imprinted Gαs expression in additional brain regions contributes to the earlier observed phenotype (Chen et al. 2012). However, it does reproduce the lower blood pressure and heart rate of the whole-brain knockout. Maternal allele-specific Gαs is also required in the PVN to mediate melanocortin-induced blood pressure increases (Li et al. 2016). In contrast to the brain knockout, the SNS stimulation of BAT, cold-induced thermoregulation as well as the melanocortin induction of energy expenditure are intact in the PVN-specific *Gnas* knockout, indicating that these physiological responses are mediated by other neurons (Chen et al. 2012).

Table 10.1 Functions of imprinted genes in the hypothalamus and related human disorders

Gene	Protein/RNA function	Expressed allele	Hypothalamus-related mouse knockout phenotype	Human disorder
Gnas	G-protein α-stimulatory subunit (Gαs); (intracellular signaling)	*Biallelic in most tissues; maternal allele in specific cell types*: PVN, DMH, and subset of MC4R-expressing neurons in the brain, proximal renal tubules, thyroid gland, pituitary somatotroph cells, and ovary	*Maternally inherited mutation*: obesity, reduced energy expenditure, reduced SNS activity, impaired melanocortin receptor-mediated stimulation of energy expenditure and blood pressure, reduced blood pressure and heart rate, decreased locomotor activity, glucose intolerance, insulin resistance *Paternally inherited mutation*: normal energy homeostasis	Albright's Hereditary Osteodystrophy (AHO) and Pseudohypoparathyroidism (PHP) Hypothalamus-associated symptoms when *maternally inherited*: obesity, reduced SNS outflow Symptoms related to other tissues when *maternally inherited*: renal parathyroid hormone resistance, hypocalcemia, hyperphosphatemia, TSH resistance, GHRH resistance *Parent-of-origin independent symptoms*: short stature, brachydactyly, heterotopic ossifications, and other variable features
Gnasxl	Extra large G-protein α-stimulatory subunit (XLαs); (intracellular signaling)	Paternal	*Postnatally*: growth retardation, hypoglycemia, reduced milk intake, reduced lipid stores in adipose tissue *Adult stage*: decreased adiposity and body weight, elevated SNS activity, elevated metabolic rate and body temperature, increased blood pressure and heart rate, increased glucose tolerance, and insulin sensitivity	*Pre- and neonatal symptoms* Small for gestational age/intrauterine growth retardation, low birth weight, postnatal growth retardation, feeding problems, low body mass index during childhood

Magel2	MAGE family ubiquitin ligase regulator (endosomal protein sorting and recycling)	Paternal	*Postnatally*: 10–50% mortality; diminished oxytocin-induced suckling initiation after birth; reduced neonatal hypothalamic levels of oxytocin, arginine-vasopressin, and orexin-A *Adult stage*: increased fat mass despite reduced food intake and orexin-A levels, reduced lean mass; hypothalamic leptin resistance; reduced POMC neurons and projections; rhythmic circadian expression in SCN and altered/reduced circadian activity patterns; and abnormal behavior	Prader–Willi syndrome
Ndn (Necdin)	MAGE family ubiquitin ligase regulator	Paternal	High rate of early neonatal mortality due to respiratory deficiencies (brainstem related); fewer hypothalamic oxytocin and gonadotropin-releasing hormone neurons; reduced thyrotropin-releasing hormone levels in PVN and hypothyroidism; abnormal axon bundles; and behavioral abnormalities	Prader–Willi syndrome
Snord116	Box C/D small nucleolar RNA	Paternal	Postnatal and adult growth retardation; reduced adiposity; increased energy expenditure; resistance to high-fat diet-induced obesity; improved body temperature maintenance at 4 °C; behavioral abnormalities Adult-onset deletion in mediobasal hypothalamus: hyperphagia and obesity	Prader–Willi syndrome

(continued)

Table 10.1 (continued)

Gene	Protein/RNA function	Expressed allele	Hypothalamus-related mouse knockout phenotype	Human disorder
Ube3a	Ubiquitin-protein ligase E3A; (protein degradation)	Biallelic in most tissues; maternal allele in brain	Sleep disturbances, circadian rhythm regulation via *Bmal1*	Angelman syndrome
Dlk1	Protein delta homolog 1, (Trans-membrane and secreted protein with similarity to Notch ligands)	Paternal	Hypothalamus-related functions uncertain, potentially involving postnatal growth retardation	Central precocious puberty; increased fat mass (Temple syndrome)
Dio3	Thyroxine 5-deiodinase (thyroid hormone catabolism)	Paternal (e.g., neonatal hypothalamus, tissue- and stage-specific)	Postnatal and adult growth retardation; reduced adiposity; abnormal thyroid hormone regulation and hypothalamus–pituitary–thyroid axis; increased energy expenditure and locomotion; changes in hypothalamic gene expression of leptin-melanocortin, oxytocin, and vasopressin pathways; abnormal circadian activity; impaired viability and fertility	Potentially involved in Temple syndrome
Kcnq1	Potassium voltage-gated channel subfamily KQT member 1	Maternal	No hypothalamus-related phenotype described, but range of other symptoms	Growth hormone deficiency and gingival fibromatosis, QT syndrome, auditory and cardiac symptoms
Peg3	Zinc finger DNA-binding protein	Paternal	Reduced body weight throughout life; increased adiposity; leptin resistance; reduced metabolic rate; lower body temperature; altered hypothalamic neuropeptide expression	

Further investigations by the group of L. S. Weinstein excluded the ventromedial hypothalamus (VMH) and identified the dorsomedial hypothalamus (DMH) as a major site of imprinted Gαs regulation of energy expenditure (Chen et al. 2017). Adenovirus-mediated Cre-deletion of *Gnas* on the maternal allele in the DMH resulted in a number of changes including obesity, reduced energy expenditure, heart rate, and locomotor activity, as well as a blunted response to melanocortin-induced O_2 consumption. Baseline activation of BAT was reduced as seen in whole-brain knockout mice, but cold-induced activation of BAT and maintenance of body temperature were intact, as was food intake (Chen et al. 2017). Blood pressure was normal in the DMH-specific deletion of Gαs, in line with previous findings attributing this *Gnas* function to the PVN. However, when *Gnas* was deleted on the maternal allele in the DMH, stimulation of MC3/4R in the DMH did not lead to phosphorylation of cAMP response element-binding protein (CREB) (Chen et al. 2017). These findings are consistent with maternal allele-specific expression of Gαs in the DMH, which is required for basic and melanocortin-induced stimulation of energy expenditure.

Recently, the importance of *Gnas* for mediating the melanocortin effects on energy homeostasis was confirmed through a specific knockout in MC4R-expressing neurons (Podyma et al. 2018). Homozygous deletion of *Gnas* in MC4R neurons results in severe obesity, while maternal allele-specific deficiency leads to a milder phenotype. Paternal heterozygous knockouts showed normal energy balance. Reduced energy expenditure and locomotor activity also featured in homozygous *Gnas* MC4R-specific knockouts, but the major contributor to the obesity in these mice was increased food intake, which was not observed in the other maternal allele-specific hypothalamus knockouts described above (Podyma et al. 2018). The milder obesity in mice with maternal allele-specific disruption of *Gnas* in MC4R neurons was attributable to reduced energy expenditure with food intake being unchanged. Furthermore, while homozygous knockouts could not maintain body temperature in a cold environment, heterozygous maternal knockouts of *Gnas* in MC4R neurons showed thermoregulation in line with previous observations in PVN- and DMH-specific deletions (Chen et al. 2012, 2017; Podyma et al. 2018).

In summary, stimulation of SNS outflow appears to be at the heart of the metabolic functions regulated by Gαs expression from the maternal allele. But it is also apparent that *Gnas* expression from the maternal allele regulates various branches of the SNS, affecting cardiovascular phenotypes via the PVN and energy expenditure via the DMH, which projects to the raphe pallidus of the medulla and other SNS-relevant brain areas. Its pathway of regulating cold-induced thermogenesis, however, remains as yet unknown (Chen et al. 2012, 2017; Podyma et al. 2018). Although disruption of MC3/4R signaling contributes substantially to the energy homeostasis phenotypes of central nervous system-specific *Gnas* knockouts, it is currently unclear whether Gαs signaling from other receptors might also contribute.

In contrast to *Gnas*, the *Gnasxl* knockout mouse phenotype, as well as case reports in human, indicate that the extra-large variant XLαs has overtly antagonistic functions to Gαs in the regulation of energy homeostasis and SNS activity (Weinstein 2014). *Gnasxl* is expressed in all major hypothalamic areas (apart from the VMH), the suprachiasmatic nucleus (SCN), preoptic area, and SNS-associated nuclei in the medulla, e.g., the raphe pallidus, nucleus of the solitary tract, as well as the intermediolateral layer of the spinal cord (Krechowec et al. 2012; Plagge et al. 2004). Expression of *Gnasxl* in peripheral tissues, for example, muscle tissues, becomes silenced postnatally with only a few exemptions like the adrenal medulla (Krechowec et al. 2012; Xie et al. 2006). Newborn XLαs-deficient pups become growth retarded, take in less milk and display hypoglycemia and lack of lipid reserves in adipose tissue, all of which results in a high mortality rate in the first week (Table 10.1; Plagge et al. 2004; Xie et al. 2006). Those *Gnasxl* knockout mice that survive to weaning age, develop into fertile adults, but remain underweight and lean throughout life, despite relatively increased food intake. Furthermore, their phenotype consists of increased energy expenditure, metabolic rate, body temperature, blood pressure, and heart rate, which is associated with elevated SNS outflow (Nunn et al. 2013; Xie et al. 2006). They also show improved glucose tolerance and insulin sensitivity (Xie et al. 2006). The precise neural pathways through which *Gnasxl* regulates SNS outflow and energy homeostasis remain unresolved, as are the neural receptors that signal via XLαs.

Human disease symptoms related to loss of *GNASXL* function have been difficult to ascertain since mutations that specifically disable the first *GNASXL* exon have not been identified. However, paternal transmission of mutations in shared downstream exons 2–13 and maternal uniparental disomies that encompass the *GNAS* locus on chromosome 20, which result in two maternal alleles and lack of a paternally inherited allele, indicate pre-/neonatal symptoms (Table 10.1). These include intrauterine growth retardation, low birth weight, postnatal growth retardation, and feeding problems (Genevieve et al. 2005; Kawashima et al. 2018; Mulchandani et al. 2016; Richard et al. 2013). Although no adult metabolic data of such patients have been reported, the neonatal symptoms are reminiscent of the *Gnasxl* knockout mouse phenotype (Plagge et al. 2004). Taken together, these findings show that XLαs exerts an important role in neonatal physiology in addition to its regulation of adult energy homeostasis, which to date has only been confirmed in mice.

In adults, XLαs acts in the brain to inhibit SNS outflow and energy expenditure, while its variant Gαs functions to stimulate SNS activity and metabolic rate. These antagonistic roles are correlated on the epigenetic level with genomic imprinting and monoallelic transcription from opposite parental alleles. *Gnasxl* is expressed exclusively from the paternal allele, while *Gnas* is only imprinted in specific hypothalamic neurons, where it is transcribed from the maternal allele. Why epigenetic regulation through genomic imprinting of crucial growth and energy homeostasis functions evolved in mammals is beyond the scope of this chapter, but has been extensively discussed in the literature (Haig 2004; Peters 2014).

10.3.2 Hypothalamic Functions of the Prader–Willi Syndrome (PWS) and Angelman Syndrome (AS) Imprinting Cluster

Another example of a cluster of imprinted genes with functions in the hypothalamus is the Prader–Willi syndrome and Angelman syndrome locus on human chromosome 15 (conserved on mouse chromosome 7) (Buiting et al. 2016; Cassidy et al. 2012; Peters 2014). The locus comprises five genes with paternal allele-specific expression (*MKRN3, MAGEL2, NDN, NPAP1*, and *SNURF/SNRPN/SNORD115-116*), and one with maternal allele-specific transcription (*UBE3A*), although the murine locus contains paternally expressed *Peg12* instead of *NPAP1* (Fig. 10.2b). Loss of function of the paternally expressed genes causes PWS, while loss of function of the maternally expressed *UBE3A*, which is imprinted in the brain only, causes AS. The monoallelic expression of all genes in the cluster is controlled by a bipartite ICR that encompasses the promoter of the *Snurf/Snrpn/Snord115–116* gene (PWS-IC) and additional sequences further upstream (AS-IC). The PWS-IC is methylated on the maternal allele, thus silencing the expression of the long *Snurf/Snrpn/Snord115–116* RNA. In neurons, this paternally expressed transcript, which is also known as *Ube3a antisense* and can be processed in different ways, extends into *Ube3a* exons and is thought to inhibit the expression of *Ube3a* in *cis* through transcriptional interference, thus resulting in intact *Ube3a* expression only from the maternal allele (Buiting et al. 2016; LaSalle et al. 2015). On the paternal allele, the PWS-IC remains unmethylated and has an activation function for all other paternally expressed genes of the cluster, which involves physical interactions via chromosome looping (Plagge 2012).

Angelman syndrome constitutes a neurodevelopmental disorder that is mainly characterized by intellectual disability, microcephaly, speech impairment, happy demeanor, EEG abnormalities and sleep disturbances (Buiting et al. 2016). Functionally, Ube3a (Ubiquitin-protein ligase 3a) is involved in labeling proteins for degradation and lack of the enzyme impairs synapse formation and remodeling. Little is known about hypothalamus-specific roles of Ube3a in the context of AS, but knockout mice show abnormalities in sleep and circadian rhythm, potentially due to impaired Ube3a-mediated degradation of the circadian clock protein Bmal1 in the hypothalamus (Ehlen et al. 2015; Shi et al. 2015). On the other hand, main symptoms of PWS are postnatal hypotonia, poor suckling, failure to thrive, respiratory disturbances, hypogonadism, childhood-onset obesity, hyperphagia, cognitive and behavioral abnormalities, short stature, and growth hormone insufficiency (Cassidy et al. 2012). Knockout mouse models indicate that all paternally expressed genes contribute to the range of these symptoms.

The functions of *Magel2* (*Mage-like protein 2*) have been characterized, and it is highly expressed in many areas of the hypothalamus (Tacer and Potts 2017). On a molecular level, Magel2 acts as a regulator of RING E3 ubiquitin ligases within a complex that controls endosomal protein sorting. Deficiency of Magel2 in a hypothalamic cell line impairs protein recycling to the plasma membrane (Tacer and Potts 2017). The physiological functions of *Magel2* are indicated by knockout mice, which show 10–50% pre-weaning mortality (Table 10.1) (Kozlov et al. 2007; Schaller et al. 2010). This mortality occurs in the first few hours after birth, due to a deficiency in oxytocin in the pups, which at this stage is required to initiate suckling activity (Schaller et al. 2010). For surviving knockout mice, a slight postnatal growth retardation is followed at the adult stage by reduced lean mass, increased fat mass despite decreased food intake and hypothalamic orexin-A

(Bischof et al. 2007; Kozlov et al. 2007). It is noteworthy that the observation of reduced food intake is contrary to PWS symptoms. Hypothalamic POMC neurons and their projections were found to be deficient in knockout mice (Maillard et al. 2016; Mercer et al. 2013), and leptin resistance might be related to reduced leptin receptor recycling from endosomes to the plasma membrane (Wijesuriya et al. 2017). The Magel2 protein is also robustly expressed in the SCN and its levels fluctuate with circadian rhythmicity (Kozlov et al. 2007). Consistent with SCN function, the knockout mice show overall reduced wheel-running activity and disturbed activity patterns under constant darkness (Kozlov et al. 2007; Tacer and Potts 2017).

Similar to *Magel2*, *Necdin* (*Ndn*) constitutes another member of the Mage family of proteins (Tacer and Potts 2017). *Necdin* is expressed in a number of brain regions, and it has roles in neuronal survival, migration, differentiation, and neurite formation (Table 10.1). *Ndn* knockout mice have neonatal respiratory problems, which lead to a substantial rate of mortality shortly after birth (Gerard et al. 1999; Muscatelli et al. 2000). Respiratory disturbances are also observed in PWS and they are due to defects in the serotoninergic system of the brainstem inspiratory rhythm generator (Matarazzo et al. 2017). In the hypothalamus, lack of *Ndn* causes reduced numbers of oxytocinergic neurons in the PVN and gonadotropin-releasing hormone neurons in the preoptic area (Miller et al. 2009; Muscatelli et al. 2000). Deficits in the latter neuron population might be responsible for the hypogonadism observed in PWS. *Necdin* knockout mice present symptoms of hypothyroidism due to reduced levels of thyrotropin-releasing hormone in the PVN (Hasegawa et al. 2012). Furthermore, Necdin has a role in axon outgrowth as indicated by aberrant axon bundles in the anterior hypothalamus and other brain regions (Lee et al. 2005). On the molecular level, Necdin appears to interact with a diverse set of binding partners, including cytoskeleton-associated proteins as well as transcription factors (Hasegawa et al. 2012; Lee et al. 2005).

The box C/D small nucleolar RNA cluster *Snord116* also has a crucial role in PWS (Cavaille 2017). Patients with microdeletions of this cluster, but intact *NECDIN* and *MAGEL2* genes, present with PWS symptoms, and knockout mouse models also show some of these disease characteristics (Bieth et al. 2015; Ding et al. 2008; Skryabin et al. 2007). Postnatal growth retardation is a shared feature in humans and mice with paternally inherited *Snord116* deletions. However, knockout mice remain lean and underweight throughout life due to increased energy expenditure, while *SNORD116*-mutant patients develop the typical PWS hyperphagic obesity toward adulthood (Bieth et al. 2015; Ding et al. 2008; Skryabin et al. 2007). Interestingly, when the *Snord116* cluster is deleted in the mediobasal hypothalamus of adult mice via an adenovirus Cre-recombinase vector injection, the mutant mice then develop hyperphagia and obesity reminiscent of PWS patients (Polex-Wolf et al. 2018). These findings highlight physiological differences between humans and mice and potentially indicate species-specific compensatory abilities to inherited gene mutations during development.

10.3.3 Other Imprinted Genes with Hypothalamic Functions

Although less well characterized, four other imprinted genes have roles in the hypothalamus, which are briefly summarized here. A cluster of imprinted genes on mouse chromosome 12 (human chromosome 14) contains two protein-coding genes with hypothalamic functions, *Dlk1* (*Protein delta homolog 1*) and *Dio3* (*Thyroxine 5-deiodinase*), both expressed from the paternal allele (Table 10.1). The Dlk1 protein has structural similarity to Notch ligands, but its precise signaling mechanism is unknown. Dlk1 is prominently expressed in several hypothalamic nuclei (arcuate nucleus, PVN, DMH, and SCN), and it has been co-localized in oxytocin, arginine–vasopressin, orexin, NPY, and dynorphin-positive neurons (Meister et al. 2013; Villanueva et al. 2012). *Dlk1* involvement in embryonic development and adiposity has been well characterized, but its hypothalamus-specific functions are poorly understood. A mutation of *DLK1* in humans leads to central precocious puberty, possibly via GnRH-neuron regulation, and increased fat mass (Dauber et al. 2017). *Dio3* shows monoallelic expression from the paternal allele in the neonatal hypothalamus (Martinez et al. 2014). Knockout mice display postnatal growth retardation, which lasts into adulthood, deregulated thyroid hormone levels and disturbances in the hypothalamic–pituitary–thyroid axis (Table 10.1). Reduced adiposity is associated with changes in hypothalamic gene expression of the leptin–melanocortin pathway, increased energy expenditure, and locomotion (Wu et al. 2017). Activity patterns of knockout mice also showed circadian changes, which might be associated with *Dio3* expression in the SCN. Additional behavioral abnormalities relating to aggression and maternal care of the knockout mice are associated with changes in the hypothalamic oxytocin and arginine–vasopressin systems (Stohn et al. 2018). The third gene is the voltage-gated potassium channel gene *Kcnq1*, which is located in a large imprinting cluster on mouse chromosome 7 (human chromosome 11) and is expressed from the maternal allele. It has a crucial role in the hypothalamic–pituitary growth hormone axis. The potassium channel is expressed in hypothalamic growth hormone-releasing hormone (GHRH) neurons and pituitary somatotropes. *KCNQ1* mutations in humans cause growth hormone deficiency and gingival fibromatosis among other symptoms (Table 10.1) (Tommiska et al. 2017). The fourth imprinted gene associated with hypothalamic functions is the *Paternally expressed gene 3* (*Peg3*), a Zinc finger-containing DNA-binding factor that is widely expressed in the hypothalamus (Li et al. 1999). *Peg3* knockout mice have reduced body weight at all stages of life, but are hypometabolic, show a proportionately increased fat mass, lower body temperature, leptin resistance and changes in hypothalamic NPY, POMC, MCH, and Orexin neuropeptide expression (Curley et al. 2005). Whether knockout females show deficiencies in maternal behavior toward their offspring (e.g., nest building, pup retrieval, nursing) that are linked to a reduced number of hypothalamic oxytocin neurons, is controversial (Li et al. 1999; Denizot et al. 2016).

10.4 Analysis of Genomic Imprinting

10.4.1 Analysis of Methylation

As mentioned in the introduction, the canonical regulatory mechanism of imprinting is the methylation of DNA at specific CGIs to form DMRs or ICRs. A breakthrough in the study of DNA methylation was achieved with the technique of bisulfite conversion of genomic DNA. The treatment with sodium bisulfite promotes the deamination of cytosine residues, which is followed by desulfonation at elevated pH, thus leading to conversion into uracil (Fig. 10.3). When a residue is methylated, however, the deamination cannot take place and the cytosine remains unchanged. Since only cytosines in single-stranded DNA are susceptible to conversion by bisulfite, complete DNA denaturation is a critical step. After bisulfite conversion, the DNA fragment of interest can be amplified by PCR using primers designed to anneal to either the sense or the antisense DNA strand, which at this point are not complementary anymore. The PCR primer design needs to be based on the expected bisulfite-converted sequence, in which methylated cytosines will be retained as C, and the unmethylated cytosines converted to uracil, i.e., such primers are usually AT-rich (Fig. 10.3). Furthermore, AT-rich bisulfite PCR amplicons (from originally unmethylated DNA) can be difficult to amplify using standard Taq-Polymerases and, therefore, amplicon sizes of less than ~400 bp are recommended. Small amplicon size is especially important when allelic differences in methylation are investigated to avoid a bias in amplification of originally methylated over unmethylated alleles, i.e., to minimize preferential amplification of templates with more balanced nucleotide composition over highly AT-rich templates. The advantages of the bisulfite method include the efficiency of the chemical conversion, relatively low cost and the speed of the procedure. Because in one reaction the whole genome can be converted and investigated in subsequent analyses, this has become the best approach for the study of DNA methylation.

Following bisulfite conversion and PCR amplification, the DNA fragment of interest can be further analyzed using several methods. Traditionally, the PCR product can be cloned into a plasmid and then a sufficiently large number of individual clones sequenced via Sanger sequencing. This procedure will produce a representation of the frequency of methylated (preserved) and unmethylated (converted) cytosines at every CpG dinucleotide position of the DNA amplicon in the cloned population (Fig. 10.4a). In the analysis of a DMR of an imprinted gene, a ratio of ~50% methylation would be expected at each CpG site whereby two types of sequences should be distinguishable: predominantly methylated and predominantly converted (Fig. 10.4a). If the DNA fragment of interest was isolated from hybrid offspring of different mouse strains and contains a single nucleotide polymorphism (SNP), the methylation can be associated with the parental origin of the allele.

An alternative to cloning and sequencing consists of the digestion of a PCR product with restriction endonucleases that contain CpGs in their recognition sites such as TaqI (T/CGA) and BstUI (CG/CG) (Fig. 10.4b). This technique, called COBRA (combined bisulfite restriction analysis) (Eads and Laird 2002), detects

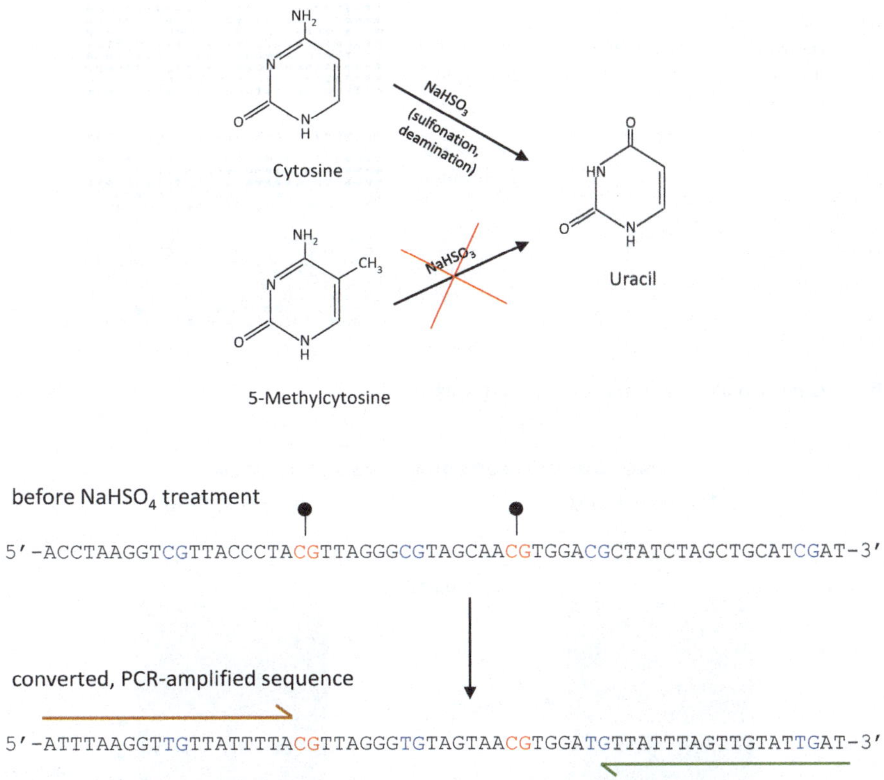

Fig. 10.3 Bisulfite conversion of genomic DNA. The treatment with $NaHSO_3$ induces the deamination of cytosine into uracil residues. This conversion will not happen, if the cytosine is methylated or hydroxy-methylated in the carbon 5 position. Following bisulfite treatment, the sense and antisense strands of the original double-stranded DNA are no longer complementary. PCR amplification of a strand will now lead to an exchange of uracil residues to thymine and the DNA sequence will then be characterized by an abundance of A and T. Only the methylated sites will retain cytosine residues. The primers for bisulfite sequencing need to be complementary to the converted sequence (orange and green arrows). Black lollipops indicate methylated CpG sites

originally methylated CpGs as they are conserved after bisulfite treatment and PCR amplification and then cut by the restriction endonuclease. By contrast, bisulfite conversion of an unmethylated CpG leads to loss of the restriction site. If a CpG is fully methylated, the PCR product will be completely digested, and vice versa (Fig. 10.4b). Quantification of partially methylated sites will require blotting of the gel, hybridization, and signal detection with a probe.

The presence of both cytosine and thymine in the same position following bisulfite treatment and PCR amplification can indicate partial methylation or potentially incomplete bisulfite conversion. In the bisulfite sequencing approach, incomplete conversion can be recognized through the presence of remaining C residues outside of CpG dinucleotides—all of these should have been converted. A potential

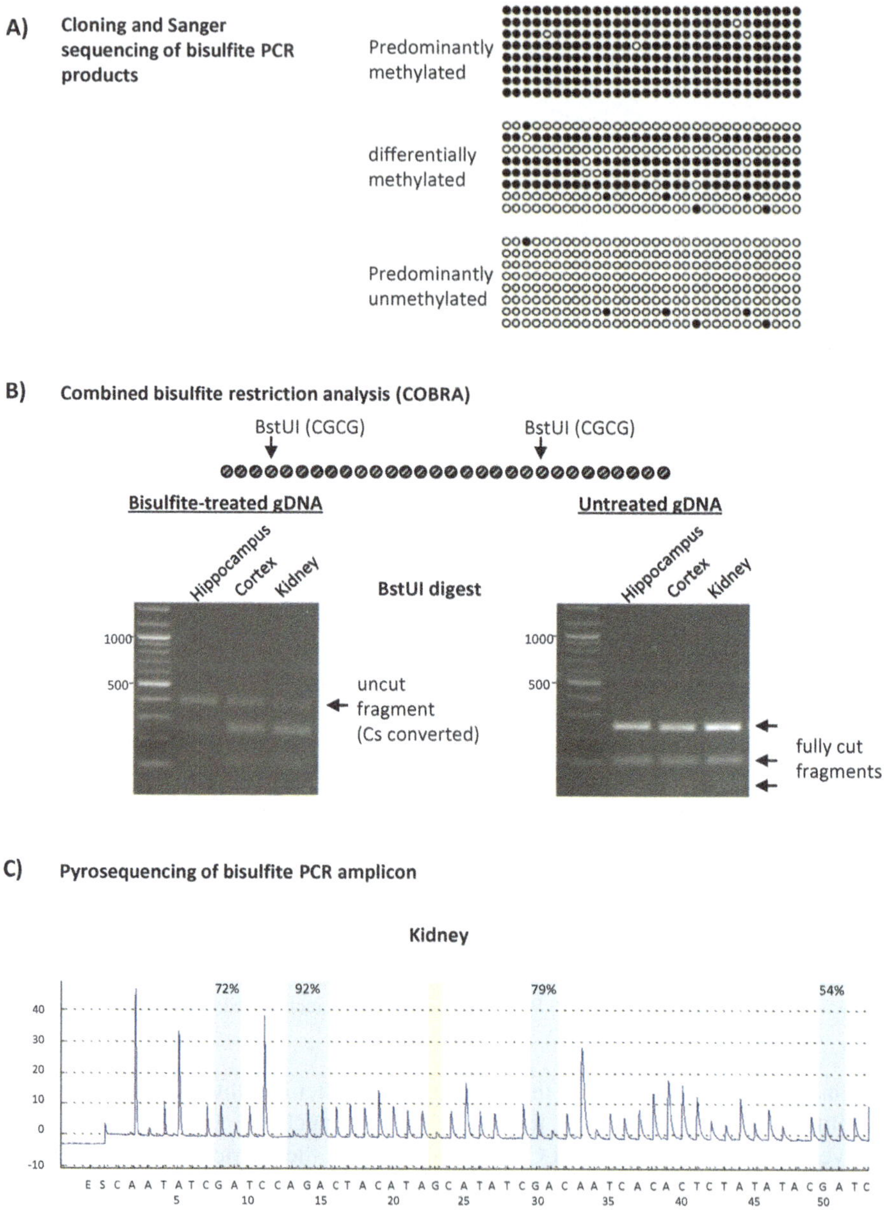

Fig. 10.4 Analysis methods of bisulfite-treated and PCR-amplified genomic DNA. (**a**) Cloning and Sanger sequencing of bisulfite PCR products. Scheme showing three examples of possible outcomes for methylation patterns in an amplicon covering a CGI. Each row of dots represents one cloned PCR product and each dot represents an individual CpG site (black = methylated, white = unmethylated). The differentially methylated example, which contains methylated and unmethylated sequences, would be typical of an imprinted DMR. Further SNP analysis would be required to associate the methylation status with the parental origin of an allele. (**b**) Combined bisulfite restriction analysis (COBRA) on untreated and bisulfite-treated genomic DNA from hippocampus, cerebral cortex, and kidney. Restriction enzyme BstUI recognizes the site CGCG. When a cytosine is unmethylated in the genomic DNA, it is converted to thymine following bisulfite treatment and PCR amplification, disrupting the integrity of the restriction sites. In contrast,

drawback of bisulfite sequencing is related to cloning biases that can occur in bacterial strains used for plasmid transformation after ligation of bisulfite PCR amplicons. In genomic imprinting, a DMR will result in bisulfite amplicons with different sequences derived from methylated and unmethylated alleles, respectively, and it is possible that the ligated amplicons are not equally propagated after transformation into *E. coli*, thus giving a wrong representation of methylation frequencies. On the other hand, in the COBRA method appropriate positive controls for restriction endonuclease functionality need to be included. For this reason, it is always recommended to confirm any results using at least two independent approaches.

The best approach to accurately quantify the methylation status is the use of pyrosequencing of bisulfite-converted genomic DNA (Dejeux et al. 2009). This method also includes PCR amplification as a first step after bisulfite treatment to generate fragments of ~100 bp. These are then sequenced by adding one nucleotide at a time and quantifying its incorporation through a cascade of enzymatic reactions that result in strictly proportional luciferase-mediated light emission. For example, at a partially methylated CpG site, incorporation of dGTP (originally methylated C) and dATP(αS) (originally unmethylated C) are tested consecutively, and the ratio of light emitted at the two nucleotide steps represents the percentage of methylated cytosine (Fig. 10.4c). Pyrosequencing is now accepted as the "gold standard" for quantitative sequencing. Apart from analyzing methylation at specific loci, bisulfite treatment has now been combined with next-generation sequencing to investigate the whole genome in specific tissues or cell types, for example, via whole-genome bisulfite sequencing (WGBS) (Harris et al. 2010). As details of the various approaches and the importance of avoiding biases at specific steps in the protocols cannot be covered here, we refer the reader to more specific, recently published work (Harris et al. 2010; Olova et al. 2018).

10.4.2 Analysis of Monoallelic Expression and Single-Cell Resolution of Imprinting

The analysis of imprinted genes also requires a proof of preferential or completely monoallelic gene expression in a parent-of-origin-specific way. To demonstrate this, exonic SNPs between different mouse strains are used, most commonly between C57BL/6J and *Mus musculus molossinus* (JF1) or *Mus musculus castaneus* (Cast/EiJ) strains. Tissues or cells from hybrid offspring of reciprocal parental crosses (e.g., C57BL/6J × JF1 and JF1 × C57BL/6J) are isolated, cDNA generated and the

Fig. 10.4 (continued) methylated sequences maintain the cytosine and the restriction enzyme will cut the PCR amplicon. This example CpG island shows the presence of both methylated and unmethylated alleles (unpublished data). (**c**) Bisulfite pyrosequencing analysis of the same CGI as in (**b**). A pyrogram sequencing trace of a bisulfite PCR amplicon is shown. The positions highlighted in light blue represent CpG sites and the degree of methylation at each site is quantified and displayed as a percentage above the highlighted peaks (unpublished data)

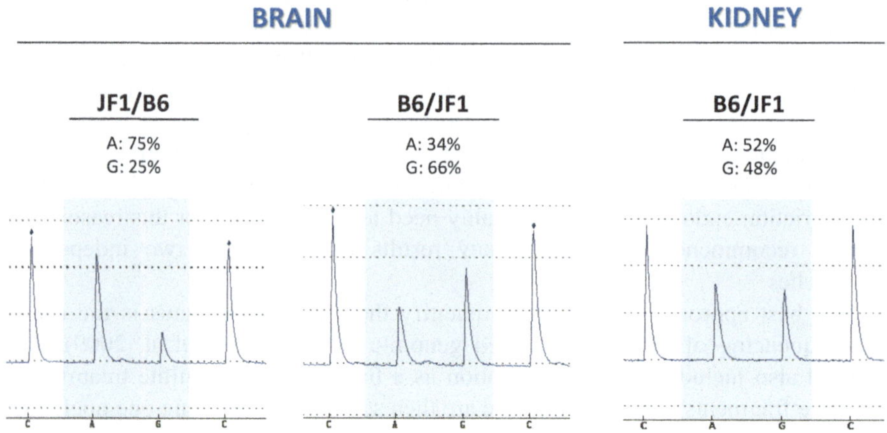

Fig. 10.5 SNP pyrosequencing to quantify allele-specific expression levels of a gene. Brain and kidney cDNAs were prepared from reciprocal crosses of C57BL/6J (B6) and Japanese fancy mouse 1 (*Mus musculus molossinus*, JF1), and a SNP-containing amplicon analyzed via pyrosequencing. Quantification of SNP prevalence indicates ~70% preferential maternal allele-specific expression of *Trappc9* in the brain, but equal biallelic expression in kidney. The reciprocal mouse crosses are shown as JF1/B6 and B6/JF1, whereby the first position conventionally indicates the maternal genotype. The SNP position is shaded in light blue

ratio of expression of the SNPs within the gene of interest determined via pyrosequencing. In the example shown in Fig. 10.5, expression of the *Trappc9* gene was found to be predominantly from the maternal allele (~70%) in brain tissue, while in kidney both SNPs show equal biallelic expression levels.

Recently, parent-of-origin biases of gene expression in the hybrid mouse brain have been characterized using RNAseq whole-transcriptome methods. In some cases, specific brain subregions, including the hypothalamus and arcuate nucleus were analyzed (Babak et al. 2015; Bonthuis et al. 2015; Huang et al. 2017; Perez et al. 2015). These transcriptome datasets, which include previously unknown transcripts, can be a good starting point for the investigation of imprinted genes in the brain. While several genes, including *Gnas*, *Dlk1*, and *Trappc9*, show tissue- or cell type-specific imprinting (Chen et al. 2009; Ferron et al. 2011; Perez et al. 2015), it is often unclear whether every cell within a population exhibits the parent-of-origin-specific, monoallelic expression. In particular, for genes with a bias toward parental allele-specific expression (e.g., 70% maternal, 30% paternal transcripts, based on whole tissue lysates) it remains unexplained whether such a ratio reflects unequal biallelic expression within each cell, or whether it is due to a mixture of cells with absolute monoallelic expression and cells with equal biallelic expression. This question is now being addressed with single-cell technologies. There are two main approaches to the single-cell analysis of allelic gene expression. On the one hand, probe hybridization onto fixed cells or tissue sections is combined with microscopy to distinguish between mono- or biallelic expression in situ (Bonthuis et al. 2015;

Ginart et al. 2016; Huang et al. 2017). On the other hand, molecular genetics or genome-wide next-generation sequencing methods are applied on isolated single cell extracts (Angermueller et al. 2016; Cheow et al. 2015; Clark et al. 2018; Deng et al. 2014; Kelsey et al. 2017). As some of these techniques have been applied to hypothalamic tissues, we briefly describe them here as promising approaches for future research.

One of the methods designed to investigate allele-specific expression at the tissue level is "single nucleotide polymorphism RNA fluorescent in situ hybridization" (SNP-FISH) (Ginart et al. 2016; Levesque et al. 2013). The method is applied on tissues or cells from hybrid mice (e.g., C57BL/6J × Cast/EiJ), which harbor several SNPs in the mRNA of the gene of interest. Three different pools of fluorescently labeled oligonucleotides are then hybridized simultaneously to fixed samples: (a) a pool of "guide" probes, which bind to the mRNA of the gene of interest outside of the SNP regions for general detection of the relevant transcript; (b) two pools of SNP-specific oligonucleotides with different fluorescent labels, which vary only in a single nucleotide position to detect the parental origin of the transcript (Fig. 10.6). These SNP-specific oligos are initially blocked by shorter "mask" oligonucleotides, leaving the SNP-containing region free to hybridize to the perfectly matching mRNA. Eventually, the SNP-specific oligos shed their "mask" oligos to fully anneal to the mRNA of interest. By analyzing co-localization of fluorescent signals from the "guide" oligos and the SNP-specific oligos, respectively, Ginart and colleagues (Ginart et al. 2016) were able to demonstrate that, within one tissue, cell populations with monoallelic maternal expression of the imprinted gene *H19* coexisted with neighboring cell populations that showed biallelic expression.

Another technique used to visualize the imprinted expression of genes in tissues is RNAscope in situ hybridization. Bonthuis and colleagues (Bonthuis et al. 2015) adapted the standard RNAscope (Advanced Cell Diagnostics, ACD) method to investigate actively transcribed alleles of specific genes in individual cell nuclei of fixed sections from the hypothalamic arcuate nucleus and dorsal raphe nucleus (Bonthuis et al. 2015). Using intronic oligonucleotide probes, which detect nascent transcripts in the nucleus before splicing occurs, they were able to determine through imaging for the presence of one or two hybridization signals ("color substrate dots") within a nucleus, whether one or both alleles of a gene were expressed. In contrast to the SNP-FISH technique, however, the RNAscope method does not provide information on the parental origin of monoallelic expression. Also, when used on tissue sections, the RNAscope approach has an inherent false detection rate problem, since the plane of sectioning might result in spatial separation of the two alleles in sequential sections. This would result in single hybridization signals within nuclei although biallelic expression occurs. This false detection rate needs to be accounted for by using appropriate control probes for biallelically expressed genes. The authors estimate that this method has an error rate of ~25% nuclei showing potentially false monoallelic expression (Bonthuis et al. 2015; Huang et al. 2017). Nevertheless, due to the high sensitivity of the hybridization probes and the straight forward protocol, this approach can provide a useful first assessment of the localization of cells with mono- and biallelic expression within tissues. Especially since it can be combined

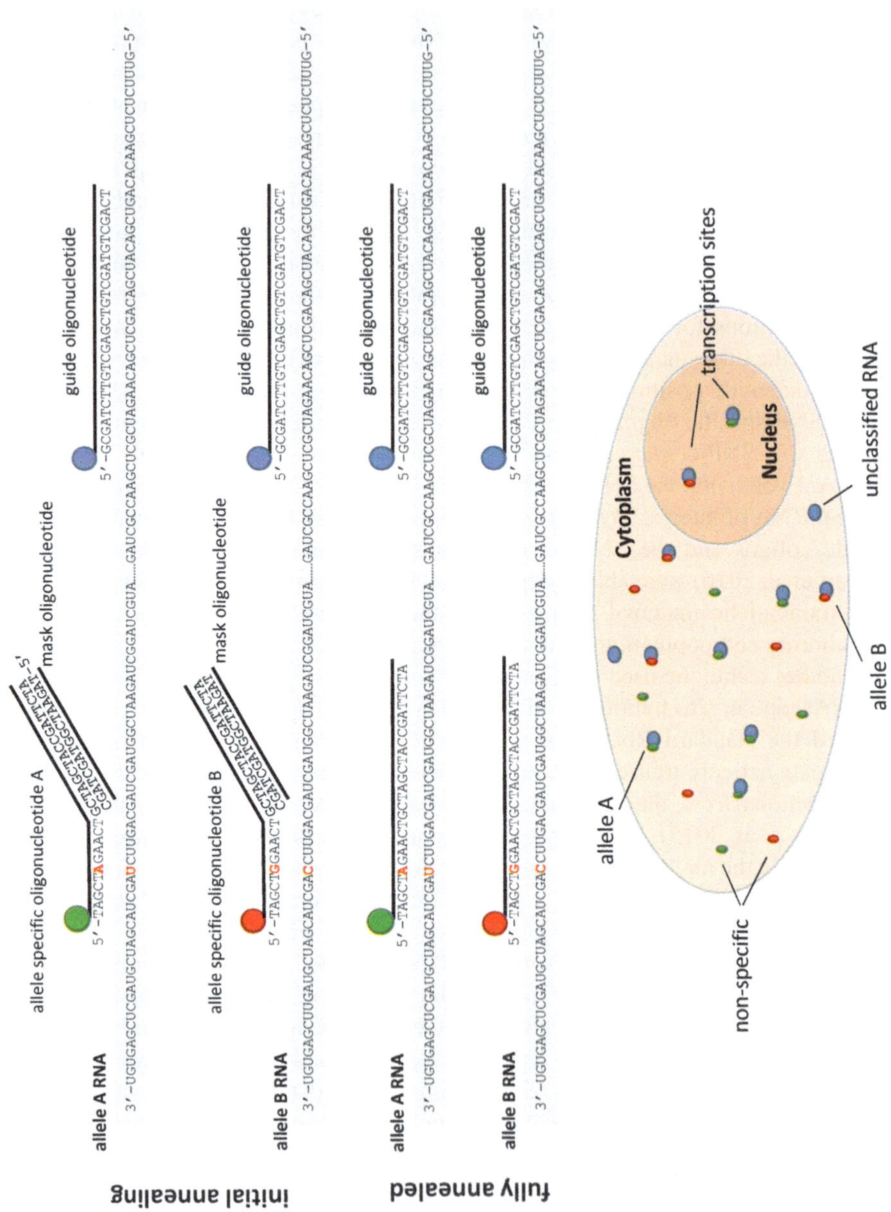
initial annealing
fully annealed
allele A RNA
allele B RNA
allele specific oligonucleotide A
allele specific oligonucleotide B
mask oligonucleotide
guide oligonucleotide
Cytoplasm
Nucleus
transcription sites
unclassified RNA
allele A
allele B
non-specific

Fig. 10.6 Schematic illustration of the SNP-FISH technique for the detection of allele-specific RNA expression in cells and tissue sections. The allelic RNA variants of a gene of interest are shown with a SNP highlighted in red. A pool of fluorescently labeled "guide" oligonucleotides, which hybridize to non-variant sequences of the RNA, serve as a general marker for the presence of the transcript. Allele-specific oligonucleotides covering SNP positions (SNV probes) are labeled with a second and third fluorescent tag, respectively. Initially, only a short stretch of the SNV probe sequence (including the actual SNP position) is free to hybridize with the target RNA, as the remainder of the oligo is bound by a short complementary "mask" oligo. This allows specific binding of the SNV probes to the allelic RNA variants. The "mask" oligos are then displaced as the allele-specific oligos fully anneal with their target RNAs. Figure adapted from Levesque et al. (2013)

with tissue RNAseq analysis (see below) for determination of parental expression biases.

Apart from these in situ techniques, molecular genetics approaches have been established to analyze epigenetic features as well as RNA expression from single-cell extracts (Angermueller et al. 2016; Cheow et al. 2015; Clark et al. 2018; Deng et al. 2014; Kelsey et al. 2017). Some methods are designed for a low-cost, focused investigation of specific loci (Cheow et al. 2015), while recent trends favor genome-wide next-generation sequencing technologies. RNAseq was first applied by Deng and colleagues (Deng et al. 2014) to investigate transcriptome-wide allelic expression in single cells using suitable strain-specific SNPs. Apart from parent-of-origin-dependent (imprinted) gene expression, they observed that a substantial proportion of genes (12–24%) are expressed monoallelically in a dynamic, random, and stochastic way (Deng et al. 2014). The most recent technological advances enable a combined analysis of DNA methylation, transcriptome, and nucleosome accessibility from single cells (scM&T-seq, scNMT-seq) (Angermueller et al. 2016; Kelsey et al. 2017). These approaches allow direct correlations to be made between epigenetic status, chromatin accessibility, and gene expression in a single cell. Although we cannot cover these methods in detail here [for a comprehensive review see Kelsey et al. (2017)], we would like to emphasize that consideration needs to be given to the way single cells can be isolated and the overall number of cells required for analysis. Most frequently, flow cytometry/fluorescence-activated cell sorting (FACS) is used to obtain single cells or nuclei (Luo et al. 2017), although microfluidics or manual collection via micromanipulators can also be applied depending on the tissue or cell type of interest (Cheow et al. 2015). Furthermore, and again depending on the complexity of the tissue to be investigated, usually hundreds or thousands of cells need to be included in an experiment to obtain a representative dataset for the cell population in question. As single-cell RNAseq and methylome analyses have already been applied on some brain regions (Luo et al. 2017; Tasic et al. 2016; Zeisel et al. 2015), it will be exciting to see the multi-omics approaches being adapted to neurobiological questions and specifically to characterize the large diversity of hypothalamic cell types in the future.

10.5 Perspectives

Within the brain, the hypothalamus has emerged as a major region in which imprinted genes are expressed. Here, they exert crucial functions in the regulation of postnatal development, energy homeostasis, hypothalamus–pituitary, and hypothalamus–thyroid neuroendocrine axes, as well as circadian rhythmicity and sleep. It is highly likely that the list provided in this chapter will be expanded in the near future, as several additional imprinted genes have been associated with a role in the hypothalamus, for example, *Neuronatin*, *Mkrn3*, *Grb10,* and *Liz.* On a molecular level, the recent developments in single-cell techniques will shed further light on the epigenetic regulation of imprinted genes. Many genes display an expression bias toward one parental allele when analyzed on a whole-tissue level. To clarify whether this is due to a mixture of cell types, some of which showing biallelic others monoallelic expression, and to understand the neurobiological relevance of this, will be an exciting field of future research.

Acknowledgment We would like to thank the Wellcome Trust PhD studentship program in Cellular and Molecular Physiology at the University of Liverpool (099795/Z/12/Z) for research support.

Key References

Buiting et al. (2016). This review provides an excellent overview about the clinical symptoms and genetics of Angelman syndrome.

Cassidy et al. (2012). A recommended review summarising the clinical symptoms as well as the genetics of Prader-Willi syndrome.

Chen et al. (2009). This paper shows for the first time genomic imprinting of *Gnas* in the brain and that loss of maternal allele-specific expression of Gnas in the brain is the cause for an obesity phenotype due to reduced energy expenditure.

Ferguson-Smith (2011) This review provides an excellent overview about the history of genomic imprinting research and well-studied imprinted genes clusters.

Kelsey et al. (2017). This recent review gives a good introduction into novel single-cell analysis techniques for epigenetics and gene expression.

Perez et al. (2015) This paper systematically analyzed imprinted gene expression within the brain transcriptome with excellent detail.

Peters (2014). This review emphasizes the physiological functions of imprinted genes.

Plagge et al. (2004). This paper identified the physiological functions of the paternally expressed transcript variant *Gnasxl* of the *Gnas* locus.

References

Angermueller C, Clark SJ, Lee HJ, Macaulay IC, Teng MJ, Hu TX, Krueger F, Smallwood S, Ponting CP, Voet T, Kelsey G, Stegle O, Reik W (2016) Parallel single-cell sequencing links transcriptional and epigenetic heterogeneity. Nat Methods 13:229–232

Babak T, DeVeale B, Tsang EK, Zhou Y, Li X, Smith KS, Kukurba KR, Zhang R, Li JB, van der Kooy D, Montgomery SB, Fraser HB (2015) Genetic conflict reflected in tissue-specific maps of genomic imprinting in human and mouse. Nat Genet 47:544–549

Bartolomei MS, Ferguson-Smith AC (2011) Mammalian genomic imprinting. Cold Spring Harb Perspect Biol 3:a002592

Bieth E, Eddiry S, Gaston V, Lorenzini F, Buffet A, Conte Auriol F, Molinas C, Cailley D, Rooryck C, Arveiler B, Cavaille J, Salles JP, Tauber M (2015) Highly restricted deletion of the SNORD116 region is implicated in Prader-Willi Syndrome. Eur J Hum Genet 23:252–255

Bischof JM, Stewart CL, Wevrick R (2007) Inactivation of the mouse Magel2 gene results in growth abnormalities similar to Prader-Willi syndrome. Hum Mol Genet 16:2713–2719

Bonthuis PJ, Huang WC, Stacher Horndli CN, Ferris E, Cheng T, Gregg C (2015) Noncanonical genomic imprinting effects in offspring. Cell Rep 12:979–991

Buiting K, Williams C, Horsthemke B (2016) Angelman syndrome—insights into a rare neurogenetic disorder. Nat Rev Neurol 12(10):584–593

Cassidy SB, Schwartz S, Miller JL, Driscoll DJ (2012) Prader-Willi syndrome. Genet Med 14:10–26

Cavaille J (2017) Box C/D small nucleolar RNA genes and the Prader-Willi syndrome: a complex interplay. Wiley Interdiscip Rev RNA 8:e1417

Chen M, Wang J, Dickerson KE, Kelleher J, Xie T, Gupta D, Lai EW, Pacak K, Gavrilova O, Weinstein LS (2009) Central nervous system imprinting of the G protein G(s)α and its role in metabolic regulation. Cell Metab 9:548–555

Chen M, Nemechek NM, Mema E, Wang J, Weinstein LS (2011) Effects of deficiency of the G protein Gsα on energy and glucose homeostasis. Eur J Pharmacol 660:119–124

Chen M, Berger A, Kablan A, Zhang J, Gavrilova O, Weinstein LS (2012) Gsα deficiency in the paraventricular nucleus of the hypothalamus partially contributes to obesity associated with Gsα mutations. Endocrinology 153:4256–4265

Chen M, Shrestha YB, Podyma B, Cui Z, Naglieri B, Sun H, Ho T, Wilson EA, Li YQ, Gavrilova O, Weinstein LS (2017) Gsα deficiency in the dorsomedial hypothalamus underlies obesity associated with Gsα mutations. J Clin Invest 127:500–510

Cheow LF, Quake SR, Burkholder WF, Messerschmidt DM (2015) Multiplexed locus-specific analysis of DNA methylation in single cells. Nat Protoc 10:619–631

Clark SJ, Argelaguet R, Kapourani CA, Stubbs TM, Lee HJ, Alda-Catalinas C, Krueger F, Sanguinetti G, Kelsey G, Marioni JC, Stegle O, Reik W (2018) scNMT-seq enables joint profiling of chromatin accessibility DNA methylation and transcription in single cells. Nat Commun 9:781

Curley JP, Pinnock SB, Dickson SL, Thresher R, Miyoshi N, Surani MA, Keverne EB (2005) Increased body fat in mice with a targeted mutation of the paternally expressed imprinted gene Peg3. FASEB J 19:1302–1304

Dauber A, Cunha-Silva M, Macedo DB, Brito VN, Abreu AP, Roberts SA, Montenegro LR, Andrew M, Kirby A, Weirauch MT, Labilloy G, Bessa DS, Carroll RS, Jacobs DC, Chappell PE, Mendonca BB, Haig D, Kaiser UB, Latronico AC (2017) Paternally inherited DLK1 deletion associated with familial central precocious puberty. J Clin Endocrinol Metab 102:1557–1567

Dejeux E, El abdalaoui H, Gut IG, Tost J (2009) Identification and quantification of differentially methylated loci by the pyrosequencing technology. Methods Mol Biol 507:189–205

Deng Q, Ramskold D, Reinius B, Sandberg R (2014) Single-cell RNA-seq reveals dynamic, random monoallelic gene expression in mammalian cells. Science 343:193–196

Denizot AL, Besson V, Correra RM, Mazzola A, Lopes I, Courbard JR, Marazzi G, Sassoon DA (2016) A novel mutant allele of Pw1/Peg3 does not affect maternal behavior or nursing behavior. PLoS Genet 12:e1006053

Ding F, Li HH, Zhang S, Solomon NM, Camper SA, Cohen P, Francke U (2008) SnoRNA Snord116 (Pwcr1/MBII-85) deletion causes growth deficiency and hyperphagia in mice. PLoS One 3:e1709

Eads CA, Laird PW (2002) Combined bisulfite restriction analysis (COBRA). Methods Mol Biol 200:71–85

Ehlen JC, Jones KA, Pinckney L, Gray CL, Burette S, Weinberg RJ, Evans JA, Brager AJ, Zylka MJ, Paul KN, Philpot BD, DeBruyne JP (2015) Maternal Ube3a loss disrupts sleep homeostasis but leaves circadian rhythmicity largely intact. J Neurosci 35:13587–13598

Ferguson-Smith AC (2011) Genomic imprinting: the emergence of an epigenetic paradigm. Nat Rev Genet 12:565–575

Ferron SR, Charalambous M, Radford E, McEwen K, Wildner H, Hind E, Morante-Redolat JM, Laborda J, Guillemot F, Bauer SR, Farinas I, Ferguson-Smith AC (2011) Postnatal loss of Dlk1 imprinting in stem cells and niche astrocytes regulates neurogenesis. Nature 475:381–385

Genevieve D, Sanlaville D, Faivre L, Kottler ML, Jambou M, Gosset P, Boustani-Samara D, Pinto G, Ozilou C, Abeguile G, Munnich A, Romana S, Raoul O, Cormier-Daire V, Vekemans M (2005) Paternal deletion of the GNAS imprinted locus (including Gnasxl) in two girls presenting with severe pre- and post-natal growth retardation and intractable feeding difficulties. Eur J Hum Genet 13:1033–1039

Gerard M, Hernandez L, Wevrick R, Stewart CL (1999) Disruption of the mouse necdin gene results in early post-natal lethality. Nat Genet 23:199–202

Ginart P, Kalish JM, Jiang CL, Yu AC, Bartolomei MS, Raj A (2016) Visualizing allele-specific expression in single cells reveals epigenetic mosaicism in an H19 loss-of-imprinting mutant. Genes Dev 30:567–578

Haig D (2004) Genomic imprinting and kinship: how good is the evidence? Annu Rev Genet 38:553–585

Harris RA, Wang T, Coarfa C, Nagarajan RP, Hong C, Downey SL, Johnson BE, Fouse SD, Delaney A, Zhao Y, Olshen A, Ballinger T, Zhou X, Forsberg KJ, Gu J, Echipare L, O'Geen H, Lister R, Pelizzola M, Xi Y, Epstein CB, Bernstein BE, Hawkins RD, Ren B, Chung WY, Gu H, Bock C, Gnirke A, Zhang MQ, Haussler D, Ecker JR, Li W, Farnham PJ, Waterland RA, Meissner A, Marra MA, Hirst M, Milosavljevic A, Costello JF (2010) Comparison of sequencing-based methods to profile DNA methylation and identification of monoallelic epigenetic modifications. Nat Biotechnol 28:1097–1105

Hasegawa K, Kawahara T, Fujiwara K, Shimpuku M, Sasaki T, Kitamura T, Yoshikawa K (2012) Necdin controls Foxo1 acetylation in hypothalamic arcuate neurons to modulate the thyroid axis. J. Neurosci 32:5562–5572

He Q, Zhu Y, Corbin BA, Plagge A, Bastepe M (2015) The G protein alpha subunit variant XLαs promotes inositol 1,4,5-trisphosphate signaling and mediates the renal actions of parathyroid hormone in vivo. Sci Signal 8:ra84

Huang WC, Ferris E, Cheng T, Horndli CS, Gleason K, Tamminga C, Wagner JD, Boucher KM, Christian JL, Gregg C (2017) Diverse non-genetic, allele-specific expression effects shape genetic architecture at the cellular level in the mammalian brain. Neuron 93:1094–1109

Ivanova E, Kelsey G (2011) Imprinted genes and hypothalamic function. J Mol Endocrinol 47: R67–R74

Kawashima S, Nakamura A, Inoue T, Matsubara K, Horikawa R, Wakui K, Takano K, Fukushima Y, Tatematsu T, Mizuno S, Tsubaki J, Kure S, Matsubara Y, Ogata T, Fukami M, Kagami M (2018) Maternal uniparental disomy for chromosome 20: physical and endocrinological characteristics of five patients. J Clin Endocrinol Metab 103:2083–2088

Kelsey G, Stegle O, Reik W (2017) Single-cell epigenomics: recording the past and predicting the future. Science 358:69–75

Kozlov SV, Bogenpohl JW, Howell MP, Wevrick R, Panda S, Hogenesch JB, Muglia LJ, Van Gelder RN, Herzog ED, Stewart CL (2007) The imprinted gene Magel2 regulates normal circadian output. Nat Genet 39:1266–1272

Krechowec SO, Burton KL, Newlaczyl AU, Nunn N, Vlatkovic N, Plagge A (2012) Postnatal changes in the expression pattern of the imprinted signalling protein XLαs underlie the changing phenotype of deficient mice. PLoS One 7:e29753

LaSalle JM, Reiter LT, Chamberlain SJ (2015) Epigenetic regulation of UBE3A and roles in human neurodevelopmental disorders. Epigenomics 7:1213–1228

Lee S, Walker CL, Karten B, Kuny SL, Tennese AA, O'Neill MA, Wevrick R (2005) Essential role for the Prader-Willi syndrome protein necdin in axonal outgrowth. Hum Mol Genet 14:627–637

Levesque MJ, Ginart P, Wei Y, Raj A (2013) Visualizing SNVs to quantify allele-specific expression in single cells. Nat Methods 10:865–867

Li L, Keverne EB, Aparicio SA, Ishino F, Barton SC, Surani MA (1999) Regulation of maternal behavior and offspring growth by paternally expressed Peg3. Science 284:330–333

Li YQ, Shrestha Y, Pandey M, Chen M, Kablan A, Gavrilova O, Offermanns S, Weinstein LS (2016) G(q/11)α and G(s)α mediate distinct physiological responses to central melanocortins. J Clin Invest 126:40–49

Luo C, Keown CL, Kurihara L, Zhou J, He Y, Li J, Castanon R, Lucero J, Nery JR, Sandoval JP, Bui B, Sejnowski TJ, Harkins TT, Mukamel EA, Behrens MM, Ecker JR (2017) Single-cell methylomes identify neuronal subtypes and regulatory elements in mammalian cortex. Science 357:600–604

Lyko F (2018) The DNA methyltransferase family: a versatile toolkit for epigenetic regulation. Nat Rev Genet 19:81–92

Maillard J, Park S, Croizier S, Vanacker C, Cook JH, Prevot V, Tauber M, Bouret SG (2016) Loss of Magel2 impairs the development of hypothalamic Anorexigenic circuits. Hum Mol Genet 25:3208–3215

Mantovani G, Bastepe M, Monk D, de Sanctis L, Thiele S, Usardi A, Ahmed SF, Bufo R, Choplin T, De Filippo G, Devernois G, Eggermann T, Elli FM, Freson K, Garcia Ramirez A, Germain-Lee EL, Groussin L, Hamdy N, Hanna P, Hiort O, Juppner H, Kamenicky P, Knight N, Kottler ML, Le Norcy E, Lecumberri B, Levine MA, Makitie O, Martin R, Martos-Moreno GA, Minagawa M, Murray P, Pereda A, Pignolo R, Rejnmark L, Rodado R, Rothenbuhler A, Saraff V, Shoemaker AH, Shore EM, Silve C, Turan S, Woods P, Zillikens MC, Perez de Nanclares G, Linglart A (2018) Diagnosis and management of pseudohypoparathyroidism and related disorders: first international Consensus Statement. Nat Rev Endocrinol 14:476–500

Martinez ME, Charalambous M, Saferali A, Fiering S, Naumova AK, Germain DS, Ferguson-Smith AC, Hernandez A (2014) Genomic imprinting variations in the mouse type 3 deiodinase gene between tissues and brain regions. Mol Endocrinol 28:1875–1886

Matarazzo V, Caccialupi L, Schaller F, Shvarev Y, Kourdougli N, Bertoni A, Menuet C, Voituron N, Deneris E, Gaspar P, Bezin L, Durbec P, Hilaire G, Muscatelli F (2017) Necdin shapes serotonergic development and SERT activity modulating breathing in a mouse model for Prader-Willi syndrome. elife 6:e32640

Mehta S, Williamson CM, Ball S, Tibbit C, Beechey C, Fray M, Peters J (2015) Transcription driven somatic DNA methylation within the imprinted Gnas cluster. PLoS One 10:e0117378

Meister B, Perez-Manso M, Daraio T (2013) Delta-like 1 homologue is a hypothalamus- enriched protein that is present in orexin-containing neurones of the lateral hypothalamic area. J Neuroendocrinol 25:617–625

Mercer RE, Michaelson SD, Chee MJ, Atallah TA, Wevrick R, Colmers WF (2013) Magel2 is required for leptin-mediated depolarization of POMC neurons in the hypothalamic arcuate nucleus in mice. PLoS Genet 9:e1003207

Miller NL, Wevrick R, Mellon PL (2009) Necdin, a Prader-Willi syndrome candidate gene, regulates gonadotropin-releasing hormone neurons during development. Hum Mol Genet 18:248–260

Mulchandani S, Bhoj EJ, Luo M, Powell-Hamilton N, Jenny K, Gripp KW, Elbracht M, Eggermann T, Turner CL, Temple IK, Mackay DJ, Dubbs H, Stevenson DA, Slattery L, Zackai EH, Spinner NB, Krantz ID, Conlin LK (2016) Maternal uniparental disomy of chromosome 20: a novel imprinting disorder of growth failure. Genet Med 18:309–315

Muscatelli F, Abrous DN, Massacrier A, Boccaccio I, Le Moal M, Cau P, Cremer H (2000) Disruption of the mouse Necdin gene results in hypothalamic and behavioral alterations reminiscent of the human Prader-Willi syndrome. Hum Mol Genet 9:3101–3110

Nunn N, Feetham CH, Martin J, Barrett-Jolley R, Plagge A (2013) Elevated blood pressure, heart rate and body temperature in mice lacking the XLαs protein of the Gnas locus is due to increased sympathetic tone. Exp Physiol 98:1432–1445

Olova N, Krueger F, Andrews S, Oxley D, Berrens RV, Branco MR, Reik W (2018) Comparison of whole-genome bisulfite sequencing library preparation strategies identifies sources of biases affecting DNA methylation data. Genome Biol 19:33

Perez JD, Rubinstein ND, Fernandez DE, Santoro SW, Needleman LA, Ho-Shing O, Choi JJ, Zirlinger M, Chen SK, Liu JS, Dulac C (2015) Quantitative and functional interrogation of parent-of-origin allelic expression biases in the brain. elife 4:e07860

Peters J (2014) The role of genomic imprinting in biology and disease: an expanding view. Nat Rev Genet 15:517–530

Plagge A (2012) Non-coding RNAs at the Gnas and Snrpn-Ube3a imprinted gene loci and their involvement in hereditary disorders. Front Genet 3:264

Plagge A, Gordon E, Dean W, Boiani R, Cinti S, Peters J, Kelsey G (2004) The imprinted signaling protein XLαs is required for postnatal adaptation to feeding. Nat Genet 36:818–826
Plagge A, Isles AR, Gordon E, Humby T, Dean W, Gritsch S, Fischer-Colbrie R, Wilkinson LS, Kelsey G (2005) Imprinted Nesp55 influences behavioral reactivity to novel environments. Mol Cell Biol 25:3019–3026
Plasschaert RN, Bartolomei MS (2014) Genomic imprinting in development, growth, behavior and stem cells. Development 141:1805–1813
Podyma B, Sun H, Wilson EA, Carlson B, Pritikin E, Gavrilova O, Weinstein LS, Chen M (2018) The stimulatory G protein Gsα is required in melanocortin 4 receptor-expressing cells for normal energy balance, thermogenesis, and glucose metabolism. J Biol Chem 293:10993–11005
Polex-Wolf J, Lam BY, Larder R, Tadross J, Rimmington D, Bosch F, Cenzano VJ, Ayuso E, Ma MK, Rainbow K, Coll AP, O'Rahilly S, Yeo GS (2018) Hypothalamic loss of Snord116 recapitulates the hyperphagia of Prader-Willi syndrome. J Clin Invest 128:960–969
Richard N, Molin A, Coudray N, Rault-Guillaume P, Juppner H, Kottler ML (2013) Paternal GNAS mutations lead to severe intrauterine growth retardation (IUGR) and provide evidence for a role of XLαs in fetal development. J Clin Endocrinol Metab 98:E1549–E1556
Schaller F, Watrin F, Sturny R, Massacrier A, Szepetowski P, Muscatelli F (2010) A single postnatal injection of oxytocin rescues the lethal feeding behaviour in mouse newborns deficient for the imprinted Magel2 gene. Hum Mol Genet 19:4895–4905
Shi SQ, Bichell TJ, Ihrie RA, Johnson CH (2015) Ube3a imprinting impairs circadian robustness in Angelman syndrome models. Curr Biol 25:537–545
Skryabin BV, Gubar LV, Seeger B, Pfeiffer J, Handel S, Robeck T, Karpova E, Rozhdestvensky TS, Brosius J (2007) Deletion of the MBII-85 snoRNA gene cluster in mice results in postnatal growth retardation. PLoS Genet 3:e235
Stohn JP, Martinez ME, Zafer M, Lopez-Espindola D, Keyes LM, Hernandez A (2018) Increased aggression and lack of maternal behavior in Dio3-deficient mice are associated with abnormalities in oxytocin and vasopressin systems. Genes Brain Behav 17:23–35
Tacer KF, Potts PR (2017) Cellular and disease functions of the Prader-Willi Syndrome gene MAGEL2. Biochem J 474:2177–2190
Tasic B, Menon V, Nguyen TN, Kim TK, Jarsky T, Yao Z, Levi B, Gray LT, Sorensen SA, Dolbeare T, Bertagnolli D, Goldy J, Shapovalova N, Parry S, Lee C, Smith K, Bernard A, Madisen L, Sunkin SM, Hawrylycz M, Koch C, Zeng H (2016) Adult mouse cortical cell taxonomy revealed by single cell transcriptomics. Nat Neurosci 19:335–346
Tommiska J, Kansakoski J, Skibsbye L, Vaaralahti K, Liu X, Lodge EJ, Tang C, Yuan L, Fagerholm R, Kanters JK, Lahermo P, Kaunisto M, Keski-Filppula R, Vuoristo S, Pulli K, Ebeling T, Valanne L, Sankila EM, Kivirikko S, Laaperi M, Casoni F, Giacobini P, Phan-Hug F, Buki T, Tena-Sempere M, Pitteloud N, Veijola R, Lipsanen-Nyman M, Kaunisto K, Mollard P, Andoniadou CL, Hirsch JA, Varjosalo M, Jespersen T, Raivio T (2017) Two missense mutations in KCNQ1 cause pituitary hormone deficiency and maternally inherited gingival fibromatosis. Nat Commun 8:1289
Turan S, Bastepe M (2013) The GNAS complex locus and human diseases associated with loss-of-function mutations or epimutations within this imprinted gene. Horm Res Paediatr 80:229–241
Villanueva C, Jacquier S, de Roux N (2012) DLK1 is a somato-dendritic protein expressed in hypothalamic arginine-vasopressin and oxytocin neurons. PLoS One 7:e36134
Weinstein LS (2014) Role of G(s)α in central regulation of energy and glucose metabolism. Horm Metab Res 46:841–844
Wijesuriya TM, De Ceuninck L, Masschaele D, Sanderson MR, Carias KV, Tavernier J, Wevrick R (2017) The Prader-Willi syndrome proteins MAGEL2 and necdin regulate leptin receptor cell surface abundance through ubiquitination pathways. Hum Mol Genet 26:4215–4230
Wu Z, Martinez ME, St Germain DL, Hernandez A (2017) Type 3 deiodinase role on central thyroid hormone action affects the leptin-melanocortin system and circadian activity. Endocrinology 158:419–430

Xie T, Plagge A, Gavrilova O, Pack S, Jou W, Lai EW, Frontera M, Kelsey G, Weinstein LS (2006) The alternative stimulatory G protein α-subunit XLαs is a critical regulator of energy and glucose metabolism and sympathetic nerve activity in adult mice. J Biol Chem 281:18989–18999

Zeisel A, Munoz-Manchado AB, Codeluppi S, Lonnerberg P, La Manno G, Jureus A, Marques S, Munguba H, He L, Betsholtz C, Rolny C, Castelo-Branco G, Hjerling-Leffler J, Linnarsson S (2015) Brain structure. Cell types in the mouse cortex and hippocampus revealed by single-cell RNA-seq. Science 347:1138–1142

Rhythmic Epigenetics in Neuroendocrine and Immune Systems

11

Christopher S. Coyle, Elisabetta Tolla, and Tyler J. Stevenson

Abstract

Biological rhythms in neuroendocrine and immune systems are pervasive. Daily and seasonal changes in day lengths regulate multiple physiological and immunological parameters in a diverse range of animals. A series of studies have shown that epigenetic modifications exhibit naturally occurring rhythms across short- and long-term timescales. In this chapter, we describe daily, estrous and seasonal oscillations in epigenetic enzymes in neuroendocrine substrates, peripheral reproductive tissues and immune cells (e.g. leukocytes). The predominant focus of the chapter includes enzymes involved in DNA methylation and histone modifications, such as DNA methyltransferase and histone deacetylases. The findings presented herein highlight that epigenetic modifications can be permanent as well as transient with long-term consequences on the timing of physiological and behavioural processes. Moreover, the bidirectional interaction between the immune system and the neuroendocrine nucleus that controls biological rhythmicity, the suprachiasmatic nucleus, emphasizes the need to understand rhythmic changes in epigenetic enzymes and the consequences of disrupted daily and seasonal rhythms.

Keywords

Methyltransferase · Acetylation · Oscillation · Circadian · Circannual

C. S. Coyle · E. Tolla · T. J. Stevenson (✉)
Institute of Biodiversity, Animal Health and Comparative Medicine, University of Glasgow, Glasgow, UK
e-mail: tyler.stevenson@glasgow.ac.uk

S. Wray, S. Blackshaw (eds.), *Developmental Neuroendocrinology*, Masterclass in Neuroendocrinology 9, https://doi.org/10.1007/978-3-030-40002-6_11

11.1 Introduction

Biochemical plasticity in epigenetic modifications has significant potential, and challenges, for our understanding of the mechanistic regulation of neuroendocrine function. It is now well established that epigenetic modifications, DNA methylation and histone acetylation, in particular, programme important genomic and chromatin changes during development (Li 2002) and can exhibit transgenerational inheritance (Heard and Martienssen 2014). Emerging research findings from multiple laboratories have revealed that epigenetic modifications exhibit naturally occurring oscillations across different periods of time. For example, daily and seasonal variation in epigenetic enzyme expression occurs across the brain and peripheral tissues (Stevenson 2018). Since epigenetic modifications are an essential component for the regulation of gene transcription across taxa, rhythmic DNA methylation and histone acetylation are likely an evolutionarily ancient system (Stevenson 2018). In this chapter, we will present the historical context that provides the necessary background information, and then describe the current research in naturally occurring rhythms in epigenetic modifications. Due to the vast amount of knowledge available, the chapter will not cover developmental epigenetics or transgenerational epigenetics and instead focus on rhythmic epigenetics. The chapter will draw from the research fields of chronobiology, reproductive neuroendocrinology and the neural control of energy balance. The last section will provide a brief overview of the research developments in the field of immunochronobiology and highlight the exciting potential for pharmaceutical reversal of epigenetic modifications for health and disease.

11.2 Historical Perspective on Epigenetics in the Neuroendocrine Axes

The study of epigenetics originates in the field of developmental biology. Early work by Conrad Waddington synthesized the concept that interactions between the environment and the genetic material during embryogenesis and early life experiences shaped the mature, complex organism (Waddington 1953). As the field advanced, it became clear that experience-dependent modifications could alter an individual's fitness, and this was subsequently passed to the next generation; commonly referred to as transgenerational inheritance. Here, the study of mitotically and/or meiotically heritable changes in gene function was observed independent of changes in the genome sequence (Riggs and Porter 1996). In parallel, the identities of enzymes that act to catalyze the biochemical modifications of genome function (i.e. DNA methylation and histone modification) were discovered to include: histone (de) acetylases (Taunton et al. 1996) and DNA methyltransferases (Yoder et al. 1997). Today, the next level of complexity focuses on RNA methylation and non-coding RNA mechanisms (Guil and Esteller 2009). For the past 20 years, there has been an exponential growth in our understanding of the environmental and hormonal regulation of epigenetic enzymes (Stevenson 2017a), particularly in the field of cancer genomics (Esteller 2008) with significantly less attention in the field of neuroendocrinology. One aim of this chapter is to present the current understanding of the role of rhythmic DNA methylation and histone acetylation in timing neuroendocrine function and outline exciting avenues for future research directions.

The intensive investigation of hormone-driven changes in epigenetic modifications in neuroendocrine substrates developed during the 1980s. Initial studies utilized pituitary tumour cell lines collected from Fischer 344 rats (Durrin et al. 1984). Rats that received chronic diethylstilbestrol treatment were shown to have reduced methylation in the prolactin promoter, and associated increase in prolactin mRNA expression. Remarkably, the effect of elevated diethylstilbestrol was localized to the pituitary as the prolactin promoter retained higher methylation in liver cells. These data indicated that hormones could induce epigenetic modifications in anatomically localized neuroendocrine substrates. One of the first studies to demonstrate a clear long-term behavioural effect of epigenetic modification within neuroendocrine circuits was the observation of experience-dependent effects of rat maternal licking-and-grooming behaviour (Weaver et al. 2004). High maternal licking and grooming behaviour was found to significantly reduced DNA methylation and increased histone acetylation of the nerve growth factor-inducible protein A (NGFI-A) consensus sequence in the hippocampal glucocorticoid receptor in offspring (Weaver et al. 2004). Over the past 35 years, the evidence has expanded to illustrate that a wide range of hormones can have cellular- and neuroendocrine nuclei-specific effects of epigenetic modifications during development (Nugent et al. 2015; also see Chap. 15) and in adulthood (Stevenson 2017a). Indeed, the technological developments that have significantly advanced genome sequencing have yielded immense insights into the identification of specific genome motifs and chromatin that are targets for hormone-driven epigenomic modifications. Since hormones are facilitators of epigenetic modifications via epigenetic enzymes, here we will focus on the hormone-dependent regulation of DNA methylation and histone acetylation epigenetic enzymes in neuroendocrine tissues.

11.3 Function and Distribution of Epigenetic Enzymes in the Neuroendocrine Axes

Epigenetic changes that impact the genome (i.e. DNA methylation) and chromatin (i.e. histone acetylation) are induced by a range of enzyme families that show tissue-specific expression profiles. DNA methylation on the genome template is regulated by DNA methyltransferases (DNMT) and Ten-Eleven-Translocation methylcytosine dioxygenase (TET) enzymes (Fig. 11.1a). DNMTs catalyze the removal of methyl from *S*-adenosyl methionine (SAM) and preferentially add to cytosine-guanine (CpG) residues, but also non-CpG residues (Jang et al. 2017). Conversely, TET enzymes oxidize 5-methylcytosine that is then actively reverted to cytosine through oxidation and thymine DNA glycosylase (TDG)-mediated base excision repair (Kohli and Zhang 2013). Both DNMT and TET enzymes have a number of different isoforms that have distinct structure–function relations that contribute to the developmental- and cell-specificity of epigenetic modifications. In adults, DNMT1, 3a, and

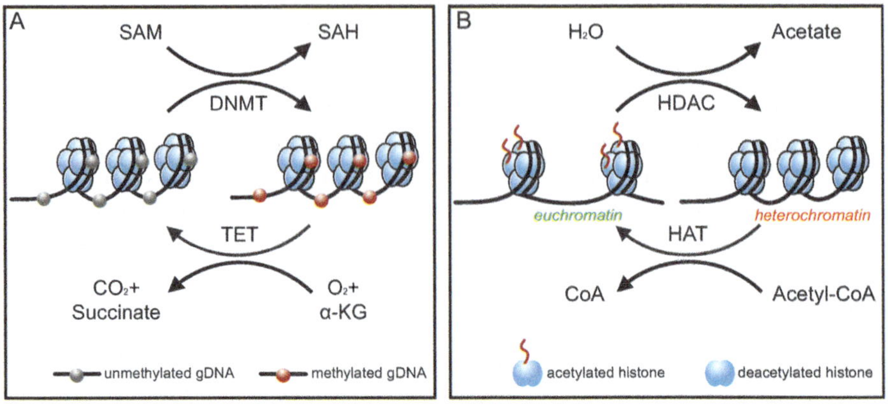

Fig. 11.1 Functional pathways of epigenetic enzymes. (**a**) DNA methyltransferase (DNMT) enzymes catalyze the addition of methyl groups, depicted by red dots, on to the genome template. Ten Eleven Translocation (TET) enzymes initiate a cascade of biochemical changes that include the oxidation of 5-methylcytosine and promote locus-specific reversal of DNA methylation indicated by grey dots. (**b**) Epigenetic enzymes involved in chromatin modifications include histone deacetylase (HDAC) and histone acetylase (HAT). HDAC triggers the removal of acetyl groups from histone. HAT enzymes catalyze the removal of acetyl groups from Acetyl-CoA and add acetyl, indicated by red curvilinear lines, on to histones. Abbreviations: *S*-adenosyl methionine (SAM), *S*-adenosyl homocysteine (SAH), carbon dioxide (CO_2), alpha-ketoglutarate (αKG), coenzyme A (CoA)

3b exhibit wide anatomical distribution with high levels of expression in the brain, reproductive tissues, and immune glands (e.g. spleen) (Fig. 11.2). The TET1-3 enzymes also show high expression levels in the brain and female reproductive tissues (i.e. ovary) (Fig. 11.2).

Chromatin modifications include histone acetylation induced by histone acetylases (HAT/KAT) and the converse histone deacetylase (HDAC) enzymes that consist of several isoforms (Klose and Bird 2006). HAT enzymes catalyze the removal of acetyl from acetyl-coenzyme A (CoA) and induce the transformation from closed heterochromatin into the transcriptional accessible euchromatin states (Fig. 11.1b). The removal of acetyl groups from histones by HDACs results in silenced gene transcription and the formation of cellular acetate. Similar to the enzymes involved in DNA methylation, both HDAC and HAT enzymes show higher levels of expression in the brain, reproductive tissues and immune glands (Fig. 11.2).

11.4 Predicable Oscillations of Epigenetic Enzymes Indicate Rhythmic Modifications

The initial evidence to indicate that epigenetic enzymes display oscillations were derived from high-throughput cDNA microarrays and RNA sequencing platforms (Stevenson 2018). The current evidence suggests that daily epigenetic enzyme oscillations are isoform-specific and tissue-dependent. For example, in the adult

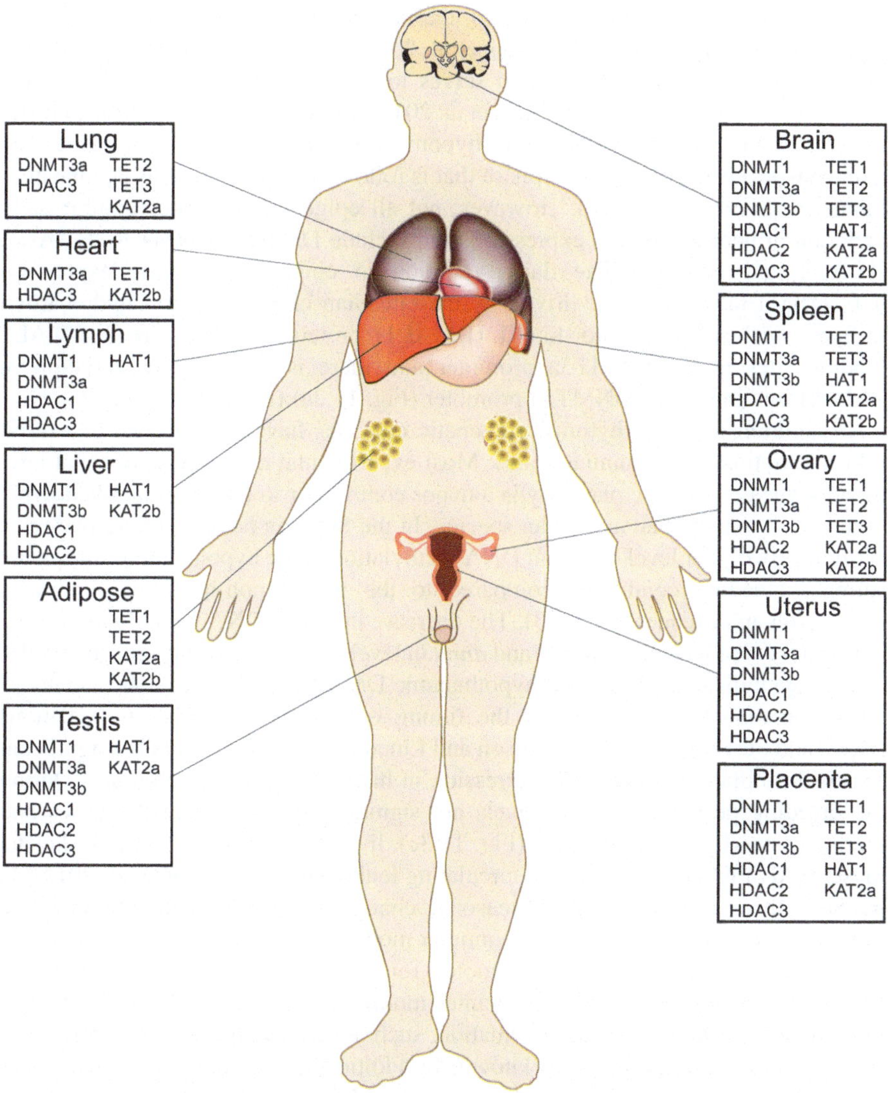

Fig. 11.2 Distribution and relative abundance of epigenetic enzymes. A representative diagram to highlight the tissue-specific and expression levels of epigenetic enzymes in multiple tissues. Epigenetic enzymes presented in tissue boxes were selected based on higher than average Reads Per Kilobase Million (RPKM) expression levels across all tissues. Most enzymes involved in DNA methylation and histone modifications are expressed in neuroendocrine substrates (i.e. brain) as well as reproductive tissues (i.e. testes, ovary and uterus). RPKM data were obtained from PUBMED and are based on Fagerberg et al. (2014)

mouse suprachiasmatic nucleus (SCN), the master neuroendocrine nucleus for the regulation of daily rhythms (Hastings et al. 2014), DNMT3a is expressed, exhibits daily rhythms in mRNA levels and serves to maintain the integrity of circadian oscillations in genome function (Azzi et al. 2014). In the adult male Siberian hamster there are robust daily oscillations in hypothalamic *dnmt3a* and *hdac4* expression with peak levels during the dark phase that is followed by a precipitous decline after lights on (Stevenson 2017b). However, not all epigenetic enzymes exhibit daily variation in neuroendocrine expression, and include DNMT1, DNMT3b, HDAC1, HDAC2 and HDAC3. The daily variation in epigenetic enzyme expression (e.g. DNMT3a) is probably driven by the circadian clock gene aryl hydrocarbon receptor nuclear translocator-like 1 (BMAL1) proteins as there are 3 BMAL1 binding sites in the DNMT3a promoter and there is robust daily variation in BMAL1 binding in the DNMT3a promoter (Fig. 11.3a) (Koike et al. 2012).

In addition to daily rhythms, epigenetic enzymes have been shown to exhibit robust variation across annual scales. Most experimental evidence shows that neuroendocrine epigenomic plasticity is a major contributor to the seasonal phenotypic variation in mammalian and avian species. In the Siberian hamster brain, there is a significantly higher level of global DNA methylation in the hypothalamus during the summer breeding conditions compared to the winter non-breeding conditions (Stevenson and Prendergast 2013). The decrease in global DNA methylation occurs in parallel to reduced daily *dnmt1* and *dnmt3a* levels (Fig. 11.3b) (Stevenson 2017b). These data indicate that reduced hypothalamic DNA methylation may be an endogenous circannual mechanism for the timing of seasonal life–history transitions (Helm and Stevenson 2015; Stevenson and Lincoln 2017). There are also significant seasonal oscillations in DNMT expression in hamster reproductive tissues. During the breeding period, DNMT3a levels are significantly reduced in the testes and uterine tissue (Lynch et al. 2016) (Fig. 11.3c), liver (Alvarado et al. 2015), skeletal muscle (Alvarado et al. 2015) and circulating leukocytes (Stevenson et al. 2014). In humans, there are also marked seasonal changes in DNA methylation. Blood samples revealed that in spring and summer months, there were significantly higher levels of DNA methylation in the promoters for cancer-related genes (RASSF1A and MGMT) compared to autumn and winter months (Ricceri et al. 2014). The exact mechanisms for how external information, such as day length, regulates epigenetic enzyme expression are not well known. In addition to day length, supplementary cues including temperature and diet (see below), likely contribute to the regulation of epigenetic enzyme expression in a tissue-dependent manner. The two primary hormonal candidates that provide an internal code of environmental cues are melatonin and gonadal-derived oestrogen (Stevenson 2017a). In human embryonic kidney cells (HEK293), the administration of melatonin induced an increase in *dnmt3a* expression (Fig. 11.3d). Given that melatonin is a reliable physiological measurement of 'night-length' a proxy for night duration, these data suggest that the annual change in day length drives melatonin-dependent entrainment of epigenetic enzyme expression, and DNMT3a in particular. In ovariectomized hamsters, a bolus injection of estradiol and progesterone (E_2P_4) significantly reduced *dnmt3a* expression (Fig. 11.3e). It is likely that oestrogen actions provide supplementary regulation

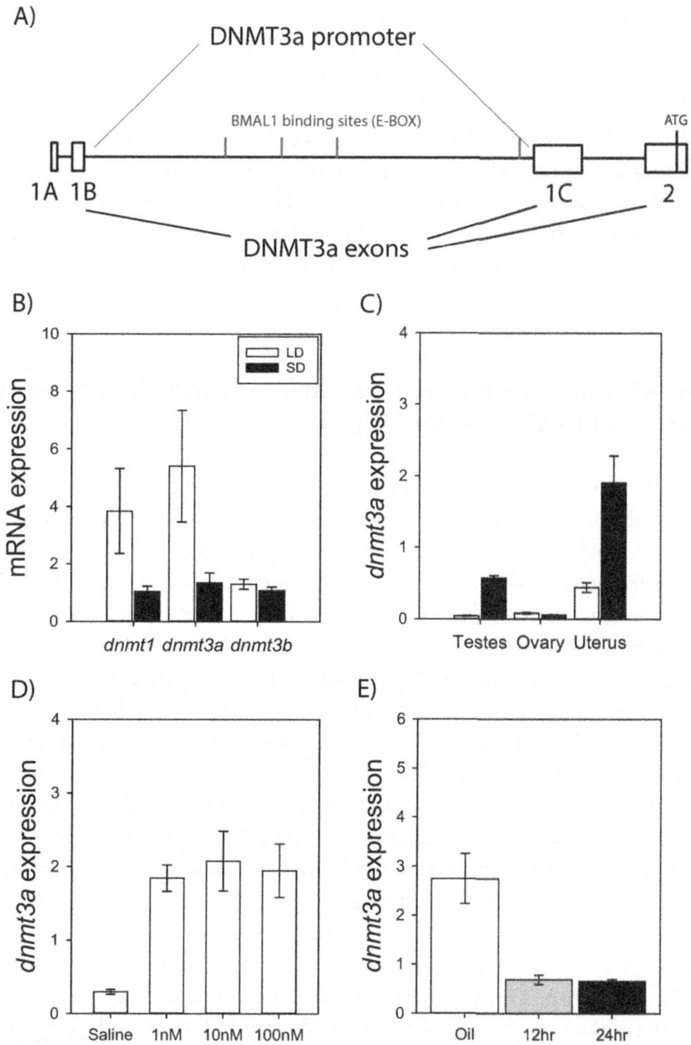

Fig. 11.3 Photoperiodic and hormonal regulation of DNA methyltransferase 3a. DNMT3a is one epigenetic enzyme that exhibits consistent rhythmic expression in multiple nuclei and tissues. (**a**) The *dnmt3a* promoter has several binding motifs for the circadian clock gene BMAL1. Koike et al. (2012) confirmed that BMAL1 directly binds to transcriptional regulatory regions for *dnmt3a* and shows daily changes with peak activity during the late dark phase. (**b**) Hypothalamic *dnmt1* and *dnmt3a* expressions display photoperiod-dependent regulation in which long summer-like day lengths (LD) stimulate higher levels and short winter-like days (SD) reduce levels. (**c**) Conversely, there is an increase in *dnmt3a* expression in the testes and uterine tissue of SD reproductively regressed hamsters. The annual change in photoperiod is encoded by high amplitude changes in the duration of melatonin and plasma oestrogen concentrations. (**d**) Melatonin was sufficient to induce *dnmt3a* expression in HEK293 cells. (**e**) A single bolus injection of E_2P_4 to ovariectomized hamsters significantly reduced uterine *dnmt3a* expression. Data are adapted from Stevenson (2017a, b) and Lynch et al. (2016)

of enzyme expression that fine-tunes the timing of epigenome modifications. The precise genomic targets for the annual change in DNA methylation have yet to be fully elucidated. In mammalian species, there are robust annual changes in the reproductive neuroendocrine peptides Kisspeptin (Greives et al. 2008; Stevenson et al. 2012) and RF-related peptide-3 (Greives et al. 2008). Whereas in avian species, the annual change in reproduction is primarily driven by the photoperiodic regulation of gonadotropin-releasing hormone (GnRH) cells (Stevenson et al. 2013). These reproductive neuropeptides are excellent candidates for rhythmic epigenetic, and in particular, the control of *seasonal* reproduction (Kurian and Terasawa 2013; also see Chap. 9).

11.5 Rhythmic Epigenetic Enzymes and the Neuroendocrine Control of the Estrous Cycle

One of the well-described neuroendocrine reproductive biological rhythms across vertebrates is the female reproductive cycle. Over several decades, the behavioural, morphological, hormonal, cellular and molecular oscillations have been characterized in mammalian species (Bronson 1990). The female reproductive cycle serves to optimize the timing of fertilization and reproductive behaviour (Pfaff et al. 2004). In rodents, there is typically a 4–5 day estrous cycle that includes complex coordination of multiple hormonal, morphological and behavioural changes. Rodentia exhibit four key estrous stages that include diestrous, proestrous, estrous and metestrous (Westwood 2008). A robust method to monitor the estrous cycle in rats and mice uses vaginal cytology. A simple vaginal smear can clearly delineate the estrous stage based on the absence, presence or proportion of neutrophils, small and large nucleated epithelial cells and anucleated keratinized epithelial cells (Cora et al. 2015). In parallel, the endometrium of the uterus exhibits large-scale programmed remodelling that involves cellular proliferation, apoptosis, angiogenesis and leukocyte infiltration (Evans et al. 1990). During the transition from diestrous–proestrous–estrous, there is an extensive growth of uterine tissue followed by a subsequent decrease in size (Lynch et al. 2016; Fig. 11.4a). These morphological and cellular changes are driven by the local production of sex steroids 17β-estradiol (E_2) and progesterone (P_4) (Wood et al. 2007) (Fig. 11.4b).

The epigenetic enzymes *dnmt1*, *dnmt3a* and *dnmt3b* are predominantly localized in the epithelial layer in ovaries (Ahluwalia et al. 2001), and endometrium cells in the uterus (Van Kaam et al. 2011), and may function to time fertility (Roy and Matzuk 2006). HDAC1 and HDAC2 are expressed in oocytes, and conditional ovarian knockouts of *hdac1/hdac2* in the ovary causes inhibition of follicle development past the secondary stage (Ma et al. 2012). Recent studies have examined cyclical changes in epigenetic enzymes during the estrous cycle and have identified marked plasticity during the transition from proestrous to estrous stages (Fig. 11.4c). For example, uterine *dnmt3a* and *dnmt3b* enzymes are significantly reduced during late proestrous and early estrous (Lynch et al. 2016). Uterine *dnmt3a* has been shown to be important for decidualization and shows a transient oestrogen-dependent decrease

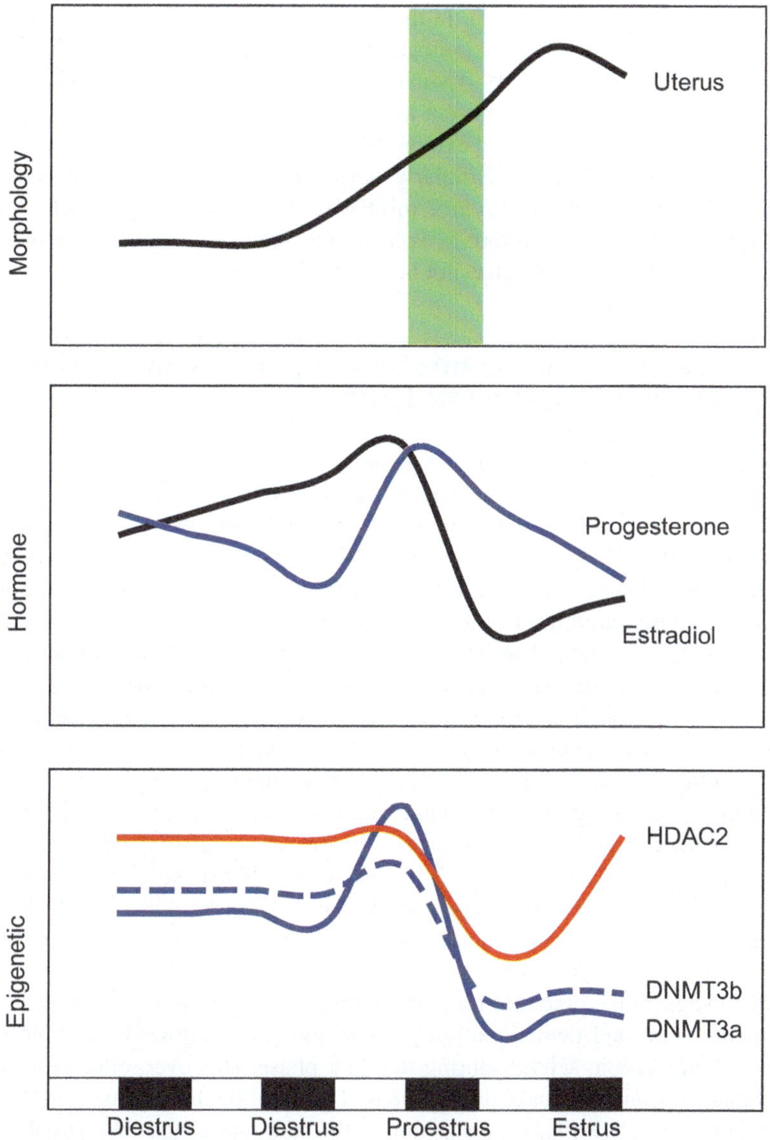

Fig. 11.4 Epigenetic enzymes oscillate in uterine tissues across the estrous cycle. The female rodent estrous cycle is characterized by robust changes in morphology, plasma hormones and uterine epigenetic enzymes. (Upper Panel) During the estrous cycle, uterine mass significantly increases during the transition from proestrous to estrous stages. On the evening of proestrous, there is a significant increase in the probability and intensity of female reproductive behaviours. The green bar depicts 'behavioural estrous', the period of high proceptivity and receptivity in females. (Middle Panel) Cycling changes in ovarian hormones estradiol and progesterone, the primary hormones that induce marked morphological changes in the uterus and transcriptional activation of several genes involved in the control of reproductive physiology. (Lower Panel) Three epigenetic enzymes in the uterus, HDAC2, DNMT3a and DNMT3b, were found to be inhibited by estradiol/ progesterone. Data are adapted from Lynch et al. (2016, 2017)

in expression (Logan et al. 2013). In the hamster, there is a significant decrease in *dnmt3a* during estrous and was significantly reduced in ovariectomized animals in response to a single bolus of E_2P_4 (Lynch et al. 2016) (Fig. 11.3e). Parallel changes in histone modifications are also present as HDAC1/2 are expressed in endometriotic and endometrial stromal cells and exhibit differential regulation by E_2 and combined E_2P_4 (Colón-Díaz et al. 2012). Similarly, Jefferson and colleagues (2013) revealed that diethylstilbestrol inhibited *hdac2* mRNA expression; yet uterine *hdac1, 2* and *3* did not change at postnatal day 5. To date, the functional significance of estrous cycling in epigenetic enzymes has not been established.

11.6 Neuroendocrine Control of Daily Energy Balance and Daily Rhythms in Epigenetic Enzymes

There are multiple systems that control energy intake and expenditure. A complex interaction of neuroendocrine substrates, adipose tissue, muscle cells, pancreatic cells and liver cells serve to maintain a homeostatic steady-state condition (Yeo and Heisler 2012). Daily rhythms in epigenetic enzymes have been characterized in discrete hypothalamic nuclei and peripheral tissues that suggest a complex epigenomic reprogramming occurs daily to ensure optimal energy balance (Asher and Sassone-Corsi 2015). The development of a freely available, online resources that contain massive datasets based on high-throughput methods (e.g. microarray, RNA sequencing), such as CircaDB (circadb.hogeneschlab.org), provide a valuable opportunity to interrogate daily rhythms in epigenetic enzyme RNA expression across multiple tissues. For example, CircaDB analyzes reveal that brown and white adipose tissues show daily rhythms in *hdac6* and *kat2b*; which coincidentally have anti-phase expression waveforms (Pizarro et al. 2013). *hdac6* expression peaked during the mid-point of a 12:12 LD cycle, whereas *kat2b* peaks during the dark phase. The enzyme expression patterns here suggest a daily switch in histone (de)acetylation occurs that is not driven by food intake and instead, reflects an endogenous timing mechanism. In the liver, *tet2* and *dnmt3b* peak prior to lights on; which suggest the potential for daily variation in *epigenetic tone*; high levels of both methylation and demethylation prior to lights on followed by a subsequent decline in epigenomic activity during the dark phase. However, other research has shown that liver *dnmt3a* peaks during the mid-point of the light phase (Hughes et al. 2009). Conversely, liver *hdac5* peaks during the dark phase, and *hdac6* peaks during the light phase.

There is growing evidence that the time of food intake has a remarkable level of bidirectional interactions between the circadian clock and energy balance (Asher and Sassone-Corsi 2015). The vast majority of research has examined the contribution of chromatin remodelling via the activity of sirtuin proteins (SIRT1) within well-defined hypothalamic nuclei involved in the regulation of metabolism, appetite and food intake (Asher and Sassone-Corsi 2015). SIRT1 removes histone acetyl groups via a nicotinamide adenine dinucleotide (NAD+)-dependent manner. Increased SIRT1 activity in the ventromedial nucleus (VMN) of the hypothalamus leads to histone deacetylation and can directly interact with the circadian clock gene CLOCK; and thereby, regulates clock-controlled genes in a histone deacetylation-dependent manner (Nakahata et al. 2008; Orozco-Solis et al. 2015). The current evidence suggests that SIRT1 expression specifically within VMN Sf1 neurons is essential to coordinate circadian food intake and daily energetic state.

11.7 Daily Rhythmic Epigenetic Modifications and Disease

Immune responses are dynamically regulated over daily and seasonal timescales (Stevenson and Prendergast 2015). This section will give an overview of the ongoing research regarding the role of epigenetic mechanisms in immune function, focusing on evidence that indicates reversible modifications. The epigenetic enzymes described above have been implicated in pathological conditions and progression, e.g. cancer, and have become a powerful tool in disease treatment (Widschwendter et al. 2018). One tantalizing conjecture is the potential loss of natural rhythmic epigenetic modifications as a contributing factor to disease progression. Therefore, the investigation of natural rhythmic variation in epigenetic modifications can give clinical researchers insight into the factors that could be involved in enzyme disruption, as well as aiding in the development of novel reversible treatment strategies.

It is becoming clear that immune function (e.g. cancer) and biological rhythms (i.e. *circadian*) have bidirectional interactions. The increased attention to disrupted circadian rhythms has resulted in the identification that entrained daily rhythms contribute to health and well-being (Savvidis and Koutsilieris 2013). Leukocyte migration from lymph nodes, lymph and plasma shows daily rhythms with peak CD4+, CD8+ T cells and B cell levels during the light phase in mice (Druzd et al. 2017) and Siberian hamsters (Prendergast et al. 2013). Disruption of circadian timing induces leukocyte arrhythmia, reduced daily amplitudes in circadian clock gene and inhibited CD11+ dendritic cells, indicating that a functional SCN is required for daily variation in immune function (Prendergast et al. 2013). Likewise, there are distinct seasonal rhythms in plasma leukocytes numbers and leukocyte transcriptomic profiles in hamsters and humans, respectively (Bilbo et al. 2002; Banks et al. 2016; Dopico et al. 2015). During winter photoperiods, there are significantly elevated circulating blood leukocytes, T cells and natural killer cells (Bilbo et al. 2002). The winter enhancement of immune function and survival is common across species and serves to increase survival (Nelson 2004). The increase in many chronic diseases and pathological states has been attributed to not only disrupted circadian rhythms as described above, but also disruption in seasonal immune rhythms (Stevenson et al. 2015).

A few studies have examined the daily or seasonal variation of epigenetic enzyme expression in immune tissues or cells. A massive transcriptomic analysis of seasonal waveforms of human leukocytes identified that *dnmt1, hdac5, -6 and -9* vary across the year (Dopico et al. 2015). In Siberian hamsters, short photoperiods resulted in higher *dnmt3b* leukocyte expression (Stevenson et al. 2014). The increased *dnmt3b* was associated with higher methylation levels in the proximal promoter of deiodinase type-3 (*dio3*), an enzyme involved in thyroid hormone catabolism. How rhythms in epigenetic enzymes relate to immunoenhancement or immunosuppression is not well known, the best evidence can be derived from the research field in autoimmune diseases and cancer.

Autoimmunity has been associated with DNA hypomethylation in a number of studies. For instance, in both animal models (Quddus et al. 1993) and human cells (Richardson et al. 1990), treatment with DNA methylation inhibitor drugs (e.g. 5-Azacytidine) causes the development of autoimmune conditions equal or comparable to Systemic Lupus Erythematosus (SLE). SLE progression has been mainly attributed to reduced DNA methylation in T cells (Richardson et al. 1990) that causes, in turn, overexpression of adhesion molecules such as lymphocyte function-associated antigen 1 (LFA-1) (Cornacchia et al. 1988). Increased LFA-1 expression appears to result in immune cell autoreactivity leading to SLE (Kaplan et al. 2000). The ability to selectively reinstate methylation of LFA-1 in T cells at specific daily phases may provide one solution to reduce the incidence and progression of autoimmune diseases such as SLE.

Once the differentiation of immune cells such as T lymphocytes was closely related to chromatin structure changes (Agarwal and Rao 1998), scientists started to investigate the precise role of DNMT1 within lymphocyte organization. Naive lymphocytes undergo a series of differentiations to become either B or T cell, to express different types of T cell surface receptors (e.g. TCRαβ or TCRγδ) and lineage markers (e.g. CD4+, CD8+) (Richter et al. 1999). CD4+ cells then are able to differentiate into regulatory (Treg) cells. An essential role of Treg cells is to alleviate the immune response and, for instance, prevent an extremely elevated immune reaction that could lead to autoimmune diseases. Lee et al. (2001) utilized a Cre/Lox system to generate *dnmt1* knockout in thymocytes, and the lack of DNMT1 enzyme activity resulted in compromised T cell development and survival. Josefowicz et al. (2009) also examined the functional role of DNMT1 for Treg cell lineage regulation and Forkhead box P3 (*foxp3*), in particular, *foxp3* is important for the stimulation of the transition of CD4+ cells into the Treg cell lineage (Fontenot et al. 2003). However, when DNMT1 is absent, the *foxp3* promoter is hypomethylated and increased FOXP3 in CD8+ cells rather than CD4+ cells. Therefore, the absence of *dnmt1* may lead to the progression of CD8+ cells into Treg cells and thus, an autoimmune response. Therefore, DNMT1 provides a system for confining *foxp3* cell specificity to the CD4+ lineage, ensuring normal Treg cell development (Josefowicz et al. 2009). It is unknown whether the loss of *dnmt1* in Treg cells leads to the reduction in daily oscillations in cells, endogenous circadian clock signalling and/or inhibition of rhythmic epigenetic modifications.

Despite the undeniable importance of DNMT1 in immune function, other methylating enzymes may also play an important role in the daily timing of essential regulatory genes. DNMT3a and DNMT3b are considered the de novo DNA methyltransferases, as these enzymes establish new epigenetic patterns in cells (Okano et al. 1999). Organisms retain functional levels of *dnmt3a* and *dnmt3b* within immune cells well into adulthood (Fitzpatrick and Wilson 2003). Their activity within the immune response has been explored both in vitro and in vivo (Okano et al. 1999; Fitzpatrick and Wilson 2003). Similar to DNMT1, their expression is required for normal development and cell survival, as their loss results in embryonic lethality (Okano et al. 1999).

It is clear that both genetics and epigenetics play a role in the development of the majority of neuroendocrine cancers. One the '10 hallmarks of cancer' outlined by Hanahan and Weinberg (2011) is genomic instability that results from alterations in the epigenome (Esteller 2008; Hanahan and Weinberg 2011), including both global hypomethylation (Gaudet et al. 2003) and gene-specific hypermethylation (Herman and Baylin 2003). High expression of *dnmts* has been shown in a variety of cancers, e.g. neuroblastoma, prostate cancer (Kautiainen and Jones 1986). Several studies have assessed the function of DNMT1 through both pharmacological inhibition and gene knockdown in vitro and in vivo. DNMT1 is required for development in mice, as *dnmt1*$^{-/-}$ is lethal (Li et al. 1992). This has been attributed to the enzyme's essential function in maintaining DNA methylation patterns in daughter cells through generations (Lei et al. 1996), a function necessary for normal survival and growth. Jackson-Grusby et al. (2001) observed that mice *dnmt1*$^{-/-}$ embryonic stem cell-derived fibroblasts undergo a p53-dependent cell death. p53 is a tumour suppressor gene, involved in inhibiting cell cycle progression when errors are detected in the genome (Symonds et al. 1994). When methylated, p53 expression is silenced and results in survival of cancerous cells (Jackson-Grusby et al. 2001). Knox et al. (2000) monitored the effects of *dnmt1*$^{-/-}$ in human lung cancer cells and found that hypomethylation due to the absence of *dnmt1* also inhibits cell proliferation through apoptosis. The transcription of other tumour suppressor genes, such as p16 and p21, was found to be upregulated after using pharmacological inhibitors of DNMT1 enzymatic activity in the human bladder (Fournel et al. 1999) and colon (Rhee et al. 2000) carcinoma cells, preventing tumour progression. It is intriguing that both expression of tumour suppressor genes (Miki et al. 2013) and epigenetic enzymes has been shown to vary in a circadian manner. p53 is able to regulate the cell cycle is via 'clocking' genes, by inhibiting the *per2* gene through interaction with BMAL1 and CLOCK (Miki et al. 2013; Fu et al. 2002). Repression of *per2* expression has been associated with overexpression of c-MYC proto-oncogene (cMyc), a gene that codes for transcription factors involved in cell cycle progression (Pelengaris et al. 2002). cMyc directly binds to DNMT3a to promote cell growth (Brenner et al. 2005). It is then possible that circadian rhythms dictate patterns of certain epigenetic components, resulting in cancer development. However, the exact mechanisms underlying the circadian control of epigenetic signals in tumourigenesis are still largely unknown.

Aberrant expression of HAT and HDAC activity also contributes to the rise and progression of specific diseases (Jones and Baylin 2007) and include cancer (Cress and Seto 2000), Alzheimer's disease (Francis et al. 2009) and mental retardation (Coffee et al. 1999). HDAC activity has been shown to be linked to the expression of certain eukaryotic cell cycle regulation factors, such as retinoblastoma (RB) protein (Brehm et al. 1998). Disrupted interaction between HDAC proteins and RB proteins prevents the apoptosis of cancerous cells, allowing tumourigenesis (Xu et al. 1993).

Direct downregulation of HDACs via chemical inhibitors is also currently being explored. Use of HDAC inhibitors have been linked to cell cycle arrest in T cell lymphoma (Olsen et al. 2007) and a reduction in cancerous cells in an acute myeloid leukaemia mouse model (Bots et al. 2014). Altogether, these data indicate that a greater understanding of the timing of epigenetic modifications, particularly daily rhythms can provide significant improvements for healthy immune function and the treatment of pathological conditions.

11.8 Future Perspectives

It has been suggested that an 'epigenetic homeostasis' exists and is needed for normal cellular function that is characterized by a balance between DNA (de) methylation as well as histone (de)acetylation (Jones and Baylin 2007). In this chapter, we have focused on 'epigenetic plasticity', the transient modifications to the genome and/or chromatin template that leads to a functional change in the neuroendocrine control of reproduction, energy balance and immune function. The predominant difference between these philosophical positions is that the disruption of epigenetic homeostasis leads to long-term consequences in cellular function, such as chronic diseases while the epigenetic plasticity model argues that the short- and long-term oscillations in epigenetic enzymes maintain cyclical cell- and tissue-function that is required for naturally occurring, normal daily, estrous and seasonal rhythms (Stevenson 2018). A greater understanding of the mechanistic and functional basis of the latter can yield significant insights into the genetic control of animal health and wellness. Below, we outlined two major perspectives for future research.

The field of epigenetic research is exciting as ongoing findings reveal the massive potential for basic, translational and applied research. The advent of high-throughput sequencing platforms such as Whole-Genome Sequencing (WGS), transcriptomic, proteomic assays has yielded immense information on the complexity of molecular and cellular signalling pathways in neuroendocrine and immune cells. Moreover, the chemical modification of genome structure using sodium bisulfite now provides the capability to identify specific nucleotide residues/motifs that show variation in epigenetic modifications (i.e. methylation). This chapter has focused on the rhythmic expression of multiple enzymes involved in epigenetic modifications. The data show tissue-, cell- and time-dependent oscillations; however, our understanding of the downstream effects is limited. The next steps are to establish the precise genomic and chromatin location of daily, estrous and seasonal epigenetic modifications. The combination of sequencing platforms, such as Illumina or PacBIO, provides a valuable opportunity to address wherein the genome epigenetic modifications exhibit rhythmic variation; and thus, novel gene regulatory mechanisms within tissues and more importantly at a single-cell resolution.

Another advantage of sequencing tools to identify epigenetic modifications is the development of individual or *personalized* approaches to medical treatments (Rasool et al. 2015). Indeed, further investigation into the exact genomic and chromatin modifications that underlie daily and seasonal plasticity in reproduction, energy balance and immune function is warranted. The ability to design epigenetic constructs that selectively (de)methylate target genome regions that is provided within an optimal daily window has the potential to make significant health and wellness advancements for both livestock animals and humans.

Key References

Azzi et al. (2014). This study shows that daily oscillations in SCN DNA methylation regulates circadian clock function.

Bilbo et al. (2002). This is an exciting report on the seasonal regulation of immune function.

Dopico et al. (2015). This study reveals large annual changes in leukocyte transcriptomes in humans.

Koike et al. (2012). One key finding identified in this paper were daily rhythms in BMAL1 binding on the *dnmt3a* promoter.

Lynch et al. (2016). This report illustrated increased *dnmt3a* expression was negatively associated with oscillations in fertility.

Nugent et al. (2015). This study discovered sex-differences in neuroendocrine DNMT3a and a lack of expression facilitated masculinization.

Stevenson and Prendergast. (2013). This study established that rhythmic neuroendocrine DNA methylation underlies seasonal timing of reproduction.

References

Agarwal S, Rao A (1998) Modulation of chromatin structure regulates cytokine gene expression during T cell differentiation. Immunity 9:765–775

Ahluwalia A, Hurteau JA, Bigsby RM, Nephew KP (2001) DNA methylation in ovarian cancer. II. Expression of DNA methyltransferases in ovarian cancer cell lines and normal ovarian epithelial cells. Gynecol Oncol 82:299–304

Alvarado S, Mak T, Liu S, Storey KB, Szyf M (2015) Dynamic changes in global and gene-specific DNA methylation during hibernation in adult thirteen-lined ground squirrels, *Ictidomys tridecemlineatus*. J Exp Biol 218:1787–1795

Asher G, Sassone-Corsi P (2015) Time for food: the intimate interplay between nutrition, metabolism and the circadian clock. Cell 161:84–92

Azzi A, Dallmann R, Casserly A, Rehrauer RH, Patrignani A, Maier B, Kramer A, Brown SA (2014) Circadian behavior is light-programmed by plastic DNA methylation. Nat Neurosci 17:377–382

Banks R, Delibegovic M, Stevenson TJ (2016) Photoperiod- and triiodothyronine-dependent regulation of reproductive neuropeptides, proinflammatory cytokines, and peripheral physiology in Siberian hamsters (*Phodopus sungorus*). J Biol Rhythm 31:299–307

Bilbo SD, Dhabhar FS, Viswanathan K, Saul A, Yellon SM, Nelson RJ (2002) Short day lengths augment stress-induced leukocyte trafficking and stress-induced enhancement of skin immune function. Proc Natl Acad Sci U S A 99:4067–4072

Bots M, Verbrugge I, Martin B, Salmon J, Ghisi M, Baker A, Stanley K, Shortt J, Ossenkoppele G, Zuber J, Rappaport A, Atadja P, Lowe S, Johnstone R (2014) Differentiation therapy for the treatment of t(8;21) acute myeloid leukemia using histone deacetylase inhibitors. Blood 123:1341–1352

Brehm A, Miska EA, McCance DJ, Reid JL, Bannister AJ, Kouzarides T (1998) Retinoblastoma protein recruits histone deacetylase to repress transcription. Nature 391:597–601

Brenner C, Deplus R, Didelot C, Loriot A, Viré E, De Smet C, Gutierrez A, Danovi D, Bernard D, Boon T, Pelicci PG (2005) Myc represses transcription through recruitment of DNA methyltransferase corepressor. EMBO J 24:336–346

Bronson FH (1990) Mammalian reproductive biology. University of Chicago Press, Chicago

Coffee B, Zhang F, Warren ST, Reines D (1999) Acetylated histones are associated with FMR1 in normal but not fragile X-syndrome cells. Nat Genet 22:98–101

Colón-Díaz M, Báez-Vega P, García M, Ruiz A, Monteiro JB, Fourquet J, Bayona M, Alvarez-Garriga C, Achille A, Seto E, Flores I (2012) HDAC1 and HDAC2 are differentially expressed in endometriosis. Reprod Sci 19:483–492

Cora MC, Looistra L, Travlos G (2015) Vaginal cytology of the laboratory rat and mouse: review and criteria for the staging of the estrous cycle using stained vaginal smears. Toxicol Pathol 43:776–793

Cornacchia E, Golbus J, Maybaum J, Strahler J, Hanash S, Richardson B (1988) Hydralazine and procainamide inhibit T cell DNA methylation and induce autoreactivity. J Immunol 140:2197–2200

Cress WD, Seto E (2000) Histone deacetylases, transcriptional control, and cancer. J Cell Physiol 184:1–16

Dopico XC, Evangelou M, Ferreira RC, Guo H, Pekalski M, Smyth DJ, Cooper N, Burren OS, Fulford AJ, Hennig BJ, Am P, Ziegler AG, Bonifacio E, Wallace C, Todd JA (2015) Wide-spread seasonal gene expression reveals annual differences in human immunity and physiology. Nat Commun 6:7000

Druzd D, Matveeva O, Ince L, Harrison U, Wenyan H, Schmal C, Herzel H, Tsang AH, Kawakami N, Leliavske A, Uhl O, Yao L, Sander LE, Chen CS, Kraus K, de Juan A, Hergenhan SM, Ehlers M, Koletzko B, Haas R, Solbach W, Oster H, Scheiermann C (2017) Lymphocyte circadian clocks control lymph node trafficking and adaptive immune responses. Immunity 46:120–132

Durrin LK, Weber JL, Gorksi J (1984) Chromatin structure, transcription, and methylation of the prolactin gene domain in pituitary tumors of Fischer 344 rats. J Biol Chem 259:7086–7093

Esteller M (2008) Epigenetics in cancer. N Engl J Med 358:1148–1159

Evans GS, Gibson DF, Roberts SA, Hind TM, Potten CS (1990) Proliferative change in the genital tissues of female ice during the oestrous cycle. Cell Tissue Kinet 23:619–635

Fagerberg L, Hallstrom BM, Oksvold P, Kampf C, Djureinovic D, Odeberg J, Habuka M, Tahmasebpoor S, Danielsson A, Edlund K, Asplund A, Sjostedt E, Lundberg E, Szigyarto CA, Skogs M, Takanen JO, Berling H, Tegel H, Mulder J, Nilsson P, Schwenk JM, Lindskog C, Danielsson F, Mardinoglu A, Sivertsson A, von Feilitzen K, Forsberg M, Zwahlen M, Olsson I, Navani S, Huss M, Nielsen J, Ponten F, Uhlen M (2014) Analysis of the human tissue specific expression by genome-wide integration of transcriptomics and antibody-based proteomics. Mol Cell Proteomics 13:397–406

Fitzpatrick DR, Wilson CB (2003) Methylation and demethylation in the regulation of genes, cells, and responses in the immune system. Clin Immunol 109:37–45

Fontenot JD, Gavin MA, Rudensky AY (2003) Foxp3 programs the development and function of CD4+ CD25+ regulatory T cells. Nat Immunol 4:330

Fournel M, Sapieha P, Beaulieu N, Besterman JM, MacLeod AR (1999) Down-regulation of human DNA-(Cytosine-5) methyltransferase induces cell cycle regulators p16 ink4A and p21WAF/Cip1 by distinct mechanisms. J Biol Chem 274:24250–24256

Francis YI, Fà M, Ashraf H, Zhang H, Staniszewski A, Latchman DS, Arancio O (2009) Dysregulation of histone acetylation in the APP/PS1 mouse model of Alzheimer's disease. J Alzheimers Dis 18:131–139

Fu L, Pelicano H, Liu J, Huang P, Lee CC (2002) The circadian gene Period2 plays an important role in tumor suppression and DNA damage response in vivo. Cell 111:41–50
Gaudet F, Hodgson JG, Eden A, Jackson-Grusby L, Dausman J, Gray JW, Leonhardt J, Jaenisch R (2003) Induction of tumors in mice by genomic hypomethylation. Science 300:489–492
Greives TJ, Kriegsfeld LJ, Bentley GE, Tsutsui K, Demas GE (2008) Recent advances in reproductive neuroendocrinology: a role for RFamide peptides in seasonal reproduction? Proc Biol Sci 275:1943–1951
Guil S, Esteller M (2009) DNA methylomes, histone codes and miRNAs: tying it all together. Int J Biochem Cell Biol 41:87–95
Hanahan D, Weinberg R (2011) Hallmarks of cancer: the next generation. Cell 144:646–674
Hastings MH, Brancaccio M, Maywood ES (2014) Circadian pacemaking in cells and circuits of the suprachiasmatic nucleus. J Neuroendocrinol 26:2–10
Heard E, Martienssen RA (2014) Transgenerational epigenetic inheritance: myths and mechanisms. Cell 157:95–109
Helm B, Stevenson TJ (2015) Circannual rhythms: history, present challenges, future directions. In: Numata H, Helm B (eds) Annual, lunar and tidal clocks: patterns and mechanisms of nature's enigmatic rhythms. Springer Japan, Tokyo, pp 213–225
Herman JG, Baylin SB (2003) Gene silencing in cancer in association with promoter hypermethylation. N Engl J Med 349:2042–2054
Hughes ME, DiTacchio L, Hayes KR, Vollmers C, Pulivarthy S, Baggs JE, Panda S, Hogenesch JB (2009) Harmonics of circadian gene transcription in mammas. PLoS Genet 5:e1000442
Jackson-Grusby L, Beard C, Possemato R, Tudor M, Fambrough D, Csankovszki G, Jaenisch R (2001) Loss of genomic methylation causes p53-dependent apoptosis and epigenetic deregulation. Nat Genet 27:31–39
Jang HS, Shin WJ, Lee JE, Tae Do J (2017) CpG and non-CpG methylation in epigenetic gene regulation and brain function. Genes (Basel) 8:148
Jefferson WN, Chevalier DM, Phelps JY, Cantor AM, Padilla-Banks E, Newbold RR, Archer TK, Kinyamu HK, Williams CJ (2013) Persistently altered epigenetic marks in the mouse uterus after neonatal estrogen exposure. Mol Endocrinol 27:1666–1677
Jones P, Baylin S (2007) The epigenomics of cancer. Cell 128:683–692
Josefowicz SZ, Wilson CB, Rudensky AY (2009) Cutting edge: TCR stimulation is sufficient for induction of Foxp3 expression in the absence of DNA methyltransferase 1. J Immunol 182:6648–6652
Kaplan MJ, Beretta L, Yung RL, Richardson BC (2000) LFA-1 overexpression and T cell autoreactivity: mechanisms. Immunol Investig 29:427–442
Kautiainen TL, Jones PA (1986) DNA methyltransferase levels in tumorigenic and nontumorigenic cells in culture. J Biol Chem 261:1594–1598
Klose RJ, Bird AP (2006) Genomic DNA methylation: the mark and its mediators. Trends Biochem Sci 31:89–97
Knox JD, Araujo FD, Bigey P, Slack AD, Price GB, Zannis-Hadjopoulos M, Szyf M (2000) Inhibition of DNA methyltransferase inhibits DNA replication. J Biol Chem 275:17986–17990
Kohli RM, Zhang Y (2013) TET enzymes, TDG and the dynamics of DNA demethylation. Nature 502:472–479
Koike N, Yoo SH, Huang HC, Kumar V, Lee C, Kim TK, Takahashi JS (2012) Transcriptional architecture and chromatin landscape of the core circadian clock in mammals. Science 338:349–354
Kurian JR, Terasawa E (2013) Epigenetic control of gonadotropin releasing hormone neurons. Front Endocrinol 4:61
Lee PP, Fitzpatrick DR, Beard C, Jessup HK, Lehar S, Makar KW, Cherry SR (2001) A critical role for Dnmtl and DNA methylation in T cell development, function, and survival. Immunity 15:763–774
Lei H, Oh SP, Okano M, Juttermann R, Goss KA, Jaenisch R, Li E (1996) De novo DNA cytosine methyltransferase activities in mouse embryonic stem cells. Development 122:3195–3205

Li E (2002) Chromatin modification and epigenetic reprogramming in mammalian development. Nat Rev Genet 3:662–673

Li E, Bestor TH, Jaenisch R (1992) Targeted mutation of the DNA methyltransferase gene results in embryonic lethality. Cell 69:915–926

Logan PC, Ponnampalam AP, Steiner M, Mitchell MD (2013) Effect of cyclic AMP and estrogen/progesterone on the transcription of DNA methyltransferase during the decidualization of human endometrial stromal cells. Mol Hum Reprod 19:302–312

Lynch EWJ, Coyle CS, Lorgen M, Campbell E, Bowman A, Stevenson TJ (2016) Cyclical DNA methyltransferase 3a expression is a seasonal and oestrus timer in reproductive tissues. Endocrinology 157:2469–2478

Lynch EWJ, Coyle CS, Stevenson TJ (2017) Light and hormonal regulation of histone deacetylase 1, 2, 3 in reproductive tissues. Gen Comp Endocrinol 246:194–199

Ma P, Pan H, Montgomery RL, Olson EN, Schultz RM (2012) Compensatory functions of histone deacetylase 1 (HDAC1) and HDAC2 regulate transcription and apoptosis during mouse oocyte development. Proc Natl Acad Sci U S A 109:E481–E489

Miki T, Matsumoto T, Zhao Z, Lee CC (2013) p53 regulates Period2 expression and the circadian clock. Nat Commun 4:2444

Nakahata Y, Kaluzova M, Grimaldi B, Sahar S, Hirayama J, Chen D, Guarente LP, Sassone-Corsi P (2008) The NAD+-dependent deacetylase SIRT1 modulates CLOCK-mediated chromatin remodelling and circadian control. Cell 134:329–340

Nelson RJ (2004) Seasonal immune function and sickness responses. Trends Immunol 25:187–192

Nugent BM, Wright CL, Shetty AC, Hodes GE, Lenz KM, Mahurkar A, Russo SJ, Devine SE, McCarthy MM (2015) Brain feminization requires active repression of masculinization via DNA methylation. Nat Neurosci 18:690–697

Okano M, Bell DW, Haber DA, Li E (1999) DNA methyltransferases Dnmt3a and Dnmt3b are essential for de novo methylation and mammalian development. Cell 99:247–257

Olsen EA, Kim YH, Kuzel TM, Pacheco TR, Foss FM, Parker S, Frankel SR, Chen C, Ricker JL, Arduino JM, Duvic M (2007) Phase IIb multicenter trial of vorinostat in patients with persistent, progressive, or treatment refractory cutaneous T-cell lymphoma. J Clin Oncol 25:3109–3115

Orozco-Solis R, Ramadori G, Coppari R, Sassone-Corsi P (2015) SIRT1 relays nutritional inputs to the circadian clock through the Sf1 neurons of the ventromedial hypothalamus. Endocrinol 156:2174–2184

Pelengaris S, Khan M, Evan G (2002) c-MYC: more than just a matter of life and death. Nat Rev Cancer 10:764

Pfaff DW, Phillips MI, Rubin RT (2004) Hormonal secretions and responses are affected by biological clocks. In: Principles of hormone/behavior relations. Elsevier Academic Press, San Diego, pp 181–196

Pizarro A, Hayer K, Lahens NF, Hogenesch JB (2013) CircaDB: a database of mammalian circadian gene expression profiles. Nucleic Acids Res 41:D1009–D1013

Prendergast BJ, Cable EJ, Patel PN, Pyter LM, Onishi KG, Stevenson TJ, Ruby NF, Bradley SP (2013) Impaired leukocyte trafficking and skin inflammatory responses in hamsters lacking a functional circadian system. Brain Behav Immun 32:94–104

Quddus J, Johnson KJ, Gavalchin J, Amento EP, Chrisp CE, Yung RL, Richardson BC (1993) Treating activated CD4+ T cells with either of two distinct DNA methyltransferase inhibitors, 5-azacytidine or procainamide, is sufficient to cause a lupus-like disease in syngeneic mice. J Clin Invest 92:38–53

Rasool M, Malik A, Naseer MI, Manan A, Ansari SA, Begum I, Qazi MH, Pushparaji PN, Abuzenadah AM, Al-Qahtani MH, Kamal MA, Pushparaji PN, Gan SH (2015) The role of epigenetics in personalized medicine: challenges and opportunities. BMC Med Genet 8:S5

Rhee I, Jair KW, Yen RWC, Lengauer C, Herman JG, Kinzler KW, Vogelstein B, Baylin SB, Schuebel KE (2000) CpG methylation is maintained in human cancer cells lacking DNMT1. Nature 404:1003–1007

Ricceri F, Trevisan M, Fiano V, Grasso C, Fasanelli F, Scoccianti C, De Marco L, Tox AG, Vineis P, Sacerdote C (2014) Seasonality modifies methylation profiles in healthy people. PLoS One 9:e106846

Richardson B, Scheinbart L, Strahler J, Gross L, Hanash S, Johnson M (1990) Evidence for impaired T cell DNA methylation in systemic lupus erythematosus and rheumatoid arthritis. Arthritis Rheum 33:1665–1673

Richter A, Löhning M, Radbruch A (1999) Instruction for cytokine expression in T helper lymphocytes in relation to proliferation and cell cycle progression. J Exp Med 190:1439–1450

Riggs AD, Porter TN (1996) Overview of epigenetic mechanisms. In: Russo VEA, Martienssen R, Riggs AD (eds) Epigenetic mechanisms of gene regulation. Cold Spring Harbor Laboratory Press, Cold Spring Harbor, NY, pp 29–45

Roy A, Matzuk MM (2006) Deconstructing mammalian reproduction: using knockouts to define fertility pathways. Reproduction 131:207–219

Savvidis C, Koutsilieris M (2013) Circadian rhythm disruption in cancer biology. Mol Med 18:1249–1260

Stevenson TJ (2017a) Environmental and hormonal regulation of neuroendocrine epigenetic enzymes. J Neuroendocrinol 29(5)

Stevenson TJ (2017b) Circannual and circadian rhythms in hypothalamic DNA methyltransferase and histone deacetylase in Siberian hamsters (*Phodopus sungorus*). Gen Comp Endocrinol 243:130–137

Stevenson TJ (2018) Epigenetic regulation of biological rhythms: an evolutionary ancient molecular timer. Trends Genet 43:90–100

Stevenson TJ, Lincoln GA (2017) Epigenetic mechanisms regulating circannual rhythms, Chap 29. In: Kumar V (ed) Biological timekeeping: clocks, rhythms and behaviour. Springer, India

Stevenson TJ, Prendergast BJ (2013) Reversible DNA methylation regulates seasonal photoperiodic time measurement. Proc Natl Acad Sci USA 110:16651–16656

Stevenson TJ, Prendergast BJ (2015) Photoperiodic time measurement and seasonal immunological plasticity. Front Neuroendocrinol 37:76–88

Stevenson TJ, Hahn TP, MacDougall-Shackleton SA, Ball GF (2012) Gonadotropin-releasing hormone plasticity: a comparative perspective. Front Neuroendocrinol 33:287–300

Stevenson TJ, Bernard DJ, McCarthy MM, Ball GF (2013) Photoperiod-dependent regulation of gonadotropin-releasing hormone 1 messenger ribonucleic acid levels in the songbird brain. Gen Comp Endocrinol 190:81–87

Stevenson TJ, Onishi KG, Bradley SP, Prendergast BJ (2014) Cell-autonomous iodothyronine deiodinase expression mediates seasonal plasticity in immune function. Brain Behav Immun 36:61–70

Stevenson TJ, Visser ME, Arnold W, Barrett P, Biello S, Dawson A, Denlinger DL, Dominoni D, Ebling FJ, Elton S, Evans N, Ferguson HM, Foster RG, Hau M, Haydon DT, Hazlerigg DG, Heideman P, Hopcraft JGC, Jonsson NN, Kronfeld-Schor N, Kumar V, Lincoln GA, MacLeod R, Martin SAM, Martinez-Bakker M, Nelson RJ, Reed T, Robinson JE, Rock D, Schwartz WJ, Steffan-Dewenter I, Tauber E, Thackeray SJ, Umstatter C, Yoshimura T, Helm B (2015) Disrupted seasonal biology impacts health food security and ecosystems. Proc R Soc B 282:1–10

Symonds H, Krall L, Remington L, Saenz-Robles M, Lowe S, Jacks T, Van Dyke T (1994) p53-dependent apoptosis suppresses tumor growth and progression in vivo. Cell 78:703–711

Taunton J, Hassig CA, Schreiber SL (1996) A mammalian histone deacetylase related to the yeast transcriptional regulator Rpd3p. Science 272:408–411

Van Kaam KJ, Delvoux B, Romano A, D'Hooghe T, Dunselman GA, Groothuis PG (2011) Deoxyribonucleic acid methyltransferase and methyl-CpG binding domain proteins in human endometrium and endometriosis. Fertil Steril 95:1421–1427

Waddington CH (1953) Epigenetics and evolution. Symp Soc Exp Biol 7:186–199

Weaver ICG, Cervoni N, Champagne FA, D'Alessio AC, Sharma S, Seckl JR, Dymov S, Szyf M, Meaney MJ (2004) Epigenetic programming by maternal behavior. Nat Neurosci 7:847–854

Westwood FR (2008) The female rat reproductive cycle: a practical histological guide to staging. Toxicol Pathol 36:375–384

Widschwendter M, Jones A, Evans I, Reisel D, Dillner J, Sundstrom K, Steyerberg EW, Vergouwe Y, Wegwarth O, Rebitschek FG, Siebert U, Sroczynski G, de Beaufort ID, Bolt I, Cibula D, Zikan M, Bjorge L, Colombo N, Harbeck N, Dudbridge F, Am T, Knoppers BM, Joly Y, Teschendorff AE, Pashayan N, FORECEE 4C Consortium (2018) Epigenome-based cancer risk prediction: rationale, opportunities and challenges. Nat Rev Clin Oncol 15:292–309

Wood GA, Fata JE, Watson KLM, Khokha R (2007) Circulating hormones and estrous stage predict cellular and stromal remodelling in murine uterus. Reproduction 133:1035–1044

Xu HJ, Cairns P, Hu SX, Knowles MA, Benedict WF (1993) Loss of RB protein expression in primary bladder cancer correlates with loss of heterozygosity at the RB locus and tumor progression. Int J Cancer 53:781–784

Yeo GS, Heisler LK (2012) Unraveling the brain regulation of appetite: lessons from genetics. Nat Neurosci 15:1343–1349

Yoder JA, Soman NS, Verdine GL, Bestor TH (1997) DNA (cytosine-T)-methyltransferase in mouse cells and tissues. Studies with a mechanism-based probe. J Mol Biol 270:385–395

Part III

Development of Neuroendocrine Circuits

Development of Limbic System Stress-Threat Circuitry

12

Newton S. Canteras, Dayu Lin, and Joshua G. Corbin

Abstract

How an animal responds to environmental stressors and threats is essential for survival. These responses are governed through an interconnected circuit in the brain dubbed the limbic system. Three main structures of the limbic system are the amygdala, bed nucleus of the stria terminalis (BNST), and hypothalamus. Together these structures define a central stress-threat circuit. This chapter describes the most up to date knowledge of how these structures function in regulating responses to stressors and how each of these structures is formed from embryonic development. This knowledge of the underlying biology of these regions is essential to design rational therapeutic approaches for conditions, such as post-traumatic stress disorder (PTSD) that are characterized by dysregulation of the stress-threat system. In this chapter, we also describe the critical unanswered questions in the field of stress-threat research and potential future research directions.

N. S. Canteras
Department of Anatomy, Institute of Biomedical Sciences, University of São Paulo; São Paulo, SP, Butanta, Brazil

D. Lin
Neuroscience Institute, New York University School of Medicine, New York, NY, USA

Department of Psychiatry, New York University School of Medicine, New York, NY, USA

Center for Neural Science, New York University, New York, NY, USA

J. G. Corbin (✉)
Center for Neuroscience Research, Department of Pediatrics, Children's Research Institute, Children's National Medical Center, Washington, DC, USA
e-mail: JCorbin@cnmcresearch.org

S. Wray, S. Blackshaw (eds.), *Developmental Neuroendocrinology*, Masterclass in Neuroendocrinology 9, https://doi.org/10.1007/978-3-030-40002-6_12

Keywords
Stress-threat system · Amygdala · Bed nucleus of stria terminalis · Hypothalamus · Embryonic development · Brain circuitry · Defense responses · Loss and gain of function · Mouse models · Circuit manipulation

12.1 Introduction

Research into the neural mechanisms underlying defensive responses in animals has a long history. Animals are naturally selected to protect themselves from threats associated with the presence of a predator or aggressive conspecifics that evoke innate fear. Evidence accumulated over the past two decades suggests that the circuitry that supports fear responses is complex and involves multiple independent circuits that process different types of fear. In particular, there is good evidence to support the existence of distinct circuits for fear of predators and fear of aggressive conspecifics (a member of the same species). In Sect. 12.2 of this chapter, we review these data and consider these segregated fear circuits and point out the crucial role of the amygdala, hypothalamus and bed nuclei of the stria terminalis. In Sect. 12.3 of this chapter, we review the current state of knowledge of how these brain regions are developmentally formed, and in Sect. 12.4 we describe the critical unanswered questions in this field and future directions.

12.2 Key Brain Structures of Fear/Threat Circuits

12.2.1 The Amygdala as a Switchboard for Fear

The mammalian amygdala comprises a heterogeneous set of distinct regions, called nuclei. Accordingly, the amygdalar nuclei have been previously broadly classified into a cortical division (cortical amygdala, basolateral amygdala (BLA), basomedial amygdala (BMA), and lateral amygdala (LA)) and a striatal division (medial amygdala (MEA) and central amygdala (CEA)) (Swanson and Petrovich 1998). However, as more recent studies have revealed a previously unrecognized level of developmental complexity as described in Sect. 12.3 of this chapter, this original simple "striatal/cortical" classification scheme has been updated.

Neural activity mapping and lesion studies in rodents show that different types of threat stimuli activate and depend on different parts of the amygdala (Fig. 12.1). In rodents, olfactory and vomeronasal cues have a crucial role in signaling the presence of a predator. These signals are conveyed through direct connections from the main and accessory olfactory bulb, respectively, to the MEA. Predator odors are thought to work as kairomones, which are semiochemicals that are released by one species and have a favorable adaptive effect on a different "receiving" species but no favorable effect on the transmitting species (Wyatt 2014). A number of studies have shown that the detection of predator odors relies on distinct olfactory

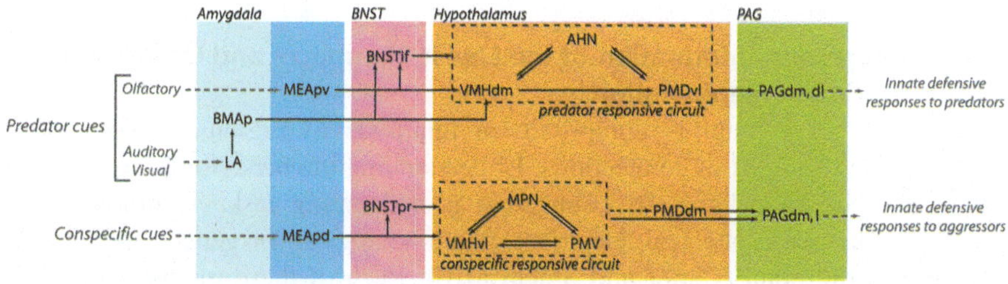

Fig. 12.1 Parallel circuits mediating innate responses to predators and aggressive conspecifics. See text for details. Abbreviations: *BMAp* basomedial amygdalar nucleus, posterior part, *BNSTif, pr* bed nuclei stria terminalis, interfascicular nucleus, principal nucleus; *LA* latereal amygdalar nucleus, *MEApv, pd* medial amygdalar nucleus, posteroventral and posterodorsal parts, *MPN* medial preoptic nucleus; PAGdm, *dl, l* periaqueductal gray, dorsomedial, dorsolateral and lateral parts, *PMDdm, vl* dorsal premammillary nucleus, dorsomedial and ventrolateral parts, *PMV* ventral preamammillary nucleus; VMHdm, *vl* ventromedial hypothalamic nucleus, dorsomedial and ventrolateral parts

subsystems for chemodetection, namely the vomeronasal organ (McGregor et al. 2004), the Grueneberg ganglion (Brechbühl et al. 2013), and subsets of sensory neurons within the main olfactory epithelium that express trace amine-associated receptors (TAARs) (Liberles 2015). Nasal detection of different predator odors has been associated with distinct olfactory subsystems, i.e., 2-phenylethylamine, which is found in carnivore urine, activates TAAR4 neurons in the main olfactory epithelium (Ferrero et al. 2011; Dewan et al. 2013), cat fur odor activates the vomeronasal organ (McGregor et al. 2004), and 2-propylthietane, which is extracted from the stoat (also known as the short-tailed weasel) anal gland, is a Grueneberg ganglion activator (see Pérez Gómez et al. 2015). Exposure to all of these different types of predator odors results in increased cFos expression in the posteroventral part of the medial amygdalar nucleus (MEApv; see Pérez-Gómez et al. 2015). Notably, rats with lesions in the MEA show a substantial reduction in innate fear responses to a cat or its odor. Exposure to a live predator also activates two other amygdala nuclei, the LA and posterior BMA (BMAp) (Martinez et al. 2011). These nuclei receive inputs from visual and auditory association areas and are likely to integrate non-olfactory predator-derived sensory cues (see Martinez et al. 2011) (Fig. 12.1). Consistent with this view, lesions in these nuclei also reduce innate fear responses to a cat (Martinez et al. 2011). Olfactory and vomeronasal inputs to MEA also have a role in the detection of conspecifics. Exposure to an aggressive conspecific predominantly activates the posterodorsal part of the MEA (MEApd; Motta et al. 2009-Key Reference), suggesting that the processing of olfactory threat cues that associate with predators and conspecifics is likely to depend on different MEA subnuclei (Fig. 12.1). Interestingly, the MEApd (responsive to an aggressive conspecific) and the MEApv (responsive to predator odor) are marked by different LIM homeodomain-containing transcription factors (Lhx6 in the MEApd and Lhx9 in the MEApv), which may form the basis for hardwiring systems which are differentially responsive to social and predatory cues (Choi et al. 2005-Key Reference).

Box 1 Interesting Information of the Latest Pharmaco- and Optogenetic Tools for Circuit Manipulation

The mammalian brain is composed on hundreds of regions, billions of neurons, and trillions of connections. Understanding the behavioral relevance of any group of cells or its connection is a daunting task. Traditionally, neuroscientists use physical, electrical, pharmacological, or genetic methods to perturb the brain (Carter and Shieh 2010). Each of these methods suffer from various drawbacks, including low spatial and temporal control and poor cell type selectivity. To achieve more precise control of the cells and their connections, novel chemogenetic and optogenetic tools were developed over the last decade by taking inspiration from nature and performing creative molecular engineering. Since the debut of optogenetic and chemogenetic probes over 10 years ago, these tools have revolutionized the field of neuroscience (Luo et al. 2018-Key Reference). They have enabled neuroscientists to investigate and dissect the neural circuits in a complex brain environment and gained incredible new insight into the neural generation of complex behaviors, such as defense.

Pharmacogenetics: As indicated by the names, pharmaco-(also known as chemo-) and opto-genetic tools use chemicals and light, respectively, to control the activity of genetically defined neuronal populations or connections. The most widely used chemogenetic probes are known as DREADDs, Designer Receptors Exclusive Activated by Designer Drugs (Roth 2016). DREADDs are virtually mutated G-protein coupled receptors (GPCRs) that no longer bind to their endogenous ligands but bind small synthetic chemicals. Depending on the secondary signaling pathways that are coupled to the DREADDs, activating the DREADDs could either reduce cell excitability by opening a GIRK channel (Gi-DREADD) or activate the cells by increasing Ca2+ concentration (Gq-DREADD) or cAMP levels (Gs-DREADD). The most commonly used excitatory DREADDs are based upon Gq-coupled human M3 muscarinic receptor (hM3Dq) while Gi-coupled human M4 muscarinic receptor (hM4Di) based DREADD is the most commonly used inhibitory type (Armbruster and Roth 2005; Armbruster et al. 2007). Both hM3Dq and hM4Di were generated via rounds of mutagenesis and screening for decreased affinity to acetylcholine (the natural ligand of muscarinic receptors) as well as increased affinity to clozapine-N-oxide (CNO), a biologically inert small molecule (Armbruster and Roth 2005: Armbruster et al. 2007). In addition to DREADDs, several other forms of ligand-controlled GPCR based receptors have also been developed, allowing orthogonal controls of multiple populations in the same animals (Coward et al. 1998; Vardy et al. 2015; Magnus et al. 2019).

Optogenetics: Nearly all optogenetic probes have been engineered from natural opsins that selectively respond to specific wavelengths of light and

(continued)

Box 1 Interesting Information of the Latest Pharmaco- and Optogenetic Tools for Circuit Manipulation (continued)

translocate ions across the membrane (Fenno et al. 2011). The best known optogenetic probe is channelrhodopsin-2 (ChR2), a nonspecific cation channel found in green algae (Boyden et al. 2005). Upon sensing blue light, ChR2 undergoes a conformational change, opening the pore, and allowing H+, Na+, K+ and Ca2+ ions to passively diffuse across the membrane and depolarize the cell. When sufficient cation ions influx through ChR2, a spike will be generated by activating sodium channels. The conformational change of ChR2 depends on the presence of retinal which happens to be present in sufficient amount in most vertebrate cells in the form of vitamin A, negating the need to supply it exogenously (Hegemann and Nagel 2013). The first opsin developed for inhibiting neural activity was halorhodopsin (NpHR), a light driven Cl-pump that is expressed by the halobacterium (Schobert and Lanyi 1982; Gradinaru et al. 2010). Upon sensing yellow light, NpHR actively pumps Cl-ions into the cells causing hyperpolarization of the membrane potential. A second class of popular inhibitory opsin is archaerhodopsin-3 (Arch), a light driven H+ pump that is found in *Halorubrum sodomense* (Chow et al. 2010). Most recently, a third class of inhibitory opsin was developed by transforming channelrhodopsin into a light-activated chloride channel (Berndt et al. 2014). To date, a wide range of optogenetic probes with diverse properties (e.g. excitation wavelength and open and close kinetics) are available, allowing neuroscientists to control neural activity with unprecedented precision.

Delivery methods: The most common way to deliver the chemogenetic or optogenetic probes is to infect cells with a replication incompetent virus, such as adeno-associated virus (AAV) that carries the transgene driven by a short promoter (Luo et al. 2008). To achieve the cell type specificity, conditional AAV vectors that carry a Cre-dependent transgene is delivered to the transgenic animals that express Cre in a specific subset of cells. The specificity can be further improved based on the connectivity. Retrograde viruses, such as herpes simplex virus or retrograde AAV, can be delivered to the downstream regions of the cells of interest so that only Cre-expressing cells with certain projection patterns are infected by the virus and express the transgene (Neve et al. 2005; Tervo et al. 2016). Additionally, several transgenic lines that express optogenetic or chemogenetic probes, either conditionally or constitutively, have been developed to achieve consistent manipulation across animals and eliminate the need for viral injection (Madisen et al. 2012; Zhu et al. 2016).

One advantage of optogenetic and chemogenetic probes is that they are controlled by specific ligands—light or small molecule. As a downside of this controllability, it requires the experimenter to deliver the ligand exogenously. Small molecules, such as CNO, are typically delivered through i.p. injection

(continued)

Box 1 Interesting Information of the Latest Pharmaco- and Optogenetic Tools for Circuit Manipulation (continued)
although recent reports suggest that CNO has poor brain-barrier penetrability and is likely metabolized to clozapine before entering the central nervous system (Gomez et al. 2017). Because DREADDs are expressed in both cell bodies and axon terminals, CNO can also be administrated directly to the regions where the axon terminals reside to block synaptic transmission of a specific projection (Stachniak et al. 2014). Light delivery is generally achieved by implantation of a light guide deep into the brain close to the target region. This is because mammalian brain tissue scatters light exponentially, with only 10% of light intensity remaining at a distance 500 μm from the light source (Aravanis et al. 2007). During an experiment, an optic fiber is connected to the implanted light guide to deliver appropriate wavelength light to the target region. The light can be delivered to either cell bodies or specific projections to achieve pathway-specific manipulation (Tye and Deisseroth 2012).

12.2.2 Parallel Hypothalamic Paths for Fear

In the hypothalamus, it has been recognized that behavioral control columns, formed by distinct hypothalamic circuits, are involved in the control of the three basic classes of goal-oriented behaviors: ingestive, reproductive, and anti-predatory defensive (Swanson 2000). This concept was mostly based on a systematic analysis of axonal projections from the medial half of the hypothalamus of the rat and was further corroborated by a number of functional studies (see Swanson 2000). Based on anatomical tract-tracing experiments of medial hypothalamic zone projections, this zone has been divided into two networks that show a high degree of interconnection among nuclei (*Fig. 12.1, and see Box 1 for description of modern tools now being employed to understand the link between circuity and behavior*). One network comprises the anterior hypothalamic nucleus (AHN), the dorsomedial part of the ventromedial nucleus (VMHdm), and the ventrolateral part of the dorsal premammillary nucleus (PMDvl), which are highly interconnected (Canteras and Swanson 1992; Canteras et al. 1994; Risold et al. 1994) (Fig. 12.1). Exposure to a predator or its odor activates the AHN–VMHdm–PMDvl circuit (i.e., the predator-responsive circuit; Fig. 12.1) (Cezario et al. 2008). Notably, the VMHdm and the AHN provide a unique type of projection to the PMdvl, with dense bilateral projection fields (Canteras et al. 1994; Risold et al. 1994). This type of synaptic arrangement enables the PMdvl to work as an amplifier for the hypothalamic neural processing of predator-related cues. Corroborating this view, the PMDvl is the most responsive brain site during exposure to a cat or its odor, and lesions or pharmacological blockade of the PMD drastically reduces fear responses to a predator or its odor (see Cezario et al. 2008). The VMHdm, in turn, receives most of the amygdalar processing of predator-related cues from the MEApv (Canteras et al. 1995) and the

BMAp (Petrovich et al. 1996) (Fig. 12.1). In this regard, of particular relevance is the finding that exposure to different types of carnivore odors converge in a pathway formed by the MEApv and VMHdm (Pérez Gómez et al. 2015). Pharmacogenetic inhibition in transgenic mice expressing hM4D in steroidogenic factor 1 (SF1)-expressing neurons in the central and dorsomedial parts of the VMH decreased the defensive responses of the mice during exposure to a rat but not during exposure to an aggressive conspecific (Silva et al. 2013). Conversely, optogenetic activation of SF1-expressing neurons in the same population of VMH cells initiates a range of behavioral responses that resemble the animal's natural defensive behaviors (Wang et al. 2015). The main targets of the VMHdm are the dorsal periaqueductal gray (PAGd) and the AHN. Accordingly, activation of the pathway from the VMHdm to the PAG induces immobility, whereas stimulation of the pathway from the VMHdm to the AHN promotes mobility, running, escape jumping, and avoidance and the animals tend to avoid the place where they had received the stimulation (Wang et al. 2015). At least part of the complex defensive behaviors that have been described for the AHN are likely to be mediated through its projections to the PMDvl. The PMDvl is the main interface between the predator-responsive circuit and the PAG (Motta et al. 2009-Key Reference). The PMDvl provides strong projections to the dorsomedial and dorsolateral parts of the PAG, both of which are activated during exposure to a live predator or its odor and is an important site for the organization of fear responses to predators (Motta et al. 2009-Key Reference) (Fig. 12.1). Notably, lesions in the dorsal PAG block the entire range of fear responses to a predator, from flight and freezing to risk assessment, and is conceivable that the degree of activation of the dorsal PAG dictates the intensity of the fear response (see Cezario et al. 2008).

The second medial hypothalamic network includes the medial preoptic nucleus (MPN), the ventrolateral part of the VMH (VMHvl), the tuberal nucleus, and the ventral premammillary nucleus (PMV), which are also highly interconnected (Simerly and Swanson 1988; Canteras et al. 1992, 1994). In contrast to predator exposure, exposure to an aggressive conspecific activates the MPN–VMHvl–PMV network (the conspecific-responsive circuit; Fig. 12.1), clearly suggesting the existence of separate circuits responsive to predators and conspecifics (see Motta et al. 2009-Key Reference). In addition, exposure to an aggressive conspecific activates the dorsomedial part of the PMD (PMDdm), which is distinct from the predator-activated PMDvl (Fig. 12.1). Of particular relevance, lesions of the PMD block freezing and passive supine responses to an aggressive conspecific but not, however, active defense, including boxing and upright defensive postures (Motta et al. 2009-Key Reference). These data suggest that separate circuits in the PMD are responsible for fear responses to predators versus fear responses to aggressive conspecifics. The link between elements of the conspecific-responsive hypothalamic circuit and the PMDdm is not clear at the moment, and the subfornical region of the lateral hypothalamic area has been suggested one possible candidate (Goto et al. 2005). As discussed for the predator-responsive medial hypothalamic circuit, the PMD serves as an important interface between the conspecific-responsive hypothalamic circuit and the PAG. However, differently from the PMDvl, the PMDdm projects to the dorsomedial PAG (PAGdm) and lateral PAG (PAGl) (Fig. 12.1), providing a

Table 12.1 Summary of gain and loss of functional studies of the defense circuit

Area (cell type)	Manipulation	Behavioral change	References
VMHdm (SF1+ cells)	ChR2 activation	Freeze, flight, jump	Wang et al. (2015) Kunwar et al. (2015)
VMHdm(SF1+ cells)	DREADDi inactivation	Reduced defense toward a predator	Silva et al. (2013)
VMHdm(SF1+ cells)	Capase3 mediated cell ablation	Reduced predator avoidance	Kunwar et al. (2015)
VMHdm (SF1)- > AHN	ChR2 activation	Freeze, flight, jump	Wang et al. (2015)
VMHdm (SF1)- > PAG	ChR2 activation	Freeze	Wang et al. (2015)
AHN (nonselective)	ChR2 activation	Flight and jump	Wang et al. (2015)
VMHvl (nonselective)	DREADDi inactivation	Reduced social defense	Silva et al. (2013)
VMHvl (Esr1+)	ChR2 activation	Increase social defense	Wang et al. (2019)
VMHvl (Esr1+)	NpHR3.0 inactivation	Decrease active social defense	Wang et al. (2019)
PMd (nonselective)	Cytotoxic lesion	Decrease passive social defense	Motta et al. (2009)-Key Reference
PAGdm (nonselective)	Cytotoxic lesion	Decrease passive and active social defense	Canteras lab, unpublished results

projection pattern that matches the PAG activation pattern in response to an aggressive conspecific (see Motta et al. 2009-Key Reference). Preliminary findings from the Canteras laboratory suggest that PAGdm cytotoxic lesions decrease both passive (freezing and the typical on-the-back position maintained after the resident leaves them alone) and active forms (an upright position with sparse boxing and dashing away from the resident) of social defensive responses. In addition to the PMD, pharmacogenetic inhibition in the VMHvl of mice locally infected with an adeno-associated virus (AAV) expressing hM4D pharmacogenetic neural inhibition tool (HA-hM4D) decreased the defensive responses of the mice during exposure to an aggressive conspecific but not during exposure to a rat (Silva et al. 2013). Moreover, estrogen receptor+ ($Esr1^+$) cells situated at the anterior portion of the VMHvl are significantly excited when animals actively defended themselves against conspecific aggressors. Furthermore, during exposure to a conspecific aggressor, inactivation of $VMHvl^{Esr1}$ cells compromised the active forms of social defensive behaviors, such as dashing away from the aggressor and assuming an upright posture (Wang et al. 2019). A summary of functional studies implicating the VMH in the defensive behaviors is shown in Table 12.1.

12.2.3 The BNST and Social and Anti-Predatory Defense

The BNST serves as an important interface between the amygdala and the hypothalamic circuits involved in social and anti-predatory defense (Fig. 12.1). However, the role the BNST plays in the context of social and anti-predatory defense remains to be fully investigated. Previous research has established the role of the BNST in sustained fear as distinct from the role of the amygdala in phasic fear (see Lebow and Chen 2016; Davis et al. 2010). Accordingly, the amygdala is likely to be generically involved in the assessment of specific cues related to fearful events, whereas the BNST would be related to the ability to differentiate between safe and non-safe situations (Sullivan et al. 2004), causing a sustained state of apprehension (see Lebow and Chen 2016; Davis et al. 2010). On hodological grounds, the elements of the posterior division of the BNST fit quite well with the parallel circuits organizing anti-predatory and social defense. Thus, in the posterior BNST, the principal nucleus is particularly targeted by the MEApd (Canteras et al. 1995), which, as previously mentioned, responds to aggressive conspecifics, and projects to elements of the conspecific-responsive hypothalamic circuit the MPN, VMHvl, tuberal nucleus, and PMV (Dong and Swanson 2004) (Fig. 12.1). Conversely, the interfascicular nucleus of the posterior division of the BNST (BNSTif) receives amygdalar inputs from regions processing predator cues, namely the MEApd and BMAp (Canteras et al. 1995; Petrovich et al. 1996) and projects to elements of the predator-responsive hypothalamic circuit, including the AHN, the VMHdm and, to a lesser degree, the PMD (Dong and Swanson 2004) (Fig. 12.1). Moreover, the interfascicular and principal nuclei of the BNST receive strong inputs from elements conspecific- and predator-responsive medial hypothalamic circuits (Simerly and Swanson 1988; Canteras et al. 1992, 1994; Risold et al. 1994), suggesting that the BNST has more integrative functions that should be investigated in the context of social and anti-predatory defense.

12.3 Genetic Regulation of Stress/Threat System Development

As described above many different brain structures and nuclei comprise the stress/threat system. In this section we cover what is currently understood regarding development of three central brain regions of this system: (1) the ventromedial hypothalamic (VMH), paraventricular (PVN), and premammilary (PMN) nuclei of the hypothalamus, (2) the MEA, and (3) the BNST. These brain structures are highly tuned to detecting and processing environmental threats, and as such make up a core interconnected circuit regulating stress/threat responses. While the knowledge of development of each system remains incomplete and lags behind that of much more studied brain regions such as the cerebral cortex, a picture is emerging of how each structure is assembled during embryogenesis and how these developmental programs may relate to how they function. Below we describe development of each brain structure individually (Sect. 12.3.1–12.3.3), followed by what is known about how connectivity is established (Sect. 12.3.4), and conclude with data and

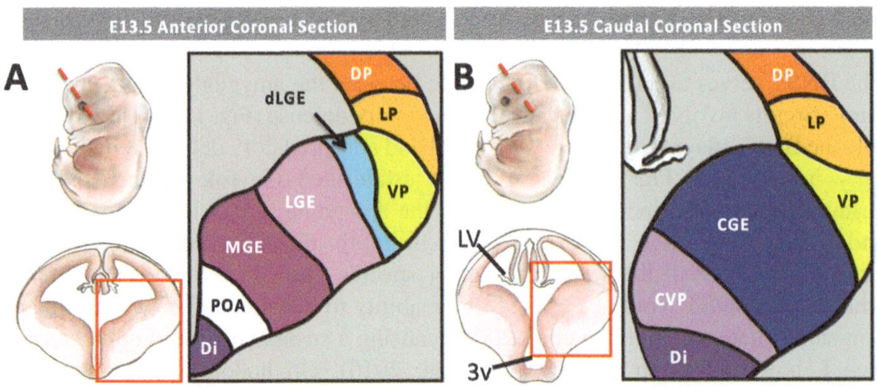

C

	DP	LP	VP	dLGE	LGE	MGE	POA	CGE	CVP	Di
MEA	*Emx1*	*Emx1*		*Tshz1*		*Nkx2.1* *Lhx6*	*Dbx1* *FoxP2*	*Coup-TFII*	*Ebf3*	*Foxb1* *Otp* *Lhx5*
BNST			*Pax6* *Dbx1*		*Islet 1* *Dlx5*	*Nkx2.1* *Lhx6*	*Dbx1*	*Coup-TFII*		

Fig. 12.2 BNST and MEA progenitor zones. Schematics of the E13.5 brain at the level of the telencephalon and diencephalon showing both the overall anatomy and regions of defined progenitor zones known to contribute to either BNST or MEA (A & B). Table below in (C) lists the transcription factors expressed in each progenitor zone and contributions to either the BNST or MEA. Abbreviations: *CGE* caudal ganglionic eminence, *CVP* caudal-ventral pallium, *Di* Diencephalon, *dLGE* dorsal lateral ganglionic eminence, *DP* dorsal pallium, *LGE* lateral ganglionic eminence, *LP* lateral pallium, *LV* lateral ventricle, *MGE* medial ganglionic eminence, *POA* preoptic area, *VP* ventral pallium, *3v* third ventricle (Illustrations by Katie Sokolowski)

speculation regarding how these developmental genetic programs may link to later stress/threat behavioral responses in adult animals (Sect. 12.3.5).

12.3.1 Development of the VMH, PVN, and PMN Hypothalamic Nuclei

The major hypothalamic subnuclei that regulate threat avoidance and detection are the VMH, the PVN, and the PMN (Gross and Canteras 2012-Key Reference; Hashikawa et al. 2016-Key Reference; Takahashi 2014) (Fig. 12.1). As development of these regions are covered in greater detail in the chapters by M. Placzek and M Towers (Development of the Neuroendocrine Hypothalamus), and J. Michaud (Development of the Paraventricular Nucleus), here we provide a general overview, as readers should refer to the above chapters as well as outstanding current reviews (Bedont et al. 2015; Alvarez-Bolado 2018; Xie and Dorsky 2017) for more details.

The vast majority of neurons in the hypothalamus are generated during embryonic development between embryonic day (E)10.5 and E16.5 in mouse (Shimada

Table 12.2 Key Transcription Factors in Hypothalamic Stress/Threat System Development

Nucleus	Key transcription factors	Neuronal cell types
VMH	*Rax, Ascl, Ngn3, Nkx2.1, Fezfl, Nr5al*	ERα+, glutamate+
PVN	*Siml, Arnt2, Otp, Brn2(Pou3f2)*	Avp+, Oxt+, Crh+
PMN	*Nkx2.1*, repression of *En1*	Da+

Abbreviations: *Avp* Arginine vasopressin, *Crh* Corticotropin releasing hormone, *DA* Dopamine, *ERα+* Estrogen receptor-alpha, *Oxt* Oxytocin

and Nakamura 1973 and reviewed in Bedont et al. 2015) (Mouse gestation is approximately 19 days). Hypothalamic progenitor cells are born in the ventricular zone surrounding the ventral aspect of the third ventricle (di on Fig. 12.2). Dividing progenitors, spatially distributed along both dorsal-ventral and rostral-caudal axes of the developing diencephalon, express different combinations of transcription factors (Ferran et al. 2015; Shimogori et al. 2010), master regulator genes that function in a combinatorial manner to determine cell fate. Differential transcription factor expression is one of the early developmental steps in which unspecified neural progenitors progressively take on neuronal subtype identity prior to achieving their mature neuronal phenotype. Table 12.2 summarizes the key transcription factors that comprise the unique "combinatorial code" for development and specification of major subclasses of neurons of the VMH, PVN, and PMN (Bedont et al. 2015; Alvarez-Bolado 2018; Xie and Dorsky 2017; Nouri and Awatramani 2017). These genes are expressed during the proliferative and/or early postmitotic stages of hypothalamic neuron development. It is important to note that Table 12.2 represents an incomplete catalog of genes required for specification of neurons in these nuclei, especially the PMN, as a number of required transcription factors remain to be identified.

12.3.2 Development of the Medial Amygdala (MEA)

At the anatomical level, the MEA can be subdivided into three distinct subdomains: Anterior (MEAA), posterio-dorsal (MePd) and posterior-ventral (MePv). At the neuronal subtype level, the MEA is composed of a mix of three major classes of neurons: local projecting interneurons (Bian 2013; Guirado et al. 2008; Real et al. 2009), and long-range projecting excitatory and inhibitory output neurons (Bian et al. 2008; Choi et al. 2005; Keshavarzi et al. 2014). Single-cell RNA-sequencing studies have revealed that the MEA is populated by at least 16 distinct molecularly identifiable neuronal subtypes (Wu et al. 2017). Much of this diversity is found across interneurons, which based on a combination of previously identified molecular, electrophysiological and morphological criteria, fall into a number of distinct subtypes (Petilla Interneuron Nomenclature Group et al. 2008). In the MEA, these include subtypes that are defined by their expression of either somatostatin (Sst), Calbindin (CB), Calretinin (CR), the serotonin receptor subunit 5Ht3a, or Neuropeptide Y (NPY). Interestingly, the MEA is devoid of interneurons marked by expression of parvalbumin (PV) (Carney et al. 2010; Guirado et al. 2008), a subtype that is found in abundance in the cerebral cortex.

In addition to these local and projection inhibitory neurons, the MEA is comprised of excitatory output neurons. This mixing of inhibitory and excitatory output neurons within the same structure is atypical for telencephalic nuclei as output neurons of telencephalic nuclei are typically either excitatory (e.g., cerebral cortex, basolateral amygdala) or inhibitory (e.g., basal ganglia), but not both. MEA neurons can also be identified based on expression of one or more neuropeptides and neurohormones, factors which mediate the internal state of the animal and appropriate behavioral responses to stressors and threats (Gross and Canteras 2012-Key Reference; Hashikawa et al. 2018; Li and Dulac 2018; Sokolowski and Corbin 2012; Yang and Shah 2014). How expression of these factors relates to the other criteria for defining neuronal identity in the MEA remains an area of active investigation. Although less well characterized than their interneuron counterparts, both inhibitory and excitatory output neurons in the MEA can also be classified according to distinct intrinsic electrophysiological profiles (Bian 2013; Bian et al. 2008; Carney et al. 2010; Keshavarzi et al. 2014).

Box 2 This Box Contains Interesting Information on the Techniques for Fate Mapping Neural Populations

With each technique, cell populations are indelibly marked by a reporter gene (for example, GFP). Fate of cells are tracked either anatomically via transplantation, via electroporation in which a cell-specific subtype is labeled via delivery of a GFP-expression vector or via genetic labeling in transgenic mice (Illustrations by Katie Sokolowski).

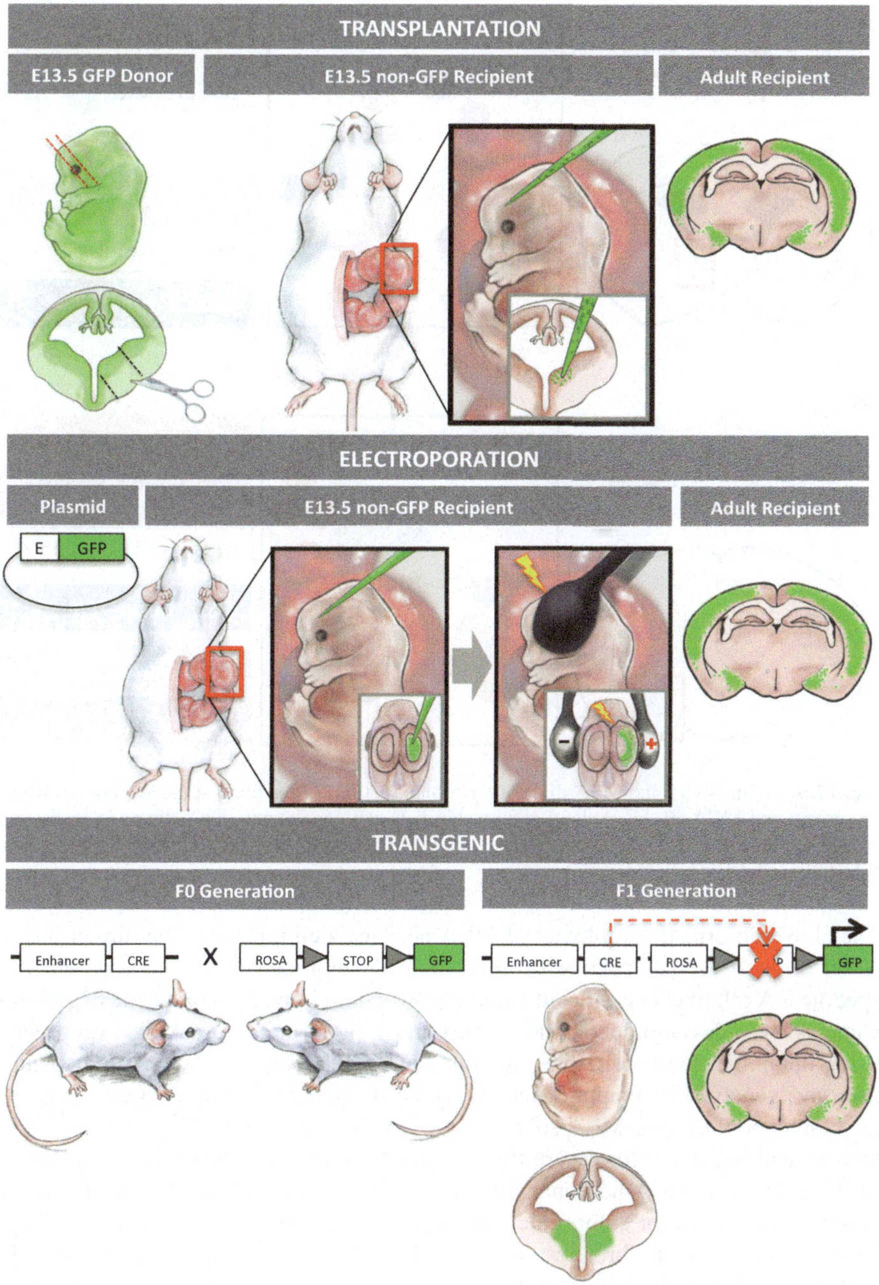
TRANSPLANTATION
E13.5 GFP Donor
E13.5 non-GFP Recipient
Adult Recipient
ELECTROPORATION
Plasmid
E
GFP
E13.5 non-GFP Recipient
Adult Recipient
TRANSGENIC
F0 Generation
F1 Generation
Enhancer
CRE
X
ROSA
STOP
GFP
Enhancer
CRE
ROSA
GFP

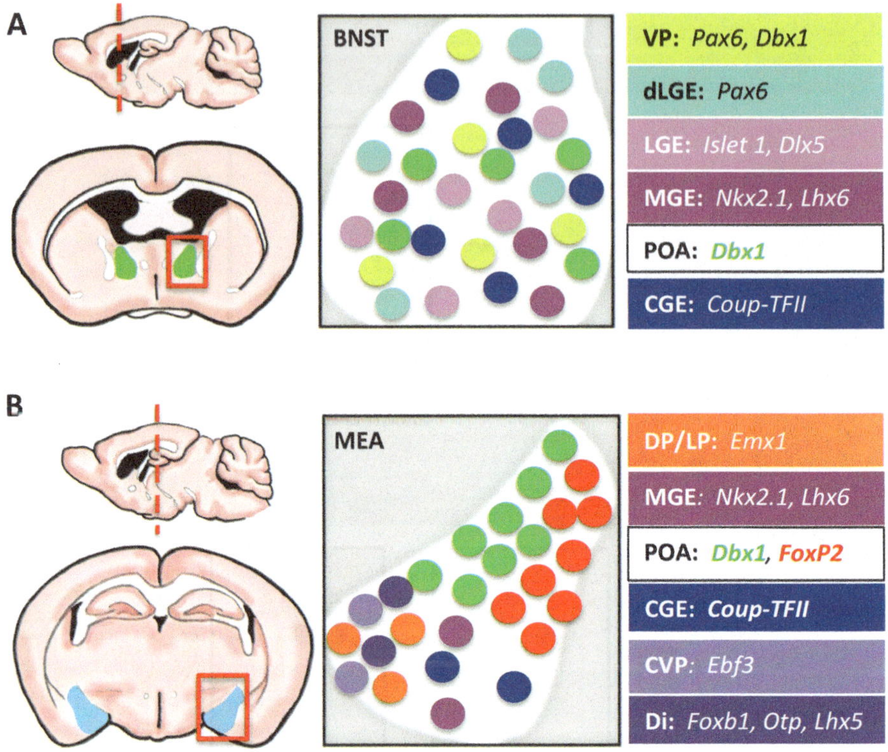

Fig. 12.3 Adult BNST and MEA diversity. Coronal sections of the brain at the level of the BNST (A, green) and MEA (B, blue) show relationship between gene expression patterns in embryonic progenitor domains and neuronal diversity in each structure. See Fig. 12.1 for abbreviations (Illustrations by Katie Sokolowski)

The above-described complexity of MEA neuronal cell types may be further understood through the perspective of embryonic development when neurons are first specified. A relatively significant amount of work studying dynamic gene expression changes over developmental time, short-and long-term cell tracking experiments using lipophilic fluorescent dyes, in vivo neuronal labeling, embryonic transplantation and genetic fate mapping has provided a relatively comprehensive map for development of diverse MEA subpopulations (Fig. 12.3, Table 12.3, and references therein, and see Box 2 for description of tracing techniques). From these studies, ~9 of the at least 16 MEA neuronal subclasses can be identified by differential developmental origin and/or embryonic transcription factor expression (Table 12.3, Figs. 12.2 and 12.3). At the broadest anatomical level, the three major MEA subnuclei (MEAA, MePD, and MePV) can be individually identified in the postnatal brain by the differential expression of members of the LIM family of homeodomain encoding transcription factors, *Lhx5*, *Lhx6*, and *Lhx9* (Choi et al. 2005-Key Reference; García-López et al. 2008). The expression of these genes is atypical of developmentally

Table 12.3 Origins of MEA neuronal diversity

Embryonic Source	Known TF markers	Mature neuronal subtype	References
MGE*	*Nkx2.1*, *Lhx6*	Sst + and Calbindin+ interneurons	Bupesh et al. (2011); Carney et al. (2010); Puelles et al. (2016)
CGE	*Coup-TFII*	Inhibitory neurons (likely including Calretinin+ and 5HT3a + interneurons)	Kanatani et al. (2015); Nery et al. (2002); Soma et al. (2009); Touzot et al. (2016)
POA	*Dbx1*	nNos+, Aromatase+ inhibitory output neurons. A subset express ERα and AR	Hirata et al. (2009); Lischinsky et al. (2017)
POA?/ Anterior diencephalon?	*Foxp2*	Inhibitory output neurons (distinct subclass from *Dbx1*-derived). A subset express ERα and to a lesser extent, AR.	Carney et al. (2010); Lischinsky et al. (2017)
POA	*Nkx5.1*	Npy + interneurons	Gelman et al. (2011)
dLGE	*Tshz1*	Presumptive inhibitory interneurons	Kuerbitz et al. (2018)
Diencephalon/ Hypothalamus	*Foxb1*	Unknown subclass of inhibitory neurons	Zhao et al. (2008)
Lateral and dorsal pallium (cerebral cortical primordium)	*Emx1*	Excitatory neurons	Cocas et al. (2009); Gorski et al. (2002)
Caudo-ventral pallium (CVP)	*Ebf3*	Excitatory neurons	Ruiz-Reig et al. (2018)
Diencephalon/ Hypothalamus	*Otp*, *Lhx5*	Excitatory neurons	Bupesh et al. (2011); Soma et al. (2009); Zhao et al. (2008)

*It has also been proposed that the source of Sst + interneurons is part of a separate subdomain in the diagonal area of the telencephalon (Dg), which has also been previously referred to alternatively as the anterior endopeduncular area (AEP), caudal or caudo-ventral MGE (cMGE, cvMGE) (Bupesh et al. 2011, 2011; Puelles et al. 2000). For simplicity we refer to this region as part of the MGE.

expressed transcription factors as their expression persists through postnatal stages, thus providing a putative fate map from embryogenesis to adult stages (with the major caveat that gene expression patterns might not be fixed within the same population over time). To overcome this inherent limitation of using gene expression studies as a proxy for fate mapping, a number of studies have employed more sophisticated genetic and nongenetic tools to follow MEA neuronal fate over time (see Box 2).

Embryonic gene expression studies have revealed that in addition to the LIM-containing transcription factor encoding genes described above, emerging MEA nuclei can also be identified by expression of a variety of other transcription factor encoding genes such as *Tbr1*, *Foxp2*, *Nkx2.1*, and *Otp* (Table 12.3, and references therein). Previous fate mapping studies using Cre-based approaches (see Box 2) have begun to uncover the relationship between neuronal progenitor cell identity as defined by transcription factor expression and ultimate MEA neuronal identity. While a full mapping of MEA neuronal diversity remains to be accomplished and subsequently linked back to developmental origin, there are clear patterns that have begun to emerge. First, fate mapping work, along with previous standard birthdating studies, revealed that MEA neuronal specification occurs within a similar epoch as the hypothalamus, roughly between E10.5 and E15.5 in the mouse (Soma et al. 2009). Second, the same ventral telencephalic domains (MGE and CGE), which generate interneurons of the cerebral cortex (Batista-Brito and Fishell 2009; Xu et al. 2003), are also major sources of interneuron populations for the MEA (Figs. 12.2, 12.3 and Table 12.3, and references therein). For example, the CGE is the source of calretinin+ neurons and the MGE is the source of Sst + and Calbindin+ interneurons in the MEA (Kanatani et al. 2015; Nery et al. 2002; Soma et al. 2009; Touzot et al. 2016). Moreover, similar to the cerebral cortex, a population of MEA excitatory neurons appears to be derived from Emx1+ progenitors originating in lateral and dorsal pallial compartments of the cerebral cortical primordium (Gorski et al. 2002) (Figs. 12.2, 12.3 and Table 12.3). In contrast to this "shared" origin with other telencephalic brain regions for populations of MEA interneurons and excitatory neurons, there appear to be multiple unique sources of origin for MEA neuronal diversity. These include progenitor pools located embryonic zones that straddle both sides of the telencephalic–diencephalic border. The telencephalic niches include the preoptic area (POA) (Bulfone et al. 1993) (not be confused with the hypothalamic preoptic area, which by all evidence does not appear to be derived from the embryonic POA), and likely a region known as the entopeduncular region (AEP) (Fig. 12.2). These niches are located ventral to the MGE and CGE, respectively, and are the ventral-most telencephalic domains. Genetic fate mapping studies have revealed contributions of Nkx5.1+ POA-derived neurons to the MEA (Although their adult identity was not established) (Gelman et al. 2011). In addition, more in depth fate mapping studies have revealed Dbx1+ and Foxp2+ contributions to the MEA from domains at the telencephalic–diencephalic boundary. As defined by molecular and electrophysiological profiles, *Dbx1*- and *Foxp2*-derived populations generate distinct subclasses of putative inhibitory output neurons that are distributed across different subnuclei of the MEA (Lischinsky et al. 2017). These developmentally defined subpopulations also contribute to known neuroendocrine-expressing populations such as those that express Aromatase, Androgen Receptor (AR), and Estrogen receptor-α (ERα) (Table 12.3). Additionally, as one moves caudally in the telencephalon there is a distinct compartment where the pallium appears to wrap underneath the CGE, dubbed the caudo-ventral pallium (CVP) (Ruiz-Reig et al. 2018) (Fig. 12.2). The CVP expresses the transcription factor encoding gene, *Ebf3*, and the secreted factor encoding genes, *Gdf10* and *Fgf15*. These populations contribute to the MEA and appear to generate a subset of excitatory neurons. This segregation of excitatory and inhibitory sources for MEA neuronal diversity follows the current dogma that the subpallial telencephalon is the exclusive (or almost exclusive) source of telencephalic inhibitory neurons (both local interneurons and long-range projecting neurons) and the pallium is the source of excitatory neurons. Interestingly, and perhaps unique for telencephalic structures, the MEA also appears to draw neurons from the more distant brain regions such as the diencephalon; the hypothalamic, and thalamic primordium. Multiple studies have uncovered waves of long-range migration and settling of diencephalic-derived populations in the emerging MEA. These populations, which appear to arise primarily from the lateral/parventricular and/or caudal aspects of the hypothalamus, cross the telencephalic–diencephalic border to reach the MEA (Bupesh et al. 2011; García-Moreno et al. 2010; Soma et al. 2009; Zhao et al. 2008). Diencephalic-derived progenitors express in combination or separately, the transcription factor encoding genes, *Otp*, *Lhx5,* or *Foxb1* (Fig. 12.2 and Table 12.3). Those expressing *Otp* or *Lhx5* appear to generate a subset of excitatory neurons of the MEA. Although not yet formally established, based on their site of origin and embryonic molecular gene expression patterns these neurons appear to be a distinct population from the dorsal telencephalic *Emx1*-derived or CVP-originating excitatory neurons. These genetic and site of origin differences suggest that excitatory neurons of the MEA, similar to inhibitory neurons, are likely a highly diverse group composed of subtypes with different projection patterns, transcriptome profiles, and biophysical properties.

In summary, work from a number of studies described above has revealed that the MEA is composed of a "patchwork" of neurons generated from spatially distinct embryonic niches derived from: (1) common sources that generate interneurons (MGE, CGE) or excitatory neurons (the dorsal telencephalon) destined for the MEA and other brain regions such as the striatum, cerebral cortex, and hippocampus and (2) unique niches highly dedicated to the MEA such as the POA and specific hypothalamic progenitor zones (Figs. 12.2, 12.3 and Table 12.3). The contribution from numerous different niches likely results in the vast diversity of functionally distinct neuronal subtypes that populate the MEA, many of which express different neuropeptides and neurohormones. In addition, neural progenitors that respond to the secreted patterning factor, Sonic Hedgehog (Shh), differentiate into a sizable portion of MEA neurons. This likely is reflective of the widespread and earlier patterning function of Shh in ventralization across the neuraxis (Ericson et al. 1995; Matise and Wang 2011). In addition, there have been a handful of studies examining the function of a subset of genes, such as *Nkx2.1*, *Otp*, and *CoupTF-II*, known to mark MEA progenitor populations (Table 12.2, and references therein). These genes are individually required for generation and/or migration of subsets of MEA interneurons (*Nkx2.1*, *CoupTF-II*) or excitatory neurons (*Otp*) (Table 12.4). Thus, perhaps not surprisingly, genes that mark distinct MEA progenitor subclasses also appear to be required for their proper development.

12.3.3 Development of the BNST

Located within the ventral telencephalon, the BNST, sometimes also referred to as "the extended amygdala" is bounded by the globus pallidus and traverses from the

Table 12.4 Genes known to function in MEA development

Gene	Gene function	References
Otp	Promotes migration of *Lhx5*+ presumptive excitatory neurons from PVH to MeA	García-Moreno et al. (2010)
Nkx2.1	Specification of subset of MePD *Sst* + neurons and regulation of *Lhx9* and *Shh* expression	Carney et al. (2010)
Coup-TFII	Via regulation of Nrp2 expression serves as a molecular switch regulating MeA versus cerebral cortical directed migration of a POA-derived presumptive local inhibitory neurons	Kanatani et al. (2015); Tang et al. (2012)

septal area to anterior to the hypothalamus (Figs. 12.2 and 12.3). It is composed of up to 18 distinct subnuclei which are broadly grouped as either part of the anterior or posterior BNST (Lebow and Chen 2016-Key Reference). As revealed largely from the work of Larry Swanson and collaborators, the BNST sends and receives projections from a variety of brain regions, most notably major subnuclei of the amygdala and hypothalamus (Dong et al. 2001; Thompson and Swanson 1998). At the neuronal level, the BNST expresses a large variety of neurohormones and neuropeptides, which points to its central role in modulation of mood, anxiety, and stress responses. Unfortunately, little work has been undertaken to understand how this complex and important structure arises, thus BNST development remains a critical understudied area. However, from a handful of studies, some insight has been gained. First, classical Brdu birthdating studies from Shirley Bayer, a pioneer along with Joseph Altman in the use of methodologies to birthdate neurons, showed that cells that will populate the BNST are born across a rostral to caudal gradient (rostral earlier) between E13 and E20 in the rat (Bayer 1987). Subsequent studies by George De Vries have revealed distinct temporal windows within this period for the sexual dimorphic development of vasopressin (Avp) + and Galanin (Gal+) cells during the earlier waves of BNST generation (Han and De Vries 1999; al-Shamma and De Vries 1996). Following Brdu labeling over time the work from Bayer (Bayer 1987) suggested that BNST cells arise from the medial edge of the MGE near the septum, and a more dorsal and posterior source located at the approximate level of the CGE (Note while the term "ganglionic eminences" were first used in the 1970s, it was not until the early 1990s that the terms MGE, LGE, and CGE were in use). Following these early studies, more recent gene expression studies carried out primarily from the laboratories of Loreta Medina and Luis Puelles have extended our understanding of BNST development (Bupesh et al. 2011; García-López et al. 2008; Puelles et al. 2000). This work, combined with work of others (Hirata et al. 2009; Nery et al. 2002; Ruiz-Reig et al. 2018; Touzot et al. 2016; Vucurovic et al. 2010), indicates that patterning of the BNST follows similar rules as MEA development with multiple sources contributing to the BNST (Table 12.5, and references therein). These sources likely include the CGE, POA, MGE (AEP), LGE, and perhaps the ventral pallium (VP) (Figs. 12.2, 12.3 and Table 12.5). However, considering the wide variety of neuronal cell types that populate the BNST, there are likely other sources including the dorsal telencephalon as well as extratelencephalic sources (Bupesh et al. 2011).

Table 12.5 Known Origins of BNST Neuronal Diversity

Embryonic origin	Cell types/markers	References
CGE	5HT3a + presumptive interneurons	Nery et al. (2002); Ruiz-Reig et al. (2018); Touzot et al. (2016); Vucurovic et al. (2010)
POA/VP?	*Dbx1*-derived neurons	Hirata et al. (2009)
MGE/AEP	*Nkx2.1*-derived and *Lhx6*+ cells	García-López et al. (2008); Puelles et al. (2000)
LGE	*Islet 1*+, *Dlx5*+ cells	Bupesh et al. (2011); García-López et al. (2008)
dLGE/VP	*Pax6*+ cells (VP origin)	Bupesh et al. (2011)

12.3.4 Development of Connectivity

The above subsections detail known mechanisms of how diverse neurons of key threat detection structures are specified during embryonic development. However, how these circuits become wired together still remains unclear. A number of lines of evidence suggest that circuitry within this system is laid down relatively early. For example, studies from the lab of Richard Simerly have shown that circuitry between the BNST and MEA is already established by postnatal day (P)5 (Hutton et al. 1998). Moreover, although not experimental evidence, reporter expression studies of limbic system expressed genes reveal that neuronal tracts of the limbic system can be observed as early as E15.5 (Fig. 12.4).

12.3.5 Linking Developmental Genetic Programs to Behavioral Regulation

One aspect of high interest is to understand how the above-described developmental programs may relate to behavioral regulation. One intriguing idea is that transcription factor developmental programs direct stereotypical patterns of circuit formation in turn regulate different and specific components of hardwired behaviors. Thus, embryonic patterning may represent a continuum linking development directly to circuit function and ultimately the behaviors regulated by these circuits. Although such a link is far from being realized, some evidence within the stress/threat detection network suggests that developmental patterning mechanisms in mice directly guide specific behavioral outputs. One such example is the regulation of stress behavior by the embryonic expressed transcription factor, *Otp*. At the developmental genetic level, as described in Sect. 12.3.1, *Otp* is a key factor implicated in development of the PVN and the MEA. In the PVN, *Otp* expression is regulated by *Fezf2,* and *Otp* directly regulates expression of corticotropin releasing hormone

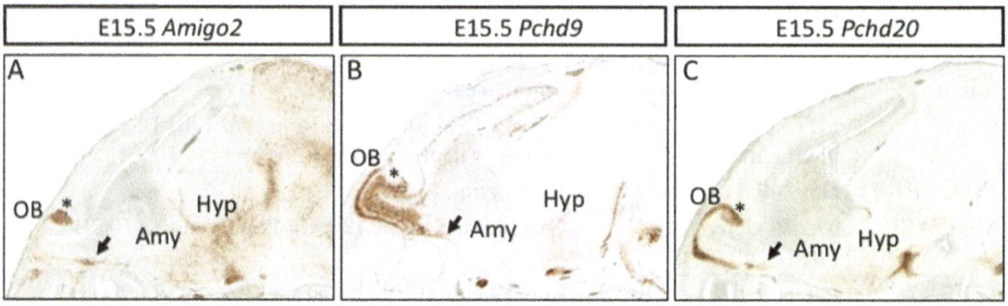

Fig. 12.4 Gensat images. DAB immunostained images of GFP expression from BAC driven cre recombination revealing gene expression of the cell adhesion encoding genes *Pchd20, Amigo2* or *Pchd9* by E15.5. Recombination is observed in the olfactory bulb (OB, with asterisk highlighting AOB), amygdala and hypothalamus, with presumptive projections observed between the OB and amygdala/hypothalamus (arrow). These images suggest establishment of limbic circuits as by as early as E15.5. Images courtesy of the Gene Expression Nervous System Atlas (GENSAT) Project, NINDS Contracts N01NS02331 & HHSN271200723701C to The Rockefeller University (New York, NY). Abbreviations: *Amy* amygdala, *AOB* accessory olfactory bulb, *Hyp* hypothalamus, *OB* olfactory bulb.

(CRH), a key molecule in modulating stress responses (Amir-Zilberstein et al. 2012; Bedont et al. 2015; Wircer et al. 2017). *Otp* mutants display anxiety-like behaviors and increased CRH expression. Thus, the embryonic patterning gene, *Otp*, functions directly in establishment of the stress (HPA) axis. A second example is the apparent modular function of *Dbx1* in development of stress/threat detection circuits in the arcuate nucleus and lateral hypothalamus. As mentioned above, *Dbx1* marks a subpopulation of neurons destined for the MEA (Figs. 12.2, 12.3 and Table 12.2). In addition, similar to many other MEA embryonic expressed genes, *Dbx1* is expressed throughout the developing hypothalamus (Sokolowski et al. 2016; Shimogori et al. 2010). Conditional deletion of *Dbx1* selectively in the hypothalamus results in a sex-specific deficit in innate fear responses to predator odor (Sokolowski et al. 2015). Correlating with this lack of innate fear response is a reduced expression of stress-induced corticosterone, and a lack of heightened neuronal activity in response to stress. *Dbx1* conditional mutant mice also do not display normal stress responses to food deprivation. These behavioral phenotypes are linked to embryonic misspecification of the lateral and arcuate nucleus hypothalamic neurons that are known to activate the PVN, without deficits in specification of the PVN, the site of CRH expression. Thus, by acting during a transient window of embryogenesis, *Dbx1* sets in motion a series of developmental events that are required for later behavioral and physiological responses to stress. Thus, work to date has revealed that at least two genes; one likely acting in a region autonomous manner to specific CRH neurons (*Otp*) and the other likely acting in a region nonautonomous manner to specify neurons that feed into the PVN (*Dbx1*), are required for specification of the stress/threat axis. This work also highlights the parallel embryonic genetic processes in which the brain's stress systems are specified.

12.4 Important Areas of Future Investigation

The above subsections describe current (early 2019) knowledge of the circuity and function of the stress/threat system (Sects. 12.2 and 12.3) and how neurons that comprise key stress/threat systems are specified from development, with specific examples of how key transcription factors establish subsystems required for select behaviors (Sect. 12.4). Despite this knowledge, there remain a number of key open questions, highlighted in Box 3 below:

Box 3 This Box Contains Interesting Information about Key Questions for Future Research

Question #1: Do embryonic developmental programs contribute to sex differences in stress/threat system behaviors, or is this solely governed by later hormonal effects?

Importance: Many stress linked disorders such as PTSD and anxiety disorders, as well as behavioral disorders characterized by altered stress responses, such as autism spectrum disorders, display a sex bias in presentation and prevalence. In addition to addressing basic biological questions of how male and female circuitry is established in sexually dimorphic brain regions such as the BNST, MEA and hypothalamus, answering Question #1 will likely inform and possibly provide better therapeutic approaches to treating these prevalent conditions.

Question #2: How do environmental stressors affect formation of these "hardwired" systems?

Importance: While formation of BNST-MEA-Hypothalamic circuits appears to be laid down primarily via developmental genetic programming mechanisms, these circuits are highly malleable by environmental influences during both fetal (embryonic) and early post-natal critical periods. Understanding the interplay between genes and environment is a critical unanswered question that has implications for a variety of human conditions described above in Question #1.

Question #3: Are there evolutionarily conserved common "blueprints" for establishing stress/threat system connectivity?

Importance: Normal responses to immediate environmental threats allow for survival of individuals across all species. This commonality may suggest that there are evolutionarily conserved mechanisms for development of threat systems across a diverse species, perhaps via utilization of similar genetic modules for neuron and circuit specification. If so, this may expand the repertoire of animal model systems in which to model human disorders affecting stress/threat systems.

Key References

Choi et al. (2005)—This article identifies subcircuits of the medial amygdala that are genetically identifiable by expression of the LIM-family of transcription factors and posits that these subcircuits control different innate behaviors.

Gross and Canteras (2012)—This review nicely describes the brain circuits controlling innate and learned fear.

Hirata et al. (2009)—Using genetic tracing techniques this study revealed that there are progenitor zones in the developing telencephalon that are dedicated to generate neuronal diversity in the amygdala.

Hashikawa et al. (2016)—This review eloquently describes the olfactory to deeper brain circuits that translate odor information to innate behavioral output.

Lebow and Chen (2016)—This review provides an excellent summary the anatomy, function and role of the bed nucleus of the stria terminalis in psychiatric disorders.

Luo, et al. (2018)—This review provides an very comprehensive description of the state of the art tools of modern neuroscience.

Motta et al. (2009)—Using a combination of approaches, this study revealed that there are distinct hypothalamic circuits for processing different fear behaviors.

References

al-Shamma HA, De Vries GJ (1996) Neurogenesis of the sexually dimorphic vasopressin cells of the bed nucleus of the stria terminalis and amygdala of rats. J Neurobiol 29:91–98

Alvarez-Bolado G (2018) Development of neuroendocrine neurons in the mammalian hypothalamus. Cell Tissue Res 375(1):23–39

Amir-Zilberstein L, Blechman J, Sztainberg Y, Norton WHJ, Reuveny A, Borodovsky N, Tahor M, Bonkowsky JL, Bally-Cuif L, Chen A et al (2012) Homeodomain protein otp and activity-dependent splicing modulate neuronal adaptation to stress. Neuron 73:279–291

Aravanis AM, Wang LP, Zhang F, Meltzer LA, Mogri MZ, Schneider MB, Deisseroth K (2007) An optical neural interface: in vivo control of rodent motor cortex with integrated fiberoptic and optogenetic technology. J Neural Eng 4:S143–S156

Armbruster B, Roth B (2005) Creation of designer biogenic amine receptors via directed molecular evolution. Neuropsychopharmacology 30:S265–S265

Armbruster BN, Li X, Pausch MH, Herlitze S, Roth BL (2007) Evolving the lock to fit the key to create a family of G protein-coupled receptors potently activated by an inert ligand. Proc Natl Acad Sci U S A 104:5163–5168

Batista-Brito R, Fishell G (2009) The developmental integration of cortical interneurons into a functional network. Curr Top Dev Biol 87:81–118

Bayer SA (1987) Neurogenetic and morphogenetic heterogeneity in the bed nucleus of the stria terminalis. J Comp Neurol 265:47–64

Bedont JL, Newman EA, Blackshaw S (2015) Patterning, specification, and differentiation in the developing hypothalamus. Wiley Interdiscip Rev Dev Biol 4:445–468

Berndt A, Lee SY, Ramakrishnan C, Deisseroth K (2014) Structure-guided transformation of Channelrhodopsin into a light-activated Chloride Channel. Science 3:420–424

Bian X (2013) Physiological and morphological characterization of GABAergic neurons in the medial amygdala. Brain Res 1509:8–19

Bian X, Yanagawa Y, Chen WR, Luo M (2008) Cortical-like functional organization of the pheromone-processing circuits in the medial amygdala. J Neurophysiol 99:77–86

Boyden ES, Zhang F, Bamberg E, Nagel G, Deisseroth K (2005) Millisecond-timescale, genetically targeted optical control of neural activity. Nat Neurosci 8:1263–1268

Brechbühl J, Moine F, Klaey M, Nenniger-Tosato M, Hurni N, Sporkert F et al (2013) Mouse alarm pheromone shares structural similarity with predator scents. Proc Natl Acad Sci U S A 110:4762–4767

Bulfone A, Puelles L, Porteus MH, Frohman MA, Martin GR, Rubenstein JL (1993) Spatially restricted expression of Dlx-1, Dlx-2 (Tes-1), Gbx-2, and Wnt-3 in the embryonic day 12.5 mouse forebrain defines potential transverse and longitudinal segmental boundaries. J Neurosci 13:3155–3172

Bupesh M, Legaz I, Abellán A, Medina L (2011) Multiple telencephalic and extratelencephalic embryonic domains contribute neurons to the medial extended amygdala. J Comp Neurol 519:1505–1525

Canteras NS, Swanson LW (1992) The dorsal premamillary nucleus: an unusual component of the mamillary body. Proc Natl Acad Sci U S A 89:10089–10093

Canteras NS, Simerly RB, Swanson LW (1992) Projections of the ventral premamillary nucleus. J Comp Neurol 324:195–212

Canteras NS, Simerly RB, Swanson LW (1994) Organization of projections from the ventromedial nucleus of the hypothalamus: a *Phaseolus vulgaris* leucoagglutinin study in the rat. J Comp Neurol 348:41–79

Canteras NS, Simerly RB, Swanson L (1995) Organization of projections from the medial nucleus of the amygdala: a PHAL study in the rat. J Comp Neurol 360:213–245

Carney RSE, Mangin J-M, Hayes L, Mansfield K, Sousa VH, Fishell G, Machold RP, Ahn S, Gallo V, Corbin JG (2010) Sonic hedgehog expressing and responding cells generate neuronal diversity in the medial amygdala. Neural Develop. 5:14

Carter M, Shieh JC (2010) Guide to research techniques in neuroscience. Elsevier, Burlington, MA, pp 1–375

Cezario AF, Ribeiro-Barbosa ER, Baldo MV, Canteras NS (2008) Hypothalamic sites responding to predator threats – the role of the dorsal premamillary nucleus in unconditioned and conditioned antipredatory defensive behavior. Eur J Neurosci 28:1003–1015

Choi GB, Dong H-W, Murphy AJ, Valenzuela DM, Yancopoulos GD, Swanson LW, Anderson DJ (2005) Lhx6 delineates a pathway mediating innate reproductive behaviors from the amygdala to the hypothalamus. Neuron 46:647–660

Chow BY, Han X, Dobry AS, Qian X, Chuong AS, Li M, Henninger MA, Belfort GM, Lin Y, Monahan PE, Boyden ES (2010) High-performance genetically targetable optical neural silencing by light-driven proton pumps. Nature 463:98–102

Coward P, Wada HG, Falk MS, Chan SD, Meng F, Akil H, Conklin BR (1998) Controlling signaling with a specifically designed Gi-coupled receptor. Proc Natl Acad Sci U S A 95:352–357

Cocas LA, Miyoshi G, Carney RS, Sousa VH, Hirata T, Jones KR, Fishell G, Huntsman MM, Corbin JG (2009) Emx1-lineage progenitors differentially contribute to neural diversity in the striatum and amygdala. J Neurosci 29(50):15933–15946. https://doi.org/10.1523/JNEUROSCI.2525-09.2009. PubMed PMID: 20016109; PubMed Central PMCID: PMC3679174

Davis M, Walker DL, Miles L, Grillon C (2010) Phasic vs sustained fear in rats and humans: role of the extended amygdala in fear vs anxiety. Neuropsychopharmacology 35:105–135

Dewan A, Pacifico R, Zhan R, Rinberg D, Bozza T (2013) Non redundant coding of aversive odours in the main olfactory pathway. Nature 497:486–489

Dong HW, Petrovich GD, Swanson LW (2001) Topography of projections from amygdala to bed nuclei of the stria terminalis. Brain Res Brain Res Rev 38:192–246

Dong HW, Swanson LW (2004) Projections from bed nuclei of the stria terminalis, posterior division: implications for cerebral hemisphere regulation of defensive and reproductive behaviors. J Comp Neurol 471(4):396–433. Erratum in: J Comp Neurol. 2004 Jul 5;474 (4):603-4. PubMed PMID: 15022261

Ericson J, Muhr J, Jessell TM, Edlund T (1995) Sonic hedgehog: a common signal for ventral patterning along the rostrocaudal axis of the neural tube. Int J Dev Biol 39:809–816

Fenno L, Yizhar O, Deisseroth K (2011) The development and application of optogenetics. Annu Rev Neurosci 34:389–412

Ferran JL, Puelles L, Rubenstein JLR (2015) Molecular codes defining rostrocaudal domains in the embryonic mouse hypothalamus. Front Neuroanat 9:46

Ferrero DM, Lemon JK, Fluegge D, Pashkovski SL, Korzan WJ, Datta SR et al (2011) Detection and avoidance of a carnivore odor by prey. Proc Natl Acad Sci U S A 108:11235–11240

García-López M, Abellán A, Legaz I, Rubenstein JLR, Puelles L, Medina L (2008) Histogenetic compartments of the mouse centromedial and extended amygdala based on gene expression patterns during development. J Comp Neurol 506:46–74

García-Moreno F, Pedraza M, Di Giovannantonio LG, Di Salvio M, López-Mascaraque L, Simeone A, De Carlos JA (2010) A neuronal migratory pathway crossing from diencephalon to telencephalon populates amygdala nuclei. Nat Neurosci 13:680–689

Gelman D, Griveau A, Dehorter N, Teissier A, Varela C, Pla R, Pierani A, Marín O (2011) A wide diversity of cortical GABAergic interneurons derives from the embryonic preoptic area. J Neurosci 31:16570–16580

Gomez JL, Bonaventura J, Lesniak W, Mathews WB, Sysa-Shah P, Rodriguez LA, Ellis RJ, Richie CT, Harvey BK, Dannals RF, Pomper MG, Bonci A, Michaelides M (2017) Chemogenetics revealed: DREADD occupancy and activation via converted clozapine. Science 357:503–507

Gorski JA, Talley T, Qiu M, Puelles L, Rubenstein JLR, Jones KR (2002) Cortical excitatory neurons and glia, but not GABAergic neurons, are produced in the Emx1-expressing lineage. J Neurosci 22:6309–6314

Goto M, Canteras NS, Burns G, Swanson LW (2005) Projections from the subfornical region of the lateral hypothalamic area. J Comp Neurol 493:412–438

Gradinaru V, Zhang F, Ramakrishnan C, Mattis J, Prakash R, Diester I, Goshen I, Thompson KR, Deisseroth K (2010) Molecular and cellular approaches for diversifying and extending optogenetics. Cell 141:154–165

Gross CT, Canteras NS (2012) The many paths to fear. Nat Rev Neurosci 13:651–658

Guirado S, Real MA, Dávila JC (2008) Distinct immunohistochemically defined areas in the medial amygdala in the developing and adult mouse. Brain Res Bull 75:214–217

Han TM, De Vries GJ (1999) Neurogenesis of galanin cells in the bed nucleus of the stria terminalis and centromedial amygdala in rats: a model for sexual differentiation of neuronal phenotype. J Neurobiol 38:491–498

Hashikawa K, Hashikawa Y, Falkner A, Lin D (2016) The neural circuits of mating and fighting in male mice. Curr Opin Neurobiol 38:27–37

Hashikawa K, Hashikawa Y, Lischinsky J, Lin D (2018) The neural mechanisms of sexually dimorphic aggressive behaviors. Trends Genet TIG 34:755–776

Hegemann P, Nagel G (2013) From Channelrhodopsins to Optogenetics. EMBO Mol Med 5:173–176

Hirata T, Li P, Lanuza GM, Cocas LA, Huntsman MM, Corbin JG (2009) Identification of distinct telencephalic progenitor pools for neuronal diversity in the amygdala. Nat Neurosci 12:141–149

Hutton LA, Gu G, Simerly RB (1998) Development of a sexually dimorphic projection from the bed nuclei of the stria terminalis to the anteroventral periventricular nucleus in the rat. J Neurosci 18:3003–3013

Keshavarzi S, Sullivan RKP, Ianno DJ, Sah P (2014) Functional properties and projections of neurons in the medial amygdala. J Neurosci 34:8699–8715

Kunwar PS, Zelikowsky M, Remedios R, Cai H, Yilmaz M, Meister M, Anderson DJ (2015) Ventromedial hypothalamic neurons control a defensive emotion state. elife 4:e06633

Kanatani S, Honda T, Aramaki M, Hayashi K, Kubo K, Ishida M, Tanaka DH, Kawauchi T, Sekine K, Kusuzawa S, Kawasaki T, Hirata T, Tabata H, Uhlén P, Nakajima K (2015) The COUP-TFII/Neuropilin-2 is a molecular switch steering diencephalon-derived GABAergic neurons in the developing mouse brain. Proc Natl Acad Sci U S A 112(36):E4985–E4994. https://doi.org/10.1073/pnas.1420701112. Epub 2015 Aug 24. PubMed PMID: 26305926; PubMed Central PMCID: PMC4568674

Kuerbitz J, Arnett M, Ehrman S, Williams MT, Vorhees CV, Fisher SE, Garratt AN, Muglia LJ, Waclaw RR, Campbell K (2018) Loss of Intercalated Cells (ITCs) in the mouse amygdala of Tshz1 mutants correlates with fear, depression, and social interaction phenotypes. J Neurosci 38 (5):1160–1177. https://doi.org/10.1523/JNEUROSCI.1412-17.2017. Epub 2017 Dec 18. PubMed PMID: 29255003; PubMed Central PMCID: PMC5792476

Lebow MA, Chen A (2016) Overshadowed by the amygdala: the bed nucleus of the stria terminalis emerges as key to psychiatric disorders. Mol Psychiatry 21:450–463

Li Y, Dulac C (2018) Neural coding of sex-specific social information in the mouse brain. Curr Opin Neurobiol 53:120–130

Liberles SD (2015) Trace amine-associated receptors: ligands, neural circuits, and behaviors. Curr Opin Neurobiol 34C:1–7

Lischinsky JE, Sokolowski K, Li P, Esumi S, Kamal Y, Goodrich M, Oboti L, Hammond TR, Krishnamoorthy M, Feldman D et al (2017) Embryonic transcription factor expression in mice predicts medial amygdala neuronal identity and sex-specific responses to innate behavioral cues. elife 6:e21012

Luo L, Callaway EM, Svoboda K (2008) Genetic dissection of neural circuits. Neuron 57 (5):634–660

Luo LQ, Callaway EM, Svoboda K (2018) Genetic dissection of neural circuits: a decade of Progress. Neuron 98:256–281

Madisen L, Mao T, Koch H, Zhuo JM, Berenyi A, Fujisawa S, Hsu YW, Garcia AJ 3rd, Gu X, Zanella S, Kidney J, Gu H, Mao Y, Hooks BM, Boyden ES, Buzsaki G, Ramirez JM, Jones AR, Svoboda K, Han X, Turner EE, Zeng H (2012) A toolbox of Cre-dependent optogenetic transgenic mice for light-induced activation and silencing. Nat Neurosci 15:793–802

Magnus CJ, Lee PH, Bonaventura J, Zemla R, Gomez JL, Ramirez MH, Hu X et al (2019) Ultrapotent Chemogenetics for research and potential clinical applications. Science 364 (6436):eaav5282. https://doi.org/10.1126/science.aav5282. [Epub ahead of print]

Martinez RC, Carvalho-Netto EF, Ribeiro-Barbosa ER, Baldo MV, Canteras NS (2011) Amygdalar roles during exposure to a live predator and to a predator-associated context. Neuroscience 172:314–328

Matise MP, Wang H (2011) Sonic hedgehog signaling in the developing CNS where it has been and where it is going. Curr Top Dev Biol 97:75–117

McGregor IS, Hargreaves GA, Apfelbach R, Hunt GE (2004) Neural correlates of cat odor-induced anxiety in rats: region-specific effects of the benzodiazepine midazolam. J Neurosci 24:4134–4144

Motta SC, Goto M, Gouveia FV, Baldo MV, Canteras NS, Swanson LW (2009) Dissecting the brain's fear system reveals the hypothalamus is critical for responding in subordinate conspecific intruders. Proc Natl Acad Sci U S A 106:4870–4875

Nery S, Fishell G, Corbin JG (2002) The caudal ganglionic eminence is a source of distinct cortical and subcortical cell populations. Nat Neurosci 5:1279–1287

Neve RL, Neve KA, Nestler EJ, Carlezon WA (2005) Use of herpes virus amplicon vectors to study brain disorders. BioTechniques 39:381–391

Nouri N, Awatramani R (2017) A novel floor plate boundary defined by adjacent En1 and Dbx1 microdomains distinguishes midbrain dopamine and hypothalamic neurons. Dev Camb Engl 144:916–927

Pérez-Gómez A, Bleymehl K, Stein B, Pyrski M, Birnbaumer L, Munger SD et al (2015) Innate predator odor aversion driven by parallel olfactory subsystems that converge in the ventromedial hypothalamus. Curr Biol 25:1340–1346

Petilla Interneuron Nomenclature Group, Ascoli GA, Alonso-Nanclares L, Anderson SA, Barrionuevo G, Benavides-Piccione R, Burkhalter A, Buzsáki G, Cauli B, Defelipe J et al (2008) Petilla terminology: nomenclature of features of GABAergic interneurons of the cerebral cortex. Nat Rev Neurosci 9:557–568

Petrovich GD, Risold PY, Swanson LW (1996) Organization of the projections of the basomedial nucleus of the amygdala: a PHAL study in the rat. J Comp Neurol 374:387–420

Puelles L, Kuwana E, Puelles E, Bulfone A, Shimamura K, Keleher J, Smiga S, Rubenstein JL (2000) Pallial and subpallial derivatives in the embryonic chick and mouse telencephalon, traced by the expression of the genes Dlx-2, Emx-1, Nkx-2.1, Pax-6, and Tbr-1. J Comp Neurol 424:409–438

Puelles L, Morales-Delgado N, Merchán P, Castro-Robles B, Martínez-de-la-Torre M, Díaz C, Ferran JL (2016) Radial and tangential migration of telencephalic somatostatin neurons originated from the mouse diagonal area. Brain Struct Funct 221(6):3027–3065. https://doi.org/10.1007/s00429-015-1086-8. Epub 2015 Jul 19. PubMed PMID: 26189100; PubMed Central PMCID: PMC4920861

Real MA, Heredia R, Labrador MDC, Dávila JC, Guirado S (2009) Expression of somatostatin and neuropeptide Y in the embryonic, postnatal, and adult mouse amygdalar complex. J Comp Neurol 513:335–348

Risold PY, Canteras NS, Swanson LW (1994) Organization of projections from the anterior hypothalamic nucleus: a Phaseolus vulgaris leucoagglutinin study in the rat. J Comp Neurol 348:1–40

Roth BL (2016) DREADDs for neuroscientists. Neuron 89:683–694

Ruiz-Reig N, Andres B, Lamonerie T, Theil T, Fairén A, Studer M (2018) The caudo-ventral pallium is a novel pallial domain expressing Gdf10 and generating Ebf3-positive neurons of the medial amygdala. Brain Struct Funct 223:3279–3295

Schobert B, Lanyi JK (1982) Halorhodopsin is a light-driven chloride pump. J Biol Chem 257:306–313

Shimada M, Nakamura T (1973) Time of neuron origin in mouse hypothalamic nuclei. Exp Neurol 41:163–173

Shimogori T, Lee DA, Miranda-Angulo A, Yang Y, Wang H, Jiang L, Yoshida AC, Kataoka A, Mashiko H, Avetisyan M et al (2010) A genomic atlas of mouse hypothalamic development. Nat Neurosci 13:767–775

Silva BA, Mattucci C, Krzywkowski P, Murana E, Illarionova A, Grinevich V, Canteras NS, Ragozzino D, Gross CT (2013) Independent hypothalamic circuits for social and predator fear. Nat Neurosci 16:1731–1733

Simerly RB, Swanson LW (1988) Projections of the medial preoptic nucleus: a Phaseolus vulgaris leucoagglutinin anterograde tract-tracing study in the rat. J Comp Neurol 270:209–242

Sokolowski K, Corbin JG (2012) Wired for behaviors: from development to function of innate limbic system circuitry. Front Mol Neurosci 5:55

Sokolowski K, Esumi S, Hirata T, Kamal Y, Tran T, Lam A, Oboti L, Brighthaupt S-C, Zaghlula M, Martinez J et al (2015) Specification of select hypothalamic circuits and innate behaviors by the embryonic patterning gene dbx1. Neuron 86:403–416

Sokolowski K, Tran T, Esumi S, Kamal Y, Oboti L, Lischinsky J, Goodrich M, Lam A, Carter M, Nakagawa Y et al (2016) Molecular and behavioral profiling of Dbx1-derived neurons in the arcuate, lateral and ventromedial hypothalamic nuclei. Neural Develop 11:12

Soma M, Aizawa H, Ito Y, Maekawa M, Osumi N, Nakahira E, Okamoto H, Tanaka K, Yuasa S (2009) Development of the mouse amygdala as revealed by enhanced green fluorescent protein gene transfer by means of in utero electroporation. J Comp Neurol 513:113–128

Stachniak TJ, Ghosh A, Sternson SM (2014) Chemogenetic synaptic silencing of neural circuits localizes a hypothalamus-->midbrain pathway for feeding behavior. Neuron 82:797–808

Swanson LW (2000) Cerebral hemisphere regulation of motivated behavior. Brain Res 886:113–164

Swanson LW, Petrovich GD (1998) What is the amygdala? Trends Neurosci 21:323–331

Sullivan GM, Apergis J, Bush DE, Johnson LR, Hou M, LeDoux JE (2004) Lesions in the bed nucleus of the stria terminalis disrupt corticosterone and freezing responses elicited by a contextual but not by a specific cue-conditioned fear stimulus. Neuroscience 128:7–14

Takahashi LK (2014) Olfactory systems and neural circuits that modulate predator odor fear. Front Behav Neurosci 8:72

Tervo DGR, Hwang BY, Viswanathan S, Gaj T, Lavzin M, Ritola KD, Lindo S, Michael S, Kuleshova E, Ojala D, Huang CC, Gerfen CR, Schiller J, Dudman JT, Hantman AW, Looger LL, Schaffer DV, Karpova AY (2016) A designer AAV variant permits efficient retrograde access to projection neurons. Neuron 92(2):372–382

Thompson RH, Swanson LW (1998) Organization of inputs to the dorsomedial nucleus of the hypothalamus: a reexamination with Fluorogold and PHAL in the rat. Brain Res Brain Res Rev 27:89–118

Touzot A, Ruiz-Reig N, Vitalis T, Studer M (2016) Molecular control of two novel migratory paths for CGE-derived interneurons in the developing mouse brain. Dev. Camb. Engl. 143:1753–1765

Tye KM, Deisseroth K (2012) Optogenetic investigation of neural circuits underlying brain disease in animal models. Nat Rev Neurosci 13:251–266

Tang K, Rubenstein JL, Tsai SY, Tsai MJ (2012) COUP-TFII controls amygdala patterning by regulating neuropilin expression. Development 139(9):1630–1639. https://doi.org/10.1242/dev.075564. PubMed PMID: 22492355; PubMed Central PMCID: PMC3317968

Vardy E, Robinson JE, Li C, Olsen RHJ, DiBerto JF, Giguere PM, Sassano FM, Huang XP, Zhu H, Urban DJ, White KL, Rittiner JE, Crowley NA, Pleil KE, Mazzone CM, Mosier PD, Song J, Kash TL, Malanga CJ, Krashes MJ, Roth BL (2015) A new DREADD facilitates the multiplexed Chemogenetic interrogation of behavior. Neuron 86:936–946

Vucurovic K, Gallopin T, Ferezou I, Rancillac A, Chameau P, van Hooft JA, Geoffroy H, Monyer H, Rossier J, Vitalis T (2010) Serotonin 3A receptor subtype as an early and protracted marker of cortical interneuron subpopulations. Cereb Cortex N Y N 1991(20):2333–2347

Wang L, Chen IZ, Lin D (2015) Collateral pathways from the ventromedial hypothalamus mediate defensive behaviors. Neuron 85:1344–1358

Wang L, Talwar V, Osakada T, Kuang A, Guo Z, Yamaguchi T, Lin D (2019) Hypothalamic control of conspecific self-defense. Cell Rep 26:1747–1758.e5

Wircer E, Blechman J, Borodovsky N, Tsoory M, Nunes AR, Oliveira RF, Levkowitz G (2017) Homeodomain protein Otp affects developmental neuropeptide switching in oxytocin neurons associated with a long-term effect on social behavior. elife 6:e22170

Wu YE, Pan L, Zuo Y, Li X, Hong W (2017) Detecting activated cell populations using single-cell RNA-Seq. Neuron 96:313–329.e6

Wyatt TD (2014) Pheromones and animal behavior, 2nd edn. Cambridge University Press, Cambridge, UK

Xie Y, Dorsky RI (2017) Development of the hypothalamus: conservation, modification and innovation. Dev Camb Engl 144:1588–1599

Xu Q, de la Cruz E, Anderson SA (2003) Cortical interneuron fate determination: diverse sources for distinct subtypes? Cereb Cortex N Y N 1991(13):670–676

Yang CF, Shah NM (2014) Representing sex in the brain, one module at a time. Neuron 82:261–278

Zhao T, Szabó N, Ma J, Luo L, Zhou X, Alvarez-Bolado G (2008) Genetic mapping of Foxb1-cell lineage shows migration from caudal diencephalon to telencephalon and lateral hypothalamus. Eur J Neurosci 28:1941–1955

Zhu H, Aryal DK, Olsen RH, Urban DJ, Swearingen A, Forbes S, Roth BL, Hochgeschwender U (2016) Cre-dependent DREADD (designer receptors exclusively activated by designer drugs) mice. Genesis 54:439–446

Organization and Postnatal Development of Visceral Sensory Inputs to the Neuroendocrine Hypothalamus

13

Linda Rinaman

Abstract

Sensory signals arising in the body's internal organs are delivered to the central nervous system by spinal and vagal afferent neurons. These signals modulate pituitary hormone release via ascending neural pathways from the caudal brainstem to the neuroendocrine hypothalamus. *Interoceptive* (i.e., visceral) feedback regarding the moment-to-moment state of gastrointestinal and other physiological systems is critical for shaping and coordinating adaptive neuroendocrine functions, including hormonal responses to real or perceived homeostatic threats. This chapter summarizes the functional organization and postnatal development of visceral sensory inputs to the neuroendocrine hypothalamus in rats and mice, the two mammalian models in which most of the relevant data have been collected. A special emphasis is placed on hypothalamic inputs arising from noradrenergic (NA) and glucagon-like peptide 1 (GLP1) neurons, whose cell bodies occupy the caudal nucleus of the solitary tract (NTS) and medullary reticular formation (RF). The axonal projections of NA and GLP1 neurons directly target the hypothalamic medial preoptic area (mPOA), arcuate nucleus (ARC), paraventricular nucleus (PVN), and supraoptic nucleus (SON), where adrenergic and GLP1 receptor signaling modulates neuroendocrine gene expression, pulsatile activity, and pituitary hormone release. NA and GLP1 inputs to the neuroendocrine hypothalamus mature postnatally in rats and mice, likely contributing to the developmental maturation of pituitary hormone release in response to both interoceptive and exteroceptive signals.

Keywords

Fos · Gastrointestinal · Glucagon-like peptide 1 · HPA axis · Noradrenergic · Nucleus of the solitary tract · Oxytocin

L. Rinaman (✉)
Department of Psychology and Program in Neuroscience, Florida State University, Tallahassee, FL, USA
e-mail: Rinaman@psy.fsu.edu

S. Wray, S. Blackshaw (eds.), *Developmental Neuroendocrinology*, Masterclass in Neuroendocrinology 9, https://doi.org/10.1007/978-3-030-40002-6_13

13.1 Introduction

Neuroendocrine systems comprise magnocellular and parvocellular neurons that control pituitary hormone release. The cell bodies of magnocellular oxytocin (OT) and vasopressin (AVP) neurons are located within the paraventricular nucleus (PVN), supraoptic nucleus (SON), and accessory cell groups distributed between these two hypothalamic nuclei. The axons of magnocellular OT and AVP neurons terminate within the posterior pituitary, where OT and AVP are released into the systemic circulation. The PVN also houses parvocellular endocrine neurons whose axon terminals secrete anterior pituitary hormone-releasing hormones [i.e., corticotropin-releasing hormone (CRH) and vasopressin, thyrotropin releasing hormone (TRH), and somatostatin (SS)] into the median eminence (ME), located at the base of the hypothalamus; dopamine (DA), which inhibits prolactin secretion, is similarly produced by tubero-infundibular (TIDA) neurons. The ME contains a specialized portal vasculature that conveys these releasing hormones directly into the anterior pituitary gland at high concentrations. Gonadotropin releasing hormone (GnRH) and growth hormone-releasing hormone (GHRH) also are secreted into the ME from the terminals of additional parvocellular endocrine neurons whose cell bodies are located within the hypothalamic medial preoptic area (mPOA) and arcuate nucleus (ARC).

Magnocellular and parvocellular endocrine neurons receive direct and relayed synaptic input from the nucleus of the solitary tract (NTS), located within the caudal brainstem (Fig. 13.1). The NTS receives *interoceptive* signals arising in gastrointestinal (GI) and other organ systems, and ascending projections from the NTS to the hypothalamus provide an interface through which visceral sensory signals influence the production and release of pituitary hormones. For example, stimulation of GI vagal sensory afferents in adult rats activates NTS neurons as well as magnocellular OT neurons and *hypophysiotropic* CRH neurons, leading to increased plasma concentrations of OT, adrenocorticotropic hormone (ACTH), and corticosterone. Conversely, similar stimulation of vagal sensory afferents in 2-day-old rat pups activates NTS neurons, as in adult rats, but fails to activate hypothalamic endocrine neurons, which is evidence that ascending projections from the NTS to the neuroendocrine hypothalamus are functionally immature. Visceral signals relayed from the NTS to the hypothalamus also impact the function of parvocellular TRH, GnRH, GHRH, SS, and DA endocrine neurons in adult rodents; however, the anatomical and functional maturation of these signaling pathways is relatively unexplored. This chapter summarizes what is known from rodent models regarding the functional organization and development of neural pathways that deliver visceral sensory signals to neuroendocrine cells of the hypothalamus.

13.1.1 Visceral Sensory Inputs to the Spinal Cord

Sensory signals arising within the body's internal organs (i.e., thoracic, abdominal, and pelvic viscera) are carried to the CNS by spinal and cranial nerve afferents. In rats, the cell bodies of spinal visceral sensory afferents occupy the dorsal root ganglia

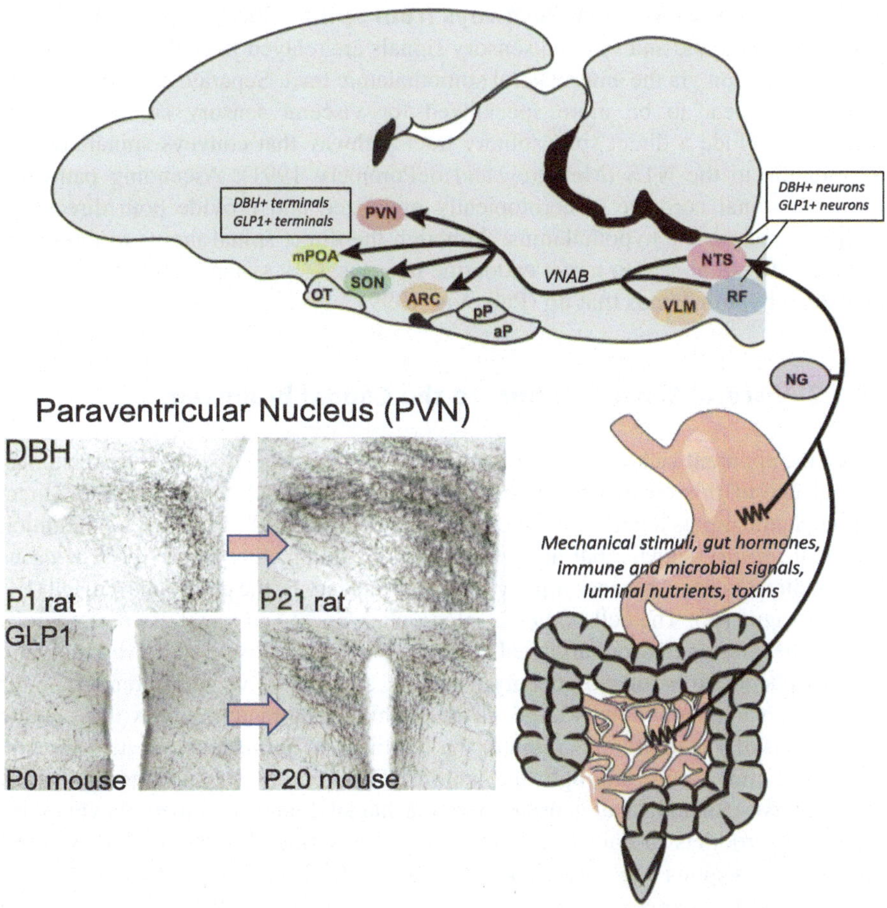

Fig. 13.1 Mechanical and chemical signals arising within the stomach, intestines, and other viscera are relayed to the caudal brainstem by vagal sensory afferents whose cell bodies occupy the nodose ganglia (NG). The central axons of vagal afferents synapse within the nucleus of the solitary tract (NTS) in the caudal medulla. Noradrenergic neurons (i.e., immunoreactive for dopamine beta hydroxylase, DBH) and peptidergic neurons expressing preproglucagon and immunoreactive for glucagon-like peptide 1 (GLP1) are located within the NTS, reticular formation (RF), and ventrolateral medulla (VLM), and carry visceral sensory signals from the caudal brainstem to the neuroendocrine hypothalamus via the ventral noradrenergic bundle (VNAB). Lower left: Postnatal changes in innervation of the neuroendocrine hypothalamus. In newborn rats and mice, DBH- and GLP1-positive axons are sparse within the paraventricular nucleus (PVN) and other hypothalamic endocrine nuclei, with innervation density increasing progressively to achieve adult-like levels by the end of the third week postnatal (pink arrow)

(DRG); their axons terminate centrally in laminae I–VII of the spinal cord dorsal horn and intermediate zone, and also in lamina X around the central canal. A subset of the second-order spinal sensory neurons that receive visceral sensory inputs also receive input from somatic sensory DRG neurons; thus, the activity of spinal sensory neurons is co-modulated by visceral, cutaneous, and proprioceptive stimuli.

13.1.1.1 Visceral Sensory Pathways from Spinal Cord to Hypothalamus

Convergent somatic and visceral sensory signals are relayed from the spinal cord to the diencephalon via the anterolateral spinothalamic tract. Separate ascending spinal pathways appear to be more specialized for visceral sensory signaling; these pathways include a direct spinosolitary tract pathway that conveys spinal sensory information to the NTS (Menétrey and dePommery 1991). Ascending pathways from the spinal cord are viscerotopically organized and provide both direct and relayed inputs to the hypothalamus. Although the direct spinal inputs to the hypothalamus do not appear to target endocrine neurons, they synapse within regions of the lateral hypothalamus that do (Palkovits 1999).

13.1.2 Visceral Sensory Inputs to the Caudal Brainstem

Given the apparent lack of direct spinal inputs to the neuroendocrine hypothalamus, the balance of this chapter will focus on the structure and development of direct hypothalamic inputs arising from the NTS and medullary RF (Fig. 13.1). In addition to spinal sensory signals that are relayed via the spinosolitary tract, NTS neurons receive direct visceral sensory input via two cranial nerves, the glossopharyngeal (IX) and the vagus (X). The cell bodies of glossopharyngeal and vagal afferent neurons occupy the inferior glossopharyngeal (aka petrosal) and nodose ganglia, respectively. The peripheral axons of glossopharyngeal and a subset of vagal afferents monitor cardiovascular and pulmonary target tissues within the thoracic cavity, while separate populations of vagal afferents monitor the GI tract and associated digestive viscera, from the oral cavity through the colon. The centrally projecting axons of glossopharyngeal and vagal neurons enter the caudal dorsolateral medulla via multiple sensory rootlets to converge within the solitary tract, which conveys visceral sensory axons along the rostrocaudal length of the NTS. Most of these sensory afferents form synapses within the caudal visceral portion of the NTS (as opposed to the more rostral gustatory NTS), but some visceral sensory axons terminate within the area postrema (AP), a dorsal midline circumventricular organ that lacks a blood-brain barrier. The AP lies in the floor of the caudal fourth ventricle, immediately adjacent to the visceral portion of the NTS. AP neurons respond to *interoceptive* signals conveyed by hormones, osmolytes, toxins, cytokines, and other circulating factors and convey these signals to postsynaptic neurons in the NTS, medullary RF, and pontine parabrachial nucleus (PBN) (Shapiro and Miselis 1985).

13.2 Anatomical Development of Vagal Afferent Inputs to the NTS

Vagal sensory neurons within the rat nodose ganglia undergo their final mitotic division between embryonic day (E)10 and E14, with peak cell production on E13. Even as neural proliferation continues, the peripheral axons of vagal sensory neurons start to innervate the GI tract, and their central axons begin to penetrate the caudal

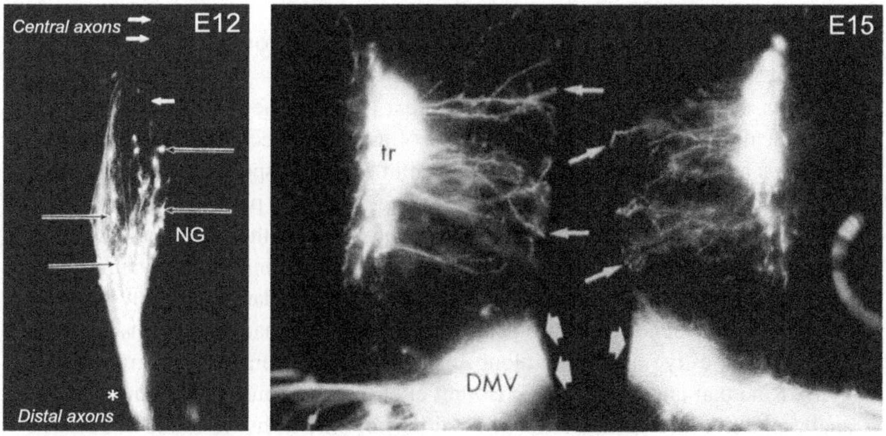

Fig. 13.2 DiI tracing of vagal neurons during embryonic development in rats. DiI crystals were placed into the developing gastric tube of formalin-fixed embryos harvested on embryonic day (E) 12 or E15. In the left panel, DiI-labeled vagal sensory neurons within the nodose ganglion (NG; long arrows) are evident, along with their distal axons (asterisk) that are labeled due to their peripheral association with the gastric tube. However, very few labeled NG neurons have axons extending centrally into the caudal medulla (short arrows). In the right panel, DiI-labeled motor neurons within the dorsal motor nucleus of the vagus (DMV) that innervate the gastric tube are evident, along with the central axons of vagal sensory neurons that bundle within the solitary tract (tr). Some of these sensory axons emerge to project into the developing NTS (nucleus tractus solitarii, nucleus of the solitary tract; long arrows). Figure adapted from Rinaman and Levitt (1993)

medulla (Rinaman and Levitt 1993). Results from a study in which crystals of DiI (1,1′-dioctadecyl-3,3,3′,3′-tetramethylindocarbocyanine perchlorate) were placed into the developing digestive tube in intact, formalin-fixed rat embryos (*see* Box 13.1) revealed that vagal sensory neurons within the nodose ganglia were first labeled in E12 specimens (Fig. 13.2, NG), evidence that their peripheral axons first reach the digestive tube at this time point. However, very few of the nodose neurons labeled at E12 had central axons that entered the caudal medulla, and labeled fibers within the solitary tract were not identifiable until E13. The density and rostrocaudal distribution of labeled vagal afferents within the solitary tract increased markedly between E13 and E14, but labeled vagal afferent axons did not emerge from the tract until E15, when labeled fibers at all rostrocaudal levels coursed directly from the solitary tract (Fig. 13.2, tr) into the proliferative ventricular zone of the alar plate (Fig. 13.2, right, long arrows). Although vagal sensory fibers were not observed to arborize within the developing NTS until E16, their distribution within the NTS was remarkably adult-like by E18, approximately 3 days before birth (Rinaman and Levitt 1993). Thus, the early embryonic development of vagal sensory inputs to the NTS provides an anatomical substrate for digestive vago-vagal reflexes (e.g., the gastric accommodation reflex) that already are operational in newborn rats.

Box 13.1 Tracing Axonal Projections During Embryonic and Early Postnatal Development

Two obstacles are immediately apparent when one considers how to anatomically trace the early development of axonal projection pathways. Many pathways are established before birth, even in precocial species such as rats and mice, and neural tracing in live developing embryos presents a formidable technical challenge. One strategic solution is provided by the use of carbocyanine dyes such as DiI (1,1′-dioctadecyl-3,3,3′,3′-tetramethylindocarbocyanine perchlorate). DiI travels by diffusion along continuous cellular lipids, fully labeling the axonal and dendritic processes of mature and immature neurons that are exposed to the dye anywhere along their plasma membrane (Godement et al. 1987; Rinaman and Levitt 1993; Bouret et al. 2004). Thus, DiI tracing permits analysis of point-to-point axonal projections not only in living tissue but also in aldehyde-fixed tissue samples, because dye diffusion along continuous lipid membranes does not require microtubule-mediated transport. DiI tracing has been used to reveal the prenatal development of vagal sensory and motor pathways between the brainstem and digestive tube in aldehyde-fixed embryonic and early postnatal rat pups (Rinaman and Levitt 1993) (Fig. 13.2). DiI also has been used to reveal the early postnatal growth of intra-hypothalamic axonal projections from the ARC to the PVN in fixed slabs of brain tissue dissected from neonatal rodent pups, and to demonstrate the effect of postnatal leptin signaling to stimulate axonal growth (Bouret et al. 2004). Generating these data in embryonic and early neonatal animals would have been an impractical (and perhaps impossible) feat using survival surgery for targeted delivery of neural tracer, as typically applied in adult rats and mice.

13.2.1 Functional Development of Vagal Afferent Inputs to the NTS

In addition to displaying functional autonomic vago-vagal reflexes, newborn rats are behaviorally responsive to at least some vagally mediated satiety signals that are conveyed by GI vagal afferents to the caudal NTS. A key example of early behavioral responsiveness to these signals is afforded by considering the effects of systemically administered cholecystokinin octapeptide (CCK) in newborn rats. *Endogenous* CCK is released from intestinal secretory cells during and after food intake and nutrient absorption, in both adult and newborn animals. CCK acts by binding to CCK-1 (aka CCK-A) receptors that are abundantly expressed in the GI tract of newborn rats (Robinson et al. 1987), including expression by GI vagal sensory afferents. In adult rats, CCK initiates and maintains various digestive processes and also acts as a satiety signal that suppresses food intake by limiting meal size through a central, vagally mediated mechanism. Low doses of *exogenous* CCK robustly suppress independent milk intake in newborn rats (Robinson et al. 1988), evidence that the neural signaling pathway underlying this behavioral effect already is functional at birth.

CCK increases the firing rate of GI vagal afferents through a receptor-mediated mechanism in both adult and newborn rats (Raybould and Tache 1988); thus, synthetic CCK is an ideal experimental tool with which to activate vagal sensory afferents and identify their central postsynaptic targets in rats at different developmental stages. CCK injected intraperitoneally evokes robust expression of *Fos* (the protein product of the immediate-early gene *c-fos*) within the caudal NTS and AP in both adult rats and newborn pups (Rinaman et al. 1994), with activated neurons located in specific NTS subregions that receive gastric vagal sensory input. Perhaps unsurprisingly, given the behavioral responsiveness of newborn rats to CCK, the distribution of CCK-induced Fos expression within the caudal NTS is virtually identical in 2-day-old and adult rats. Notably, surgical mid-collicular decerebration studies in adult rats have revealed that brainstem circuits alone are sufficient for feeding responses to CCK and other vagal satiety signals (Grill and Norgren 1978). Thus, the available data indicate that CCK receptor-mediated activation of vagal afferent inputs to the NTS and subsequent processes for recruitment of brainstem circuits that underlie CCK-induced *hypophagia* already are functional in 2-day-old rats. However, (*discussed further, below*), newborn rats are decidedly immature in their hypothalamic Fos and pituitary hormone responses to CCK treatment, reflecting delayed maturation of synaptic connections between visceral sensory neurons within the caudal NTS and their hypothalamic targets (Rinaman and Koehnle 2010).

13.3 Visceral Sensory Stimulation of the Neuroendocrine Hypothalamus

13.3.1 Responses to Vagal Stimuli in Adult Animals

The ability of visceral sensory stimuli to modulate pituitary hormone release in adult animals has been appreciated for many years. Early studies demonstrated posterior pituitary release of AVP and OT in rats after electrical stimulation of the central end of the severed vagus nerve (Chang et al. 1939; Thieblot et al. 1957), and later work demonstrated that gastrointestinal or vagal nerve stimulation increases the firing activity of magnocellular and parvocellular endocrine neurons within the PVN and SON in anesthetized rats (Jin et al. 1993; Ueta et al. 1993). Some of these experiments involved mechanical gastric distension or electrical stimulation of vagal afferents, while others involved pharmacological manipulation of vagal afferents via systemic administration of CCK or other gut peptides in adult rats. For example, systemically administered ghrelin (a hormone that is *endogenously* secreted from gastric tissue when the stomach is empty) suppresses the firing activity of GI vagal afferents, whereas CCK increases vagal afferent firing. In adult rodents, increased gastric release of ghrelin promotes increased release of growth hormone (GH) from the anterior pituitary. Ghrelin binds to receptors expressed by vagal sensory neurons, which transport the receptor to their peripheral axon terminals within the stomach wall (see Fig. 13.1). Blockade of vagal afferent signaling to the

NTS abolishes ghrelin-induced GH secretion and also blocks the ability of ghrelin to activate parvocellular GHRH endocrine neurons (Date et al. 2002).

13.3.2 Ascending Visceral Sensory Inputs to the Hypothalamus in Adult Rats and Mice

Visceral sensory signals arising within the GI tract and other peripheral tissues are relayed to the neuroendocrine hypothalamus via direct and indirect ascending neural projection pathways from the NTS and AP in adult rats and mice (Rinaman 2007). The indirect pathways include projections from AP and NTS neurons that are relayed through the medullary RF (including the VLM) and pontine PBN before ascending to the hypothalamus. Projections from AP neurons to the subjacent NTS and also directly to the VLM and PBN provide a neural route for blood borne signals to alter central visceral sensory processing. However, while endocrine neurons within the hypothalamic PVN, SON, mPOA, and ARC receive direct synaptic input from catecholaminergic and peptidergic NTS and VLM neurons in adult rats, they appear to receive little or no direct synaptic input from the pontine PBN (Saper and Loewy 1980). The PBN instead appears to target primarily pre-autonomic hypothalamic neurons and also provides robust input to the central nucleus of the amygdala (CeA) and bed nucleus of the stria terminalis (BNST), two highly interconnected limbic forebrain regions which also receive direct input from the NTS and VLM (Bienkowski and Rinaman 2013).

13.3.3 Stimulus-Induced Activation of Visceral Sensory Inputs to the Hypothalamus

As described in previous sections (Sects. 13.2.1 and 13.3.1), *exogenous* (synthetic) CCK provides a useful experimental tool with which to manipulate ascending vagal sensory pathways. In adult rats, systemic administration of CCK increases pituitary release of ACTH, OT, GH, and thyroid stimulating hormone (TSH) through vagally mediated, CCK-1 receptor-dependent activation of NTS neurons that provide direct and relayed input to neuroendocrine cells of the hypothalamus (Fig. 13.1) (Rinaman 2007). Plasma OT levels also are increased by gastric distension, which synergizes with *exogenous* CCK to further activate GI vagal afferents (Verbalis et al. 1986). Systemic CCK increases plasma levels of AVP rather than OT in adult humans, ferrets, and rhesus macaque monkeys. While the neural circuitry underlying species differences in CCK-induced posterior pituitary hormone release remains unclear, the difference is consistent with known species differences in endocrine responses to *stress*. Rats and mice (which lack an *emetic reflex*) release OT along with ACTH during *stress responses*, whereas primates and other animals with an emetic reflex release AVP and ACTH (Verbalis et al. 1987).

Voluntary intake of a large meal by adult rats activates only a subset of the NTS visceral sensory neurons that are activated after mechanical gastric distension, supraphysiological doses of systemic CCK, and other "visceral stressors" that increase ACTH and OT release (Rinaman 1999a). Nevertheless, the ability of these visceral stimuli to alter pituitary hormone secretion indicates that visceral sensory signals impact neuroendocrine hypothalamic cells in adult rats; conversely, the lack of hypothalamic neural Fos activation and hormonal responses to vagal stimulation in neonatal rats indicates that the central pathways conveying visceral signals from the NTS to the hypothalamus are functionally immature at birth (Rinaman and Koehnle 2010). The following section reviews what is currently known regarding the chemical phenotypes, anatomical organization, and postnatal maturation of neural projections from the caudal brainstem NTS and VLM to the neuroendocrine hypothalamus.

13.3.4 Chemical Phenotypes of Hindbrain Neurons Conveying Visceral Sensory Signals to the Neuroendocrine Hypothalamus

13.3.4.1 Noradrenergic (NA) Signaling Pathways

In adult rats, visceral sensory inputs to the hypothalamus feature axonal projections that arise from medullary NA and adrenergic neurons within the NTS (A2/C2 cell groups) and VLM (A1/C1 cell groups) (Sawchenko and Swanson 1982; Rinaman 2010). For convenience, these NTS and VLM neurons are hereafter referred to collectively as NA neurons, because they all are immunoreactive for the NA synthetic enzyme, dopamine β hydroxylase (DBH). Medullary NA neurons appear to be glutamatergic, based on their expression of mRNA for vesicular glutamatergic transporter 2, which can also be visualized immunocytochemically in DBH-positive terminals within the neuroendocrine hypothalamus (Zheng et al. 2015). Subpopulations of NA neurons within the NTS and VLM that project to the hypothalamus also express multiple peptides, including prolactin-releasing peptide (PrRP) and neuropeptide Y (NPY) (Rinaman 2007).

Endocrine neurons within the mPOA, ARC, PVN, and SON express high levels of adrenergic receptors (Kaminski and Watts 2012). The caudal NTS and VLM provide most of the NA input to these hypothalamic nuclei, with a notably smaller contribution from the locus coeruleus (A6 cell group) (Day et al. 1980). NA terminals form both symmetric- and asymmetric-type synapses with the dendrites and somata of OT, AVP, CRF, and TRH neurons. NA signaling promotes activation of these neurons, thereby increasing hormone release (Nillni 2010). For example, adrenergic signaling pathways from the caudal medulla to the hypothalamus provide critical control over vagally mediated activation of the hypothalamic–pituitary–adrenal (HPA) axis in adult rats, due to synaptic stimulation of CRF neurons (Kamilaris et al. 1992). In addition to direct synaptic inputs onto the somata and dendrites of identified endocrine neurons, NA axon terminals form close appositions with axons in the external layer of the ME in adult rats, suggesting an additional site for regulating hormone release into the portal system supplying the anterior pituitary.

In contrast to the stimulatory effects of adrenergic signaling on endocrine CRH, TRH, AVP, and OT neurons, adrenergic receptor signaling directly and robustly suppresses the activity of GnRH neurons within the mPOA in adult male and female mice, in which NA signaling may serve a permissive role in the ability of other inputs to regulate pulsatile GnRH and LH secretion (Szawka et al. 2013; Briski and Shrestha 2016). During periods of caloric deficit, for example, NA inputs to the neuroendocrine hypothalamus provide the primary source of glucoprivic stimulation that inhibits pituitary LH secretion (Ohkura et al. 2000). Metabolic status is monitored within the NTS, and NA signaling from the NTS suppresses the pre-ovulatory LH surge (Ibrahim and Briski 2014). Indeed, energy deficiency is a key cause of reduced frequency or cessation of ovulation in women.

NA projections from the NTS and VLM to the hypothalamus course primarily through the ventral noradrenergic ascending bundle (VNAB) (Fig. 13.1). Electrical stimulation of the VNAB elicits CRF secretion into the portal vasculature serving the anterior pituitary through an adrenergic receptor-mediated mechanism (Plotsky 1987), thereby increasing plasma levels of ACTH. Similarly, direct stimulation of the NTS or VLM increases PVN neuronal firing and pituitary hormone secretion, and these effects depend on NA inputs to the PVN (Day et al. 1985). Extracellular NA content increases in the PVN after systemic administration of CCK, which, as reviewed above, activates vagal sensory inputs to the NTS and stimulates pituitary release of several hormones. Systemic CCK activates NA neurons within the NTS and VLM in rats, ferrets, and rhesus macaques (Rinaman 2007). In rats, NA neurons activated after CCK administration include those that project directly to the PVN (Rinaman et al. 1995), and immunotoxin-induced destruction of these NA neurons attenuates the ability of systemic CCK to activate Fos expression within PVN endocrine neurons (Rinaman 2010). A CCK-induced increase in NA content measured within the PVN directly parallels increased plasma levels of ACTH and OT in rats, consistent with the predominantly excitatory effects of systemic CCK and synaptic effects of NA on hypothalamic CRF and OT neurons (Verbalis et al. 1991).

13.3.4.2 Glucagon-like Peptide 1 (GLP1) Projections

Separate populations of non-NA hypothalamic-projecting neurons within the caudal NTS and VLM express immunoreactivity for a variety of neuropeptides (Rinaman 2007), including CCK and GLP1. A subset of CCK-expressing NTS neurons have ascending axonal projections to the PVN in adult mice (Roman et al. 2017), but the synaptic targets remain to be determined. Conversely, GLP1 circuits are more well defined. GLP1 neurons within the caudal NTS and adjacent medullary RF are glutamatergic (Zheng et al. 2015), similar to NA neurons (Stornetta et al. 2002), and provide the sole source of GLP1-positive axon terminals within the hypothalamus. Similar to ascending NA projections from the NTS and VLM, the axons of GLP1-positive neurons ascend to the hypothalamus within the VNAB (Fig. 13.1) to target parvocellular CRH and TRH neurons in the PVN, magnocellular OT neurons in the PVN and SON, and parvocellular GnRH neurons in the mPOA. GLP1 binding sites and GLP1 receptor (GLP1-R) mRNA expression is evident in these and other hypothalamic endocrine nuclei (Goke et al. 1995; Shughrue et al. 1996; Zueco et al.

1999), and GLP1 terminals synapse on kisspeptin-expressing neurons within the ARC that regulate GnRH neurons (Heppner et al. 2017). The available data indicate that GLP1 signaling increases the excitability of endocrine neurons that express GLP1-R and also acts presynaptically to modulate glutamatergic and GABAergic synaptic inputs to endocrine neurons (Farkas et al. 2016).

GLP1 neurons have been implicated in neuroendocrine responses to acute stress. Early studies found that central (but not systemic) infusion of GLP1 or GLP1-R agonists activates Fos in hypothalamic endocrine neurons and increases plasma levels of OT, ACTH, and corticosterone in adult rodents (Larsen et al. 1997). Subsequent studies found that a variety of interoceptive stressors [i.e., systemic administration of lithium chloride (LiCl), lipopolysaccharide (LPS), or high doses of CCK] and cognitive stressors (i.e., elevated platform exposure, restraint/immobilization) activate Fos expression in the majority of GLP1 neurons in adult rats, including GLP1 neurons that project to the hypothalamus (Rinaman 1999a). Further, central antagonism of GLP1-R suppresses the ability of visceral/interoceptive or cognitive stressors to activate the HPA axis (Rinaman 1999b; Kinzig et al. 2003). Thus, both physiological and cognitive threats to homeostasis recruit ascending GLP1 neural pathways to the neuroendocrine hypothalamus, similar to threat-induced recruitment of ascending NA pathways (Rinaman 2007). Notably, mechanical gastric distension, supraphysiological doses of CCK, and rapid intake of an extremely large liquid meal can activate Fos expression in GLP1 neurons (Rinaman 1999a; Kreisler et al. 2014), consistent with evidence that GLP1 neurons respond to unusually strong (i.e., homeostatically threatening) GI sensory signals (Maniscalco et al. 2013). Thus, recruitment of hindbrain GLP1 neurons, including those that innervate hypothalamic neuroendocrine cells, may depend on the intensity, modality, and/or temporal features of visceral stimulation, and may even differentiate stressful from non-stressful stimuli.

13.3.5 Postnatal Development of NA and GLP1 Axonal Inputs to the Neuroendocrine Hypothalamus

Immunocytochemical techniques have demonstrated that NA projections from the NTS and VLM to the hypothalamus are present at birth. However, the density of DBH-immunopositive NA terminals increases markedly in rats during the first 3 weeks of postnatal development (Fig. 13.1) (Khachaturian and Sladek 1980; Rinaman 2001), in concert with increasing levels of hypothalamic norepinephrine (Coyle and Axelrod 1971). Additional evidence for delayed neurochemical maturation of ascending NA projection pathways is offered by analysis of fiber immunolabeling for PrRP, which is co-expressed by caudal subsets of NA neurons within the NTS (A2 cell group) and VLM (A1 cell group) that project to the hypothalamus and limbic forebrain (Chen et al. 1999; Morales et al. 2000). In rats, PrRP-positive fibers are first visible within the anterior ventral BNST (which is the main axonal target of hindbrain PrRP neurons) on postnatal day (P)3 (Yano et al.

2001). By P6, PrRP-positive fibers also are present within the PVN and other hypothalamic nuclei, but fiber density is markedly sparser at P6 than in adulthood.

It is important to note that an absence or scarcity of neurotransmitter-positive fibers should not be interpreted as evidence that the axons of NTS and VLM neurons are not present within the hypothalamus. In one study in which retrograde neural tracer was microinjected into the PVN in rat pups on the day of birth (P0), the number of PVN-projecting neurons within the NTS and VLM quantified 24 h later (on P1) was similar to the number of retrogradely labeled, PVN-projecting neurons in adult rats (Rinaman 2001). However, a smaller proportion of PVN-projecting NTS and VLM neurons were DBH-positive in newborn rats compared with adults, consistent with reduced DBH immunostaining in the neonatal PVN. Conversely, fiber immunolabeling for phenylethanolamine N-methyltransferase (PNMT, the enzyme that converts norepinephrine to epinephrine) within the PVN was most dense at P1, gradually decreasing to adult-like levels by P21. The dynamic temporal shift in the density of immunolabeling for these catecholaminergic synthetic enzymes reflects a gradual maturation of noradrenergic and adrenergic signaling to the hypothalamus in rats during the first 3 weeks of life (Rinaman 2001). It remains unclear whether the shift in immunolabeling patterns represents parallel increases and reductions in axonal terminal density arising from PNMT-negative versus PNMT-positive NA neurons, or whether it represents altered enzyme content within individual neurons whose axons innervate the neuroendocrine hypothalamus.

13.3.5.1 Postnatal Development of Visceral Sensory Pathways Assessed in Transgenic Rodent Models

Recent advances in transgenic technology have provided additional tools with which to assess the development and distribution of axonal projections arising from phenotypically identified neurons. In one transgenic mouse model, neurons that express *preproglucagon* (*PPG*, which is cleaved to GLP1, GLP2, glucagon, glicentin, oxyntomodulin, and other peptides) were engineered to also express yellow fluorescent protein (YFP). In these "PPG-YFP" mice, YFP expression completely fills the somatic, dendritic, and axonal compartments of PPG-expressing neurons, providing a convenient and chemically specific marker that does not depend on immunolabeling of GLP1 itself (Llewellyn-Smith et al. 2011). In previously unpublished work, we observed abundant YFP-positive cell bodies, dendrites, and axons within the caudal NTS in newborn PPG-YFP mice (i.e., P0), and dual immunolabeling confirmed that YFP-expressing neurons are immunopositive for GLP1 on P6 (Fig. 13.3). However, YFP-positive fibers are nearly absent within the neuroendocrine hypothalamus in newborn mice, with YFP immunolabeling increasing gradually within the PVN over the first 3 weeks of postnatal development (Fig. 13.4). GLP1 neurons express leptin receptors in mice, in which leptin signaling stimulates the postnatal development of axonal outgrowth from the ARC to the PVN (Bouret and Simerly 2007). An ongoing study is investigating whether the postnatal surge in leptin is also required for normal development of GLP1 signaling pathways from the NTS to the PVN (J. Biddinger and R. Simerly, Vanderbilt University Medical Center; personal communication). GLP1 neurons in rats do not express leptin receptors; instead, leptin receptors in rats

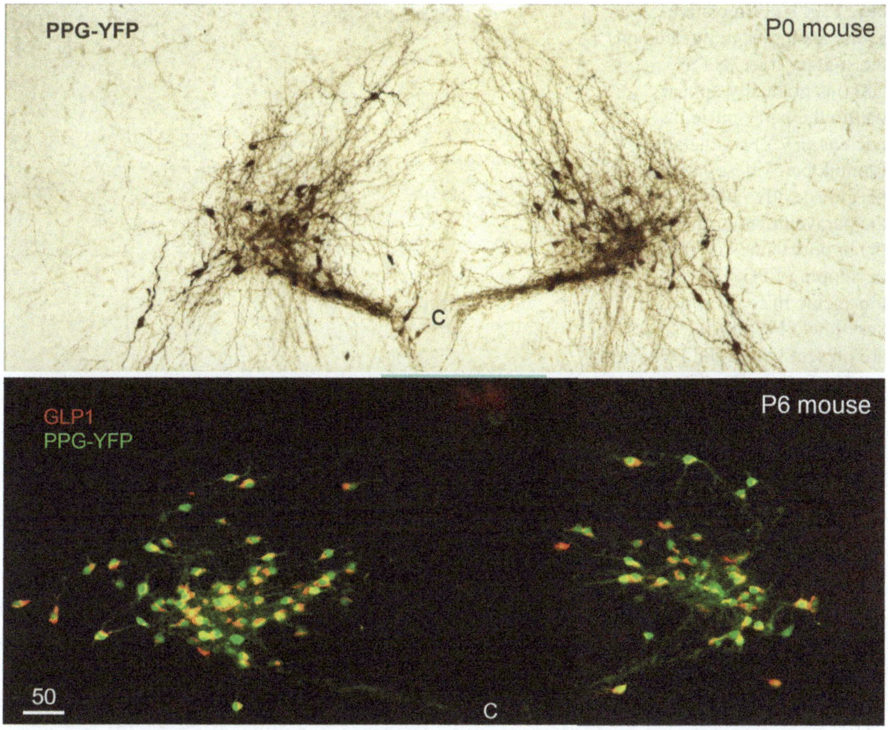

Fig. 13.3 Immunolabeling for yellow fluorescent protein (YFP) expressed under control of the preproglucagon (PPG) promoter in transgenic PPG-YFP mice. Top panel: YFP immunoperoxidase labeling of PPG neurons and processes within the caudal NTS in a newborn mouse (P0). Bottom panel: green YFP immunofluorescence is co-localized with red GLP1 immunofluorescence to generate yellow double-labeled neurons within the caudal NTS in a 6-day old PPG-YFP mouse (P6). C: central canal. Scale bar: 50 μm

are expressed by NA neurons within the caudal NTS, which do not express leptin receptors in mice (Huo et al. 2008). The potential significance of these species differences in expression of leptin receptors in the caudal NTS remains unclear, but it presumably is linked to differential effects of leptin on hindbrain-hypothalamic circuits in both developing and mature animals.

13.3.5.2 Functional Immaturity of Visceral Sensory Inputs to the Hypothalamus in Neonatal Rats

The apparent immaturity of NA and GLP1 projections from the caudal medulla to the hypothalamus in neonatal rats predicts that endocrine neurons are less sensitive to visceral sensory signals early in life. Indeed, CCK does not activate Fos expression in the PVN (Fig. 13.5) and does not increase plasma levels of OT in newborn rat pups (Rinaman et al. 1994), whereas plasma OT levels *are* increased in response to stimulation of gastric vagal afferents in 10-day-old rat pups (Nelson et al. 1998).

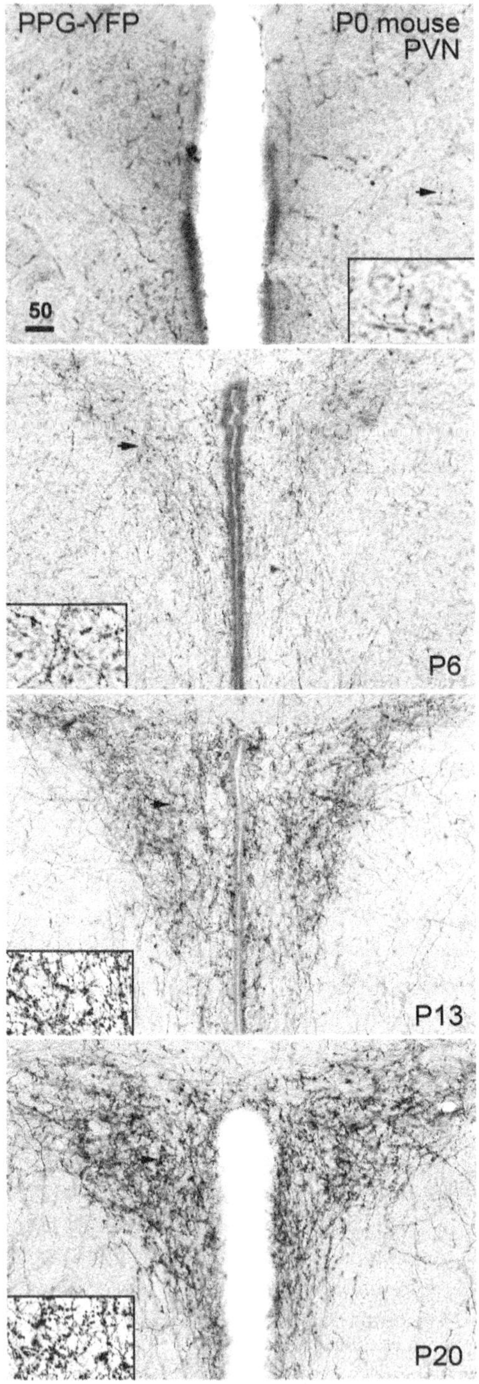

Fig. 13.4 YFP-positive axons arising from nucleus of the solitary tract (NTS) neurons gradually appear within the paraventricular nucleus (PVN) postnatally. Immunoperoxidase labeling of YFP-positive fibers within the paraventricular nucleus of the hypothalamus (PVN) in developing PPG-YFP transgenic mice. Top panel: very few YFP-positive fibers are present within the PVN in newborn mice (P0). The density of YFP-positive fibers increases within the PVN at later postnatal ages (P6, P13), with adult-like labeling evident by the time of weaning (P20, bottom panel). Scale bar: 50 μm

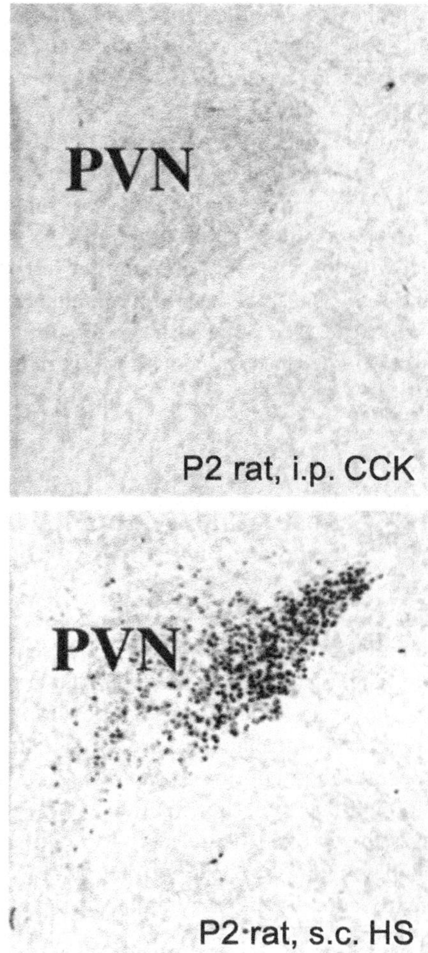

Fig. 13.5 Responses to visceral signals mature differentially. Immunocytochemical labeling of Fos within the hypothalamic paraventricular nucleus (PVN) in neonatal rats. Top panel: i.p. injection of cholecystokinin (CCK) does not increase Fos labeling within the PVN in 2-day-old rat pups, although it does so in adult rats. Figure adapted from Rinaman et al. (1994). *Bottom panel:* s.c. injection of hypertonic saline (HS) robustly activates Fos expression in 2-day-old rat pups, similar to activation in adult rats. Scale bar: 50 μm. Figure adapted from Rinaman et al. (1997)

Another study investigated the postnatal maturation of central neural Fos responses to i.p. injection of LiCl, a malaise-inducing agent (Koehnle and Rinaman 2007). Compared to Fos activation in neonatal control rats injected with physiological saline, LiCl did not increase Fos in the PVN or other forebrain regions at P0, but did so at P7 and later, with maximal PVN Fos responses to LiCl achieved by P14. Comparable results have been obtained examining central Fos responses to an acute LPS challenge in developing rats (Oladehin and Blatteis 1995). These findings provide additional evidence that vagal sensory signaling to the neuroendocrine hypothalamus undergoes significant functional maturation in rats during the first 2–3 weeks of postnatal life (Rinaman and Koehnle 2010).

Given the apparent inability of systemic CCK, LiCl, or LPS to activate hypothalamic neurons in neonatal rats, one might hypothesize that hypothalamic endocrine neurons are refractory to all sources of excitatory input early in development. This hypothesis is challenged by evidence that acute osmotic dehydration (induced by subcutaneous injection of hypertonic saline) robustly activates Fos expression in osmosensitive regions of the hypothalamus (Fig. 13.5), including magnocellular PVN and SON neurons, in both neonatal and adult rats (Rinaman et al. 1997). The important difference in the ability of *interoceptive* challenge to activate neuroendocrine cells of the hypothalamus seems to be whether or not hypothalamic responses require ascending visceral sensory inputs from the caudal brainstem. In adult rats, these ascending inputs are unnecessary for hypothalamic responses to osmotic dehydration, but they are required for responses to gastric distension, CCK, LPS, and LiCl. Thus, the lack of hypothalamic Fos activation in neonatal rats treated with CCK, LiCl, and other visceral stimuli is most likely due to functional immaturity of ascending projection pathways that relay visceral sensory signals from the NTS and VLM.

13.4 Postnatal Development and Potential Plasticity of Visceral Sensory Signaling to the Neuroendocrine Hypothalamus

The findings summarized above (Sect. 13.3.5) support the view that ascending neural pathways conveying visceral sensory signals from hindbrain to hypothalamus are functionally immature in neonatal rats. Thus, these circuits may exhibit plasticity as they assemble during a developmentally defined sensitive period. Highly specialized mechanisms are crucial for the initial establishment of postsynaptic specializations during synaptogenesis, and for activity-related changes in synaptic strength that underlie experience-dependent plasticity. By analogy with other CNS systems, evoked neural activity within ascending visceral sensory pathways may shape ongoing synapse formation and stabilization of visceral sensory inputs to the neuroendocrine hypothalamus during the first few weeks of postnatal life.

13.4.1 The Stress-Hyporesponsive Period (SHRP)

In rats, the first 2 weeks of postnatal life correspond to the so-called SHRP, characterized by reduced HPA axis responsiveness to stressful stimuli (Walker et al. 1991). Most of the peripheral and central components of stress-responsive circuits are capable of being activated in neonatal rats, given sufficient experimental or natural conditions (Levine 2001). However, HPA axis responsiveness in neonatal rats is apparently suppressed by maternal factors that generate interoceptive signals in their developing offspring, including thermal, chemical, and mechanoreceptive components (Rinaman and Koehnle 2010).

13.4.2 Early Life Experience Alters Visceral Sensory Inputs to the Neuroendocrine Hypothalamus

Plasma levels of GI-derived CCK and pituitary-derived GH fall when rat pups are separated from their dam and increase during and after maternal reunion and feeding (Kuhn et al. 1979; Weller et al. 1992). Experimental paradigms that reduce/fragment the early life maternal care received by rat pups exert a significant impact on adult measures of stress-evoked norepinephrine release within the PVN, modulating stimulus-induced activation of CRH endocrine neurons and anterior pituitary release of ACTH (Liu et al. 2000; Walker et al. 2017). Further, CCK-1 receptors are especially abundant in the upper GI tract in newborn rats (Robinson et al. 1987). Based on the results of experiments in which early life maternal care was manipulated together with manipulation of CCK-1 receptor signaling, we have proposed that the impact of maternal care on central visceral circuits is based, at least in part, on the activity of GI vagal sensory input to the NTS (Weber et al. 2009; Rinaman and Koehnle 2010), which in turn regulates the neuroendocrine hypothalamus through ascending NA, GLP1, and potentially other neurochemical signaling pathways.

13.5 Perspectives

Pituitary hormone secretion is modulated by *interoceptive* feedback signals that are relayed primarily through the caudal brainstem NTS. NA and GLP1-expressing NTS neurons project to the hypothalamus, where adrenergic and GLP1R-mediated signaling regulates gene expression and secretory activity in endocrine neurons. Given the fundamental importance of the caudal NTS as a sensory interface between internal body organs and the brain, it is striking that we still know relatively little regarding the developmental maturation of neural signaling pathways from the NTS to the neuroendocrine hypothalamus or the means by which these circuits might be programmed by early life experience and other epigenetic influences. In humans, visceral hypersensitivity and neuroendocrine dysfunctions are recognized as common symptoms of stress-related disorders such as anxiety and depression, which feature inappropriate physiological responses to stressful, emotionally triggering stimuli. Aberrant neuroendocrine and other physiological responses may arise, at least in part, as a product of postnatal life events that shape the developmental assembly of brainstem-hypothalamic visceral sensory pathways.

Key References

Bouret and Simerly (2007)—This review article summarizes evidence that the intrinsic hypothalamic circuits that integrate interoceptive signals (here, leptin signaling from visceral adipose tissue) undergo a significant amount of postnatal development in rodents. The authors present evidence that a natural postnatal surge in leptin promotes neural outgrowth and synapse formation within the neuroendocrine hypothalamus.

Coyle and Axelrod (1971)—One of the first publications providing evidence that noradrenergic inputs to the hypothalamus (which originate primarily from the hindbrain NTS and VLM) develop gradually during the first 3 weeks postnatal in rats.

Day et al. (1985)—This report provides compelling evidence in adult rats that noradrenergic projections from the hindbrain NTS and VLM to the neuroendocrine hypothalamus facilitate the activity of neuroendocrine neurons that control anterior pituitary hormone release.

Kamilaris et al. (1992)—A comprehensive set of experiments demonstrating that systemic CCK activates vagal sensory inputs to the NTS in rats, promoting activation of neuroendocrine CRH neurons at the apex of the HPA axis.

Rinaman and Koehnle (2010)—This book chapter summarizes what was known at the time regarding the anatomical and functional development of ascending visceral sensory pathways from the hindbrain NTS to the hypothalamus and other forebrain regions in rats.

Sawchenko and Swanson (1982)—A classic paper and required reading for anyone interested in understanding the anatomical organization of ascending hindbrain-hypothalamic circuits. The authors focus on inputs to the neuroendocrine hypothalamus.

Walker et al. (1991)—A comprehensive analysis of what was previously called the "stress nonresponsive period" of postnatal development. The authors conclude that the HPA axis is responsive to certain types of stimuli (including those that do not depend on ascending visceral sensory pathways to the hypothalamus), and effectively argue that this early developmental period should instead be called the "stress hyporesponsive period" (SHRP).

References

Bienkowski MS, Rinaman L (2013) Common and distinct neural inputs to the medial central nucleus of the amygdala and anterior ventrolateral bed nucleus of stria terminalis in rats. Brain Struct Funct 218:187–208

Bouret SG, Simerly RB (2007) Development of leptin-sensitive circuits. J Neuroendocrinol 19:575–582

Bouret SG, Draper SJ, Simerly RB (2004) Formation of projection pathways from the arcuate nucleus of the hypothalamus to hypothalamic regions implicated in the neural control of feeding behavior in mice. J Neurosci 24:2797–2805

Briski KP, Shrestha PK (2016) Hindbrain estrogen receptor-beta antagonism normalizes reproductive and counter-regulatory hormone secretion in hypoglycemic steroid-primed ovariectomized female rats. Neuroscience 331:62–71

Chang HC, Chia KF, Huang JJ, Lim RKS (1939) Vagus-post-pituitary reflex; antidiuretic effect. Clin J Physiol 14:161–172

Chen C-T, Dun SL, Dun NJ, Chang J-K (1999) Prolactin-releasing peptide-immunoreactivity in A1 and A2 noradrenergic neurons of the rat medulla. Brain Res 822:276–279

Coyle JT, Axelrod J (1971) Development of the uptake and storage of L-[^{3}H]norepinephrine in the rat brain. J Neurochem 18:2061–2075

Date Y, Murakami N, Toshinai K, Matsukura S, Niijima A, Matsuo H, Kangawa K, Nakazato M (2002) The role of the gastric afferent vagal nerve in ghrelin-induced feeding and growth hormone secretion in rats. Gastroenterology 123:1120–1128

Day TA, Blessing W, Willoughby JO (1980) Noradrenergic and dopaminergic projections to the medial preoptic area of the rat. A combined horseradish peroxidase/catecholamine fluorescence study. Brain Res 193:543–548

Day T, Ferguson AV, Renaud L (1985) Noradrenergic afferents facilitate the activity of tuberoinfundibular neurons of the hypothalamic paraventricular nucleus. Neuroendocrinology 41:17

Farkas I, Vastagh C, Farkas E, Balint F, Skrapits K, Hrabovszky E, Fekete C, Liposits Z (2016) Glucagon-like peptide-1 excites firing and increases GABAergic miniature postsynaptic currents (mPSCs) in gonadotropin-releasing hormone (GnRH) neurons of the male mice via activation of nitric oxide (NO) and suppression of endocannabinoid signaling pathways. Front Cell Neurosci 10:214

Godement P, Vanselow J, Thanos S, Bonhoeffer F (1987) A study in developing visual systems with a new method of staining neurones and their processes in fixed tissue. Development 101:697–713

Goke R, Larsen PJ, Mikkelsen JD, Sheikh SP (1995) Distribution of GLP-1 binding sites in the rat brain: evidence that exendin-4 is a ligand of brain GLP-1 binding sites. Eur J Neurosci 7:2294–2300

Grill HJ, Norgren R (1978) Chronically decerebrate rats demonstrate satiation but not bait shyness. Science 201:267–269

Heppner KM, Baquero AF, Bennett CM, Lindsley SR, Kirigiti MA, Bennett B, Bosch MA, Mercer AJ, Ronnekleiv OK, True C, Grove KL, Smith MS (2017) GLP-1R signaling directly activates arcuate nucleus kisspeptin action in brain slices but does not rescue luteinizing hormone inhibition in ovariectomized mice during negative energy balance. eNeuro 4(1): ENEURO.0198-16

Huo L, Gamber KM, Grill HJ, Bjorbaek C (2008) Divergent leptin signaling in proglucagon neurons of the nucleus of the solitary tract in mice and rats. Endocrinology 149:492–497

Ibrahim BA, Briski KP (2014) Role of dorsal vagal complex A2 noradrenergic neurons in hindbrain glucoprivic inhibition of the luteinizing hormone surge in the steroid-primed ovariectomized female rat: effects of 5-thioglucose on A2 functional biomarker and AMPK activity. Neuroscience 269:199–214

Jin Y, Ueta Y, Kannan H, Yamashita H (1993) Synaptic inputs from the stomach to tuberoinfundibular neurons in the paraventricular nucleus of the hypothalamus in rats. Brain Res 617:151–154

Kamilaris TC, Johnson EO, Calogero AE, Kalogeras KT, Bernardini R, Chrousos GP, Gold PW (1992) Cholecystokinin-octapeptide stimulates hypothalamic-pituitary-adrenal function in rats: role of corticotropin-releasing hormone. Endocrinology 130:1764–1774

Kaminski KL, Watts AG (2012) Intact catecholamine inputs to the forebrain are required for appropriate regulation of corticotrophin-releasing hormone and vasopressin gene expression by corticosterone in the rat paraventricular nucleus. J Neuroendocrinol 24:1517–1526

Khachaturian H, Sladek JR (1980) Simultaneous monoaminehistofluorescence and neuropeptide immunocytochemistry: III. Ontogeny of catecholamine varicosities and neurophysin neurons in the rat supraoptic and paraventricular nuclei. Peptides 1:77–95

Kinzig KP, D'Alessio DA, Herman JP, Sakai RR, Vahl TP, Figueiredo HF, Murphy EK, Seeley RJ (2003) CNS glucagon-like peptide-1 receptors mediate endocrine and anxiety responses to interoceptive and psychogenic stressors. J Neurosci 23:6163–6170

Koehnle T, Rinaman L (2007) Progressive postnatal increases in Fos immunoreactivity in the forebrain and brainstem of rats after viscerosensory stimulation with lithium chloride. Am J Physiol Regul Integr Comp Physiol 292:R1212–R1223

Kreisler AD, Davis EA, Rinaman L (2014) Differential activation of chemically identified neurons in the caudal nucleus of the solitary tract in non-entrained rats after intake of satiating vs. non-satiating meals. Physiol Behav 136:47–54

Kuhn CM, Evoniuk G, Schanberg SM (1979) Loss of tissue sensitivity to growth hormone during maternal deprivation in rats. Life Sci 25:2090–2097

Larsen PJ, Tang-Christensen M, Jessop DS (1997) Central administration of glucagon-like peptide-1 activates hypothalamic neuroendocrine neurons in the rat. Endocrinology 138:4445–4455

Levine S (2001) Primary social relationships influence the development of the hypothalamic-pituitary-adrenal axis in the rat. Physiol Behav 73:255–260

Liu D, Caldji C, Sharma S, Plotsky PM, Meaney MJ (2000) Influence of neonatal rearing conditions on stress-induced adrenocorticotropin responses and norepinephrine release in the hypothalamic paraventricular nucleus. J Neuroendocrinol 12:5–12

Llewellyn-Smith IJ, Reimann F, Gribble FM, Trapp S (2011) Preproglucagon neurons project widely to autonomic control areas in the mouse brain. Neuroscience 180:111–121

Maniscalco JW, Kreisler AD, Rinaman L (2013) Satiation and stress-induced hypophagia: examining the role of hindbrain neurons expressing prolactin-releasing peptide or glucagon-like peptide 1. Front Neurosci 6:199

Menétrey D, dePommery J (1991) Origins of spinal ascending pathways that reach central areas involved in visceroception and visceronociception in the rat. Eur J Neurosci 3:249–259

Morales T, Hinuma S, Sawchenko PE (2000) Prolactin-releasing peptide is expressed in afferents to the endocrine hypothalamus, but not in neurosecretory neurones. J Neuroendocrinol 12:131–140

Nelson EE, Alberts JR, Tian Y, Verbalis JG (1998) Oxytocin is elevated in plasma of 10-day-old rats following gastric distension. Dev Brain Res 111:301–303

Nillni EA (2010) Regulation of the hypothalamic thyrotropin releasing hormone (TRH) neuron by neuronal and peripheral inputs. Front Neuroendocrinol 31:134–156

Ohkura S, Tanaka T, Nagatani S, Bucholtz DC, Tsukamura H, Maeda K, Foster DL (2000) Central, but not peripheral, glucose-sensing mechanisms mediate glucoprivic suppression of pulsatile luteinizing hormone secretion in the sheep. Endocrinology 141:4472–4480

Oladehin A, Blatteis CM (1995) Lipopolysaccharide-induced fos expression in hypothalamic nuclei of neonatal rats. Neuroimmunomodulation 2:282–289

Palkovits M (1999) Interconnections between the neuroendocrine hypothalamus and the central autonomic system. Front Neuroendocrinol 20:270–295

Plotsky P (1987) Facilitation of immunoreactive corticotropin-releasing factor secretion into the hypophysial-portal circulation after activation of catecholaminergic pathways or central norepinephrine injection. Endocrinology 121:924–930

Raybould HE, Tache Y (1988) Cholecystokinin inhibits gastric motility and emptying via a capsaicin-sensitive vagal pathway in rats. Am J Phys 255:G242–G246

Rinaman L (1999a) Interoceptive stress activates glucagon-like peptide-1 neurons that project to the hypothalamus. Am J Phys 277:R582–R590

Rinaman L (1999b) A functional role for central glucagon-like peptide-1 receptors in lithium chloride-induced anorexia. Am J Phys 277:R1537–R1540

Rinaman L (2001) Postnatal development of catecholamine inputs to the paraventricular nucleus of the hypothalamus in rats. J Comp Neurol 438:411–422

Rinaman L (2007) Visceral sensory inputs to the endocrine hypothalamus. Front Neuroendocrinol 28:50–60

Rinaman L (2010) Hindbrain noradrenergic A2 neurons: diverse roles in autonomic, endocrine, cognitive, and behavioral functions. Am J Physiol Regul Integr Comp Physiol 00556:02010

Rinaman L, Koehnle TJ (2010) Development of central visceral circuits. In: Blumberg MS, Freeman JH, Robinson SR (eds) Oxford handbook of developmental behavioral neuroscience. Oxford University Press, New York, pp 298–321

Rinaman L, Levitt P (1993) Establishment of vagal sensorimotor circuits during fetal development in rats. J Neurobiol 24:641–659

Rinaman L, Hoffman GE, Stricker EM, Verbalis JG (1994) Exogenous cholecystokinin activates cFos expression in medullary but not hypothalamic neurons in neonatal rats. Dev Brain Res 77:140–145

Rinaman L, Hoffman GE, Dohanics J, Le WW, Stricker EM, Verbalis JG (1995) Cholecystokinin activates catecholaminergic neurons in the caudal medulla that innervate the paraventricular nucleus of the hypothalamus in rats. J Comp Neurol 360:246–256

Rinaman L, Stricker EM, Hoffman GE, Verbalis JG (1997) Central c-fos expression in neonatal and adult rats after subcutaneous injection of hypertonic saline. Neuroscience 79:1165–1175

Robinson PH, Moran TH, Goldrich M, McHugh PR (1987) Development of cholecystokinin binding sites in the rat upper gastrointestinal tract. Am J Phys 252:G529–G534

Robinson PH, Moran TH, McHugh PR (1988) Cholecystokinin inhibits independent ingestion in neonatal rats. Am J Phys 255:R14–R20

Roman CW, Sloat SR, Palmiter RD (2017) A tale of two circuits: CCK NTS neuron stimulation controls appetite and induces opposing motivational states by projections to distinct brain regions. Neuroscience 358:316–324

Saper CB, Loewy AD (1980) Efferent connections of the parabrachial nucleus in the rat. Brain Res 197:291–317

Sawchenko PE, Swanson LW (1982) The organization of noradrenergic pathways from the brainstem to the paraventricular and supraoptic nuclei in the rat. Brain Res Rev 4:275–325

Shapiro RE, Miselis RR (1985) The central neural connections of the area postrema of the rat. J Comp Neurol 234:344–364

Shughrue PJ, Lane MV, Merchenthaler I (1996) Glucagon-like peptide-1 receptor (GLP1-R) mRNA in the rat hypothalamus. Endocrinology 137:5159–5162

Stornetta RL, Sevigny CP, Guyenet PG (2002) Vesicular glutamate transporter DNPI/VGLUT2 mRNA is present in C1 and several other groups of brainstem catecholaminergic neurons. J Comp Neurol 444:191–206

Szawka RE, Poletini MO, Leite CM, Bernuci MP, Kalil B, Mendonca LB, Carolino RO, Helena CV, Bertram R, Franci CR, Anselmo-Franci JA (2013) Release of norepinephrine in the preoptic area activates anteroventral periventricular nucleus neurons and stimulates the surge of luteinizing hormone. Endocrinology 154:363–374

Thieblot L, Duchene-Marullaz P, Berthelay J (1957) Nouvelle preuve de la libération de principes posthypophysaires par l'excitation centripète du vague. Ann Endocrinol 18:651–653

Ueta Y, Kannan H, Higuchi T, Negoro H, Yamashita H (1993) CCK-8 excites oxytocin-secreting neurons in the paraventricular nucleus in rats - possible involvement of noradrenergic pathway. Brain Res Bull 32:453–459

Verbalis JG, McCann MJ, McHale CM, Stricker EM (1986) Oxytocin secretion in response to cholecystokinin and food: differentiation of nausea from satiety. Science 232:1417–1419

Verbalis JG, Richardson DW, Stricker EM (1987) Vasopressin release in response to nausea-producing agents and cholecystokinin in monkeys. Am J Phys 252:R749–R753

Verbalis JG, Stricker EM, Robinson AG, Hoffman GE (1991) Cholecystokinin activates cFos expression in hypothalamic oxytocin and corticotropin-releasing hormone neurons. J Neuroendocrinol 3:205–213

Walker CD, Scribner VA, Cascio CS, Dallman MF (1991) The pituitary-adrenocortical system of neonatal rats is responsive to stress throughout development in a time dependent and stress-specific fashion. Endocrinology 128:1385–1396

Walker CD, Bath KG, Joels M, Korosi A, Larauche M, Lucassen PJ, Morris MJ, Raineki C, Roth TL, Sullivan RM, Tache Y, Baram TZ (2017) Chronic early life stress induced by limited bedding and nesting (LBN) material in rodents: critical considerations of methodology, outcomes and translational potential. Stress 20:421–448

Weber BC, Manfredo HN, Rinaman L (2009) A potential gastrointestinal link between enhanced postnatal maternal care and reduced anxiety-like behavior in adolescent rats. Behav Neurosci 123:1178–1184

Weller A, Corp ES, Tyrka A, Ritter RC, Brenner L, Gibbs J, Smith GP (1992) Trypsin inhibitor and maternal Reunion increase plasma cholecystokinin in neonatal rats. Peptides 13:939–941
Yano T, Iijima N, Kataoka Y, Hinuma S, Tanaka M, Ibata Y (2001) Developmental expression of prolactin releasing peptide in the rat brain: localization of messenger ribonucleic acid and immunoreactive neurons. Dev Brain Res 128:101–111
Zheng H, Stornetta RL, Agassandian K, Rinaman L (2015) Glutamatergic phenotype of glucagon-like peptide 1 neurons in the caudal nucleus of the solitary tract in rats. Brain Struct Funct 220:3011–3022
Zueco JA, Esquifino AI, Chowen JA, Alvarez E, Castrillon PO, Blazquez E (1999) Coexpression of glucagon-like peptide-1 (GLP-1) receptor, vasopressin, and oxytocin mRNAs in neurons of the rat hypothalamic supraoptic and paraventricular nuclei: effect of GLP-1(7-36)amide on vasopressin and oxytocin release. J Neurochem 72:10–16

Astrocytes and Development of Neuroendocrine Circuits

14

Lydia L. DonCarlos and Julie A. Chowen

Abstract

Astrocytes, one of the most abundant cell types in the hypothalamus, perform a myriad of functions throughout all stages of development. Radial glia, astroglia-like progenitor cells, are fundamental for neurogenesis, neuronal migration, axon extension, and synaptic formation throughout the brain. These progenitor cells also give rise to mature astrocytes, which are essential for the proper functioning of the brain at all ages. In the hypothalamus, astrocytes play an important role in all neuroendocrine systems. Together with tanycytes, specialized glial cells surrounding the third ventricle of the hypothalamus, astrocytes maintain some of their developmental potential into adulthood, including their progenitor competence and ability to modulate synaptic plasticity. In this chapter, we will address what is known regarding the role of astrocytes in hypothalamic development. We will briefly cover some of the functions that astrocytes perform in specific neuroendocrine systems, as well as how the maintenance of certain developmental capacities during later stages of life is fundamental for systemic adaptation and homeostatic control. Although the roles of astrocytes in neuroendocrine control have received increasing attention in recent years, there is still much to be learned regarding these fascinating cells. Identification of functional astrocytic populations with specific markers will allow these cells to be targeted

L. L. DonCarlos
Department of Cell and Molecular Physiology, Stritch School of Medicine, Loyola University Chicago, Maywood, IL, USA

J. A. Chowen (✉)
Department of Pediatrics and Pediatric Endocrinology, Hospital Infantil Universitario Niño Jesús, Princesa Research Institute La Princesa, Madrid, Spain

Centro de Investigación Biomédica en Red de Fisiopatología de la Obesidad y Nutrición (CIBEROBN), Instituto de Salud Carlos III, IMDEA Food Institute, CEIUAM+CSIC, Madrid, Spain
e-mail: julieann.chowen@salud.madrid.org

S. Wray, S. Blackshaw (eds.), *Developmental Neuroendocrinology*, Masterclass in Neuroendocrinology 9, https://doi.org/10.1007/978-3-030-40002-6_14

and more easily manipulated, so that we can begin to learn details about how astroglia participate in the development of specific hypothalamic circuits.

Keywords
Arcuate nucleus · Astrocytes · Glia · Gliogenesis · Neuroendocrine circuits · Paraventricular nucleus · Tanycytes · Suprachiasmatic nucleus · Supraoptic nucleus

14.1 Introduction

Astrocytes are among the most abundant glial cells in the central nervous system, with the proportion of astrocytes to other glial cells or to neurons varying depending on the brain region (Ero et al. 2018). Although astrocytes were early on recognized as having processes contacting multiple different neurons and communicating with each other as well (Fig. 14.1), they were relegated to a supporting role for some years. Research beginning in the 1970s led to increasing recognition of the prominent role of astrocytes in the function of the mature and damaged nervous system, and, for the purpose of this review, as being key to the development and honing of neural circuits (Allen and Lyons 2018) including the hypothalamus and neuroendocrine systems.

Because they do not generate action potentials, many aspects of astrocytic action have been difficult to study, and research into their function has often been restricted by their physiologically inaccessible nature. Overall, astrocytes typically develop and mature later than neurons, rendering it difficult to study them independent of developing neuronal circuits. Therefore, our understanding and characterization of the development and diversity of astrocytes, and how that diversity is determined, is still in early stages relative to what we know about neurons. However, among the most informative studies on astrocyte function and plasticity are those that have examined astrocytes in the hypothalamus, both in development and in response to manipulations of, for example, gonadal steroid hormones or metabolism. Moreover, although underappreciated by neuroscientists in general, these studies of astrocytes in the developing hypothalamus have been both particularly illuminating and yet particularly enigmatic, as discussed in more detail below.

14.2 Astrocytes in Development

Glial cells include astrocytes, microglia, oligodendrocytes, tanycytes, and NG2 cells. They are distinguished by expression of specific proteins and based on their morphology. This group of cells is involved in many of the aspects of brain development that follow early patterning of the nervous system, including neurogenesis, neuronal migration, and synaptogenesis (Allen and Lyons 2018; Bosworth and Allen 2017). These cells are intimately involved in neuroprotection (Belanger and Magistretti

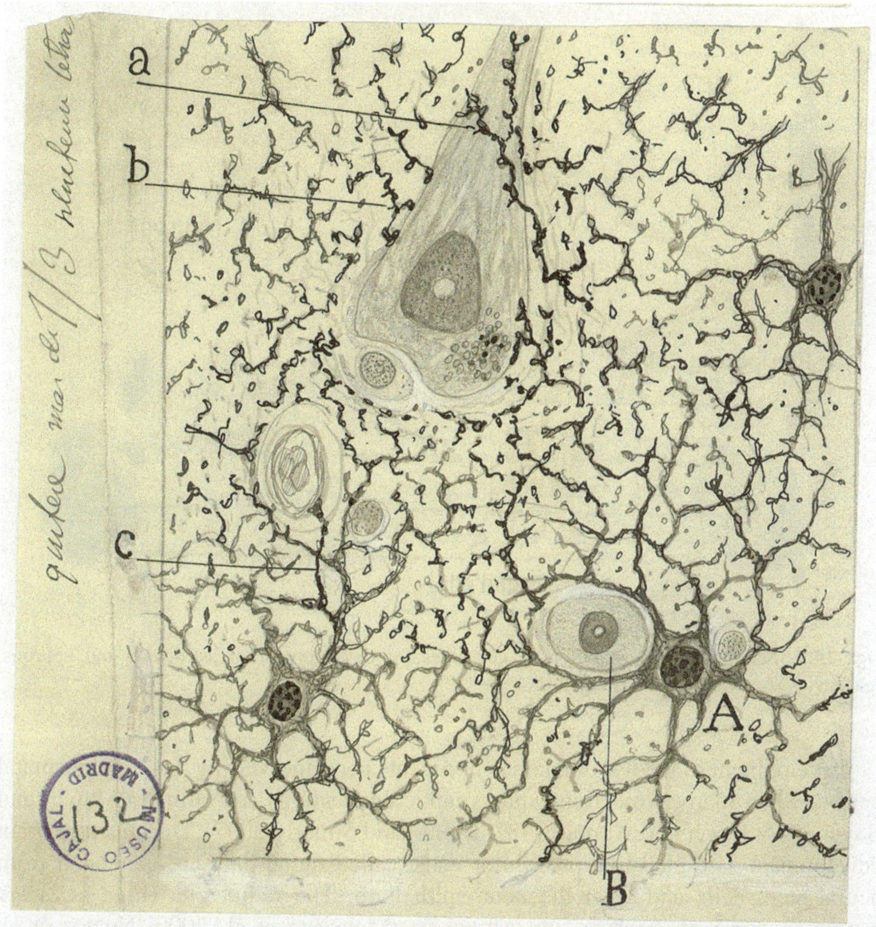

Fig. 14.1 Drawing of astrocytes by Santiago Ramón y Cajal. Courtesy of the Cajal Institute, Cajal Legacy, Spanish National Research Council (CSIC), Madrid, reprinted with permission. This detailed drawing by Santiago Ramón y Cajal demonstrates how his keen eye observed the intricate relationship between astrocytes and neurons. Astrocytes (**A**) can ensheath neuronal cell bodies (**B**) with their projections, as well as contacting neuronal processes and cell bodies (**a**, **b**) as well as blood vessels (**c**)

2009), neurotransmission (Chung et al. 2015; Perea et al. 2009), and adaptation of the CNS to physiological changes by participating in synaptic reorganization and modifications in synaptic function (Chowen et al. 2016; Hatton 1997; Sims et al. 2015). They also sustain neurons and neuronal circuits by supplying nutrients and maintaining extracellular homeostasis (Leloup et al. 2015; Magistretti and Allaman 2018) with intimate contact and communication between astrocytes and neurons throughout development.

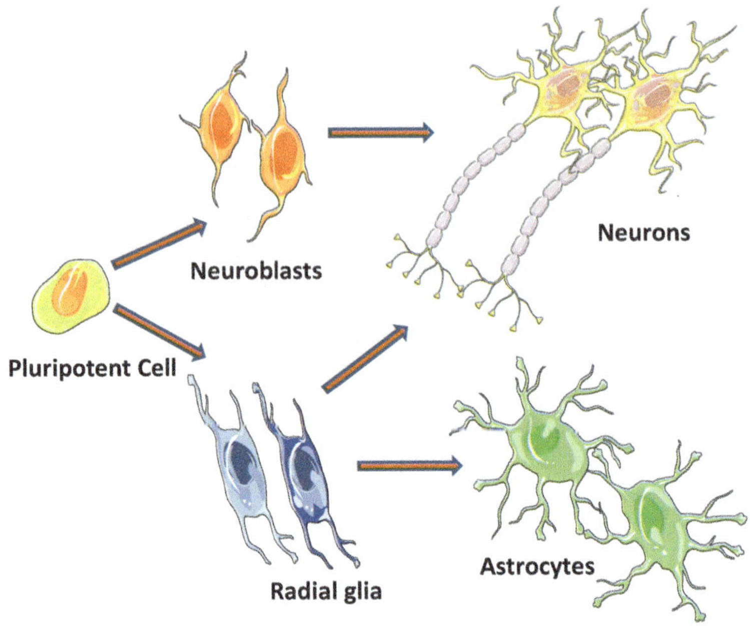

Fig. 14.2 Schematic representation of cell lines for development of astrocytes and neurons. Astrocytes are derived from radial glia

Development of the hypothalamus in rodents continues during the early postnatal period, when many projections and synaptic connections are formed (Altman and Bayer 1986; Bayer and Altman 1987; Bouret and Simerly 2007). During early brain development astroglial-like progenitor cells, called radial glia, are generated from neural stem cells and span the neuroepithelium. The radial glia (Fig. 14.2) are precursors for both neurons and astrocytes (Malatesta et al. 2000; Noctor et al. 2001), but not all radial glia have the same potential for neuronal or glial production, and part of this differential potential is likely dependent on their local environment (Malatesta et al. 2003). In rats, vimentin-reactive radial glia are found widely distributed throughout the hypothalamus and preoptic area at embryonic day 17, with this vimentin staining beginning to decrease at postnatal day (PND) 0 and dramatically decreasing by PND3. At PND6, vimentin staining is reported to be observed only in ependymal cells surrounding the third ventricle and dispersed in the optic chiasm (Munekawa et al. 2000). Glial fibrillary acidic protein (GFAP) has been widely used as a marker of astrocytes, although during development radial glial cells can also express this structural protein, as can tanycytes. Moreover, not all mature astrocytes express GFAP. Nevertheless, most of the studies to date have used markers for this protein to follow astroglial development and plasticity in the hypothalamus. Between gestational day 17.5 and PND0, both GFAP-positive and GFAP/Ki67 (Ki67 is a protein associated with proliferation) positive cells increase in the arcuate and supraoptic nuclei (Kim et al. 2016), indicating that both their number and proliferation increase.

The first GFAP-positive cells in the suprachiasmatic nucleus (SCN) are found on E15, suggesting the appearance of maturing astrocytes. The number of GFAP-positive glia increases until P21, when an adult-like pattern of GFAP immunoreactivity is observed (Botchkina and Morin 1995). In the ventrolateral portion of the SCN, Munekawa and colleagues (Munekawa et al. 2000) reported that GFAP-positive cells appear at embryonic day 20, increase at PND0, and then rapidly increase between PND3 and PND4. These developmental changes are influenced by light, with neuronal inputs from the retinohypothalamic tract being fundamental.

Although GFAP-positive cells are present in the ependymal layer of the arcuate nucleus by the first postnatal week, consistent with these cells being developing tanycytes, it is not until the second postnatal week when some GFAP-positive cells are also located in more lateral regions of the arcuate nucleus, where astrocytes are known to reside. Later, around the sixth postnatal week, GFAP-positive cells are found close to arcuate neurons and the number of these cells increases through the eighth postnatal week (Suarez et al. 1987). This study suggests that in the arcuate nucleus, GFAP-positive cells that comprise the astrocytic population do not have a mature cellular organization until after the eighth week of postnatal life. Although gliogenesis in the hypothalamus (i.e., the proliferation of astrocytes, in this case) continues through puberty and adulthood (Ahmed et al. 2008; Mohr et al. 2016), the rate of incorporation decreases with age.

14.2.1 Postnatal Maturation of Astrocytes in the Hypothalamus

Postmitotic maturation of astrocytes includes the onset of, and an increase in, expression of proteins important for astrocytic functions (e.g., glutamate transporters) and increased arborization of glial projections. Maturation of astrocytes is suggested to occur, at least in part, in response to surrounding neuronal activity, which could be involved in the differential phenotype/functionality of astrocytes in distinct brain regions. Maturation of astrocytes in areas such as the hippocampus (Bushong et al. 2004) and cortex (Morel et al. 2014) has been shown to occur from early postnatal life until approximately one month of age. Postnatal maturation of astrocytes has been documented in some hypothalamic nuclei, but much is yet to be learned regarding this process in the hypothalamus as it appears to differ to some extent from that in other brain areas. For example, in the mouse cerebral cortex, arborization of astrocytes increases between PND4 and PND26, with this rise being most significant between PND14 and 26 (Morel et al. 2014). In contrast, in the ventromedial or dorsomedial nuclei of the hypothalamus, this modification in astrocyte morphology was not observed during this time period. Moreover, loss of synaptic inputs from vesicular glutamate transporter 1 (VGluT1) positive neurons during development had no effect on astrocyte arborization in these hypothalamic nuclei, while in the cortex astrocyte morphology was altered and resulted in a reduction in the glial ensheathment of synapses in this brain area, suggesting that glutamate signaling is important for these morphological changes in the cortex (Morel et al. 2014).

In the median eminence, electron microscopic examination of gonadotropin releasing hormone (GnRH) neuron-containing transplants in adult hypogonadal mice, which have a spontaneous deletion of GnRH expression, demonstrated that the transplants sent out GnRH axons that generally followed specialized glial "channels," although the presence of astrocytes did not fully account for axon targeting specificity since the GnRH axons did not home to reactive astrocytes present at the transplantation site (Silverman et al. 1991). This may suggest that at least some preoptic/hypothalamic astroglial-independent signals beyond the channels provided cues to the GnRH axons to home to the ME or that reactive and nonreactive astrocytes produce different factors (Zamanian et al. 2012).

The proliferation and maturation of astrocytes may differ between males and females, although studies comparing these processes between the sexes in all areas of the hypothalamus remain to be performed. The number of GFAP-positive astrocytes in the SCN differs between males and females during development and this is suggested to possibly participate in the establishment of the sex differences in the neuroanatomy of this nucleus (Collado et al. 1995). Glutamate can arrest the cell cycle of astrocytes and astrocytes in the prenatal preoptic area of male rats are reported to be more susceptible to inhibition by L-glutamate than those from females (Hsu et al. 2001). The morphology of astrocytes in the preoptic area differs between the sexes as early as PND0, with astrocytes from males having more and longer primary projections than those from females (Amateau and McCarthy 2002). Astrocyte morphology in the arcuate nucleus also differs between males and females as early as PND0, with female astrocytes exhibiting a more biopolar morphology and those from males are more complex and stellate shaped. Early estrogen actions underlie these sex differences in both the preoptic area and arcuate nucleus (Amateau and McCarthy 2002; Mong et al. 1999, 1996; Mong and McCarthy 2002). In contrast, the response to early sex steroids differs in the VMH, as astrocyte morphology at PND2 was found to be unmodified by differences in sex steroids (Mong et al. 1999). Moreover, in this study, a clear difference between the detection of GFAP-positive astrocytes in the arcuate nucleus and VMH at PND2 was reported, with few cells being found in the VMH compared to the arcuate nucleus. Thus, future studies are necessary to determine where sex differences in astrocyte number and morphology occur in the hypothalamus and how this affects the function of these areas. Moreover, identifying the signals involved in determining these early sex differences is of great interest, particularly since expression of gonadal steroid receptors in developing astrocytes appears to occur after most sex differences in astrocyte morphology and responses have already been established (Lorenz et al. 2005) (see below).

Box 14.1 Does GFAP Serve as a Proxy for Events Occurring at the Terminals of Astrocytic Processes?
Large-scale studies of astrocyte dynamics in the hypothalamus have necessarily relied on data from light-microscopic investigations, often using changes in cellular GFAP distribution and expression to infer the impact of the changing hormonal milieu on the physical modulation of interneuronal communication. However, GFAP is an intermediate filament protein that is confined to more proximal processes and to endfeet of those astrocytes that express GFAP, meaning that GFAP does not demonstrate the entire extent of any given astrocyte. Therefore, it is unclear at present to what extent GFAP serves to delineate events taking place at the distal ends of astrocytes. Other methods illustrating astrocyte morphology have revealed the potential short comings of GFAP as a proxy for events occurring at astrocytic process terminals, although each of these other methods has its own shortcomings. Other markers used for labeling, genetically manipulating, or purifying astrocytes include glutamate transporter-1 (Glt1), S100 calcium-binding protein B (S100b), aldehyde dehydrogenase family 1 member L1 (Aldh1L1), aldolase C (AldoC), and aquaporin 4.

14.3 Astrocyte Diversity

Our understanding and characterization of the development and diversity of astrocytes, and how that diversity is determined, remains limited relative to what we know about neurons. Nevertheless, informative studies on hypothalamic astrocytes have emerged over the years, both in development and in response to manipulations of gonadal steroid hormones or metabolism. An electron microscopic study using the H222 antibody [which recognizes estrogen receptor (ER) alpha] demonstrated that a subpopulation of hypothalamic astrocytes expresses ERs, indicating molecular heterogeneity (Langub and Watson 1992). A later study, using confocal microscopy to examine the immunolocalization of androgen receptors in GFAP-positive cells, confirmed astrocyte heterogeneity, adding that the expression of steroid receptors changes with age. A subpopulation of astrocytes in the adult hypothalamus, but not the early postnatal hypothalamus, was shown to be androgen receptor immunoreactive (AR-ir) (Johnson et al. 2012; Lorenz et al. 2005). Astrocytic responses to metabolic signals are also anatomically specific, as expression of leptin receptors in these glial cells differs among hypothalamic nuclei, as well as throughout the brain (Garcia-Caceres et al. 2012). Studies described below show that morphological responses to hormonal signals vary by region as well. Therefore, the development of astrocytes in the hypothalamus is likely to be affected by both intrinsic and local signals that change with age, across regions, and within regions.

New transcriptional profiling methods are paving the way for a more precise study of the developmental changes and physiologic responses in potentially heterogeneous populations of astrocytes in the neuroendocrine system. One such method is Act-seq, a modified single cell RNA-seq method of high throughput transcriptional profiling (Wu et al. 2017). Act-seq was designed to prevent the artificial upregulation of immediate early genes (IEGs) that typically occurs in response to proteases used during conventional cell dissociation. The method employs the transcription inhibitor, actinomycin-D, along with modifications in dissociation methods, to inhibit transcription during cell dissociation prior to running a single cell RNA-seq assay and cluster analysis. Three distinct clusters of astrocytes were found in the medial amygdala. While all three clusters expressed known astrocytic markers, each cluster also had a distinct transcriptional profile, suggestive of unique functions. By suppressing dissociation-related upregulation of IEGs, the group was able to show that the three subsets of astrocytes in the medial amygdala differentially upregulate IEGs during a physiologic response, in this case, drug-induced seizure.

Similar approaches would be advantageous in studying hypothalamic astrocytes, in particular because the data obtained would assist the identification of specific genes of interest in neuroendocrine studies, allowing a better understanding of astrocyte heterogeneity and what circumstances drive that heterogeneity. These studies would further provide information to design and use additional methods to study astrocytes, such as translating ribosome affinity purification (TRAP) for translational profiling of the different subsets of astrocytes, or regional and morphological studies following environmental, genetic, or pharmacological manipulations. In a study by Morel et al. (2017) TRAP analysis followed by RNA-Seq on samples of somatosensory cortex, hippocampus, thalamus, hypothalamus, nucleus accumbens, and the caudate-putamen identified both a dorsoventral gradient and regional differences in expression of translating mRNAs. Hypothalamic astrocytes were reported to most closely resemble thalamic astrocytes, with a 60% overall similarity in transcriptional profile, and the transcriptional profiles of astrocytes in these diencephalic regions were different from those in the telencephalic regions sampled. It will be of great interest to determine the similarities, or differences, between astrocytes from specific hypothalamic nuclei under physiologically relevant paradigms. In the next section, we review the actions of astrocytes in the regulation of neuroendocrine function, and then discuss what is known about the participation of astrocytes in the development of neuroendocrine systems, and what molecular mechanisms mediate these effects.

Box 14.2 Astrocyte Diversity Within Regions

Astrocytes are not a homogeneous population within the central nervous system or, indeed, within a given region of the nervous system. This is the case even though astrocytes appear to have properties that are intrinsic to these cells at the species level (Oberheim et al. 2009). It is known, for example, that

(continued)

Box 14.2 (continued)
human astrocyte precursors transplanted into the cerebral cortex of mice develop the cellular morphology and functional impact on neurotransmission that would be expected of astrocytes developing in the human cortex. From region to region, though, there is a marked difference in astrocytes that can be demonstrated on multiple levels (Zhang and Barres 2010).

Within a given region of the adult nervous system, astrocytes are heterogeneous as well. Illustrating this heterogeneity of neighboring astrocytes are data on the expression of gonadal steroid hormone receptors. In neurons, only a subset typically express steroid-specific gonadal steroid hormone receptors in limbic regions (Morrell et al. 1986). Similarly, steroid receptors are not expressed in all astrocytes across regions, or within regions. The proportion of astrocytes within a region that express gonadal steroid receptors or other proteins may depend on multiple factors, but regardless, the population is heterogeneous. This constitutes an impediment to understanding how astrocytes operate within particular regions since biochemical or molecular methods to date are limited with respect to recognizing that heterogeneity within a small area. Future studies will focus on identifying these subpopulations and their specific, related, functions using genetic cell labeling methods, single cell transcriptomic and proteomic methods, among others, as discussed in the body of the chapter.

14.4 Astrocytes in Neuroendocrine Function

It has been four decades since hypothalamic astrocytes were first reported to participate in the response to hormones and the control of specific neuroendocrine functions (Tweedle and Hatton 1977). These early studies indicated that glial cells were involved in the neuroendocrine control of osmotic homeostasis, lactation, and reproduction (Olmos et al. 1989; Perlmutter et al. 1985; Theodosis et al. 1986; Tweedle and Hatton 1977). Despite these intriguing early observations, glial cells have remained in the background in neuroendocrine investigation for many years. It is only recently that they have begun to receive their due attention, and important advances have been made in understanding how they affect the functioning of neuroendocrine systems, although there is still much to be learned. Ultimately, astrocyte development in the preoptic area and hypothalamus may provide a powerful model for future investigation into the development of heterogeneity in astrocytes in general.

14.4.1 Osmotic Homeostasis

Magnocellular neurons of the paraventricular nucleus (PVN) and supraoptic nucleus (SON) are fundamental in controlling osmotic homeostasis by regulating the intake and clearance of water and salts (Stern 2015). These neurons synthesize and release the neurohypophysial hormones oxytocin and vasopressin into the blood. The secretion of oxytocin and vasopressin from the neurohypophysis is stimulated by an increase in plasma osmotic pressure and inhibited by decreased osmotic stimuli. In addition to receiving information from both peripheral and central osmoreceptors, these magnocellular neurons also have intrinsic osmosensing capabilities (Bourque and Oliet 1997; Bourque et al. 1994).

The somas and dendrites of oxytocin and vasopressin neurons are in very close proximity to one another (Fig. 14.3) and, under basal conditions, thin lamella-like astrocytic processes separate the dendritic and perikaryal membranes (Tweedle and Hatton 1977). Oxytocin stimulates the retraction of astrocytic processes in the SON and PVN, resulting in increased contact between plasma membranes from adjacent neuronal somas and adjacent dendrites (Marzban et al. 1992; Perlmutter et al. 1985; Theodosis et al. 1986; Theodosis and Poulain 1984). Structural changes resulting from magnocellular oxytocin release during lactation or parturition are thought to participate in the coordinated activation of oxytocin neurons (Wang and Hatton 2007, 2009). These studies demonstrated an inverse relationship between astroglial coverage of neuronal cell bodies and synaptic inputs to these neurons.

The decrease in astroglial coverage of oxytocin neurons during lactation results in an increase in extracellular glutamate levels in the space surrounding these neurons, which switches the function of the presynaptic kainate receptors that control GABA release from inhibitory synapses on oxytocin neurons (Bonfardin et al. 2010). Bonfardin and colleagues reported that in non-lactating rats, activation of kainate receptors facilitates inhibitory GABAergic transmission. The opposite occurs in lactating rats, where activation of kainate receptors inhibits GABA release. The restructuring of glial-neuronal contacts results in modifications in synaptic inputs to oxytocin neurons and in synaptic transmission that increases oxytocin neuronal activity, culminating in milk ejection in lactating rats (Wang and Hatton 2007, 2009). Dehydration and rehydration also affect modifications in astroglial–neuronal interactions of oxytocin and vasopressin neurons in the SON and PVN (Marzban et al. 1992; Perlmutter et al. 1985; Tweedle and Hatton 1977). Moreover, these glial cells release taurine in response to dehydration, which then activates glycine receptors on nearby neurons (Deleuze et al. 1998) to regulate extrasynaptic glutamate excitation of *N*-methyl-D-aspartate (NMDA) receptors (Joe et al. 2014).

Glial cells in the hypothalamic neurohypophysial system are reported to maintain embryonic features that are important for these structural synaptic modifications. These astrocytes possess features, such as expression of vimentin and high levels of polysialyated NCAM, which are found in radial glia. In addition, in the adult, astrocytes in the SON are also reported to maintain a radial glia-like morphology (Bonfanti et al. 1993; Theodosis and Poulain 1999).

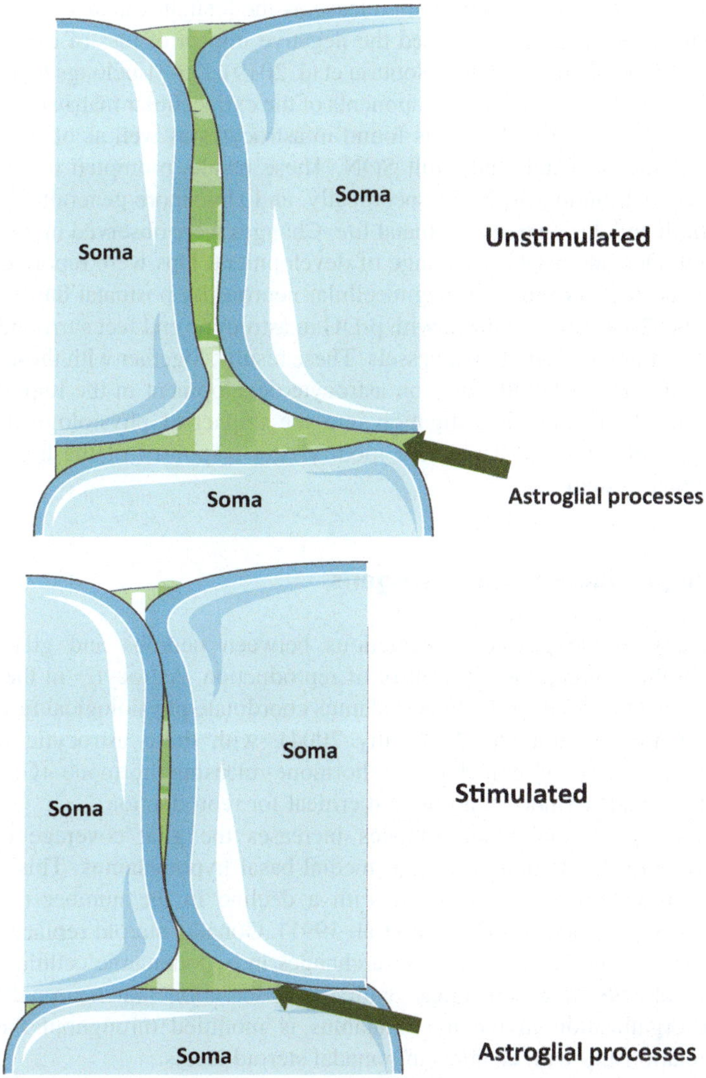

Fig. 14.3 Schematic representation of magnocellular neurons of the paraventricular and supraoptic nucleus. Under basal or unstimulated conditions, the somas (blue) of these neurons are separated by astrocytic processes (green). When stimulated (e.g., by oxytocin for oxytocin neurons), astrocytic processes are retracted and contact between the plasma membranes of neurons is increased

Our understanding of the development of astrocytic function in the neurohypophysial system is largely restricted to inference based on changes in astrocyte actions in adult physiology. It is known, however, that both maternal obesity and interleukin 6 increase astrocyte proliferation in the fetal and neonatal SON (Kim et al. 2016) as discussed below, but the functional impact of these manipulations on astrocytic modulation of oxytocin or vasopressin release by the fetal, neonatal, or adult SON is not known. A recent study addressed the negative impact of loss of dystrophin-71 (Dp-71) on SON function in mice (Souttou et al. 2019). Dp-71 belongs to a family of dystrophins (Dps) that comprise components of the extracellular matrix cell receptor, beta-dystroglycan (β-DG). Dp-71 is found in astrocytes as well as other cell types within the fetal, neonatal, and adult SON. These results prompted the authors to evaluate the distribution of Dp-71 specifically, and Dps, more generally, as well as β-DG at birth and during early postnatal life. Changes were observed in the distribution of both Dps and β-DG with stage of development. Dps were reported to be in astrocytic end feet surrounding magnocellular neurons by postnatal day 10, and by postnatal day 20 were colocalized with β-DG in astrocytic end feet surrounding both magnocellular neurons and blood vessels. These results, together with those showing an impact of early metabolic cues on astrocyte development in the hypothalamus, support a need for more investigation into the impact of physiological cues on developing astrocytes and how these modify later responses of the hypothalamic magnocellular system.

14.4.2 Reproduction and Estrogens

Modifications in the physical interactions between neurons and glia are also involved in the neuroendocrine control of reproduction. Astrocytes in the preoptic area and arcuate nucleus of the hypothalamus coordinate physiological responses to estradiol (Garcia-Segura and McCarthy 2004), with these astrocytic responses regulating the surge of gonadotropin hormone-releasing hormone (GnRH) that precipitates female ovulation, a function critical for reproduction.

Ovariectomy of nonhuman primates increases the glial coverage of GnRH neurons in both the preoptic area and medial-basal hypothalamus. This increased neuronal ensheathment is associated with a decline in the number of synaptic contacts onto these neurons (Witkin et al. 1991). Gonadal steroid replacement was shown to at least partially reverse these changes in hypothalamic cellular structure (Witkin et al. 1991), which agrees with the observation that neuronal-astrocyte structural organization in the hypothalamus is modified throughout the estrous cycle in relationship with changes in gonadal steroid levels.

In rodents, hypothalamic astrocytes also undergo morphological modifications in association with the physiological hormonal variations that occur throughout the estrous cycle and this is associated with coordinated changes in the number of synapses (Fernandez-Galaz et al. 1997; Garcia-Segura et al. 2008). Between the morning and afternoon of proestrus, astroglial processes increase their coverage of neuronal cell surfaces and GABAergic synaptic terminals are removed from the soma of arcuate neurons, and this coincides with the luteinizing hormone surge (Csakvari et al. 2008). These glial processes remain in place, separating the presynaptic terminals from the postsynaptic membrane until the morning of estrus. At this time the glial processes retract so that by the morning of metestrus, GABAergic synaptic contacts are reestablished in the arcuate nucleus (Csakvari et al. 2007).

During proestrus when GABAergic inputs to the cell soma are decreased there is a rise in the number of excitatory synapses on dendritic spines and together this results in increased firing of arcuate neurons (Csakvari et al. 2007). The synaptic changes observed in the arcuate nucleus during the estrous cycle are most likely driven by the proestrus rise in circulating 17β-estradiol levels, as administration of this sex steroid to ovariectomized rats can mimic the changes observed in cycling rats (Naftolin et al. 1996). A subpopulation of neurons in the arcuate nucleus projects to the median eminence and controls pituitary hormone secretion and this estrogen-induced synaptic remodeling most likely induces functional changes in some neurosecretory neurons to modulate pituitary hormone secretion (Csakvari et al. 2008), as well as reorganization of synaptic circuits that control GnRH neurons (Garcia-Segura et al. 2008) and those involved in metabolic control (Gao et al. 2007).

The astrocytic responses in the preoptic area and arcuate nucleus are fascinating because the estradiol-induced plasticity in morphology of astrocytic processes is entirely different in the preoptic area from that in the arcuate nucleus, and yet completely coordinated in ensuring appropriate regulation of female reproduction (Garcia-Segura and McCarthy 2004). Astrocytes in both regions respond to estradiol with alterations in extension of processes that affect neurotransmission, but in the preoptic area, astrocytic processes respond by retracting processes that are interposed between excitatory glutamatergic inputs and GnRH neurons, stimulating GnRH release. In contrast, the astrocytic processes in the arcuate nucleus respond to estradiol with interposition of processes that cut off inhibitory GABAergic synapses, and further encourage GnRH release in response to high levels of estradiol and then progesterone. The result is a coordinated increase in excitatory inputs and decrease in inhibitory inputs to GnRH neurons that supports a sustained and powerful release of GnRH. But, many questions remain. For example, how do the astrocytes in two different regions develop the perfectly coordinated, but opposite, morphological responses to the same stimulus? Does this differential response depend on the developmental environment or the neuronal environment, since the preoptic area is derived from telencephalic anlage, whereas the arcuate nucleus is derived from the diencephalic anlage? How does local heterogeneity of astrocyte gene expression develop, and how does this affect function?

Evidence indicates that the developmental environment can alter the number of axosomatic synapses in the arcuate nucleus. In rodents, the number of axosomatic synapses increases between postnatal days 10 and 20 in both males and females (Perez et al. 1990). However, from postnatal day 20 onward females have more axosomatic synapses than males, with this sex difference at least partially dependent on the neonatal sex steroid environment (Perez et al. 1990). Astrocyte morphology and expression of GFAP are also modified by the sex steroid environment during this responsive period of hypothalamic development (Chowen et al. 1995), allowing astrocytes to intervene in critical synaptic interactions (Garcia-Segura et al. 1995).

14.4.3 Metabolism

Information regarding the role of astrocytes in metabolic control has expanded rapidly in recent years, due in part to the increased interest in the search for treatments of the obesity epidemic. Research on the neuronal circuits involved in appetite and metabolic control had been largely neglected until the need for therapeutics to combat obesity and its comorbidities became a prime interest for drug development. The increased attention that has been focused on this dramatic health problem has led to rapid advances in our understanding of how the CNS controls systemic metabolism. The mechanisms of homeostatic energy control are mainly coordinated by the hypothalamus, with the primary neuronal populations found in the arcuate nucleus (Morton et al. 2006). Here, the majority of neuropeptide Y (NPY) neurons co-express Agouti-related peptide (AgRP) and GABA and exert an orexigenic effect (Krashes et al. 2013). In opposition, the anorexigenic proopiomelanocortin (POMC) neurons in this brain area produce melanocyte-stimulating hormone (MSH) to inhibit appetite and stimulate energy expenditure (Boston et al. 1997). Astrocyte ensheathment of the somas of these important metabolic neuronal populations is inversely related to the number of synaptic inputs that the neurons receive, with this organization being modulated by metabolic hormones such as leptin and ghrelin (Horvath et al. 2010; Pinto et al. 2004). Moreover, the basal astroglial/synaptic arrangement in the arcuate nucleus is suggested to influence the metabolic response of an individual, including their propensity to become obese on a high fat diet (HFD) (Horvath et al. 2010). That is, animals resistant to diet-induced obesity have a different synaptic organization/astrocytic coverage than those animals that become obese on HFD. This hypothalamic architecture involving glial-neuronal contact may result from both genetic and environmental influences.

Notably, astrocytes in development encourage the formation of functional synapses (Pfrieger and Barres 1997). For example, postnatal mice undergo a switch in their method of nutrient intake, initially suckling, and then increasingly beginning to eat solid food until weaning occurs. This switch involves the development of synapses onto the key sets of neurons involved in feeding behavior in the arcuate nucleus. Although the critical neurons are present early on, functional inhibitory synapses onto neurons that regulate feeding behavior and energy metabolism are not present during the early postnatal period. The development of astrocytes may encourage incursion of the necessary glutamatergic axons into the region and promote synaptogenesis.

14.5 Developmental Influences on Glial Organization

The number of GFAP-positive astrocytes in the arcuate nucleus is greater in rats exposed to increased energy intake during neonatal life (Fig. 14.4), and these rats have an increase in the number of primary astrocytic projections (Fuente-Martin et al. 2012). In this study, no changes in the number of GFAP-positive astrocytes

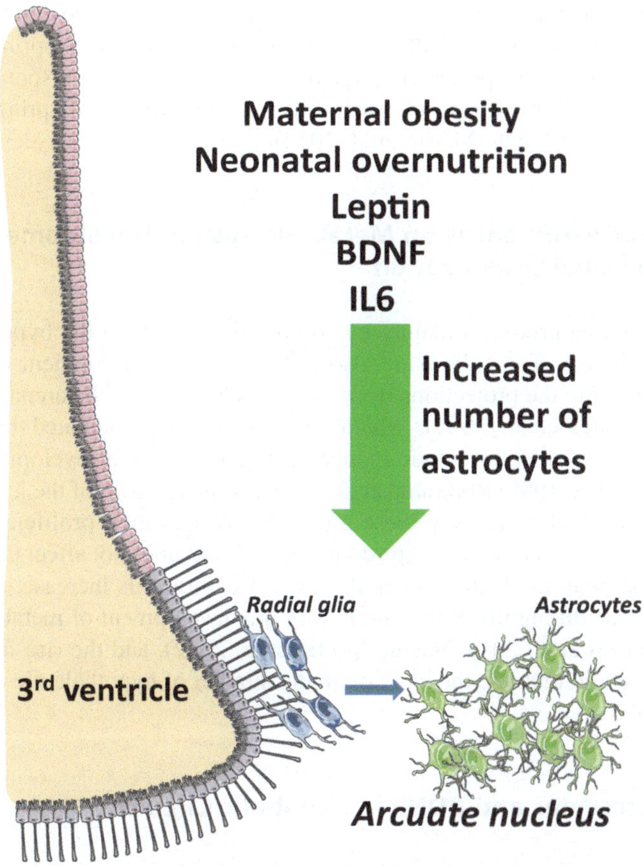

Fig. 14.4 Factors changing the number of astrocytes in the arcuate nucleus during fetal and neonatal development. Several different physiological signals can alter astrocyte numbers at different times in development. *BDNF* Brain derived neurotrophic factor, *IL6* Interleukin 6

were observed in other brain areas such as the hippocampus or cerebellum, suggesting a region-specific effect of this early overnutrition.

Maternal obesity increases astrocyte proliferation in both the fetal and neonatal mouse hypothalamus (Fig. 14.4), with specific increases found in the arcuate nucleus and the SON (Kim et al. 2016). These authors suggest that the obesity-associated rise in interleukin (IL) 6 in the maternal circulation could be at least one of the factors involved in increased numbers of astrocytes in the fetal hypothalamus, as well as the increased proliferative capacity of hypothalamic astrocytes in the offspring in later life (Kim et al. 2016). The astroglial coverage and function of POMC neurons in offspring are modified by maternal high fat intake. In the offspring of dams fed HFD, the astroglial coverage of POMC neurons in the arcuate nucleus is significantly increased compared with the offspring of mothers on a normal chow diet (Fuente-Martin et al. 2012). These highly-ensheathed POMC neurons display a decreased frequency of miniature inhibitory postsynaptic currents. The glucose-sensing capacity of highly ensheathed POMC neurons in HFD-exposed offspring was also affected, as their action potential frequency was decreased in response to lower glucose concentrations, a response that was not observed in offspring of normal chow fed mothers (Fuente-Martin et al. 2012).

14.5.1 Leptin Influences on Metabolic System Development and Glial Organization

Leptin, a cytokine produced mainly by adipose tissue, acts in the hypothalamus to inhibit food intake. This adipokine also influences the development of metabolic circuits, increasing the projections from metabolic neurons in the arcuate nucleus to the PVN (Bouret et al. 2004), which may be partially mediated by astrocyte-dependent actions of leptin. Leptin induces astrogenesis in the developing hypothalamus (Ahima et al. 1999; Rottkamp et al. 2015), while removal of the leptin receptor during early postnatal life from these glial cells reduces their proliferation and the number of GFAP-positive cells at postnatal day 21, but does not affect the number of hypothalamic neurons (Rottkamp et al. 2015). Leptin levels increase in neonatally overfed animals during the critical period for the development of metabolic circuits (Argente-Arizon et al. 2016; Fuente-Martin et al. 2012), and the rise in circulating leptin levels could underlie the increase in the number of hypothalamic astrocytes in these animals.

14.5.2 Astrocytes and BDNF in Metabolic Circuits

Brain derived neurotrophic factor (BDNF) is important for the development of projections of POMC and AgRP neurons from the arcuate nucleus to the PVN and dorsomedial nucleus of the hypothalamus (DMH), although the number of POMC or AgRP neurons in the arcuate nucleus were unchanged in BDNF genetic knockout mice (Liao et al. 2015). The lack of BDNF specifically reduced projections from POMC neurons of the arcuate nucleus to neurons of the DMH and from AgRP neurons of the arcuate nucleus to neurons located in the lateral portion of the PVN. Astrocytes produce BDNF (Wu et al. 2004) and activation of the melanocortin 4 receptor (MC4R) induces production of this growth factor (Caruso et al., 2012; Ramirez et al. 2015). BDNF is also involved in synaptic plasticity in metabolic circuits (Zhang et al. 2017) and regulates inhibitory synapse formation during brain development (Elmariah et al. 2005).

Other growth factors, such as ciliary neurotrophic factor, (CNTF) have been reported to induce generation of metabolically active neurons in the arcuate nucleus of the adult mouse (Kokoeva et al. 2005), although these results are controversial (Borg et al. 2016). Astrocytes produce and respond to CNTF (Muller et al. 2007; Poluch and Juliano 2007), with CNTF treatment inducing radial glia to differentiate into astrocytes, disrupting neuron migration (Poluch and Juliano 2007).

Many questions remain regarding the interaction of astrocytes and neurons and how they control appetite and systemic metabolism. For example, astrocytic contact with both POMC and NPY neurons is increased in obese subjects, in association with a decrease in the number of somatic synaptic inputs in both populations of neurons (Horvath et al. 2010). However, on POMC neurons this decrease is due to a decline in inhibitory inputs, whereas on NPY neurons it is the result of a reduction in stimulatory inputs, which together promote an overall balance toward an anorexigenic signal. How is this specific rewiring coordinated? What is the mechanism of cross talk between astroglia and neurons that generates this specific synaptic rewiring, both during development and in adults? Are the astrocytes that surround these two neuronal populations different, even though they both respond to leptin? Is there cross talk between astrocytes and neurons during development that ensures that the required functional circuits in each hypothalamic area are formed, although the signals involved remain elusive? The hypothalamic circuits controlling each neuroendocrine system or axis are often studied and spoken of as if they function independently, but there must also be precise communication between these circuits to maintain systemic homeostasis. Astrocytes and radial glia are involved in the generation of neurons and their projections. How they determine the precise organization of the neuroendocrine hypothalamus remains largely unknown.

Studies using genetically modified mice can be used to provide evidence for how neurons affect astrocytes. For example, *Gnasxl*-deficient mice, which lack a G-protein alpha stimulatory subunit that regulates energy balance, are lean and hypermetabolic (Holmes et al. 2016). In these animals, expression of GFAP in the adult hypothalamus was greatly diminished and the numbers of astrocytes and a subpopulation of tanycytes (alpha) were also diminished. In juvenile animals as late as postnatal day 15, however, the numbers of astrocytes and tanycytes were not different in mutants. Since *Gnasxl* is expressed only in neurons, this study provides tentative evidence that late development of astrocytes in the hypothalamus is modulated by neuronal activity, although it is also possible that any effect of these mutants on astrocytes may be secondary to undernutrition.

14.6 The Role of Tanycytes in Neuroendocrine Regulation

Tanycytes, specialized radial glia located in the ventral wall of the third ventricle, have been proposed to be a source of neural progenitor cells in the adult hypothalamus (Haan et al. 2013; Lee et al. 2012; Robins et al. 2013). Tanycytes are classified into subtypes depending on their location and function, with some studies indicating that all subtypes have neural progenitor properties. However, other studies suggest that α2-tanycytes generate β-tanycytes in vivo and that only α2-tanycytes are capable of generating proliferative neurospheres in vitro (Robins et al. 2013).

Tanycytes have also been proposed to participate in the neuroendocrine control of reproduction and metabolism (Ojeda et al. 2003; Prevot et al. 2018) and to modulate hypothalamic control of the thyroid axis (Muller-Fielitz et al. 2017). Tanycytes have end feet that extend into the VMH, arcuate nucleus, and median eminence

(Rodriguez et al. 2005) and can be retracted. They contribute to the regulation of GnRH release, as their retractable end feet are interlaced with GnRH synaptic terminals and the portal vasculature in the median eminence (Baroncini et al. 2007; Ojeda et al. 2008; Prevot et al. 2018). Extension of tanycytic processes to ensheath GnRH terminals prevents release of GnRH into the portal vasculature. These tanycytic processes retract during the preovulatory stage of the estrous cycle, which permits the contact of GnRH terminals with portal capillaries for neuropeptide release (Bellefontaine et al. 2011; Prevot et al. 1999). More than 30 years ago, Sergio Ojeda hypothesized that glial cells released substances that modulated the release of GnRH to the portal vasculature (Ojeda 1994). These signals are now known to include prostaglandin E2, transforming growth factor beta, and nitric oxide (NO), which stimulate and coordinate GnRH neuronal firing (Bellefontaine et al. 2011; Clasadonte et al. 2011; Melcangi et al. 1995). More recently, tanycytic were shown to produce semaphorin7A which binds to plexinC1 and β1-integrin (Parkash et al. 2015). The levels of semaphorin7A are modulated by ovarian steroids, and the signaling of semaphorin 7A is necessary for normal estrous cycling and fertility. It is proposed that semaphorin7A acts through is receptors in a coordinated fashion to induce expansion of tanycytic end feet and promote the retraction of GnRH nerve terminals from the pericapillary space (Parkash et al. 2015).

14.7 Perspectives

In sum, astrocytes have been difficult to study because: (1) they are not electrically active and do not generate action potentials; (2) they have divergent morphologies (including small nuclei and irregular processes); and (3) there is a lack of specific antibodies or other markers that identify astrocytic subpopulations.

The lack of astrocyte-specific reagents is certainly problematic. For example, GFAP is found in astrocytes and progenitor cells and not all astrocytes express GFAP. Moreover, GFAP immunoreactivity does not allow full visualization of astrocyte morphology, and the extent of GFAP staining within astrocytes may be less indicative of morphological changes than of changes in GFAP expression. Other difficulties relate to the problem of visualizing astrocytes, especially at the same time as neurons since they are often found in different planes, making it difficult to simultaneously label and visualize the nuclei of both cell types.

Controversies in studies of astrocytes have also arisen recently, even for issues that were previously considered to be somewhat resolved. For example, the idea that there are co-transmitters released by astrocytes that assist neuronal neurotransmission has been called into question (Wolosker et al. 2016, 2017). Does the concept of the tripartite synapse hold true, or is this functionally restricted to developmental stages? Are calcium waves between astrocytes an artifact of culture conditions, rather than an in vivo event? Should the umbrella term "astrocyte" be eliminated in favor of more specific terms that refer to distinct subtypes of this class of glia? Do astrocytes deliver glucose, lactate or both to neurons?

Few studies to date have used optogenetic techniques to interrogate astrocyte function in the hypothalamus. In one study, channelrhodopsin-2 was activated specifically in astrocytes of the posterior hypothalamus of mice following injections of a recombinant adeno-associated virus in which the GFAP promoter drove expression of channelrhodopsin and a reporter gene. Optogenetic activation of astrocytes increased sleep during the active (lights-off) phase (Pelluru et al. 2016). Optogenetic stimulation of medial-basal hypothalamic astrocytes was shown to suppress food intake in a frequency-dependent manner (Sweeney et al. 2016). These methods will likely provide additional tools for dissecting the role of astrocytes in neuroendocrine development, especially once more specific markers of astrocyte subpopulations have been identified.

As of this writing, many perspectives about astrocytes remain in flux partly because significant developmental changes in their competence are now recognized. Astrocytes are typically derived from young animals for in vitro studies and these cells do not necessarily behave as do those from adults (Verkhratsky and Nedergaard 2018), or as young astrocytes in vivo for that matter. The renewed interest in astrocytes in neuroendocrine systems will undoubtedly advance our understanding of these fascinating cells, both in their developmental capacities and in their lifelong contribution to regulating physiological processes.

Key References

Ahmed et al. (2008). Challanged the dogma of that pubertal hormones, as opposed to sex steroids during embryological/neonatal development did not affect neurogenesis.

Bouret et al. (2004). Seminal study in understanding how early metabolic/hormonal changes can have long-term effects on metabolism.

Malatestaet al. (2003). Important contributions in order to understand radial glia.

Mong et al. (1996). One of the first studies to demonstrate sex differences in glial cells in the hypothalamus.

Morel et al. (2017). Important advances in understanding differences in astrocytes in different brain regions.

Naftolin et al. (1996). An early study associating synaptic remodeling in the hypothalamus with a physiological outcome.

Perlmutter et al. (1985). A seminal study in understanding the physiological outcome of interactions between neurons and glia in the hypothalamus.

Pfrieger et al. (1997). A clear advance in undestanding the role of astrocytes in synaptic transmission.

Prevot et al. (2018). A recent update on the implications of tanycytes in neuroendocrine control.

Theodosis et al. (1986). A seminal study of neuron-glial interactions in physiology.

References

Ahima RS, Bjorbaek C, Osei S, Flier JS (1999) Regulation of neuronal and glial proteins by leptin: implications for brain development. Endocrinology 140:2755–2762

Ahmed EI, Zehr JL, Schulz KM, Lorenz BH, DonCarlos LL, Sisk CL (2008) Pubertal hormones modulate the addition of new cells to sexually dimorphic brain regions. Nat Neurosci 11:995–997

Allen NJ, Lyons DA (2018) Glia as architects of central nervous system formation and function. Science 362:181–185

Altman J, Bayer SA (1986) The development of the rat hypothalamus. Adv Anat Embryol Cell Biol 100:1–178

Amateau SK, McCarthy MM (2002) Sexual differentiation of astrocyte morphology in the developing rat preoptic area. J Neuroendocrinol 14:904–910

Argente-Arizon P, Ros P, Diaz F, Fuente-Martin E, Castro-Gonzalez D, Sanchez-Garrido MA, Barrios V, Tena-Sempere M, Argente J, Chowen JA (2016) Age and sex dependent effects of early overnutrition on metabolic parameters and the role of neonatal androgens. Biol Sex Differ 7:26

Baroncini M, Allet C, Leroy D, Beauvillain JC, Francke JP, Prevot V (2007) Morphological evidence for direct interaction between gonadotrophin-releasing hormone neurones and astroglial cells in the human hypothalamus. J Neuroendocrinol 19:691–702

Bayer SA, Altman J (1987) Development of the preoptic area: time and site of origin, migratory routes, and settling patterns of its neurons. J Comp Neurol 265:65–95

Belanger M, Magistretti PJ (2009) The role of astroglia in neuroprotection. Dialog Clin Neurosci 11:281–295

Bellefontaine N, Hanchate NK, Parkash J, Campagne C, de Seranno S, Clasadonte J, d'Anglemont de Tassigny X, Prevot V (2011) Nitric oxide as key mediator of neuron-to-neuron and endothelia-to-glia communication involved in the neuroendocrine control of reproduction. Neuroendocrinology 93:74–89

Bonfanti L, Poulain DA, Theodosis DT (1993) Radial glia-like cells in the supraoptic nucleus of the adult rat. J Neuroendocrinol 5:1–5

Bonfardin VD, Fossat P, Theodosis DT, Oliet SH (2010) Glia-dependent switch of kainate receptor presynaptic action. J Neurosci 30:985–995

Borg ML, Reichenbach A, Lemus M, Oldfield BJ, Andrews ZB, Watt MJ (2016) Central administration of the ciliary neurotrophic factor analogue, axokine, does not play a role in long-term energy homeostasis in adult mice. Neuroendocrinology 103:223–229

Boston BA, Blaydon KM, Varnerin J, Cone RD (1997) Independent and additive effects of central POMC and leptin pathways on murine obesity. Science 278:1641–1644

Bosworth AP, Allen NJ (2017) The diverse actions of astrocytes during synaptic development. Curr Opin Neurobiol 47:38–43

Botchkina GI, Morin LP (1995) Ontogeny of radial glia, astrocytes and vasoactive intestinal peptide immunoreactive neurons in hamster suprachiasmatic nucleus. Brain Res Dev Brain Res 86:48–56

Bouret SG, Simerly RB (2007) Development of leptin-sensitive circuits. J Neuroendocrinol 19 (8):575–582

Bouret SG, Draper SJ, Simerly RB (2004) Trophic action of leptin on hypothalamic neurons that regulate feeding. Science 304:108–110

Bourque CW, Oliet SH (1997) Osmoreceptors in the central nervous system. Annu Rev Physiol 59:601–619

Bourque CW, Oliet SH, Richard D (1994) Osmoreceptors, osmoreception, and osmoregulation. Front Neuroendocrinol 15:231–274

Bushong EA, Martone ME, Ellisman MH (2004) Maturation of astrocyte morphology and the establishment of astrocyte domains during postnatal hippocampal development. Int J Dev Neurosci 22:73–86

Caruso C, Carniglia L, Durand D, Gonzalez PV, Scimonelli TN, Lasaga M (2012) Melanocortin 4 receptor activation induces brain-derived neurotrophic factor expression in rat astrocytes through cyclic AMP-protein kinase A pathway. Mol Cell Endocrinol 348:47–54

Chowen JA, Busiguina S, Garcia-Segura LM (1995) Sexual dimorphism and sex steroid modulation of glial fibrillary acidic protein messenger RNA and immunoreactivity levels in the rat hypothalamus. Neuroscience 69:519–532

Chowen JA, Argente-Arizon P, Freire-Regatillo A, Frago LM, Horvath TL, Argente J (2016) The role of astrocytes in the hypothalamic response and adaptation to metabolic signals. Prog Neurobiol 144:68–87

Chung WS, Welsh CA, Barres BA, Stevens B (2015) Do glia drive synaptic and cognitive impairment in disease? Nat Neurosci 18:1539–1545

Clasadonte J, Poulain P, Hanchate NK, Corfas G, Ojeda SR, Prevot V (2011) Prostaglandin E2 release from astrocytes triggers gonadotropin-releasing hormone (GnRH) neuron firing via EP2 receptor activation. Proc Natl Acad Sci U S A 108:16104–16109

Collado P, Beyer C, Hutchison JB, Holman SD (1995) Hypothalamic distribution of astrocytes is gender-related in Mongolian gerbils. Neurosci Lett 184:86–89

Csakvari E, Hoyk Z, Gyenes A, Garcia-Ovejero D, Garcia-Segura LM, Parducz A (2007) Fluctuation of synapse density in the arcuate nucleus during the estrous cycle. Neuroscience 144:1288–1292

Csakvari E, Kurunczi A, Hoyk Z, Gyenes A, Naftolin F, Parducz A (2008) Estradiol-induced synaptic remodeling of tyrosine hydroxylase immunopositive neurons in the rat arcuate nucleus. Endocrinology 149:4137–4141

Deleuze C, Duvoid A, Hussy N (1998) Properties and glial origin of osmotic-dependent release of taurine from the rat supraoptic nucleus. J Physiol 507(Pt 2):463–471

Elmariah SB, Oh EJ, Hughes EG, Balice-Gordon RJ (2005) Astrocytes regulate inhibitory synapse formation via Trk-mediated modulation of postsynaptic GABAA receptors. J Neurosci 25:3638–3650

Ero C, Gewaltig MO, Keller D, Markram H (2018) A cell atlas for the mouse brain. Front Neuroinform 12:84

Fernandez-Galaz MC, Morschl E, Chowen JA, Torres-Aleman I, Naftolin F, Garcia-Segura LM (1997) Role of astroglia and insulin-like growth factor-I in gonadal hormone-dependent synaptic plasticity. Brain Res Bull 44:525–531

Fuente-Martin E, Garcia-Caceres C, Granado M, de Ceballos ML, Sanchez-Garrido MA, Sarman B, Liu ZW, Dietrich MO, Tena-Sempere M, Argente-Arizon P, Diaz F, Argente J, Horvath TL, Chowen JA (2012) Leptin regulates glutamate and glucose transporters in hypothalamic astrocytes. J Clin Invest 122:3900–3913

Gao Q, Mezei G, Nie Y, Rao Y, Choi CS, Bechmann I, Leranth C, Toran-Allerand D, Priest CA, Roberts JL, Gao XB, Mobbs C, Shulman GI, Diano S, Horvath TL (2007) Anorectic estrogen mimics leptin's effect on the rewiring of melanocortin cells and Stat3 signaling in obese animals. Nat Med 13:89–94

Garcia-Caceres C, Fuente-Martin E, Argente J, Chowen JA (2012) Emerging role of glial cells in the control of body weight. Mol Metab 1:37–46

Garcia-Segura LM, McCarthy MM (2004) Minireview: Role of glia in neuroendocrine function. Endocrinology 145:1082–1086

Garcia-Segura LM, Duenas M, Busiguina S, Naftolin F, Chowen JA (1995) Gonadal hormone regulation of neuronal-glial interactions in the developing neuroendocrine hypothalamus. J Steroid Biochem Mol Biol 53:293–298

Garcia-Segura LM, Lorenz B, DonCarlos LL (2008) The role of glia in the hypothalamus: implications for gonadal steroid feedback and reproductive neuroendocrine output. Reproduction 135(4):419–429

Haan N, Goodman T, Najdi-Samiei A, Stratford CM, Rice R, El Agha E, Bellusci S, Hajihosseini MK (2013) Fgf10-expressing tanycytes add new neurons to the appetite/energy-balance regulating centers of the postnatal and adult hypothalamus. J Neurosci 33:6170–6180

Hatton GI (1997) Function-related plasticity in hypothalamus. Annu Rev Neurosci 20:375–397
Holmes AP, Wong SQ, Pulix M, Johnson K, Horton NS, Thomas P, de Magalhaes JP, Plagge A (2016) Reductions in hypothalamic Gfap expression, glial cells and alpha-tanycytes in lean and hypermetabolic Gnasxl-deficient mice. Mol Brain 9:39
Horvath TL, Sarman B, Garcia-Caceres C, Enriori PJ, Sotonyi P, Shanabrough M, Borok E, Argente J, Chowen JA, Perez-Tilve D, Pfluger PT, Bronneke HS, Levin BE, Diano S, Cowley MA, Tschop MH (2010) Synaptic input organization of the melanocortin system predicts diet-induced hypothalamic reactive gliosis and obesity. Proc Natl Acad Sci U S A 107:14875–14880
Hsu C, Hsieh YL, Ho ML, Hsu HK, Yu JY (2001) Sexually dimorphic effect of glutamate treatment on cell cycle arrestment of astrocytes from the preoptic area of neonatal rats. Dev Neurosci 23:399–405
Joe N, Scott V, Brown CH (2014) Glial regulation of extrasynaptic NMDA receptor-mediated excitation of supraoptic nucleus neurones during dehydration. J Neuroendocrinol 26:35–42
Johnson RT, Schneider A, DonCarlos LL, Breedlove SM, Jordan CL (2012) Astrocytes in the rat medial amygdala are responsive to adult androgens. J Comp Neurol 520:2531–2544
Kim DW, Glendining KA, Grattan DR, Jasoni CL (2016) Maternal obesity leads to increased proliferation and numbers of astrocytes in the developing fetal and neonatal mouse hypothalamus. Int J Dev Neurosci 53:18–25
Kokoeva MV, Yin H, Flier JS (2005) Neurogenesis in the hypothalamus of adult mice: potential role in energy balance. Science 310:679–683
Krashes MJ, Shah BP, Koda S, Lowell BB (2013) Rapid versus delayed stimulation of feeding by the endogenously released AgRP neuron mediators GABA, NPY, and AgRP. Cell Metab 18:588–595
Langub MC Jr, Watson RE Jr (1992) Estrogen receptor-immunoreactive glia, endothelia, and ependyma in guinea pig preoptic area and median eminence: electron microscopy. Endocrinology 130:364–372
Lee DA, Bedont JL, Pak T, Wang H, Song J, Miranda-Angulo A, Takiar V, Charubhumi V, Balordi F, Takebayashi H, Aja S, Ford E, Fishell G, Blackshaw S (2012) Tanycytes of the hypothalamic median eminence form a diet-responsive neurogenic niche. Nat Neurosci 15 (6):700–702
Leloup C, Allard C, Carneiro L, Fioramonti X, Collins S, Penicaud L (2015) Glucose and hypothalamic astrocytes: more than a fueling role? Neuroscience 323:110–120
Liao G-Y, Bouyer K, Kamitakahara A, Sahibzada N, Wang C-H, Rutlin M, Simerly RB, Xu B (2015) Brain-derived neurotrophic factor is required for axonal growth of selective groups of neurons in the arcuate nucleus. Mol Metab 4(6):471–482
Lorenz B, Garcia-Segura LM, DonCarlos LL (2005) Cellular phenotype of androgen receptor-immunoreactive nuclei in the developing and adult rat brain. J Comp Neurol 492:456–468
Magistretti PJ, Allaman I (2018) Lactate in the brain: from metabolic end-product to signalling molecule. Nat Rev Neurosci 19:235–249
Malatesta P, Hartfuss E, Gotz M (2000) Isolation of radial glial cells by fluorescent-activated cell sorting reveals a neuronal lineage. Development 127:5253–5263
Malatesta P, Hack MA, Hartfuss E, Kettenmann H, Klinkert W, Kirchhoff F, Gotz M (2003) Neuronal or glial progeny: regional differences in radial glia fate. Neuron 37:751–764
Marzban F, Tweedle CD, Hatton GI (1992) Reevaluation of the plasticity in the rat supraoptic nucleus after chronic dehydration using immunogold for oxytocin and vasopressin at the ultrastructural level. Brain Res Bull 28:757–766
Melcangi RC, Galbiati M, Messi E, Piva F, Martini L, Motta M (1995) Type 1 astrocytes influence luteinizing hormone-releasing hormone release from the hypothalamic cell line GT1-1: is transforming growth factor-beta the principle involved? Endocrinology 136:679–686
Mohr MA, Garcia FL, DonCarlos LL, Sisk CL (2016) Neurons and glial cells are added to the female rat anteroventral periventricular nucleus during puberty. Endocrinology 157:2393–2402
Mong JA, McCarthy MM (2002) Ontogeny of sexually dimorphic astrocytes in the neonatal rat arcuate. Brain Res Dev Brain Res 139:151–158

Mong JA, Kurzweil RL, Davis AM, Rocca MS, McCarthy MM (1996) Evidence for sexual differentiation of glia in rat brain. Horm Behav 30:553–562
Mong JA, Glaser E, McCarthy MM (1999) Gonadal steroids promote glial differentiation and alter neuronal morphology in the developing hypothalamus in a regionally specific manner. J Neurosci 19:1464–1472
Morel L, Higashimori H, Tolman M, Yang Y (2014) VGluT1+ neuronal glutamatergic signaling regulates postnatal developmental maturation of cortical protoplasmic astroglia. J Neurosci 34:10950–10962
Morel L, Chiang MSR, Higashimori H, Shoneye T, Iyer LK, Yelick J, Tai A, Yang Y (2017) Molecular and functional properties of regional astrocytes in the adult brain. J Neurosci 37:8706–8717
Morrell JI, Krieger MS, Pfaff DW (1986) Quantitative autoradiographic analysis of estradiol retention by cells in the preoptic area, hypothalamus and amygdala. Exp Brain Res 62: 343–354
Morton GJ, Cummings DE, Baskin DG, Barsh GS, Schwartz MW (2006) Central nervous system control of food intake and body weight. Nature 443:289–295
Muller A, Hauk TG, Fischer D (2007) Astrocyte-derived CNTF switches mature RGCs to a regenerative state following inflammatory stimulation. Brain 130:3308–3320
Muller-Fielitz H, Stahr M, Bernau M, Richter M, Abele S, Krajka V, Benzin A, Wenzel J, Kalies K, Mittag J, Heuer H, Offermanns S, Schwaninger M (2017) Tanycytes control the hormonal output of the hypothalamic-pituitary-thyroid axis. Nat Commun 8:484
Munekawa K, Tamada Y, Iijima N, Hayashi S, Ishihara A, Inoue K, Tanaka M, Ibata Y (2000) Development of astroglial elements in the suprachiasmatic nucleus of the rat: with special reference to the involvement of the optic nerve. Exp Neurol 166:44–51
Naftolin F, Mor G, Horvath TL, Luquin S, Fajer AB, Kohen F, Garcia-Segura LM (1996) Synaptic remodeling in the arcuate nucleus during the estrous cycle is induced by estrogen and precedes the preovulatory gonadotropin surge. Endocrinology 137:5576–5580
Noctor SC, Flint AC, Weissman TA, Dammerman RS, Kriegstein AR (2001) Neurons derived from radial glial cells establish radial units in neocortex. Nature 409:714–720
Oberheim NA, Takano T, Han X, He W, Lin JHC, Wang F, Qiwu Xu Q, Wyatt JD, Pilcher W, Ojemann JG, Ransom BR, Goldman SA, Nedergaard M (2009) Uniquely hominid features of adult human astrocytes. J Neurosci 29(10):3276
Ojeda SR (1994) The neurobiology of mammalian puberty: has the contribution of glial cells been underestimated? J NIH Res 6:51–56
Ojeda SR, Prevot V, Heger S, Lomniczi A, Dziedzic B, Mungenast A (2003) Glia-to-neuron signaling and the neuroendocrine control of female puberty. Ann Med 35:244–255
Ojeda SR, Lomniczi A, Sandau US (2008) Glial-gonadotrophin hormone (GnRH) neurone interactions in the median eminence and the control of GnRH secretion. J Neuroendocrinol 20:732–742
Olmos G, Naftolin F, Perez J, Tranque PA, Garcia-Segura LM (1989) Synaptic remodeling in the rat arcuate nucleus during the estrous cycle. Neuroscience 32:663–667
Parkash J, Messina A, Langlet F, Cimino I, Loyens A, Mazur D, Gallet S, Balland E, Malone SA, Pralong F, Cagnoni G, Schellino R, De Marchis S, Mazzone M, Pasterkamp RJ, Tamagnone L, Prevot V, Giacobini P (2015) Semaphorin7A regulates neuroglial plasticity in the adult hypothalamic median eminence. Nat Commun 6:6385
Pelluru D, Konadhode RR, Bhat NR, Shiromani PJ (2016) Optogenetic stimulation of astrocytes in the posterior hypothalamus increases sleep at night in C57BL/6J mice. Eur J Neurosci 43:1298–1306
Perea G, Navarrete M, Araque A (2009) Tripartite synapses: astrocytes process and control synaptic information. Trends Neurosci 32:421–431
Perez J, Naftolin F, Garcia Segura LM (1990) Sexual differentiation of synaptic connectivity and neuronal plasma membrane in the arcuate nucleus of the rat hypothalamus. Brain Res 527:116–122

Perlmutter LS, Tweedle CD, Hatton GI (1985) Neuronal/glial plasticity in the supraoptic dendritic zone in response to acute and chronic dehydration. Brain Res 361:225–232

Pfrieger FW, Barres BA (1997) Synaptic efficacy enhanced by glial cells in vitro. Science 277:1684–1687

Pinto S, Roseberry AG, Liu H, Diano S, Shanabrough M, Cai X, Friedman JM, Horvath TL (2004) Rapid rewiring of arcuate nucleus feeding circuits by leptin. Science 304(5667):110–115

Poluch S, Juliano SL (2007) A normal radial glial scaffold is necessary for migration of interneurons during neocortical development. Glia 55:822–830

Prevot V, Croix D, Bouret S, Dutoit S, Tramu G, Stefano GB, Beauvillain JC (1999) Definitive evidence for the existence of morphological plasticity in the external zone of the median eminence during the rat estrous cycle: implication of neuro-glio-endothelial interactions in gonadotropin-releasing hormone release. Neuroscience 94:809–819

Prevot V, Dehouck B, Sharif A, Ciofi P, Giacobini P, Clasadonte J (2018) The versatile tanycyte: a hypothalamic integrator of reproduction and energy metabolism. Endocr Rev 39:333–368

Ramirez D, Saba J, Carniglia L, Durand D, Lasaga M, Caruso C (2015) Melanocortin 4 receptor activates ERK-cFos pathway to increase brain-derived neurotrophic factor expression in rat astrocytes and hypothalamus. Mol Cell Endocrinol 411:28–37

Robins SC, Stewart I, McNay DE, Taylor V, Giachino C, Goetz M, Ninkovic J, Briancon N, Maratos-Flier E, Flier JS, Kokoeva MV, Placzek M (2013) alpha-Tanycytes of the adult hypothalamic third ventricle include distinct populations of FGF-responsive neural progenitors. Nat Commun 4:2049

Rodriguez EM, Blazquez JL, Pastor FE, Pelaez B, Pena P, Peruzzo B, Amat P (2005) Hypothalamic tanycytes: a key component of brain-endocrine interaction. Int Rev Cytol 247:89–164

Rottkamp DM, Rudenko IA, Maier MT, Roshanbin S, Yulyaningsih E, Perez L, Valdearcos M, Chua S, Koliwad SK, Xu AW (2015) Leptin potentiates astrogenesis in the developing hypothalamus. Mol Metab 4:881–889

Silverman RC, Gibson MJ, Silverman AJ (1991) Relationship of glia to GnRH axonal outgrowth from third ventricular grafts in hpg hosts. Exp Neurol 114:259–274

Sims RE, Butcher JB, Parri HR, Glazewski S (2015) Astrocyte and neuronal plasticity in the somatosensory system. Neural Plast 2015:732014

Souttou S, Benabdesselam R, Siqueiros-Marquez L, Sifi M, Deliba M, Vacca O, Charles-Messance-H, Vaillend C, Rendon A, Guillonneau X, Dorbani-Mamine L (2019) Expression and localization of dystrophins and beta-dystroglycan in the hypothalamic supraoptic nuclei of rat from birth to adulthood. Acta Histochem 121:218–226

Stern JE (2015) Neuroendocrine-autonomic integration in the paraventricular nucleus: novel roles for dendritically released neuropeptides. J Neuroendocrinol 27:487–497

Suarez I, Fernandez B, Bodega G, Tranque P, Olmos G, Garcia-Segura LM (1987) Postnatal development of glial fibrillary acidic protein immunoreactivity in the hamster arcuate nucleus. Brain Res 465:89–95

Sweeney P, Qi Y, Xu Z, Yang Y (2016) Activation of hypothalamic astrocytes suppresses feeding without altering emotional states. Glia 64:2263–2273

Theodosis DT, Poulain DA (1984) Evidence for structural plasticity in the supraoptic nucleus of the rat hypothalamus in relation to gestation and lactation. Neuroscience 11:183–193

Theodosis DT, Poulain DA (1999) Contribution of astrocytes to activity-dependent structural plasticity in the adult brain. Adv Exp Med Biol 468:175–182

Theodosis DT, Montagnese C, Rodriguez F, Vincent JD, Poulain DA (1986) Oxytocin induces morphological plasticity in the adult hypothalamo-neurohypophysial system. Nature 322:738–740

Tweedle CD, Hatton GI (1977) Ultrastructural changes in rat hypothalamic neurosecretory cells and their associated glia during minimal dehydration and rehydration. Cell Tissue Res 181:59–72

Verkhratsky A, Nedergaard M (2018) Physiology of astroglia. Physiol Rev 98:239–389

Wang YF, Hatton GI (2007) Interaction of extracellular signal-regulated protein kinase 1/2 with actin cytoskeleton in supraoptic oxytocin neurons and astrocytes: role in burst firing. J Neurosci 27:13822–13834

Wang YF, Hatton GI (2009) Astrocytic plasticity and patterned oxytocin neuronal activity: dynamic interactions. J Neurosci 29:1743–1754

Witkin JW, Ferin M, Popilskis SJ, Silverman AJ (1991) Effects of gonadal steroids on the ultrastructure of GnRH neurons in the rhesus monkey: synaptic input and glial apposition. Endocrinology 129:1083–1092

Wolosker H, Balu DT, Coyle JT (2016) The rise and fall of the D-serine-mediated gliotransmission hypothesis. Trends Neurosci 39:712–721

Wolosker H, Balu DT, Coyle JT (2017) Astroglial versus neuronal D-serine: check your controls! Trends Neurosci 40:520–522

Wu H, Friedman WJ, Dreyfus CF (2004) Differential regulation of neurotrophin expression in basal forebrain astrocytes by neuronal signals. J Neurosci Res 76:76–85

Wu YE, Pan L, Zuo Y, Li X, Hong W (2017) Detecting activated cell populations using single-cell RNA-seq. Neuron 96:313–329.e316

Zamanian JL, Xu L, Foo LC, Nouri N, Zhou L, Giffard RG, Barres BA (2012) Genomic analysis of reactive astrogliosis. J Neurosci 32(18):6391–6410

Zhang Y, Barres BA (2010) Astrocyte heterogeneity: an underappreciated topic in neurobiology. Curr Opin Neurobiol 20:588–594

Zhang Y, Reichel JM, Han C, Zuniga-Hertz JP, Cai D (2017) Astrocytic process plasticity and IKKbeta/NF-kappaB in central control of blood glucose, blood pressure, and body weight. Cell Metab 25:1091–1102.e1094

Origins of Sex Differentiation of Brain and Behavior

15

Margaret M. McCarthy

Abstract

Sex differences in adult brain and behavior are often established early in development when the brain is remarkably immature. In adults sex differences take on many forms including latent variables, dimorphisms, frequency, and more. Androgens and estrogens from the developing male fetal testis masculinize the brain, so that physiology and behavior are in synchrony and harmony with gonadal phenotype. This differentiation process occurs during a critical window in males, but females remain sensitive to exogenous steroid treatment for a longer developmental period. The cellular mechanisms of sexual differentiation are diverse but often involve inflammatory signaling molecules and immune cells. This may have consequences for the higher vulnerability of males to neurological and neuropsychiatric disorders of development.

Keywords

Amygdala · Androgens · Estrogens · Hypothalamus · Neuroimmune · Preoptic area

15.1 Introduction

The brain is a reproductive organ, essential for both the control of gamete production by the gonads and the physiology and behaviors required for achieving fertilization, gestation, parturition, and parenting. By definition, these are different in males and females since sex is defined by gamete size, production, and delivery. But determining how the brain drives the different physiologies and behaviors of males and females is confounded by the inherent complexity of the nervous system and its wide

M. M. McCarthy (✉)
Department of Pharmacology, University of Maryland School of Medicine, Baltimore, MD, USA
e-mail: mmccarth@umaryland.edu

S. Wray, S. Blackshaw (eds.), *Developmental Neuroendocrinology*, Masterclass in Neuroendocrinology 9, https://doi.org/10.1007/978-3-030-40002-6_15

range of functions, from regulating heart beat to composing a symphony. Many of these functions are the same, or essentially the same, in males and females, but many are not. Moreover, many things influence how a brain develops, including nutrition, genetics, injury, nurturing, stress, and environmental complexity. Sex differences may appear under some circumstances or diminish under others. Males and females may exhibit the same response, but the underlying neural mechanisms of that response may be profoundly different. What is true in one species might not be true in another. For these reasons, it is essential to be precise in discussions of sex differences in physiology and behavior, but it is also possible to strive to identify general principles that broadly apply.

This chapter will focus on the role of testicular hormones in driving masculinization of brain and behavior, with an emphasis on rodents, both because it is what we know the most about as it has been the preferred model system for over 50 years, and because it provides a frame work for the appropriate study of more complex organisms, including humans.

15.2 Historical Overview

The story of sex differences in the brain is one of on-again off-again. Unbeknownst to many students of behavioral neuroendocrinology, in the 1960s and 1970s it was considered radical to say there was any difference in the brains of males and females, so radical that the first two studies in mammals were published in Science (Pfaff 1966; Raisman and Field 1971), and both found only very small and seemingly insignificant differences, thereby confirming that male and female brains were largely the same. This was prior to a full understanding of the neural control of gonadotropin secretion from the pituitary, and indeed even the relationship between the brain and the "master gland" was misunderstood until the 1950s (Melmed 2011). In the 1960s, Charles Barraclough made the surprising observation that treating newborn female rat pups with high doses of androgen resulted in a sterile adult (Barraclough and Gorski 1961). He assumed this was the result of a direct effect on the immature ovary, but through systematic investigation determined that it was actually the result of dysregulation in the brain, and resulted from the loss of the ability to release a surge of luteinizing hormone from the pituitary. This illustrated for the first time that hormones early in development could "program" the brain differently in males and females, but there were no clear neuroanatomical underpinnings to the effect.

The story changed dramatically when studies on song birds revealed robust differences in the nuclei in the brain that control song. This was intuited by the fact that males sing complex melodies while females mostly tweet, but the fact that this behavioral difference so clearly mapped onto neuroanatomy had not been previously demonstrated (Nottebohm and Arnold 1976). The observation in birds reopened investigation into mammals and several neuroanatomical sex differences related to reproduction were subsequently identified. Among these are the sexually dimorphic nucleus of the preoptic area (SDN-POA), the spinal nucleus of the bulbocavernosus (SNB), and the anteroventral periventricular nucleus (AVPV). Because each of these differences was directly tied to a reproductive endpoint, either behavior or gonadotropin secretion, everyone was comfortable. Yes, male and female brains differ, the thinking was, but only within the narrow confines of reproduction. A paradigm shift came in the 1990s when Bruce McEwen and colleagues published a series of papers showing effects of estrogens on the density of dendritic spines on hippocampal pyramidal neurons (Gould et al. 1990). At that time the hippocampus was the focus of an enormous swath of neuroscience for its role in learning and memory, as well as regulating the stress axis. As a result, the effect that McEwen and colleagues reported drew tremendous interest, and not without some initial skepticism. Eventually the evidence of hormonal modulation of hippocampal function was irrefutable and while not a sex difference per se, the profound effect of female hormones on a nonreproductive related neuroanatomical endpoint suggested that the influence of sex could be more widespread in the nervous system than previously thought. This may or may not be the reason that about the same time there was a constriction in neurophysiological studies to almost exclusive use of male subjects (Beery and Zucker 2010). As a result, there was little thought given to sex differences by most neuroscientists, and our increased understanding of the brain over the next two decades has been largely built around the model male.

Despite the neglect of sex as an important variable in the majority of neuroscience research, the field of behavioral neuroendocrinology continued to flourish with steady progress made in understanding the cellular and molecular mechanisms by which steroids impact both the developing brain and adult physiology and behavior.

Box 15.1 Scientific Societies Dedicated to Hormones and Behavior

In response to the increasing awareness of the widespread impact of hormones on the brain, a new scientific society was formed in 1996 called the Society for Behavioral Neuroendocrinology (SBN). The journal *Hormones and Behavior* was adopted as its official journal, a publication founded in 1969 and predating the era of neuroscience. Ten years on from the founding of SBN saw formation of a second scientific society, the Organization for the Study of Sex Differences (OSSD), which broadly addresses all aspects of biology that differ in males and females and formed a new journal, Biology of Sex Differences.

Box 15.2 NIH Policy on SABV

The last few years have seen a paradigm shift, the inclusion of sex as a biological variable in all NIH- funded preclinical research in the US (Clayton and Collins 2014), as well as efforts at Canadian and European funding agencies to increase the inclusion of female subjects in research, now referred

(continued)

Box 15.2 (continued)
to as SABV for Sex as a Biological Variable. The intent was to remove the impact of gender, which includes the impact of cultural and societal expectations, and focuses only on that which is biological. Whether and how this will transform our understanding of the brain and all its mysteries is unknown, but early signs suggest that dogmas will fall, and new and unexpected vistas will appear.

15.3 All Sex Differences Are Not Created Equal

In science, semantics matter. Precision in word use is essential for accurately conveying meaning in our principle mode of communication, writing. Often the greatest slippage in the precise use of language is with the most common words. In the realm of sex differences research, there are two mistakes that are particularly frequent. First is the use of the word gender in discussion of research on animals. Gender is a human construct that combines self-perception and others' perceptions of one's sex. We cannot ask animals what sex they think they are, nor what sex they think others are, and so they do not have gender. Thus, any discussion of gender differences should be restricted to humans. The second common mistake is the overuse of the term "sexually dimorphic" to describe any and all sex differences.

(a) Dimorphic is a compound word consisting of di = two, and morph = form, so two forms. If something is sexually dimorphic this means it has two different forms in males and females, which does happen, but not very often. The most notable would be the control of LH release from the pituitary which is not only pulsatile in males and females but also cyclical in females (and in seasonally breeding males and females), including a large surge for inducing ovulation. Mating behavior is also sexually dimorphic, involving mounting and intromitting in males and solicitations and lordosis in females. But many people use the term sexually dimorphic for what is really a sex difference, often a very small one. The descriptor "sex difference" is the appropriate term to apply to any continuous variable that on average differs between males and females. In many instances, there is overlap in the response or endpoint in males and females, but the average is significantly different. Ironically, the sexually dimorphic nucleus of the POA, or SDN, is arguably the most historically important sex difference in the brain, but it is really just a sex difference. Both males and females have an SDN, it is just much larger in males. Indeed, it is so much larger there is very little overlap between the sexes, making it reasonable to call it "dimorphic."
(b) Sometimes sex differences are latent, meaning they are not readily visible or only appear after a challenge or in a specific context. Learning is one such example, with males performing better under stress and female performance

deteriorating. Tracey Shors and colleagues determined this was correlated with a change in the synaptic profile of hippocampal neurons known to be associated with that particular form of learning (Shors 2006). An additional means by which sex differences may be hidden is by appearing the same in presentation in the two sexes but differing in prevalence. Alzheimer's disease is the same disease in men and women, but more frequently diagnosed in women, and not just because they live longer. Conversely, stuttering is more common in boys than girls. This type of sex difference has been referred to as a population sex difference.

(c) Yet another form that sex differences take is in the existence of convergence, also referred to as compensation and first articulated by Geert De Vries (2004). In this instance, males and females are at a different baseline or set point, but under the appropriate circumstance can converge on the same endpoint. Sometimes this occurs because of a lack of the necessary physiology in one sex or the other. For instance, nurturing behavior toward young by adult males is not a general feature of the behavioral repertoire of most rodent species, including toward their own progeny, whereas females who have given birth universally care for their offspring. Pregnancy and parturition involve marked physiological and hormonal changes that are critical to inducing maternal behavior. In mammals, males are deprived of this experience, both physiologically and hence behaviorally as well, with one notable exception. Males in one particular species of monogamous voles have evolved a unique vasopressinergic circuitry that promotes paternal behavior, meaning they have compensated for the lack of exposure to the hormones of pregnancy by generating a new independent regulatory circuit. Catherine Woolley has articulated a related but distinct principle that she also refers to as latent sex differences. A latent sex difference is one that is not immediately apparent. For instance, measures of synapse function of hippocampal pyramidal neurons are the same in males and females, but the underlying cellular mechanisms mediating synaptic plasticity are highly distinct, involving separate signal transduction pathways or different receptors all together (Huang and Woolley 2012). It is essential to conceptualize the various ways a sex difference can manifest when attempting to unravel and understand the origins of either a developmental or adult response profile (Fig. 15.1).

15.4 Critical and Sensitive Periods for Sexual Differentiation

The brain is designed to respond to the environment and reprogram itself in accordance with what it encounters, i.e., learn. The reprogramming that occurs during development is particularly important as it is anticipatory of future events and circumstances. This often means that one developmental path is followed at the

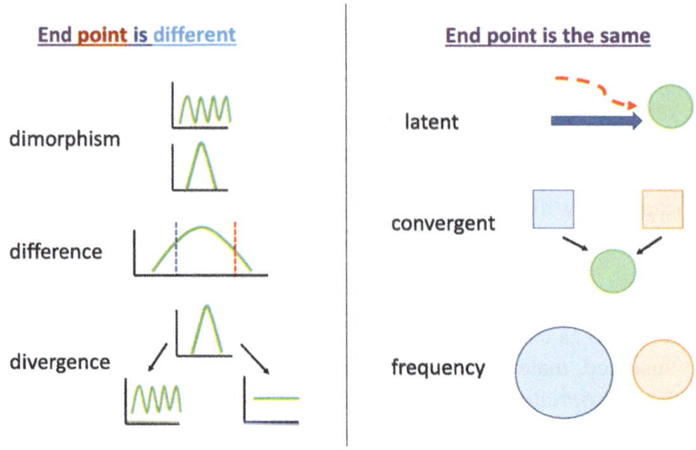

Fig. 15.1 Categories of sex differences. There are multiple ways in which the sexes can differ and these begin with whether the dependent variable (endpoint) is the same or different. Left. When the endpoint is different, it may be sexually dimorphic, meaning it takes on a different form in males versus females, or it may be the average measure of the endpoint is significantly different in the two sexes. A third possibility is that the two sexes are similar at baseline but diverge to different endpoints upon a challenge or with aging. Right. When the endpoint is the same it can be the result of different pathways or mechanisms, which are referred to as latent variables. Alternatively, the sexes may be divergent at baseline but converge to the same endpoint following challenge or evolve a mechanism to achieve the same endpoint via distinct pathway (therefore involving latent variables). Lastly, the endpoint may be the same in males and females but occur more frequently in the population in one sex versus the other

exclusion of another. It also usually means the decision point is restricted to a critical period during which the relevant environmental stimulus is more salient and the biological processes mediating fate determination are in synchrony. One of the best examples is the coinciding events of eye opening and innervation of the lateral geniculate in altricial mammals. Light is the essential external stimulus, and coordinated neuronal activity is the essential internal biological response. The result is a functioning visual system. If light is withheld during the critical period, the animal will be forever blind. Going one step further, if an animal is raised during the critical period in an environment consisting of only vertical lines, no horizontal lines, it will be forever blind to horizontal lines. In this way, the nervous system fine tunes its responses during the critical period, and those responses are then permanently embedded in the brain.

In the case of the visual system, the environmental signal is external light. But internal signals can also drive critical periods of developmental programming. The amount and type of nutrition experienced prenatally programs future metabolomics. Similarly, the amount and type of stress before and shortly after birth impact adult behavioral and physiological phenotypes associated with anxiety, emotionality, and activation of the hypothalamic–pituitary–adrenal axis. In both these instances, males and females may differ in sensitivity or other aspects of critical period programming,

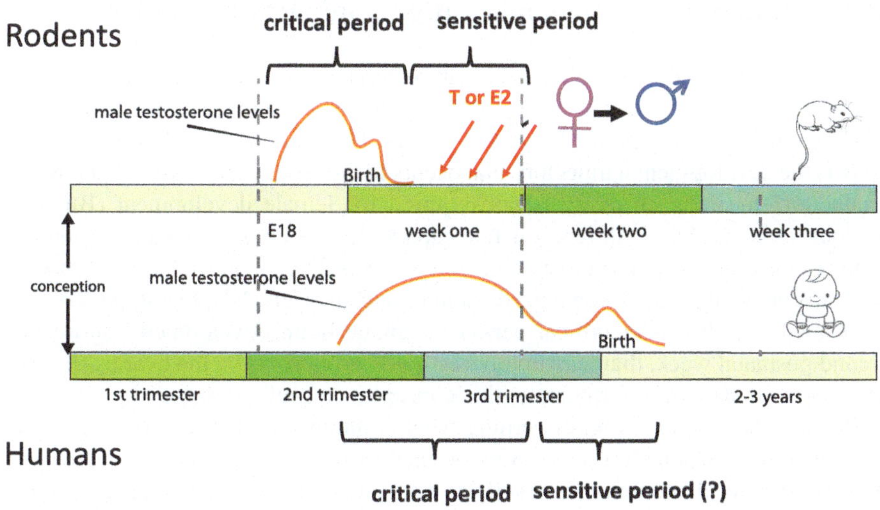

Fig. 15.2 Critical and sensitive periods for sexual differentiation in rodents and humans. The critical period for sexual differentiation begins with the onset of testicular androgen production by the developing fetus in rodents and humans. This is deemed a critical period because if androgens levels do not rise at this time in males, the brain will fail to be masculinized. The sensitive period is defined as that time during which a female can be treated with exogenous steroid (testosterone or estrogen, or other masculinizing agents) and has the potential to be phenotypically converted to the male for a particular endpoint. In rodents this extends into the first postnatal week but may vary by endpoint. In humans, we do not know when or if there is an analogous sensitive period

but they both are exposed to the same salient internal or external stimuli. The case of sexual differentiation of the brain is unique, however, in that only one sex is exposed to the essential agent that drives the developmental program down a particular path. That essential agent is testosterone from the testis of the fetal male (Fig. 15.2). Here the goal is to differentiate male from female in order to assure that adult neurophysiology and behavior is in synchrony with gonadal phenotype. Thus, males need to continuously release small pulses of LH from the pituitary to maintain spermatogenesis and they need to seek out and mate with females and, in some species, compete with other males and/or defend territories and vital resources. Females, by contrast, must be capable of a massive surge in LH release in order to induce ovulation. They must also seek out and mate with males, but usually preferably only one, and then gestate, deliver, and nurture the offspring. Nurturing is a complex combination of nesting, lactating, retrieving, grooming, and defending the young, each of which involves distinct neural circuits and remain incompletely understood. This complex divergence in physiology and behavior is programmed into the brain beginning prior to birth of the individual, despite the actual endpoints not being manifest until adulthood.

15.5 Masculinization, Feminization, and Defeminization

Up to this point, the process of masculinization has been emphasized because it is the key "differentiation" event in sexual differentiation, meaning it is the process by which the male is differentiated from the female. The feminization of both brain and body is the developmental trajectory that occurs in the absence of testis and androgen production, meaning an ovary is not required for female development (But it is obviously essential to adult reproductive capacity). This does not mean that feminization of the brain is not an active process, it surely is, but it is much harder to discern what it is in the absence of some third "neither male nor female" phenotype. There is some evidence of a later critical period in female brain development, during the second postnatal week, that may involve estrogen production by the ovaries (Bakker and Baum 2008), but unfortunately, this concept is not fully developed.

Perhaps the best angle on discerning what is feminization, is its polar opposite, defeminization, which is an active process whereby the female phenotype is removed (Fig. 15.3). This phenomenon is best illustrated, and perhaps limited to, the sex-typic mating behavior seen in rats and mice where males show mounting of sexually receptive females which respond with lordosis, a posture that allows the male to intromit his penis. Since feminization is the default, the neural circuitry of lordosis comes as preinstalled software. Removing that programming is achieved via defeminization which is also driven by androgens aromatized to estrogens in developing

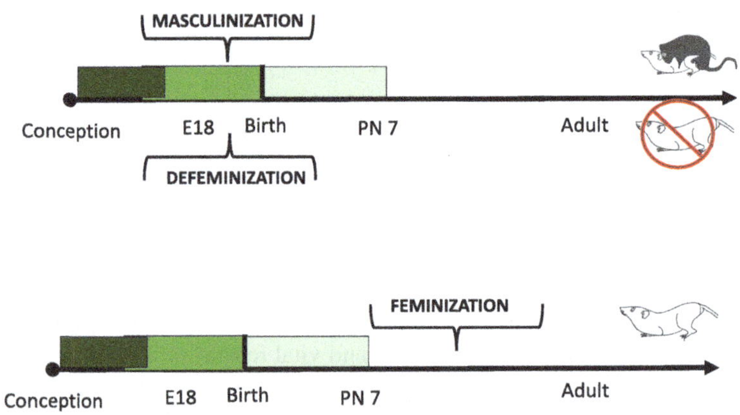

Fig. 15.3 Masculinization, feminization and defeminization. There are three distinct processes that occur during sexual differentiation of the brain to regulate adult mating behavior. Upper. Masculinization is induced by gonadal steroids of the fetal male testis to organize the neural substrate to be responsive to androgens in adulthood and promote male mating behavior. Defeminization is also a steroid induced process occurring during the same time as masculinization, and occurs via distinct cellular mechanisms that result in erasing the innate program of feminization. As a result, sex differences in adult mating behavior are maximized. Lower. Feminization occurs in the absence of gonadal steroids and programs the neural substrate to respond to cyclical estrogens and progestins in adulthood to promote female mating behavior. These events occur during distinct periods of development, although the parameters surrounding a sensitive period for feminization remain much less well defined than those for masculinization

males but via distinct cellular mechanisms (Schwarz and McCarthy 2008). Why such a system has evolved is a mystery, and whether it applies outside the context of sex behavior is debatable, but it tells us there are multiple independent processes occurring simultaneously in the developing brain that assure as little overlap as possible between males and females for certain key reproductive functions. Notably, there is no parallel process of demasculinization in females, and when masculinization is blocked in males it is best referred to as dysmasculinization, since it is the disruption of a normal process, not itself a normal process.

15.6 Testicular Steroids Drive Sexual Differentiation

Everyone is aware that puberty is a time of marked divergence for males and females in terms of the hormonal milieu. But many are not as aware that as early as fetal life there is a similar marked divergence created when the fetal testis becomes steroidogenic and synthesizes sufficiently high levels of androgens that the steroid enters the circulation and gains access to the fetal brain. In rodents, this process begins during the last days of an approximately 3-week gestation and extends into the first hours after birth (Weisz and Ward 1980). In human males, fetal steroidogenesis starts early in the second trimester and is waning by birth but followed by a later active period postnatally that is sometimes referred to as "mini-puberty" (Kuiri-Hanninen et al. 2014). The opening of the critical period for sexual differentiation of the brain is operationally defined by the onset of testicular androgen production. Artificially elevating androgens prior to the naturally occurring period of steroidogenesis is largely without effect, as the essential biological processes are not yet in play. The closing of the critical period is another matter. In males, the critical period closes when the surge of androgen production ends. Once the system is fully exposed to testosterone the programming process is initiated and cannot be undone, meaning that within one day of birth masculinization is largely complete and the critical period closed. Females, however, remain sensitive to the programming potential of testosterone well into the postnatal period, up to a week (Fig. 15.2). If injected with either testosterone or estradiol, the masculinization process is initiated, reflecting a sensitive period as opposed to a critical period (McCarthy et al. 2018). This seemingly lucky accident of timing in rodents provides an excellent tool for studying the mechanisms of steroid-mediated programming, by essentially using the female as a platform for exogenously induced masculinization. In nonhuman primates, however, the end of the critical and the sensitive period are both prenatal (Wallen and Baum 2002), confounding our ability to replicate mechanistic findings in rodents and instead forcing us to consider other routes for translational insight, which will be discussed below. Moreover, since we cannot experiment on humans, we do not know if there is a postnatal sensitive period in addition to the prenatal critical period.

The rodent model has many advantages, but there are confounds just as there are in any "model." The ability to induce the masculinization process in newborn females is an important strength but since the naturally occurring process in males begins in utero, the experimental females are in fact older when treatment occurs. At this stage of development, even a few days is a lot of time for changes to occur. It is possible to treat pregnant dams with steroids and thereby begin the exposure of females in utero, but the endocrine milieu of late pregnancy is both robust and dynamic in preparation for birth. Introducing exogenous steroids against this backdrop is unavoidably confounded. In early studies, beginning with the iconic Phoenix et al. (1959) study, androgen treatment of pregnant guinea pigs not only masculinized the sexual behavior of female offspring but it also masculinized the genitalia, an obvious confound that needs to be avoided if one wishes to understand the masculinization of the brain in isolation from the body. Lastly, rodents give birth to litters, meaning multiple offspring develop in two uterine horns where they are packed in like peas in a pod. Evidence of just how sensitive the developing brain is to gonadal steroids is found in the intrauterine position (IUP) phenomenon whereby females who develop between two males are slightly masculinized and males who develop between two females are slightly less masculinized relative to littermates that sit between mixed-sex or same sex pups. The shift in masculinization is due to androgen production of fetal males being so profuse that some is transferred to nearest neighbors. This is a naturally occurring source of individual variation in phenotype that is independent of genetics and is often overlooked. Endpoints that are impacted by IUP include adult hormonal levels, aggression, sexual behavior, and susceptibility to endocrine disruption. Limited evidence in humans suggests that females from mixed-sex twins exhibit some evidence of masculinization (Ryan and Vandenbergh 2002).

15.7 Impacts of Critical and Sensitive Periods Are Delayed

The famous geneticist Dobszhansky is often quoted for saying "Nothing in biology makes sense except in the light of evolution." Equally true would be "Nothing in neuroscience makes sense except in the light of development." The brain is the slowest developing organ of the body and is designed to be impacted by the local environment and experience as it matures. A fundamental goal of neuroscience is to understand how the brain controls complex adult behaviors. We want to know how the brain determines that now is the time to feed, now to run, now to mate, now to flee, now to fight, and now to sleep. All of these decisions are made against a backdrop of current context and past experience, but they are also impacted by how the brain was wired from the beginning, during development. Development is a simple word for a dynamic multifactorial robust process that is also sensitive to perturbations, sometimes only during a specific window or time period. More importantly, in development what happens first affects what happens next, which affects what happens next and so on in an increasingly constrained one-way trajectory. Development cannot be undone. It is the pervasive underpinning of all that follows.

Sexual differentiation of the brain is a developmental process but one of the more astonishing things about it is the timing, meaning how early. In humans our best estimate of the critical period for sexual differentiation is early second trimester to mid- to late-third trimester. This means the brain is being organized for behaviors that will not be expressed for a decade or more, with all the intervening maturation of this most complex organ. In rodents the critical period is more clearly defined due to our ability to experiment, beginning with the onset of testicular hormones during the last days of gestation and lasting until the first few days of life. But here too the brain is remarkably immature during the process of differentiation. So how is it that these early organizational events endure until adulthood?

The conceptualization of sexual differentiation has evolved along with our expanding views of the brain and its capabilities. In the 1970s and 1980s the mature brain was still considered essentially immutable, with no capacity for neurogenesis, regeneration, or even new synapse formation. The brain was often interpreted as analogous to a computer, being hard wired and programmed. Consistent with that, the impact of sexual differentiation was referred to as building the "neuroarchitecture," establishing a "blue print" and "constructing" a circuit. But we now know the brain is highly mutable, even plastic, with new neurons being born well into adulthood and synapses coming and going with astonishing frequency. At first this view makes it even more challenging to fathom how a developmental process in a brain that is essentially a gelatinous mass can be programmed to regulate later complex social behaviors. Here a second revolution in our thinking about the brain comes into play, the role of epigenetic modifications.

15.8 Neuroepigenetics

Epigenetics literally means "above the genome" and thus refers to modifications to the genetic code that impact gene expression. The dominant forms of epigenetic modifications are changes to the histones that are integral to the chromatin around the DNA and the addition of methyl groups to the nucleotides themselves. There are a variety of changes to the histones but the most prevalent is the addition or removal of acetyl and/or methyl groups to lysine residues. Changes to the DNA itself are largely restricted to the addition of methyl groups and this occurs mostly but not exclusively on cytosines proximal to guanines, so-called CpGs. The notion of epigenetics has been with us for a long time, dating back to the geneticist Waddington in the 1940s who invoked the term to explain how the same DNA in every cell could lead to divergent cell states. There needed to be a mechanism by which some genes would be permanently repressed while others were expressed. This is required in order for a hepatocyte to remain a hepatocyte while an osteoclast stays an osteoclast.

Epigenetics was rediscovered in the nervous system as a means to explain how behavioral phenotypic variation could be transferred from dam to pup. More specifically, Michael Meaney and colleagues determined that rat dams that were highly attentive to their pups produced adult females that were also highly attentive to their pups. That this was not genetic in the classic sense was proven by cross fostering pups born to low attentive dams to highly attentive dams and observing that they were highly attentive as adults. Epigenetic modifications in the promoter region of the glucocorticoid receptor were identified as a mediator of this transfer in behavioral phenotype (Szyf et al. 2007). Many subsequent studies have revealed important roles for epigenetics in behaviors ranging from learning and memory to drug addiction. A central principle emerged that when it comes to epigenetics, the brain is different. Rather than the tight irreversible epigenetic control exerted during cell fate determination, epigenetics of the nervous system is far more pliable and appears designed to respond to experience and the local environment. This is presumably a means for the brain to predict the future and thereby prepare itself. The term neuroepigenetics was coined in recognition of the seemingly unique rules and regulations regarding epigenetics in the brain (Sweatt 2013).

Sexual differentiation of the brain is also a case of predicting the future, as adult physiology and behavior need to be in synch with gonadal phenotype. Since gonadal differentiation is determined by the presence or absence of the gene Sry, and NOT the Y chromosome per se, the brain matches itself to the gonads instead of chromosome complement. This can be achieved epigenetically. Moreover, steroid hormone receptors are transcription factors that interact with the DNA both directly and indirectly and form complexes with other proteins that possess enzymatic activity that regulates histone modifications. These are often in the category of co-activators, such as SRC-1, and co-repressors, such as NCor (Smith and O'Malley 2004). Given the intimate relationship between steroids and the genome, it was natural to postulate epigenetic programming as the basis for the enduring impact of sexual differentiation.

There are two general approaches to exploring epigenetic contributions to sex differences in the brain. First is to pick a key enzyme or epigenetic modification and then measure it in males versus females. This approach has detected sex differences in some histone modifications, as well as activity of the enzymes methylating DNA, the DNA methyltransferases (DNMTs) (McCarthy and Rissman 2014). Second is to use pharmacological or genetic manipulations of the same key enzymes and see if they impact outcomes differently in males versus females. The impacted outcomes may be neuroanatomical, physiological, or behavioral. This approach was used to detect a critical role for histone modifications in the onset of puberty (Lomniczi et al. 2013) and for DNA methylation in masculinization of the preoptic area (Nugent et al. 2015).

The POA is a key brain region regulating male sexual behavior (Hull and Dominguez 2007). If the POA is lesioned or pharmacologically inactivated, males lose all interest in sex. If the POA is stimulated, males lose all interest in anything but sex. The synaptic profile of POA dendrites is established within the first few days of life, endures to adulthood and correlates with the frequency and intensity of male mating behavior in a standardized test (Wright et al. 2008). In an effort to understand how the synaptic density pattern is maintained across the lifespan, the methylation of DNA in the POA was quantified and found to be greater in females (Nugent et al. 2015). This was determined to be the result of greater activity of the DNMTs. Males have lower enzymatic activity due to inhibition from estrogens derived from testicular androgens. To determine if this methylation was functionally significant, neonatal females were injected with a drug into the POA that demethylates the DNA and briefly prevents further methylation during the first few postnatal days and then raised to adulthood. In order to determine if their behavior was masculinized, the females were ovariectomized and implanted with silastic capsules that release testosterone. Once the circulating levels of testosterone in these females were on par with that normally seen in males, they were tested for mating behavior in a controlled setting using test females who were sexually receptive.

Intriguingly, these females were fully masculinized, behaving just as normal males do, and examination of their brains at completion of the experiment revealed the synaptic pattern was the same as that seen in males as well. Collectively these data suggest that epigenetic modifications produced by steroid-mediated DNMT activity during the critical period help to maintain a dendritic architecture that is conducive to male-typic mating behavior in adulthood (Fig. 15.4). Of particular interest, this is a steroid driven process but does not involve interaction of the steroid and its receptor with the DNA that is being modified, it is more of an epigenetic modification. The mechanism by which estradiol modulates DNMT activity is unknown but there is no change in either the mRNA or the protein levels of DNMT1, DNMT3, and DNMT3a. Steroids never cease to amaze.

Critical periods have a beginning and an end. The beginning is usually associated with the onset of some stimulus, such as in the case of sexual differentiation, the onset of steroidogenesis by the fetal testis. But the ends of critical periods are much more elusive, usually being defined as that developmental time point at which the brain no long responds to the requisite stimulus. But what determines the loss of responsiveness? The observation of a change in DNA methylation offered a clue. If hormone-responsive genes are epigenetically silenced at a particular developmental time point, that could close the window of the critical/sensitive period. Indeed, this does appear to be the case since in the aforementioned study. If females were treated with the DNA demethylating drug after the normal close of the critical/sensitive period, they were still masculinized as evidenced by adult brain and behavior. Importantly, if females were treated with a masculinizing dose of steroid at the same time point, there was no effect, indicating a loss of steroid responsiveness due to DNA methylation. Thus, the close of the critical/sensitive period for sexual differentiation may be mediated by epigenetic repression of hormone-responsive genes that promote masculinization (Nugent et al. 2015).

In addition to the epigenetic effects of steroids described above, there are examples of direct gene transcription mediated sexual differentiation. The amygdala is a brain region critical to social behavior in general and a mediator of sex differences in juvenile play behavior in particular. In males the cells of the amygdala express more of the neuropeptide vasopressin, a sex difference that is programmed developmentally by androgens. If a particular DNA methyl-binding protein called MeCP2 is reduced during the critical period in males, the adult vasopressin level is also reduced, indicating a neonatal epigenetic programming of adult gene expression (Forbes-Lorman et al. 2012).

Given the central role of steroid receptors in both developmental and adult sensitivity to gonadal hormones, it is reasonable to also ask if the genes coding for estrogen receptors alpha and beta or progesterone receptor are also epigenetically

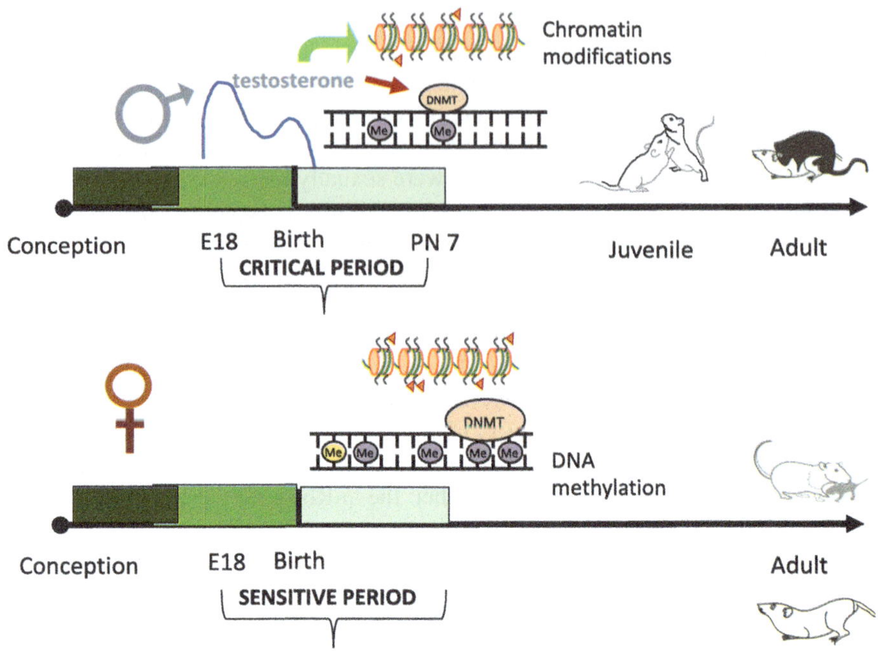

Fig. 15.4 Epigenetic programming of masculinization. Upper, males. The process of masculinization of brain and behavior occurs weeks, months to decades prior to the expression of the phenotype. One mechanism by which this early life programming is maintained is via epigenetic modifications of the DNA in which methyl groups are added to cytosine nucleotides or via changes to the surrounding chromatin, both of which impart long-term regulatory functions on gene expression. In the developing POA elevated gonadal steroids in males inhibit the DNMT enzymes that transfer a methyl group to cytosines, resulting in reduced DNA methylation compared to females. Adult male mounting behavior is illustrated. Lower, females. The higher level of methylation of the female genome in the POA inhibits the gene expression profile essential for adult expression of male sex behavior and may also provide the means by which the sensitive period is closed. Adult female lordosis behavior and maternal behavior are illustrated

modified. There are studies of the promoter regions of all these receptors and there are sex differences, but there is little agreement about the direction and magnitude of the amount of methylation or, even in some cases, precisely which CpGs are differentially methylated (McCarthy et al. 2017). As more studies are conducted, it is becoming increasingly clear that the epigenetics of the brain truly are highly malleable and that a sex difference can exist at one point in time in one brain region, disappear and reappear at another time at different CpG's within the same gene. Moreover, some hormonally induced epigenetic changes do not appear until long after hormone exposure. In what might be called an "epigenetic echo," two independent studies have found hormonally induced changes in DNA methylation that were not apparent within days of steroid treatment in neonates but did appear in juveniles or adults (Ghahramani et al. 2014; Nugent et al. 2010).

How this happens at a mechanistic level remains entirely unknown but highlights one of the biggest challenges in our quest to understand epigenetic underpinnings of sexual differentiation—our inability to repeatedly monitor changes in the same animal. Many things that we seek to measure as neuroendocrinologists can be measured repeatedly, for example, circulating corticosterone, or LH release, body weight, even neuronal activity via indwelling electrodes. But epigenetic modifications to the DNA and surrounding chromatin can only be assessed ex vivo. Moreover, most techniques involve a mixture of cell types and thereby contribute noise to the output and uncertainty to the interpretation. Lastly, the approaches with the highest fidelity, such as deep sequencing, are not as widely available for most laboratories as other techniques. Fortunately, all of these problems are improving in the right direction and so while neuroepigenetics is still at the frontier of neuroscience, it is a worthy frontier to explore.

15.9 Neuroinflammation as a Mediator of Sexual Differentiation

Steroids drive masculinization of the developing brain, but how? Many investigators have explored the mechanistic basis of sexual differentiation for many years, but it is fair to say that no clear unifying theory emerged from the effort and a general conclusion of "it's complicated" pervaded the field. The majority of work focused on neurotransmitter systems for the obvious reason that they are central to the control of behavior. Disruption of a particular neurotransmitter system will often impair normal masculinization, but there was no documented incidence of being able to induce masculinization in the absence of steroids. This stalemate was ended with the discovery that the key masculinizing agent in at least one case—the dendritic synaptic profile of the POA and adult mating behavior—was not a neurotransmitter at all, but instead the membrane derived inflammatory signaling molecule, prostaglandin E2, also called PGE2 (Amateau and McCarthy 2004). Prostaglandins are synthesized from arachidonic acid in the cell membrane by the cyclooxygenase enzymes, COX1 and COX2. Males have higher levels of both enzymes in the POA, leading to higher PGE2 production, which binds to the EP2 and EP4 receptors, G-protein coupled receptors that activate adenylate cyclase and cAMP production. Subsequent activation of protein kinase A (PKA) and phosphorylation of key subunits of glutamate receptors induces clustering of glutamate receptors at the membrane and the formation and stabilization of dendritic spines (Lenz et al. 2011). Many cells throughout the body are capable of synthesizing prostaglandins, but of particular note are microglia, the brain's innate immune system. These macrophage-lineage cells are resident in the brain and they tile throughout, meaning they spread themselves out so that each cell occupies a limited domain that does not overlap with other microglia but the entire brain is in contact with at least one microglial cell. They also respond to and produce prostaglandins, and, do so more in males than females when located within the POA. Indeed, the microglia of the POA are essential for normal masculinization to occur. If they are either depleted or inhibited, the male brain develops without the capacity for male sexual behavior (Lenz et al. 2013; VanRyzin et al. 2018).

The discovery that an innate immune cell is essential for normal masculinization of brain and behavior was a surprise. But this surprise was surpassed with the additional discovery that another innate immune cell, the mast cell, which is derived from precursors made in the bone marrow, is rather the true mastermind behind the immune-driven process of masculinization (Lenz et al. 2018). Mast cells are not ubiquitous throughout the brain, instead being clustered in the meninges, and located deep in the neuropil of the POA. Importantly, there are more in the male POA than in the female. Remarkably, these cells number only in the tens, but when stimulated to release histamine they signal to neighboring microglia which then increase production of prostaglandins and the entire cascade leading to masculinization of brain and behavior is initiated (Fig. 15.5). If the mast cells are prevented from releasing histamine, the train is stopped and the brain dysmasculinized. Blocking the degran ulation of mast cells requires pharmacological treatment. Evidence for a more nuanced role for mast cells came from inducing an allergic reaction in pregnant dams and observing that the mast cells of the POA in the offspring were more activated. Moreover, in adulthood the females were shifted towards a masculinized behavioral profile while the male offspring were less masculinized (Lenz et al. 2019). In both cases the behavior of the males and females was within the normal range, if that range includes both sexes, suggesting this is a potential source of natural variation in behavior. In summary, there are multiple cell types involved in the process of masculinizing the POA, and at least two of them are non-neuronal cells; microglia and mast cells.

Microglia have proven to be equally important to another sexually differentiated behavior, juvenile social play, but involving a completely different mechanism. Play in rodents, sometimes called "rough-and-tumble" play is a complex social behavior that may involve dyads of the same or different sex, or could involve large groups of six or more animals (Argue and McCarthy 2015). The medial amygdala has long been known as a key brain region in the neural circuitry of play and appears to be the critical node for driving a sex difference in which males play more intensely and frequently than females. How this occurs, however, was unknown until recently when it was discovered that the number of astrocytes in the medial amygdala is tightly controlled and is different between males and females, there being more in females. This is not because more astrocytes are born in the female amygdala, however, instead it is because the microglia of this brain region in males phagocytose recently born astrocytes, thereby eliminating a portion of the population before it has time to mature (VanRyzin et al. 2019). In this instance, there is no role for prostaglandins or histamine but instead another class of membrane derived signaling molecules, the endocannabinoids. These short-lived local signaling molecules are found at a higher concentration in the developing amygdala of males and promote phagocytosis by microglia. The end result is a neural circuitry that is more conducive to rough-and-tumble play in males.

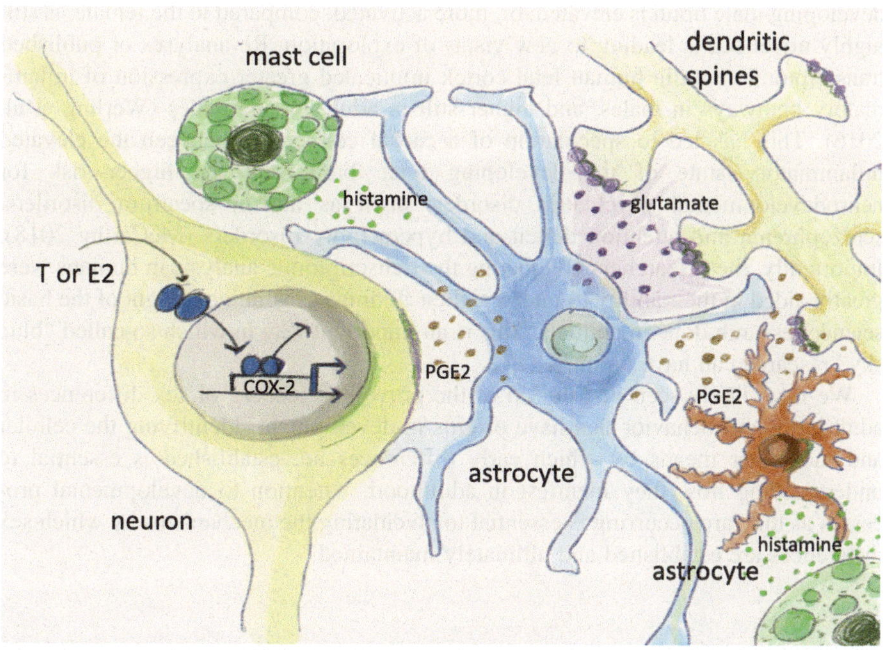

Fig. 15.5 Inflammatory signaling molecules and masculinization. In the preoptic area (POA), sexual differentiation of synaptic density begins with hormonal regulation (by testosterone, T, or estradiol, E2) of gene expression for the cyclooxygenase enzymes (COX-2) which make prostaglandin E2 (PGE2), an inflammatory signaling molecule. However additional PGE2 from the microglia (brown cells), the brain's own immune system, is also required for the full masculinization process. The production of PGE2 by microglia is initiated by histamine release from resident mast cells which are more numerous in males and thereby drive higher PGE2 production which induces glutamate release from neighboring astrocytes. The glutamate binds to AMPA receptors which have been induced to cluster at the membrane surface by the actions of protein kinase A which is activated by PGE2 binding to G-protein coupled receptors on neurons. The activation of the AMPA receptors causes the formation of dendritic spine synapses. Thus at least four cell types are involved in the masculinization of the preoptic area, and two of them are of non-neuronal origin but instead are components of the immune system

15.10 Perspectives

Sex differences are established during the process of sexual differentiation which occurs weeks, months and in some cases years before the physiology and behavior are manifest. Epigenetic programming both establishes and maintains sex differences but also provides for reversibility across the life span. Sex differences come in a variety of forms and understanding the nature of any particular one is essential to proper interpretation and experimental design. The mechanisms of sexual differentiation are diverse but often include inflammatory signaling molecules and immune cells. The discovery that the immune profile of portions of the developing male brain is elevated, or, more activated, compared to the female, is still highly novel but is leading to new vistas of exploration. Re-analyses of published transcriptomic data in human fetal cortex implicated greater expression of inflammatory pathways in males, and higher still in adult autistic males (Werling et al. 2016). This has led to speculation of a causal connection between the elevated inflammatory state of the developing male brain and the higher risk for neurodevelopmental psychiatric disorders such as autism spectrum disorders, schizophrenia and attention deficit and hyperactivity disorders (McCarthy 2018). Importantly, the researchers conducting the transcriptomic analysis in humans were greatly aided in their ability to interpret their findings as genuine in light of the basic science research done in animals. This is an important way in which so-called "blue sky" research can have an impact.

We have likely seen just the tip of the proverbial iceberg of sex differences in adult brain and behavior that have origins in development. Identifying the cellular and molecular means by which early differences are established is essential to understanding how they manifest in adulthood. Attention to developmental processes as they are occurring is essential to elucidating the mechanisms by which sex differences are established and ultimately maintained.

Key References

Amateau and McCarthy (2004)—This was the first report in which masculinization of a female brain was achieved by a downstream signaling molecule from steroid action, as opposed to treating directly with steroids. This opened a new era in understanding of the novel mechanisms by which steroids can impact the brain, including through the enhancement of immune system signaling.

Beery and Zucker (2010)—An in depth analyses of a large number of publications across a wide swatch of biomedical research revealed the over reliance on male animals, especially in neuroscience. Almost six times as many studies used exclusively male rats and mice. Subsequent to this publication the NIH mandated the incorporation of sex as a biological variable in all preclinical research.

De Vries (2004)—This minireview proposed the novel idea that sometimes the two sexes evolve mechanisms to try and be more similar, not more different. Using the example of the biparental vole, DeVries highlights how males have evolved a new neural circuit that promotes parenting and is active in the absence of the hormones of pregnancy and parturition that are so important to females.

Nottebohm and Arnold (1976)—Arguably the first robust neuroanatomical sex difference discovered in a vertebrate brain, this land mark publication led to increased scrutiny of the mammalian brain and the discovery of multiple sex differences in the size of various brain regions and subnuclei.

Nugent et al. (2015)—One of a small number of studies that revealed sex differences in the epigenome and in this case demonstrates functional significance for the sexual differentiation of sexual behavior.

Werling et al. (2016)—Secondary analyses of the transcriptome from human fetal postmortem cortical tissue found that more genes associated with inflammation, activated astrocytes and microglia are up regulated in the male compared to the female. Preclinical studies in animals were essential to interpretation of these findings as indicative of normal healthy brain development in males.

References

Amateau SK, McCarthy MM (2004) Induction of PGE(2) by estradiol mediates developmental masculinization of sex behavior. Nat Neurosci 7(6):643–650

Argue KJ, McCarthy MM (2015) Characterization of juvenile play in rats: importance of sex of self and sex of partner. Biol Sex Differ 6:16

Bakker J, Baum MJ (2008) Role for estradiol in female-typical brain and behavioral sexual differentiation. Front Neuroendocrinol 29(1):1–16. https://doi.org/10.1016/j.yfrne.2007.06.001

Barraclough CA, Gorski RA (1961) Evidence that the hypothalamus is responsible for androgen-induced sterility in the female rat. Endocrinology 68:68–79

Beery AK, Zucker I (2010) Sex bias in neuroscience and biomedical research. Neurosci Biobehav Rev 35:565–572. https://doi.org/10.1016/j.neubiorev.2010.07.002

Clayton JA, Collins FS (2014) Policy: NIH to balance sex in cell and animal studies. Nature 509 (7500):282–283

De Vries GJ (2004) Minireview: sex differences in adult and developing brains: compensation, compensation, compensation. Endocrinology 145(3):1063–1068

Forbes-Lorman RM, Rautio JJ, Kurian JR, Auger AP, Auger CJ (2012) Neonatal MeCP2 is important for the organization of sex differences in vasopressin expression. Epigenetics 7 (3):230–238. https://doi.org/10.4161/epi.7.3.19265

Ghahramani NM, Ngun TC, Chen PY, Tian Y, Krishnan S, Muir S et al (2014) The effects of perinatal testosterone exposure on the DNA methylome of the mouse brain are late-emerging. Biol Sex Differ 5:8. https://doi.org/10.1186/2042-6410-5-8

Gould E, Woolley CS, Frankfurt M, McEwen BS (1990) Gonadal steroids regulate dendritic spine density in hippocampal pyramidal cells in adulthood. Neuroscience 10:1286–1291

Huang GZ, Woolley CS (2012) Estradiol acutely suppresses inhibition in the hippocampus through a sex-specific endocannabinoid and mGluR-dependent mechanism. Neuron 74(5):801–808. https://doi.org/10.1016/j.neuron.2012.03.035

Hull EM, Dominguez JM (2007) Sexual behavior in male rodents. Horm Behav 52(1):45–55. https://doi.org/10.1016/j.yhbeh.2007.03.030

Kuiri-Hanninen T, Sankilampi U, Dunkel L (2014) Activation of the hypothalamic-pituitary-gonadal axis in infancy: minipuberty. Horm Res Paediatr 82(2):73–80. https://doi.org/10.1159/000362414

Lenz KM, Wright CL, Martin RC, McCarthy MM (2011) Prostaglandin E regulates AMPA receptor phosphorylation and promotes membrane insertion in preoptic area neurons and glia during sexual differentiation. PLoS One 6(4):e18500. https://doi.org/10.1371/journal.pone.0018500

Lenz KM, Nugent BM, Haliyur R, McCarthy MM (2013) Microglia are essential to masculinization of brain and behavior. J Neurosci 33(7):2761–2772

Lenz KM, Pickett LA, Wright CL, Davis KT, Joshi A, McCarthy MM (2018) Mast cells in the developing brain determine adult sexual behavior. J Neurosci 38(37):8044–8059. https://doi.org/10.1523/JNEUROSCI.1176-18.2018

Lenz KM, Pickett LA, Wright CL, Galan A, McCarthy MM (2019) Prenatal allergen exposure perturbs sexual differentiation and programs lifelong changes in adult social and sexual behavior. Sci Rep 9:4837

Lomniczi A, Loche A, Castellano JM, Ronnekleiv OK, Bosch M, Kaidar G et al (2013) Epigenetic control of female puberty. Nat Neurosci 16(3):281–289. https://doi.org/10.1038/nn.3319
McCarthy MM (2018) Sex differences in neuroimmunity as an inherent risk factor. Neuropsychopharmacology 44(1):38-44. https://doi.org/10.1038/s41386-018-0138-1
McCarthy MM, Rissman EF (2014) Epigenetics of reproduction. In: Plant T, Zeleznik AJ (eds) Knobil & Neill's physiology of reproduction, vol 2, 4th edn. Academic, London, pp 2439–2501
McCarthy MM, Nugent BM, Lenz KM (2017) Neuroimmunology and neuroepigenetics in the establishment of sex differences in the brain. Nat Rev Neurosci 18(8):471–484. https://doi.org/10.1038/nrn.2017.61
McCarthy MM, Herold K, Stockman SL (2018) Fast, furious and enduring: sensitive versus critical periods in sexual differentiation of the brain. Physiol Behav 187:13–19. https://doi.org/10.1016/j.physbeh.2017.10.030
Melmed S (2011) The pituitary, 3rd edn. Academic, London
Nottebohm F, Arnold AP (1976) Sexual dimorphism in vocal control areas of the songbird brain. Science 194:211–213
Nugent BM, Schwarz JM, McCarthy MM (2010) Hormonally mediated epigenetic changes to steroid receptors in the developing brain: implications for sexual differentiation. Horm Behav. https://doi.org/10.1016/j.yhbeh.2010.08.009
Nugent BM, Wright CL, Shetty AC, Hodes GE, Lenz KM, Mahurkar A et al (2015) Brain feminization requires active repression of masculinization via DNA methylation. Nat Neurosci 18(5):690–697. https://doi.org/10.1038/nn.3988
Pfaff DW (1966) Morphological changes in the brains of adult male rats after neonatal castration. J Endocrinol 36:415–416
Phoenix CH, Goy RW, Gerall AA, Young WC (1959) Organizing action of prenatally administered testosterone proprionate on the tissues mediating mating behavior in the female guinea pig. Endocrinol 65:369–382
Raisman G, Field PM (1971) Sexual dimorphism in the preoptic area of the rat. Science 173:731–733
Ryan BC, Vandenbergh JG (2002) Intrauterine position effects. Neurosci Biobehav Rev 26 (6):665–678
Schwarz JM, McCarthy MM (2008) Steroid-induced sexual differentiation of the brain: multiple pathways, one goal. J Neurochem 105:1561–1572
Shors TJ (2006) Stressful experience and learning across the lifespan. Annu Rev Psychol 57:55–85
Smith CL, O'Malley BW (2004) Coregulator function: a key to understanding tissue specificity of selective receptor modulators. Endocr Rev 25(1):45–71
Sweatt JD (2013) The emerging field of neuroepigenetics. Neuron 80(3):624–632. https://doi.org/10.1016/j.neuron.2013.10.023
Szyf M, Weaver I, Meaney M (2007) Maternal care, the epigenome and phenotypic differences in behavior. Reprod Toxicol 24(1):9–19. https://doi.org/10.1016/j.reprotox.2007.05.001
VanRyzin JW, Pickett LA, McCarthy MM (2018) Microglia: driving critical periods and sexual differentiation of the brain. Dev Neurobiol 78(6):580–592. https://doi.org/10.1002/dneu.22569
VanRyzin JW, Marquardt AE, Argue KJ, Vecchiarelli HA, Ashton SE, Arambula SE, Hill MN, McCarthy MM (2019) Microglial phagocytosis of newborn cells is induced by endocannabinoids and sculpts sex differences in juvenile rat social play. Neuron 102:435–449
Wallen K, Baum MJ (2002) Masculinization and defeminization in altricial and precocial mammals: comparative aspects of steroid hormone action. In: Pfaff D (ed) Hormones brain and behavior, vol 5. Academic, London, pp 385–424
Weisz J, Ward IL (1980) Plasma testosterone and progesterone titers of pregnant rats, their male and female fetuses and neonatal offspring. Endocrinology 106:306–313
Werling DM, Parikshak NN, Geschwind DH (2016) Gene expression in human brain implicates sexually dimorphic pathways in autism spectrum disorders. Nat Commun 7:10717. https://doi.org/10.1038/ncomms10717
Wright CL, Burks SR, McCarthy MM (2008) Identification of prostaglandin E2 receptors mediating perinatal masculinization of adult sex behavior and neuroanatomical correlates. Dev Neurobiol 68:1406–1419

Development and Modulation of Female Reproductive Function by Circadian Signals 16

Neta Gotlieb, Jacob Moeller, and Lance J. Kriegsfeld

Abstract

Female reproductive success requires multifaceted, temporally coordinated hormone secretion. The circadian timing system is central to this complex neuroendocrine regulation, with circadian disruption associated with irregular ovulatory cycles, reduced fertility, increased miscarriage rates, and anomalous fetal development. Because living in the modern world is associated with relatively chronic circadian disruption due to limited sunlight exposure during the day and exposure to artificial light at night, this issue is of broad concern. The master mammalian circadian pacemaker in the suprachiasmatic nucleus communicates monosynaptically to neuroendocrine cells in the brain to mediate the hormonal events required for ovulation and pregnancy maintenance. Likewise, maternal hormones cross the placenta to drive rhythmic fetal processes critical for typical development. The present overview describes the means by which the circadian timing system integrates with the reproductive axis to regulate female reproductive functioning. Likewise, the negative consequences of circadian disruptions on female reproductive health, and the mechanisms underlying these deleterious outcomes, are considered.

N. Gotlieb
Department of Psychology, University of California, Berkeley, Berkeley, CA, USA

J. Moeller
Graduate Group in Endocrinology, University of California, Berkeley, Berkeley, CA, USA

L. J. Kriegsfeld (✉)
Department of Psychology, University of California, Berkeley, Berkeley, CA, USA

Department of Integrative Biology, University of California, Berkeley,, Berkeley, CA, USA

Graduate Group in Endocrinology, University of California, Berkeley, Berkeley, CA, USA

The Helen Wills Neuroscience Institute, University of California, Berkeley, Berkeley, CA, USA
e-mail: kriegsfeld@berkeley.edu

S. Wray, S. Blackshaw (eds.), *Developmental Neuroendocrinology*, Masterclass in Neuroendocrinology 9, https://doi.org/10.1007/978-3-030-40002-6_16

Keywords
Biological rhythms · Circadian · Hormone · Reproduction

16.1 Introduction

Coordinated timing of neuroendocrine events is fundamental for successful female reproduction across mammalian species, including humans (Mahoney 2010; Williams and Kriegsfeld 2012; Kriegsfeld et al. 2018; Boden et al. 2013; Gamble et al. 2013; Simonneaux et al. 2017). Each phase of the female reproductive cycle, from ovulation, pregnancy, and fetal development, to parturition, requires specifically timed patterns of hormone secretion regulated by the circadian system (McEwen et al. 1987; Blaustein et al. 1994; Kriegsfeld et al. 2002; Egli et al. 2004; Mong and Pfaff 2003). Disruptions to circadian timing have marked, negative consequences for female reproductive health. For example, women with irregular work schedules or frequent travel across time zones experience abnormal menstrual cycles (Lawson et al. 2011; Wang et al. 2016), reduced fertility (Ahlborg et al. 1996; Fernandez et al. 2016), and increased miscarriage rates (Nurminen 1998; Lawson et al. 2012; Fernandez et al. 2016). Analogously, in rodents, ablation of the master circadian clock in the brain, blocking relevant clock output signals, or disrupting the genes driving circadian clock function at the cellular level, lead to pronounced deficits in ovulation and reproductive success (Miller et al. 2004; Nunez and Stephan 1977; Wiegand and Terasawa 1982; van der Horst et al. 1999; Chu et al. 2013).

The circadian system confers a selective advantage by allowing organisms to anticipate rhythmic and predictable environmental change and adjust their physiology and behavior accordingly. Although circadian rhythms are endogenously generated, to effectively synchronize internal timing with the external environment, exposure to sunlight during the day and darkness at night entrains (synchronizes) these rhythms to environmental time (Hughes et al. 2015). Unfortunately, a major consequence of contemporary lifestyles and technological advancements omnipresent in the modern world is increased exposure to sun-free environments during the day and artificial lighting at night. This combination results in an incongruence between the endogenous circadian timing system and the external environment, leading to chronic and pervasive "jet lag" in the modern world (Bedrosian and Nelson 2013; Fonken and Nelson 2014). Such concerns have attracted the attention of the medical community, with the American Medical Association adopting a policy statement on the dangers of light at night for health and reproductive functioning (Stevens et al. 2013).

Because the majority of studies to date have concentrated on the role of circadian rhythms in female reproductive function, the present chapter focuses on this sex. Although, it is worth noting that circadian disruption negatively affects semen quality and sperm numbers (Cagnacci et al. 1999; Xie et al. 2018) and compromises fertility in men (Deng et al. 2018) with analogous results seen in mouse models in which the genes regulating the circadian clockwork are knocked out (Alvarez et al.

2008; Schoeller et al. 2016). Herein, we consider all phases of the female reproductive cycle, including ovulation, mating, pregnancy and fetal development, and parturition and describe how the circadian system integrates with the reproductive axis to mediate reproductive success. Likewise, we consider development in relevant systems and circuits where applicable and the negative consequences of circadian disruption during each stage of reproduction.

16.2 The Circadian Timing System

The circadian system consists of a master brain clock in the suprachiasmatic nucleus (SCN) of the anterior hypothalamus that is synchronized to environmental time via a direct retinal pathway. As indicated previously, although circadian rhythms are endogenously generated, to be adaptive for an organism and anticipate environmental changes across the day, these endogenous cycles are entrained to environmental time. Light entrains the SCN via retinal communication from rods and cones that, in turn, target specialized, intrinsically photosensitive retinal ganglion cells containing the photopigment, melanopsin (Hattar et al. 2002; Berson et al. 2002; Lucas et al. 2003; Panda et al. 2002). Successively, the SCN uses neural, diffusible, and autonomic communication to convey timing information to the whole organism (Butler et al. 2010; Buijs et al. 2013).

Circadian rhythms are a cell-autonomous property, with circadian rhythms being generated by an autoregulatory transcription–translation feedback loop consisting of clock genes and their protein products (Takahashi 2015; Honma 2018). The core feedback loop begins in the morning with the clock protein, CLOCK, binding to BMAL1 to drive the transcription of the Period (*Per1* and *Per2*) and Cryptochrome (*Cry1* and *Cry2*) genes. Over the course of the day, *Per* and *Cry* transcripts are translated into their respective proteins that inevitably feed back to the cell nucleus to repress CLOCK: BMAL1-mediated transcription until the next morning when transcription resumes. Whereas the *Clock* gene is constitutively expressed, an additional feedback loop driven by the CLOCK: BMAL1 complex regulates *Bmal1* transcription through repression by *Rev-erbα* and transcriptional activation via retinoic acid receptors (RORs) (Fig. 16.1). Circadian timekeeping is a ubiquitous property of cells throughout the brain and body, with virtually all cells exhibiting circadian timekeeping (Honma 2018). Even when the SCN is isolated in culture, the master clock maintains indefinite circadian rhythms at the tissue level due to unique coupling among independent oscillators. In contrast, in the absence of master clock communication or other entraining stimuli, extra-SCN brain loci and peripheral organs exhibit loss of rhythmicity after several cycles (Yamazaki et al. 2000; Abe et al. 2002). This loss of rhythmicity in extra-SCN systems results from loss of coupling among cellular oscillators having slightly different periods (Welsh et al. 2004).

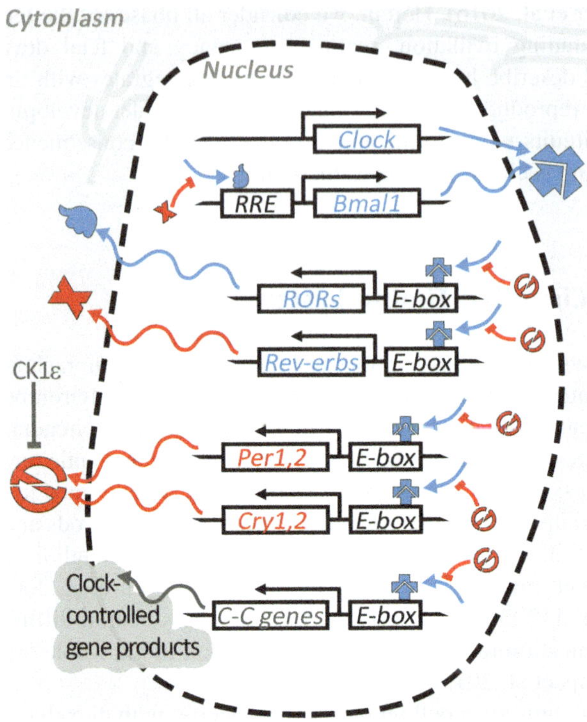

Fig. 16.1 The molecular clockwork. A simplified model of the intracellular mechanisms responsible for mammalian circadian rhythm generation. The process begins when CLOCK and BMAL1 proteins dimerize to drive the transcription of the *Per* (*Per1* and *Per2*) and Cry (*Cry1* and *Cry2*) genes. Throughout the day, PER and CRY proteins rise within the cell cytoplasm. When levels of PER and CRY reach a threshold, they form heterodimers, feed back to the cell nucleus, and negatively regulate CLOCK: BMAL1-mediated transcription of their own genes. Levels of PER are regulated by casein kinase 1 epsilon (CK1ε) which phosphorylates these proteins and marks them for degradation, thereby appropriately delaying negative feedback. Whereas *Clock* is constitutively expressed, a secondary feedback loop drives the transcription of ROR and Rev-Erbα that, in turn, induce rhythms in *Bmal1* transcription through stimulatory and inhibitory actions on ROR response elements (RRE) in the *Bmal1* promotor, respectively. Clock-controlled genes are tissue-specific genes that are produced rhythmically by the CLOCK:BMAL1 complex but are not part of the clockwork mechanism (i.e., do not feed back onto the clockwork)

16.3 The Hypothalamic–Pituitary–Gonadal Axis (HPG)

At the pinnacle of the HPG axis, hypothalamic gonadotropin-releasing hormone (GnRH) neurons send axonal projections to the median eminence. Here, GnRH is released from nerve terminals into the portal vasculature to trigger the secretion of the gonadotropins, follicle stimulating hormone (FSH) and luteinizing hormone (LH), from the anterior pituitary. In turn, LH and FSH act on the gonads to stimulate sex steroid (e.g., progesterone, estradiol, and testosterone) synthesis and secretion and gametogenesis, respectively. Sex steroids and gonadotropins feed back to the hypothalamus and anterior pituitary to regulate HPG axis activity. Upstream of the GnRH system, two key neuropeptides were identified around the turn of the millennium that have pronounced inhibitory and stimulatory actions on GnRH neurons, namely RFamide-related peptide 3 and kisspeptin, respectively. The discovery and significance of these neuropeptides are described below.

16.3.1 RFamide-Related Peptide 3

In a search for novel peptides in the Arg-Phe-NH_2 (i.e., R=Arginine, F=Phenylalanine; termed RFamide peptides) family, Tsutsui and colleagues discovered a novel RFamide peptide that inhibited gonadotropin release from cultured quail pituitaries (Tsutsui et al. 2000). Because this hypothalamic neuropeptide specifically inhibited the gonadotropins without affecting other pituitary hormones, they named this peptide gonadotropin-inhibitory hormone (GnIH) (Tsutsui et al. 2000). In birds, the GnIH precursor encodes one GnIH and two GnIH-related peptides (GnIH-RP-1 and GnIH-RP-2) (Satake et al. 2001; Tsutsui et al. 2009). In mammals, the neuropeptide precursor cDNA encodes three peptides (known as RFamide-related peptides [RFRPs]), RFRP-1, -2 and -3), with RFRP-3 thought to be the ortholog of avian GnIH (Tsutsui et al. 2009). Studies across mammalian species have found pronounced roles for this neuropeptide in regulating reproductive function.

The receptor for GnIH/RFRP-3 is a G-protein coupled receptor (GPR), the formerly orphaned GPR147 [also called NPFF1 (Bonini et al. (2000)]. GPR147 is most commonly coupled to an inhibitory G-protein (Gαi), with GnIH/RFRP3 suppressing adenylate cyclase activity (Hinuma et al. 2000; Shimizu and Bedecarrats 2010). However, in some instances, GPR147 is coupled to Gαs or Gαq proteins, where this differential coupling may account for disparity in the effects of RFRP3 based on sex or reproductive status [described further below (Gouarderes et al. 2007)].

In most rodent species, RFRP3-cell bodies are localized exclusively to the dorsomedial hypothalamus (DMH; reviewed in (Henningsen et al. 2017; Kriegsfeld et al. 2018; Ubuka and Tsutsui 2018). In mammals, RFRP-3-immunoreactive (–ir) neuronal fibers are widely distributed in the diencephalon, mesencephalon, and limbic structures (Smith et al. 2008; Kriegsfeld et al. 2006; Yano et al. 2003), providing a direct synaptic mechanism for broadly affecting neurophysiology and behavior. Across mammalian species, including humans, RFRP-3 generally suppresses GnRH and gonadotropin secretion via direct actions on GnRH cells and potentially at the level of the pituitary (Clarke et al. 2008; Johnson et al. 2007; Kriegsfeld et al. 2006; George et al. 2017; Ducret et al. 2009; Wu et al. 2009); reviewed in Ubuka and Tsutsui (2018).

Under some circumstances, RFRP-3 stimulates the reproductive axis. For example, male Siberian hamsters (*Phodopus sungorus*) that typically breed during long, summer-like days exhibit an increase in LH concentrations following central administration of RFRP-3 when maintained in short-day conditions (Ubuka et al. 2012). Additionally, male Syrian hamsters maintained in long days exhibit elevated LH and FSH in response to RFRP-3 administration (Ancel et al. 2012). Likewise, in striped hamsters (*Cricetulus barabensis*), the relationship between RFRP-3 and GnRH expression depends on sex and developmental/reproductive status (Zhao et al. 2014). Finally, RFRP3 stimulates LH secretion in male mice but inhibits LH release in female mice when estradiol concentrations are high at the time of the LH surge, but has no effect during diestrus or in ovariectomized females with low estradiol concentrations provided exogenously (Ancel et al. 2017). These findings suggest that the effects of RFRP-3 on gonadotropins may depend on the species, season, sex, and developmental status. Likewise, these findings suggest that GPR147 couples differentially with Gαi, Gαs, or Gαq based on these same variables (Gouarderes et al. 2007).

16.3.2 Kisspeptin

Kisspeptin is also a member of the RFamide family of peptides. Kisspeptin was initially known as metastin, a tumor metastasis suppressor gene in human melanoma and breast carcinoma, discovered in 1996 in Hershey, PA (Lee et al. 1996; Lee and Welch 1997; Kotani et al. 2001). The gene for metastin was named *Kiss1*, with the "SS" representing suppressor sequence and "Ki" added as an homage for the home town's famous Hershey kisses (Lee et al. 1996). The role of kisspeptin as a positive regulator of the HPG axis was discovered in 2003 in individuals exhibiting hypophysiotropic hypogonadism. These individuals were found to have a mutation in G-protein coupled receptor (GPR54), the cognate receptor for *Kiss1* gene protein products (de Roux et al. 2003; Seminara et al. 2003). These individuals exhibit low or absent circulating LH, fail to undergo puberty and are reproductively incompetent as adults. When the authors created GPR54-deficient mice, these animals exhibited an analogous phenotype (de Roux et al. 2003; Seminara et al. 2003). The protein product of the *Kiss1* gene is now commonly referred to as kisspeptin.

The *Kiss1* gene encodes a family of kisspeptin peptides, beginning with a precursor polypeptide of 145 amino acids, that is cleaved into a 54 amino acid protein (named kisspeptin-54) that can be further cleaved into 10, 13, and 14 amino acid proteins, all of which share an RFamide sequence on their C-terminus, are biologically active, and share a similar affinity to GPR54 (Kotani et al. 2001; Muir et al. 2001; Ohtaki et al. 2001).

Kisspeptin directly stimulates the secretion of GnRH and LH across mammalian species, including humans, via direct actions on GnRH neurons (Trevisan et al. 2018; Saedi et al. 2018; Yeo and Colledge 2018). In women with hypothalamic amenorrhea, acute administration of kisspeptin stimulates gonadotropin release (Jayasena et al. 2009, 2010), indicating that failure of kisspeptin signaling contributes to reproductive cessation is these women. Both pituitary cells and GnRH neurons express GPR54; however, whether kisspeptin directly stimulates gonadotropin release from the pituitary requires further investigation. For example, although prominent expression of GPR54 is detected in human pituitary (Muir et al. 2001; Ohtaki et al. 2001), and peripheral administration of kisspeptin increases plasma LH concentrations in some cases (Thompson et al. 2004), the effects of kisspeptin on cultured pituitary cells are equivocal (Matsui et al. 2004; Thompson et al. 2004; Navarro et al. 2005a, b).

Kisspeptin cell bodies are concentrated in the anteroventral periventricular (AVPV) and the arcuate (Arc) nuclei of the hypothalamus in rodents (Smith et al. 2005a, b; Clarkson et al. 2008; Clarkson and Herbison 2006) and in the preoptic area (POA) and infundibulum in humans (Hrabovszky et al. 2010). A large percentage of kisspeptin cells express estrogen receptor (ER)-α (Smith et al. 2005a), thereby mediating positive (AVPV) and negative (Arc) feedback effects of estradiol in females (Smith et al. 2005a, 2006; Adachi et al. 2007; Williams et al. 2011; Mittelman-Smith et al. 2012).

A number of studies suggest that Arc kisspeptin cells serve as critical components of estradiol negative feedback and the GnRH pulse generator. Arc kisspeptin neurons co-express two additional peptides, neurokinin B (NKB) and dynorphin (Burke et al. 2006; Goodman et al. 2007, 2014; Navarro et al. 2009; Rance 2009; Wakabayashi et al. 2010; Okamura et al. 2017). Together, this population is referred to as KNDy (pronounced "candy") neurons. These triple phenotype cells communicate in an autosynaptic feedback manner in which KNDy cells project to other KNDy cells (Foradori et al. 2002; Burke et al. 2006; Krajewski et al. 2010). Local secretion of NKB stimulates kisspeptin release from KNDy cells onto GnRH neurons (Wakabayashi et al. 2013). Mutations in the NKB gene (i.e., *Tac3*) or its receptor (*Tacr3*) lead to hypogonadotropic hypogonadism (Guran et al. 2009; Topaloglu et al. 2009; Young et al. 2010). In contrast to NKB, dynorphin inhibits Arc kisspeptin secretion and is thought to participate in the termination of each pulse (Goodman et al. 2013; Ruka et al. 2013).

16.4 Circadian Control of Ovulation

In species that ovulate spontaneously, the preovulatory LH surge that initiates ovulation is ubiquitous [rodents (Levine and Ramirez 1982; Mahoney et al. 2004; Moline et al. 1981; Sarkar et al. 1976), sheep (Moenter et al. 1992), rhesus macaques (Pau et al. 1993), and women (Cahill et al. 1998; Edwards et al. 1980; Elkind-Hirsch et al. 1984)]. The LH surge occurs in early morning in women (Kerdelhue et al. 2002) and diurnal rodents (Mahoney et al. 2004) but in early evening in nocturnal rodents (Christian and Moenter 2010), coordinating the time of maximal fertility with activity. During most of the rodent estrous cycle, gonadotropin concentrations are low due to the negative feedback effects of gonadal steroids (reviewed in Christian and Moenter 2010). At the time of ovulation, however, positive feedback effects of estradiol serve a permissive role in initiating the LH surge by a timed, neural signal from the SCN (i.e., positive feedback) (Karsch et al. 1997; Levine 1997; Levine and Ramirez 1982; Williams and Kriegsfeld 2012). This "switch" in estradiol action results from the coordinated timing of positive and negative regulators of the reproductive axis by the circadian timing system. The requirement for both high estradiol concentrations and a circadian timing signal to initiate the LH surge ensures appropriate oocyte maturation and maximal sexual motivation coincide with the time of ovulation.

As indicated previously, lesions of the SCN or the severing of connections between the SCN and the preoptic area results in acyclicity in rodents (Brown-Grant and Raisman 1977; Wiegand et al. 1980; Wiegand and Terasawa 1982). Furthermore, genetic impairments of the molecular clockwork also disrupt estrous cycles and the LH surge in mice (Khan and Kauffman 2012; Chu et al. 2013; Dolatshad et al. 2006; Miller et al. 2004). Although the SCN can support behavioral rhythms via a diffusible signal, the SCN-derived signal that generates the LH surge is neural; SCN-lesioned female hamsters receiving fetal SCN transplants that do not form neural connections with the host brain exhibit rhythmic behavior but not an LH surge (Lehman et al. 1987; Silver et al. 1996; Meyer-Bernstein et al. 1999). As described further below, both direct SCN-GnRH neuronal communication and indirect connections from the SCN to estrogen receptor-α (ERα)-expressing cell phenotypes that positively and negatively regulate the GnRH system integrate circadian and estradiol signaling to initiate the preovulatory LH surge and ovulation (Fig. 16.2).

16.4.1 Monosynaptic SCN Regulation of the GnRH System in LH Surge Generation

Two major neuropeptidergic SCN cell phenotypes have been implicated in the generation of the preovulatory LH surge and ovulation in rodents, neurons synthesizing the neuropeptide vasoactive intestinal peptide (VIP) in the retinorecipient SCN core and vasopressin (AVP) in the dorsal SCN (Moore et al. 2002). Both VIP and AVP exhibit peak release around the time of the LH surge (Schwartz et al. 1983; Kalsbeek et al. 1995; Francl et al. 2010) and central injections of either VIP or AVP receptor antagonists attenuate LH surge amplitude (Funabashi et al. 1999; Palm et al. 1999; Harney et al. 1996; Sun et al. 2012). SCN VIPergic neurons project monosynaptically to GnRH neurons in the POA (Van der Beek et al. 1997; Horvath et al. 1998). Several lines of evidence suggest a role for this direct pathway in LH surge generation. First, concomitant with the LH surge, FOS, is preferentially expressed in GnRH neurons receiving VIPergic input on the afternoon of proestrus (van der Beek et al. 1994). Furthermore, in vitro, VIP stimulates GnRH secretion (Zhao and Kriegsfeld 2009; Samson et al. 1981; Ohtsuka et al. 1988), neural firing (Christian and Moenter 2008; Piet et al. 2016), and intracellular calcium release (Piet et al. 2016). Finally, VIPergic projections from the SCN to GnRH cells are sexually dimorphic, emerging during the pubertal transition into adulthood (Kriegsfeld et al. 2002), with female rats demonstrating higher VIPergic innervation than males (Horvath et al. 1998). Notably, GnRH neurons do not express ERα, the estrogen receptor mediating positive and negative feedback effects of estradiol. This led researchers to search for estradiol-responsive systems upstream of the GnRH

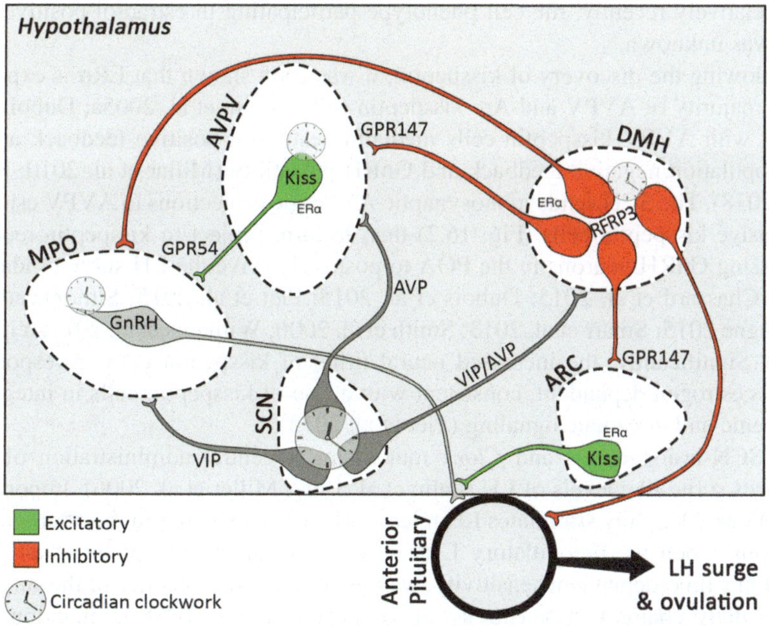

Fig. 16.2 Circadian control of the preovulatory LH surge and ovulation. Model of interactions among the circadian, RFRP-3, and kisspeptin systems in control of the preovulatory LH surge and ovulation. At the time of the LH surge, the SCN coordinates estradiol positive feedback (through kisspeptin cell activation) with removal of estradiol negative feedback (through suppression of RFRP-3 cells in the DMH). GnRH, RFRP-3, and AVPV kisspeptin cells all exhibit rhythms in clock protein or gene expression to further temporal precision in this circuit. See text for further details

system that integrate circadian and estradiol signaling (Herbison and Theodosis 1992; Hrabovszky et al. 2000; Wintermantel et al. 2006). As described below, RFRP-3 and kisspeptin have emerged as key, estrogen-responsive cell phenotypes upstream of the GnRH system that mediate circadian-controlled estrogen negative and positive feedback, respectively.

16.4.2 Integration of Circadian and Estrogenic Signaling Upstream of the GnRH System

Early findings pointed to the AVPV as a likely neural locus regulating estradiol positive feedback and the LH surge. ERα expressing neurons the AVPV send monosynaptic projections to GnRH neurons, express FOS at the time of the LH surge, and AVPV lesions result in the loss of estrous cyclicity in intact and ovariectomized/estradiol-treated rats (Wiegand and Terasawa 1982; Wiegand et al. 1980; Ronnekleiv and Kelly 1988; Gu and Simerly 1997; Le et al. 1999). However, until relatively recently, the cell phenotype participating in estradiol positive feedback was unknown.

Following the discovery of kisspeptin, it was soon shown that ERα is expressed in the majority of AVPV and Arc kisspeptin cells (Smith et al. 2005a; Dubois et al. 2015), with AVPV kisspeptin cells mediating estradiol positive feedback and the Arc population negative feedback and GnRH pulsatility (Millar et al. 2010; Moore et al. 2018). The SCN sends monosynaptic AVPergic projections to AVPV estrogen-responsive kisspeptin cells (Fig. 16.2) that, in turn, project to kisspeptin-receptor-expressing GnRH neurons in the POA to positively drive the LH surge (Vida et al. 2010; Chassard et al. 2015; Dubois et al. 2015; Piet et al. 2015; Simonneaux and Bahougne 2015; Smarr et al. 2013; Smith et al. 2006; Williams et al. 2011; Yip et al. 2015). Significantly, the increased neural firing of kisspeptin cells in response to AVP is estrogen dependent, consistent with a role of kisspeptin cells in integrating estrogenic and circadian signaling (Piet et al. 2015).

In SCN-lesioned rats and *Clock* mutant mice, central administration of AVP produces surge-like levels of LH (Palm et al. 1999; Miller et al. 2006). Importantly, exogenous AVP only stimulates LH release when administered in the afternoon, the timepoint when the preovulatory LH surge occurs in rodents (Palm et al. 2001). Given this time-dependent sensitivity, we asked if time dependence of the surge was due to daily changes in sensitivity of kisspeptin cells to AVP stimulation, daily changes in GnRH cell sensitivity to kisspeptin stimulation, or a combination of both processes. In Syrian hamsters, AVPV kisspeptin cells do not demonstrate time-dependent changes in response to AVP stimulation, whereas GnRH neurons exhibit daily changes in their responsiveness to kisspeptin, suggestive of autonomous circadian timekeeping in cells downstream of the SCN (Williams et al. 2011). Indeed, circadian oscillations in core clock genes that drive circadian rhythms at the cellular level are found both in vitro and in vivo in GnRH cells (Chappell et al. 2003; Olcese et al. 2003; Zhao and Kriegsfeld 2009). Immortalized GnRH neurons exhibit circadian rhythms in responsiveness to VIP and kisspeptin stimulation, further indicating that the GnRH system maintains circadian timing potentially as a mechanism mediating daily changes in responsiveness to upstream signaling (Zhao and Kriegsfeld 2009). Despite not exhibiting daily changes to AVP stimulation, kisspeptin cells express the clock gene *Per1* and the AVPV exhibits sustained circadian rhythms in clock gene expression in cultured AVPV explants (Jacobs et al. 2016; Chassard et al. 2015). Whether these sustained rhythms in AVPV kisspeptin cells confer daily changes in responsiveness to upstream neurochemicals other than AVP remains to be determined.

A specific role for ARC kisspeptin cells in the generation of the preovulatory LH surge has not been established. However, several findings suggest a role for Arc kisspeptin cells in this process. For example, ablation of ARC kisspeptin cells results in abnormal estrous cycles and LH surges, indicating a potential role in generating the LH surge (Helena et al. 2015; Mittelman-Smith et al. 2016). Furthermore, this population expresses receptors for both AVP and VIP (Lukas et al. 2010; Mounien et al. 2006; Ronnekleiv et al. 2014), and AVP and VIP increase intracellular calcium in subsets of Arc kisspeptin neurons in sexually dimorphic manner (Schafer et al. 2018).

As mentioned previously, prior to the LH surge and ovulation, estradiol acts via negative feedback to maintain gonadotropins at low concentrations. Several lines of evidence indicate a role for RFRP-3 cells in integrating circadian and estrogenic signals to mediate estradiol negative feedback. First, RFRP-3 cells exhibit a high activational state (as measured by FOS expression) during diestrus, reduced activity around the time of the LH surge, and increased activity soon thereafter (Gibson et al. 2008). This pattern of timing is mediated by the SCN, with both AVP/VIPergic SCN cell terminals forming close appositions with RFRP-3 cells in Syrian hamsters (Gibson et al. 2008). Likewise, RFRP-3 neurons express ERα and increase their activity in response to estradiol injections at the time of the LH surge (Kriegsfeld et al. 2006; Molnar et al. 2011; Iwasa et al. 2012; Poling et al. 2012). Additionally, analogous to effects seen for GnRH cells, VIP suppresses RFRP-3 cellular activity around the time of the LH surge, but not prior to the surge, suggesting that removal of RFPR-3-mediated estradiol negative feedback is accomplished via time-dependent sensitivity to VIP signaling (Russo et al. 2015). This time-dependent sensitivity is associated with rhythmic clock protein expression in RFRP-3 cells (Russo et al. 2015). Additionally, RFRP-3 cell terminals form close appositions with GnRH cell soma and terminals at median eminence, both of which express Gpr147 (Kriegsfeld et al. 2006; Johnson et al. 2007; Bentley et al. 2010; Kriegsfeld et al. 2010). The same impact of RFRP-3 is observed in both an estradiol surge implant model (Gibson et al. 2008) and during the afternoon of proestrus (Henningsen et al. 2017). Finally, in mice, RFRP-3 similarly inhibits LH secretion when administered at the time of the preovulatory LH surge, but not during diestrus (Ancel et al. 2017). Together, these findings suggest that RFRP-3 cells integrate estrogen and circadian signaling to time the removal of estradiol negative feedback with stimulation of the LH surge.

16.5 Implications for Circadian Rhythms in Pregnancy Success and Fetal Development

16.5.1 Circadian Timing and Pregnancy Maintenance

The reproductive system is exposed to an array of rhythmic hormonal secretions during pregnancy (Magiakou et al. 1996; Tamura et al. 2008) and disruptions to maternal timing negatively impact pregnancy success and fetal development. In one early study using a chronic jet lag model in which mice were repeatedly subjected to a 6 h phase advance or delay of the light:dark cycle, jet-lagged animals exhibited a dramatic reduction in full-term pregnancy success (Summa et al. 2012). Likewise, female mice maintained in 22 or 26 h light:dark cycles, cycles to which they cannot entrain, exhibit decreased mating behavior and experience higher rates of fetal resorption, reduced embryo weights, and delayed development (Endo and Watanabe 1989).

Prolactin, a key hormone regulating pregnancy maintenance, is under strict circadian control in rodents. In mice and rats, for example, prolactin exhibits twice daily surges during pregnancy (a diurnal and nocturnal surge) that maintains the viability of the corpora lutea (CL) and the secretion of progesterone in the first half of pregnancy. Around midpregnancy, these surges cease and placental lactogens maintain progesterone secretion for the remainder of gestation (Robertson and Friesen 1981; Bertram et al. 2010; Freeman et al. 1974; Shiu et al. 1973; Strauss et al. 1996). The SCN regulates prolactin release through the pacing of inhibiting and stimulating factors for this hormone, principally dopamine (DA) and oxytocin, respectively (Bertram et al. 2010; Freeman et al. 2000; Mai et al. 1994). This regulation likely occurs via VIPergic projections from the SCN, as both arcuate tuberoinfundibular DA (TIDA) neurons and paraventricular nucleus (PVN) oxytocin neurons are innervated by VIPergic fibers originating in the SCN (Arey and Freeman 1992; Teclemariam-Mesbah et al. 1997; Gerhold et al. 2001; Egli et al. 2004; Poletini et al. 2010). In addition, VIP antisense oligonucleotides aimed at the SCN abolish prolactin surges in rats (Harney et al. 1996). As would be expected given this mechanism of control, SCN lesions abolish prolactin surges (Kawakami et al. 1980; Pan and Gala 1985). Likewise, prolactin surges entrain to light:dark cycles and free run in constant darkness (Bethea and Neill 1979; Yogev and Terkel 1980). Finally, knockdown of essential clock genes (*Per1, Per2*, and *Clock*) in the SCN abolish prolactin surges (Poletini et al. 2007), and mice lacking a functional *Clock* gene exhibit reduced concentrations of progesterone and marked pregnancy failure, suggestive of disrupted prolactin timing (Miller et al. 2004). These findings point to the critical role of precisely timed prolactin secretion for the maintenance of pregnancy.

In addition to SCN regulation of TIDA and oxytocin neurons, kisspeptin cells project to TIDA neurons and kisspeptin administration increases prolactin release via dopaminergic cell inhibition (Szawka et al. 2010; Sawai et al. 2012, 2014; Saedi et al. 2018). Furthermore, kisspeptin neurons express prolactin receptors, and hyperprolactinemia causes reduced activity of hypothalamic kisspeptin cells, indicating reciprocal communication between kisspeptin and TIDA neurons (Saedi et al. 2018; Sonigo et al. 2012). As with TIDA and oxytocin regulation by the SCN, arcuate kisspeptin cells express VIP receptor and VIP administration alters their cellular activity (Schafer et al. 2018), suggesting that the SCN likely regulates TIDA neurons directly and indirectly, via kisspeptin cells through VIPergic signaling. Whether or not disruptions to kisspeptin cell timing affect prolactin release and pregnancy success remains to be determined.

In contrast to rodents, the pituitary gland (or pituitary hormones) is not required for the initiation and maintenance of pregnancy in humans. In the first 8 weeks of pregnancy, the CL is maintained by human chorionic gonadotropin (hCG) produced by the trophoblast (a layer of tissue that later forms part of the placenta), with placental progesterone sufficient to maintain pregnancy thereafter (Malassine et al. 2003). Although humans do not exhibit a twice daily prolactin surge and, progesterone secretion is maintained via the placenta, circadian disruptions profoundly reduce pregnancy success as seen in rodents (Gamble et al. 2013; Bonzini et al. 2011; Knutsson 2003; Lin et al. 2011). Given that the placenta is responsible for fetal-maternal nutrient exchange and maintains rhythmic clock gene and hormonal expression, it is likely that circadian disruption negatively influences pregnancy success in humans via disruptions to this structure, a hypothesis that has not yet been explored.

16.5.2 Molecular Clockwork and Pregnancy Success

Molecular clockwork plays a key role in regulating pregnancy success, with anomalies in core clock genes resulting in marked female reproductive deficits. For example, particular single nucleotide polymorphisms in the *Bmal1* or *Clock* genes are associated with a higher rate of miscarriage in women (Kovanen et al. 2010; Hodzic et al. 2018). In animal models, mice with mutations or genetic ablation of clock genes (*Clock*, *Bmal1*, *Per1*, and *Per2*) exhibit deficiencies in implantation, pregnancy maintenance, parturition, and have higher rates of fetal resorptions (i.e., miscarriage) (Pilorz and Steinlechner 2008; Alvarez et al. 2008; Boden et al. 2010; Ratajczak et al. 2009; Miller and Takahashi 2013). These mouse models also exhibit shorter duration prolactin surges or failure to initiate prolactin release following mating and pregnancy (Poletini et al. 2010). Likewise, *Clock* mutant mice exhibit reduced concentrations of progesterone, higher rates of fetal resorptions, and fewer full-term pregnancies (Miller et al. 2004; Miller and Takahashi 2013; Csapo and Wiest 1969), all likely due to altered prolactin secretion.

Notably, although clock gene knockout and mutant mice exhibit reduced fertility, these animals lack the gene of interest systemically and permanently (i.e., clock gene expression is absent in all tissues throughout development). Hence, it is not possible to determine whether the reproductive deficits originate through circadian disruption in specific reproductive tissues and/or during specific time points. For example, it is possible that lacking essential clock genes throughout development leads to abnormal establishment of circuits required for mating and pregnancy maintenance (i.e., organizational effects). It is equally possible that a dysfunctional cellular clockwork results in reproductive deficits through abnormal timing in these same circuits in adulthood (i.e., activational effects). Moreover, even when reproductive deficits do not manifest in knockout mice, it is possible that compensatory mechanisms permit typical mating and pregnancy success. Studies applying approaches that allow temporal and spatial specificity of genetic manipulations would help to disambiguate developmental and postpubertal roles of the circadian clockwork (e.g., *Bmal1* knockdown specifically in the SCN following puberty).

16.5.3 Circadian Timing in Early in Fetal Development

The developing embryo is exposed to a circadian environment as early as its initial migration from the oviduct into the uterus. The oviduct provides essential nutrients and growth factors to the developing embryo, and although studies examining oviduct rhythmicity are sparse, findings suggest that the oviducts regulate the embryonic environment via the rhythmic expression of clock genes as well as *plasminogen activator inhibitor 1* (Kennaway et al. 2003; Johnson et al. 2002;

Hastings et al. 2007), a clock-controlled gene that is involved in oviduct activity during peri-implantation and is thought to play a role in embryo protection (Kouba et al. 2000). The uterus also exhibits rhythmic expression of clock genes, in both nonpregnant and pregnant females, throughout different stages of pregnancy and under constant conditions (Hastings et al. 2007; Johnson et al. 2002; Ratajczak et al. 2010, 2012; Resuehr et al. 2007; Nakamura et al. 2010; Akiyama et al. 2010). Finally, another key structure in pregnancy maintenance and regulation, the placenta, exhibits rhythmic expression of clock genes as well as glucocorticoid receptor and glucocorticoid metabolic enzyme expression (Akiyama et al. 2010; Waddell et al. 2012; Wharfe et al. 2011). The placenta is responsible for maternal–fetal exchanges, secreting hormones to maintain gestation and promote the health of the fetus. Taken together, the developing conceptus is exposed to a rhythmic environment from its first day, long before developing its own autonomous clock. As suggested below, this rhythmic environment is likely critical for normal fetal and postnatal development (Fig. 16.3).

16.5.4 Maternal–Fetal Rhythm Synchronization

Circadian disruptions that alter maternal endocrine timing signals can impair maternal–fetal synchronization and fetal development (Fig. 16.3). A major maternal signal providing photoperiodic information to the embryo is melatonin, a hormone secreted from the maternal pineal gland at darkness. Melatonin crosses the placenta and influences the embryo's physiological rhythms and development (Schenker et al. 1998; Williams et al. 1991; Okatani et al. 1998). Melatonin receptors are widespread in the fetal nervous system (including the SCN), as well as in peripheral organs, beginning early in fetal development (Drew et al. 1998; Thomas et al. 2002; Rivkees and Reppert 1991; Vanecek 1988; Torres-Farfan et al. 2008). Likewise, the placenta expresses melatonin receptors, and melatonin plays a role in placenta development (Tamura et al. 2008). Nighttime melatonin secretion increases throughout pregnancy (Nakamura et al. 2001), and melatonin rhythms in maternal blood are mirrored in fetal circulation (Kennaway et al. 1996; Okatani et al. 1998). In addition to melatonin, fetal rhythms are also entrained by maternal cortisol and body temperature rhythms (Seron-Ferre et al. 2001), and possibly feeding times (Novakova et al. 2010) (Fig. 16.3).

Because the pineal gland matures only after birth and the developing fetus and newborn do not produce their own melatonin, maternal melatonin signaling is conveyed to offspring through the placenta (in utero) and milk (after birth) and is required for fetal/newborn rhythms. Thus, it is not surprising that maternal circadian disruptions are reflected in fetal rhythms. In nonhuman primates, for example, maternal light exposure disrupts rhythmic expression of fetal clock genes that can be rescued by maternal melatonin administration (Torres-Farfan et al. 2006). Likewise, suppression of maternal melatonin by constant light exposure during pregnancy is associated with intrauterine growth restriction (IUGR), lower concentrations and altered rhythms of cortisol, modified mRNA expression of

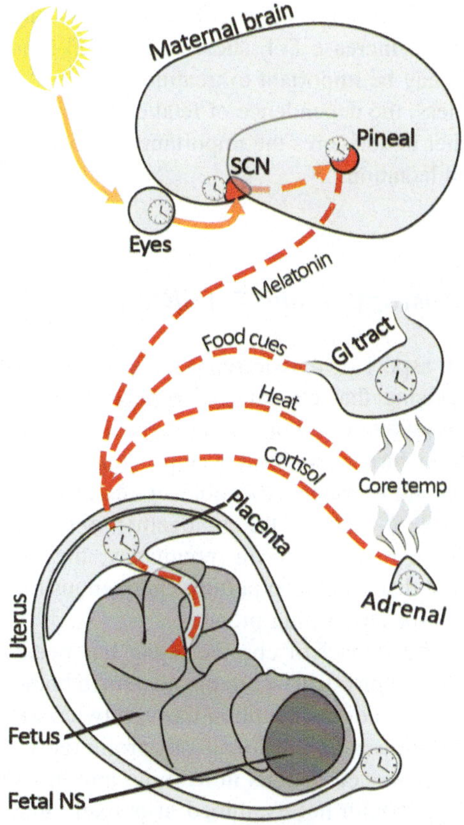

Fig. 16.3 Maternal–fetal rhythm synchronization. The developing fetus is exposed to an array of time cues from its mother. The main signal providing rhythmic information is melatonin, secreted from the mother's pineal gland at darkness and crossing the placenta to the embryo. Additional entraining signals include maternal cortisol, body temperature rhythms, and feeding times. In addition, the developing embryo is exposed to a rhythmic environment via clock genes that are rhythmically expressed in maternal reproductive tissues, including the oviduct, uterus, and placenta. This rhythmic environment is likely critical for normal fetal and postnatal development. Circadian disruptions that alter maternal endocrine timing signals can impair maternal–fetal synchronization and fetal development. Clocks indicate rhythmic expression of clock genes

clock genes and clock-controlled genes in the fetal adrenal gland, and aberrant adrenal response to adrenocorticotropic hormone (ACTH) in rats. Melatonin administration during the subjective night rescues all of the above (Mendez et al. 2012). In in vitro fertilization (IVF) treatments, melatonin promotes embryo development (Ishizuka et al. 2000; Tian et al. 2010; Sampaio et al. 2012). Likewise, melatonin administration prior to IVF treatments and throughout pregnancy is associated with improved pregnancy outcomes; fertility rates are 50% higher in melatonin treated IVF cycles (Rizzo et al. 2010; Unfer et al. 2011). These findings raise questions regarding the static nature of the environment that the conceptus is typically exposed to in cultured IVF conditions and whether a rhythmic environment that better mimics the in vivo milieu would increase IVF success rates. Likewise, the timing of IVF embryo implantation may be important to treatment success, but this remains to be determined. Nonetheless, the dependence of fetal/early postnatal rhythms on maternal environment further underscores the importance of maintaining circadian health during pregnancy and lactation.

16.5.5 Circadian Disruption and Fetal/Offspring Development

Because almost every aspect of female reproduction is regulated by the circadian system, it is not surprising that circadian disruption is associated with a host of negative reproductive outcomes. For example, shift work during pregnancy is associated with preterm birth, low birth weight, small for gestational age births, and increased risk of miscarriages (Gamble et al. 2013; Bonzini et al. 2011; Knutsson 2003; Lin et al. 2011). Taken together with findings indicating the prominence of circadian rhythms during pregnancy, these findings indicate that environmental temporal disruption can perturb endogenous timing and have marked negative consequences for developing progeny.

In animal models, the impact of chronic jetlag has been shown to negatively impact a host of mental and physical health parameters in both adults and developing offspring. For example, circadian disruption through twice weekly, 6 h advances of the light:dark cycle, decreases hippocampal cell proliferation and neurogenesis by >50% and leads to pronounced deficits in learning and memory in adult hamsters (Gibson et al. 2010). Although not examined, it is likely that animals can recover from these deficits after a period of time. However, in developing offspring, the impact of circadian disruption appears to be permanent, with the impact of in utero circadian disruption lasting until adulthood likely by modifying neural development. For example, fetal hippocampal clock and clock-controlled gene rhythms are suppressed by maternal exposure to constant light, leading to deficits in spatial memory in these offspring as adults (Vilches et al. 2014). These adult deficits are associated with dampened circadian rhythms of hippocampal clock and clock-controlled genes. Intriguingly, these deficits can be rescued by providing pregnant dams with rhythmic, exogenous melatonin (Vilches et al. 2014), pointing to a potential for clinical interventions for women unable to escape circadian disruption during pregnancy. Additionally, in utero and/or early life circadian disruption impairs the ability of mice to elicit maternal care, even when crossed-fostered to non-disrupted dams, and leads to adult deficits in social behavior and anxiety (Smarr et al. 2017). Taken together, these findings indicate that circadian disruption can have long lasting, likely permanent, impact on offspring neurobehavioral development.

16.6 Circadian Control of Birth

The length of gestation is determined by multiple clocks, including a principal clock monitoring fetal development and additional clock-regulated mechanisms governing labor onset and parturition timing. Fetal membrane senescence is thought to initiate the signaling cascades leading to parturition via inflammatory processes that increase uterine sensitivity to uterotonins, including prostaglandins and oxytocin (reviewed in Menon et al. 2016). For the sake of the present review, we focus on the circadian regulation of parturition via maternal neuroendocrine pathways, although parturition is a process regulated by multiple clocks at multiple levels (i.e., mother, fetus, and placenta) (Fig. 16.3).

As different species adapted to specific temporal niches over the course of evolution, a selective advantage was gained by initiating parturition during the daytime or nighttime dependent of selective pressures. Because delivering offspring in the home den is safer than in the open areas for prey species, many diurnal species have evolved to initiate parturition at nighttime and nocturnal species in daytime. Circadian timing of labor onset or parturition has been reported in mice, rats, hamsters, sheep, pigs, horses, and primates, including humans (Cooperstock et al. 1987a, b; Ngwenya and Lindow 2004; Boer et al. 1975; Jensen and Bobbitt 1967; Zahn and Hattensperger 1993; Germain et al. 1993; Moore et al. 1994; Takayama et al. 2003; Olcese 2012; Olcese et al. 2013; Chaney et al. 2018; Reppert et al. 1987). In rats, the length of pregnancy is influenced by the light regimen in which animals are maintained, with longer day lengths leading to later parturition (i.e., gestational day 23 instead of 22) (Bosc and Nicolle 1980). Clock genes are likely involved in timing parturition, as mice lacking a functional molecular clockwork fail to either enter labor or have prolonged and nonproductive parturition, resulting in resorption of the fully developed embryos (Miller et al. 2004). Furthermore, as described previously, *Clock* mutant mice experience higher rates of fetal resorptions and fewer pregnancies reaching term.

In humans, labor onset and parturition tend to cluster between the late night and early morning, exhibiting a trough in the late afternoon (Kaiser and Halberg 1962; Olcese 2012; Olcese et al. 2013; Gamble et al. 2013). A similar circadian pattern has been reported in the timing of parturition among indigenous populations living in rural areas (Chaney et al. 2018) and in the onset of preterm labors (Vatish et al. 2010). Underlying the circadian rhythm of labor and birth may be a rhythm in uterine myometrial activity, with contractions peaking at nighttime (Germain et al. 1993; Moore et al. 1994; Zahn and Hattensperger 1993). Moreover, the nocturnal surge in uterine activity in the last trimester of gestation can predict preterm deliveries, with women who deliver prematurely losing these nocturnal surges weeks before birth (Germain et al. 1993). Together, these findings suggest that the mechanisms underlying parturition are under circadian regulation across species.

The species-specific difference in circadian phase of parturition likely results from antiphase uterine cell responses to endocrine signals. In humans, uterine contractions exhibit a peak at night and a trough in the morning (Zahn and Hattensperger 1993; Germain et al. 1993). This pattern may result from motility-enhancing factors, such as oxytocin, estrogens, and prostaglandins, being higher at nighttime (Zahn and Hattensperger 1993; Seron-Ferre et al. 1993), and from the actions of nighttime melatonin secretion from the pineal gland (Tamura et al. 2008; Kivela et al. 1989). In a proof of principle study, late-term pregnant women exposed to bright light at night exhibited suppression in plasma melatonin concentrations that was associated with a reduction in contraction intensity (Olcese et al. 2013), likely resulting in the potentiation of oxytocin's actions on the uterus (Sharkey et al. 2009). Whether nighttime melatonin potentiates the actions of Pitocin (synthetic oxytocin) in women induced into labor, potentially achieving a faster and safer process, has not been explored.

In contrast to humans, melatonin exhibits a negative (tocolytic) effect on oxytocin-induced uterine contractility in nocturnal rodents (Dominguez Rubio et al. 2014; Takayama et al. 2003; Olcese 2012; Reppert et al. 1987). Thus, although oxytocin stimulates uterine contractions in both diurnal and nocturnal species, and melatonin peaks at night in both cases, this hormone exhibits opposite effects on the myometrium. Rats in which the source of melatonin is eliminated via pinealectomy deliver their pups independent of time of day. However, when melatonin is administered to pinealectomized females at the onset of the dark phase, the circadian rhythm of parturition is rescued. If melatonin is administered at the onset of the light phase or in a constant release manner (i.e., via capsules), the circadian pattern of deliveries is not regained (Takayama et al. 2003). This finding suggests that melatonin signaling is required at the appropriate time to initiate birth, likely due to interactions with other regulators of this complex process (e.g., oxytocin).

16.7 Conclusions and Considerations

The circadian timing system participates in all aspects of female reproduction, from ovulation to childbirth. Disruptions to circadian timing have marked negative consequences for ovulation, pregnancy success and maintenance, and offspring development. The modern world makes circadian disruption virtually inescapable, underscoring the importance of developing safe and effective strategies to maximize circadian health in the face of pervasive circadian disruption. Likewise, educating couples about the importance of circadian health represents an important consideration for health professionals providing counseling during family planning.

Key References

de Roux et al. (2003)—The authors investigated individuals with hypophysiotropic hypogonadism and mice lacking the GPR54 gene, establishing a critical role for this G-protein coupled receptor, the cognate receptor for *Kiss1* gene protein products (i.e., kisspeptin), as critical for reproductive function.

Endo and Watanabe (1989)—This article describes the effects of non-24 h days on female reproduction in mice. The authors found that mice maintained on a 22 or 26 h light:dark cycle exhibit decreased mating behavior and experience higher rates of fetal resorption, reduced embryo weights, and delayed development.

Kriegsfeld et al. (2006)—Based on the discovery of gonadotropin-inhibitory hormone (GnIH) in birds, the authors provided the first, broad characterization of GnIH (also known as RFamide-related peptide-3) in mammals (i.e., rats, mice and hamsters), establishing a more general modulatory role for this neuropeptide across species.

Mendez et al. (2012)—This article demonstrated a role for melatonin in fetal development and rhythm generation/synchronization. The authors established that exposure of pregnant rats to constant light during the second half of gestation negatively affects fetal development, with embryos exhibiting intrauterine growth retardation and disrupted adrenal clock genes and clock-controlled genes as well as altered corticosterone rhythms. However, when mothers received melatonin during their subjective night, reproductive function was rescued.

Miller et al. (2006)—This article examined reproduction of *Clock* mutant mice, establishing an important role for circadian rhythms in female reproductive function. The authors found that *Clock* mutant mice exhibit disruptions in estrous cyclicity and pregnancy maintenance, including higher rates of fetal resorptions and fewer pregnancies reaching full term.

Seminara et al. (2003)—This article employed complementary approaches in humans and mice to investigate the onset of puberty. The authors found that mutations in the G-protein coupled receptor, GPR54, the cognate receptor kisspeptin, caused idiopathic hypophysiotropic hypogonadism in both mice and humans. This finding was key to identifying a critical role for kisspeptin in reproductive functioning.

Takayama et al. (2003)—This article demonstrated a role for melatonin in timing parturition. The authors found that female rats lacking melatonin deliver pups throughout the day and night, unlike melatonin-proficient rats who enter parturition during the day. Melatonin administration to rats exclusively during the dark phase restored parturition to daytime.

Tsutsui et al. (2000)—This article identified a hypothalamic peptide that inhibits gonadotropin release in a vertebrate pituitary and named it gonadotropin-inhibitory hormone (GnIH). This finding represented a discovery with broad implications for understanding reproductive axis regulation.

References

Abe M, Herzog ED, Yamazaki S, Straume M, Tei H, Sakaki Y, Menaker M, Block GD (2002) Circadian rhythms in isolated brain regions. J Neurosci 22(1):350–356

Adachi S, Yamada S, Takatsu Y, Matsui H, Kinoshita M, Takase K, Sugiura H, Ohtaki T, Matsumoto H, Uenoyama Y, Tsukamura H, Inoue K, Maeda K (2007) Involvement of anteroventral periventricular metastin/kisspeptin neurons in estrogen positive feedback action on luteinizing hormone release in female rats. J Reprod Dev 53(2):367–378

Ahlborg G Jr, Axelsson G, Bodin L (1996) Shift work, nitrous oxide exposure and subfertility among Swedish midwives. Int J Epidemiol 25(4):783–790

Akiyama S, Ohta H, Watanabe S, Moriya T, Hariu A, Nakahata N, Chisaka H, Matsuda T, Kimura Y, Tsuchiya S, Tei H, Okamura K, Yaegashi N (2010) The uterus sustains stable biological clock during pregnancy. Tohoku J Exp Med 221(4):287–298

Alvarez JD, Hansen A, Ord T, Bebas P, Chappell PE, Giebultowicz JM, Williams C, Moss S, Sehgal A (2008) The circadian clock protein BMAL1 is necessary for fertility and proper testosterone production in mice. J Biol Rhythm 23(1):26–36. https://doi.org/10.1177/0748730407311254

Ancel C, Bentsen AH, Sebert ME, Tena-Sempere M, Mikkelsen JD, Simonneaux V (2012) Stimulatory effect of RFRP-3 on the gonadotrophic axis in the male Syrian hamster: the exception proves the rule. Endocrinology 153(3):1352–1363. https://doi.org/10.1210/en.2011-1622

Ancel C, Inglis MA, Anderson GM (2017) Central RFRP-3 stimulates LH secretion in male mice and has cycle stage-dependent inhibitory effects in females. Endocrinology 158(9):2873–2883. https://doi.org/10.1210/en.2016-1902

Arey BJ, Freeman ME (1992) Activity of vasoactive intestinal peptide and serotonin in the paraventricular nucleus reflects the periodicity of the endogenous stimulatory rhythm regulating prolactin secretion. Endocrinology 131(2):736–742. https://doi.org/10.1210/endo.131.2.1639019

Bedrosian TA, Nelson RJ (2013) Influence of the modern light environment on mood. Mol Psychiatry 18(7):751–757. https://doi.org/10.1038/mp.2013.70mp201370

Bentley GE, Tsutsui K, Kriegsfeld LJ (2010) Recent studies of gonadotropin-inhibitory hormone (GnIH) in the mammalian hypothalamus, pituitary and gonads. Brain Res 1364:62–71. https://doi.org/10.1016/j.brainres.2010.10.001

Berson DM, Dunn FA, Takao M (2002) Phototransduction by retinal ganglion cells that set the circadian clock. Science 295(5557):1070–1073. https://doi.org/10.1126/science.1067262

Bertram R, Helena CV, Gonzalez-Iglesias AE, Tabak J, Freeman ME (2010) A tale of two rhythms: the emerging roles of oxytocin in rhythmic prolactin release. J Neuroendocrinol 22(7):778–784. https://doi.org/10.1111/j.1365-2826.2010.02012.x

Bethea CL, Neill JD (1979) Prolactin secretion after cervical stimulation of rats maintained in constant dark or constant light. Endocrinology 104(4):870–876. https://doi.org/10.1210/endo-104-4-870

Blaustein JD, Tetel MJ, Ricciardi KH, Delville Y, Turcotte JC (1994) Hypothalamic ovarian steroid hormone-sensitive neurons involved in female sexual behavior. Psychoneuroendocrinology 19 (5–7):505–516

Boden MJ, Varcoe TJ, Voultsios A, Kennaway DJ (2010) Reproductive biology of female Bmal1 null mice. Reproduction 139(6):1077–1090. https://doi.org/10.1530/REP-09-0523

Boden MJ, Varcoe TJ, Kennaway DJ (2013) Circadian regulation of reproduction: from gamete to offspring. Prog Biophys Mol Biol 113(3):387–397. https://doi.org/10.1016/j.pbiomolbio.2013.01.003

Boer K, Lincoln DW, Swaab DF (1975) Effects of electrical stimulation of the neurohypophysis on labour in the rat. J Endocrinol 65(2):163–176

Bonini JA, Jones KA, Adham N, Forray C, Artymyshyn R, Durkin MM, Smith KE, Tamm JA, Boteju LW, Lakhlani PP, Raddatz R, Yao WJ, Ogozalek KL, Boyle N, Kouranova EV, Quan Y, Vaysse PJ, Wetzel JM, Branchek TA, Gerald C, Borowsky B (2000) Identification and characterization of two G protein-coupled receptors for neuropeptide FF. J Biol Chem 275 (50):39324–39331. https://doi.org/10.1074/jbc.M004385200

Bonzini M, Palmer KT, Coggon D, Carugno M, Cromi A, Ferrario MM (2011) Shift work and pregnancy outcomes: a systematic review with meta-analysis of currently available epidemiological studies. BJOG 118(12):1429–1437. https://doi.org/10.1111/j.1471-0528.2011.03066.x

Bosc MJ, Nicolle A (1980) Influence of photoperiod on the time of parturition in the rat. I. Effect of the length of daily illumination on normal or adrenalectomized animals. Reprod Nutr Dev 20(3A):735–745

Brown-Grant K, Raisman G (1977) Abnormalities in reproductive function associated with the destruction of the suprachiasmatic nuclei in female rats. Proc R Soc Lond B Biol Sci 198 (1132):279–296

Buijs RM, Escobar C, Swaab DF (2013) The circadian system and the balance of the autonomic nervous system. Handb Clin Neurol 117:173–191. https://doi.org/10.1016/B978-0-444-53491-0.00015-8

Burke MC, Letts PA, Krajewski SJ, Rance NE (2006) Coexpression of dynorphin and neurokinin B immunoreactivity in the rat hypothalamus: morphologic evidence of interrelated function within the arcuate nucleus. J Comp Neurol 498(5):712–726. https://doi.org/10.1002/cne.21086

Butler MP, Kriegsfeld LJ, Silver R (2010) Circadian regulation of endocrine functions. In: Pfaff D, Etgen A (eds) Hormones, brain and behavior, 2nd edn. Academic, New York, pp 473–507

Cagnacci A, Maxia N, Volpe A (1999) Diurnal variation of semen quality in human males. Hum Reprod 14(1):106–109

Cahill DJ, Wardle PG, Harlow CR, Hull MG (1998) Onset of the preovulatory luteinizing hormone surge: diurnal timing and critical follicular prerequisites. Fertil Steril 70(1):56–59

Chaney C, Goetz LG, Valeggia C (2018) A time to be born: variation in the hour of birth in a rural population of Northern Argentina. Am J Phys Anthropol 166(4):975–978. https://doi.org/10.1002/ajpa.23483

Chappell PE, White RS, Mellon PL (2003) Circadian gene expression regulates pulsatile gonadotropin-releasing hormone (GnRH) secretory patterns in the hypothalamic GnRH-secreting GT1-7 cell line. J Neurosci 23(35):11202–11213

Chassard D, Bur I, Poirel VJ, Mendoza J, Simonneaux V (2015) Evidence for a putative circadian Kiss-clock in the hypothalamic AVPV in female mice. Endocrinology 156(8):2999–3011. https://doi.org/10.1210/en.2014-1769

Christian CA, Moenter SM (2008) Vasoactive intestinal polypeptide can excite gonadotropin-releasing hormone neurons in a manner dependent on estradiol and gated by time of day. Endocrinology 149(6):3130–3136. https://doi.org/10.1210/en.2007-1098

Christian CA, Moenter SM (2010) The neurobiology of preovulatory and estradiol-induced gonadotropin-releasing hormone surges. Endocr Rev 31(4):544–577. https://doi.org/10.1210/er.2009-0023

Chu A, Zhu L, Blum ID, Mai O, Leliavski A, Fahrenkrug J, Oster H, Boehm U, Storch KF (2013) Global but not gonadotrope-specific disruption of Bmal1 abolishes the luteinizing hormone surge without affecting ovulation. Endocrinology 154(8):2924–2935. https://doi.org/10.1210/en.2013-1080

Clarke IJ, Sari IP, Qi Y, Smith JT, Parkington HC, Ubuka T, Iqbal J, Li Q, Tilbrook A, Morgan K, Pawson AJ, Tsutsui K, Millar RP, Bentley GE (2008) Potent action of RFamide-related peptide-3 on pituitary gonadotropes indicative of a hypophysiotropic role in the negative regulation of gonadotropin secretion. Endocrinology 149(11):5811–5821. https://doi.org/10.1210/en.2008-0575

Clarkson J, Herbison AE (2006) Postnatal development of kisspeptin neurons in mouse hypothalamus; sexual dimorphism and projections to gonadotropin-releasing hormone neurons. Endocrinology 147(12):5817–5825. https://doi.org/10.1210/en.2006-0787

Clarkson J, d'Anglemont de Tassigny X, Moreno AS, Colledge WH, Herbison AE (2008) Kisspeptin-GPR54 signaling is essential for preovulatory gonadotropin-releasing hormone neuron activation and the luteinizing hormone surge. J Neurosci 28(35):8691–8697. https://doi.org/10.1523/JNEUROSCI.1775-08.2008

Cooperstock M, England JE, Wolfe RA (1987a) Circadian incidence of labor onset hour in preterm birth and chorioamnionitis. Obstet Gynecol 70(6):852–855

Cooperstock M, England JE, Wolfe RA (1987b) Circadian incidence of premature rupture of the membranes in term and preterm births. Obstet Gynecol 69(6):936–941

Csapo AI, Wiest WG (1969) An examination of the quantitative relationship between progesterone and the maintenance of pregnancy. Endocrinology 85(4):735–746. https://doi.org/10.1210/endo-85-4-735

de Roux N, Genin E, Carel JC, Matsuda F, Chaussain JL, Milgrom E (2003) Hypogonadotropic hypogonadism due to loss of function of the KiSS1-derived peptide receptor GPR54. Proc Natl Acad Sci USA 100(19):10972–10976. https://doi.org/10.1073/pnas.1834399100

Deng N, Kohn TP, Lipshultz LI, Pastuszak AW (2018) The relationship between shift work and Men's health. Sex Med Rev 6(3):446–456. https://doi.org/10.1016/j.sxmr.2017.11.009
Dolatshad H, Campbell EA, O'Hara L, Maywood ES, Hastings MH, Johnson MH (2006) Developmental and reproductive performance in circadian mutant mice. Hum Reprod 21(1):68–79. https://doi.org/10.1093/humrep/dei313
Dominguez Rubio AP, Sordelli MS, Salazar AI, Aisemberg J, Bariani MV, Cella M, Rosenstein RE, Franchi AM (2014) Melatonin prevents experimental preterm labor and increases offspring survival. J Pineal Res 56(2):154–162. https://doi.org/10.1111/jpi.12108
Drew JE, Williams LM, Hannah LT, Barrett P, Abramovich DR (1998) Melatonin receptors in the human fetal kidney: 2-[125I]iodomelatonin binding sites correlated with expression of Mel1a and Mel1b receptor genes. J Endocrinol 156(2):261–267
Dubois SL, Acosta-Martinez M, DeJoseph MR, Wolfe A, Radovick S, Boehm U, Urban JH, Levine JE (2015) Positive, but not negative feedback actions of estradiol in adult female mice require estrogen receptor alpha in kisspeptin neurons. Endocrinology 156(3):1111–1120. https://doi.org/10.1210/en.2014-1851
Ducret E, Anderson GM, Herbison AE (2009) RFamide-related peptide-3, a mammalian gonadotropin-inhibitory hormone ortholog, regulates gonadotropin-releasing hormone neuron firing in the mouse. Endocrinology 150(6):2799–2804. https://doi.org/10.1210/en.2008-1623
Edwards RG, Steptoe PC, Fowler RE, Baillie J (1980) Observations on preovulatory human ovarian follicles and their aspirates. Br J Obstet Gynaecol 87(9):769–779
Egli M, Bertram R, Sellix MT, Freeman ME (2004) Rhythmic secretion of prolactin in rats: action of oxytocin coordinated by vasoactive intestinal polypeptide of suprachiasmatic nucleus origin. Endocrinology 145(7):3386–3394. https://doi.org/10.1210/en.2003-1710
Elkind-Hirsch K, Ravnikar V, Tulchinsky D, Schiff I, Ryan KJ (1984) Episodic secretory patterns of immunoreactive luteinizing hormone-releasing hormone (IR-LH-RH) in the systemic circulation of normal women throughout the menstrual cycle. Fertil Steril 41(1):56–61
Endo A, Watanabe T (1989) Effects of non-24-hour days on reproductive efficacy and embryonic development in mice. Gamete Res 22(4):435–441. https://doi.org/10.1002/mrd.1120220409
Fernandez RC, Marino JL, Varcoe TJ, Davis S, Moran LJ, Rumbold AR, Brown HM, Whitrow MJ, Davies MJ, Moore VM (2016) Fixed or rotating night shift work undertaken by women: implications for fertility and miscarriage. Semin Reprod Med 34(2):74–82. https://doi.org/10.1055/s-0036-1571354
Fonken LK, Nelson RJ (2014) The effects of light at night on circadian clocks and metabolism. Endocr Rev 35, 648–670. https://doi.org/10.1210/er.2013-1051
Foradori CD, Coolen LM, Fitzgerald ME, Skinner DC, Goodman RL, Lehman MN (2002) Colocalization of progesterone receptors in parvicellular dynorphin neurons of the ovine preoptic area and hypothalamus. Endocrinology 143(11):4366–4374. https://doi.org/10.1210/en.2002-220586
Francl JM, Kaur G, Glass JD (2010) Regulation of vasoactive intestinal polypeptide release in the suprachiasmatic nucleus circadian clock. Neuroreport 21(16):1055–1059. https://doi.org/10.1097/WNR.0b013e32833fcba4
Freeman ME, Smith MS, Nazian SJ, Neill JD (1974) Ovarian and hypothalamic control of the daily surges of prolactin secretion during pseudopregnancy in the rat. Endocrinology 94(3):875–882. https://doi.org/10.1210/endo-94-3-875
Freeman ME, Kanyicska B, Lerant A, Nagy G (2000) Prolactin: structure, function, and regulation of secretion. Physiol Rev 80(4):1523–1631. https://doi.org/10.1152/physrev.2000.80.4.1523
Funabashi T, Aiba S, Sano A, Shinohara K, Kimura F (1999) Intracerebroventricular injection of arginine-vasopressin V1 receptor antagonist attenuates the surge of luteinizing hormone and prolactin secretion in proestrous rats. Neurosci Lett 260(1):37–40
Gamble KL, Resuehr D, Johnson CH (2013) Shift work and circadian dysregulation of reproduction. Front Endocrinol 4:92. https://doi.org/10.3389/fendo.2013.00092
George JT, Hendrikse M, Veldhuis JD, Clarke IJ, Anderson RA, Millar RP (2017) Effect of gonadotropin-inhibitory hormone on luteinizing hormone secretion in humans. Clin Endocrinol 86(5):731–738. https://doi.org/10.1111/cen.13308

Gerhold LM, Horvath TL, Freeman ME (2001) Vasoactive intestinal peptide fibers innervate neuroendocrine dopaminergic neurons. Brain Res 919(1):48–56

Germain AM, Valenzuela GJ, Ivankovic M, Ducsay CA, Gabella C, Seron-Ferre M (1993) Relationship of circadian rhythms of uterine activity with term and preterm delivery. Am J Obstet Gynecol 168(4):1271–1277

Gibson EM, Humber SA, Jain S, Williams WP 3rd, Zhao S, Bentley GE, Tsutsui K, Kriegsfeld LJ (2008) Alterations in RFamide-related peptide expression are coordinated with the preovulatory luteinizing hormone surge. Endocrinology 149(10):4958–4969. https://doi.org/10.1210/en.2008-0316

Gibson EM, Wang C, Tjho S, Khattar N, Kriegsfeld LJ (2010) Experimental 'jet lag' inhibits adult neurogenesis and produces long-term cognitive deficits in female hamsters. PLoS One 5(12):e15267. https://doi.org/10.1371/journal.pone.0015267

Goodman RL, Lehman MN, Smith JT, Coolen LM, de Oliveira CV, Jafarzadehshirazi MR, Pereira A, Iqbal J, Caraty A, Ciofi P, Clarke IJ (2007) Kisspeptin neurons in the arcuate nucleus of the ewe express both dynorphin A and neurokinin B. Endocrinology 148(12):5752–5760. https://doi.org/10.1210/en.2007-0961

Goodman RL, Hileman SM, Nestor CC, Porter KL, Connors JM, Hardy SL, Millar RP, Cernea M, Coolen LM, Lehman MN (2013) Kisspeptin, neurokinin B, and dynorphin act in the arcuate nucleus to control activity of the GnRH pulse generator in ewes. Endocrinology 154(11):4259–4269. https://doi.org/10.1210/en.2013-1331

Goodman RL, Coolen LM, Lehman MN (2014) A role for neurokinin B in pulsatile GnRH secretion in the ewe. Neuroendocrinology 99(1):18–32. https://doi.org/10.1159/000355285

Gouarderes C, Mazarguil H, Mollereau C, Chartrel N, Leprince J, Vaudry H, Zajac JM (2007) Functional differences between NPFF1 and NPFF2 receptor coupling: high intrinsic activities of RFamide-related peptides on stimulation of [35S]GTPgammaS binding. Neuropharmacology 52(2):376–386. https://doi.org/10.1016/j.neuropharm.2006.07.034

Gu GB, Simerly RB (1997) Projections of the sexually dimorphic anteroventral periventricular nucleus in the female rat. J Comp Neurol 384(1):142–164

Guran T, Tolhurst G, Bereket A, Rocha N, Porter K, Turan S, Gribble FM, Kotan LD, Akcay T, Atay Z, Canan H, Serin A, O'Rahilly S, Reimann F, Semple RK, Topaloglu AK (2009) Hypogonadotropic hypogonadism due to a novel missense mutation in the first extracellular loop of the neurokinin B receptor. J Clin Endocrinol Metab 94(10):3633–3639. https://doi.org/10.1210/jc.2009-0551

Harney JP, Scarbrough K, Rosewell KL, Wise PM (1996) In vivo antisense antagonism of vasoactive intestinal peptide in the suprachiasmatic nuclei causes aging-like changes in the estradiol-induced luteinizing hormone and prolactin surges. Endocrinology 137(9):3696–3701. https://doi.org/10.1210/endo.137.9.8756535

Hastings M, O'Neill JS, Maywood ES (2007) Circadian clocks: regulators of endocrine and metabolic rhythms. J Endocrinol 195(2):187–198. https://doi.org/10.1677/JOE-07-0378

Hattar S, Liao HW, Takao M, Berson DM, Yau KW (2002) Melanopsin-containing retinal ganglion cells: architecture, projections, and intrinsic photosensitivity. Science 295(5557):1065–1070. https://doi.org/10.1126/science.1069609

Helena CV, Toporikova N, Kalil B, Stathopoulos AM, Pogrebna VV, Carolino RO, Anselmo-Franci JA, Bertram R (2015) KNDy neurons modulate the magnitude of the steroid-induced luteinizing hormone surges in ovariectomized rats. Endocrinology 156(11):4200–4213. https://doi.org/10.1210/en.2015-1070

Henningsen JB, Ancel C, Mikkelsen JD, Gauer F, Simonneaux V (2017) Roles of RFRP-3 in the daily and seasonal regulation of reproductive activity in female Syrian hamsters. Endocrinology 158(3):652–663. https://doi.org/10.1210/en.2016-1689

Herbison AE, Theodosis DT (1992) Localization of oestrogen receptors in preoptic neurons containing neurotensin but not tyrosine hydroxylase, cholecystokinin or luteinizing hormone-releasing hormone in the male and female rat. Neuroscience 50(2):283–298

Hinuma S, Shintani Y, Fukusumi S, Iijima N, Matsumoto Y, Hosoya M, Fujii R, Watanabe T, Kikuchi K, Terao Y, Yano T, Yamamoto T, Kawamata Y, Habata Y, Asada M, Kitada C, Kurokawa T, Onda H, Nishimura O, Tanaka M, Ibata Y, Fujino M (2000) New neuropeptides containing carboxy-terminal RFamide and their receptor in mammals. Nat Cell Biol 2 (10):703–708. https://doi.org/10.1038/35036326

Hodzic A, Lavtar P, Ristanovic M, Novakovic I, Dotlic J, Peterlin B (2018) Genetic variation in the CLOCK gene is associated with idiopathic recurrent spontaneous abortion. PLoS One 13(5): e0196345. https://doi.org/10.1371/journal.pone.0196345

Honma S (2018) The mammalian circadian system: a hierarchical multi-oscillator structure for generating circadian rhythm. J Physiol Sci 68(3):207–219. https://doi.org/10.1007/s12576-018-0597-5

Horvath TL, Cela V, van der Beek EM (1998) Gender-specific apposition between vasoactive intestinal peptide-containing axons and gonadotrophin-releasing hormone-producing neurons in the rat. Brain Res 795(1–2):277–281

Hrabovszky E, Shughrue PJ, Merchenthaler I, Hajszan T, Carpenter CD, Liposits Z, Petersen SL (2000) Detection of estrogen receptor-beta messenger ribonucleic acid and 125I-estrogen binding sites in luteinizing hormone-releasing hormone neurons of the rat brain. Endocrinology 141(9):3506–3509. https://doi.org/10.1210/endo.141.9.7788

Hrabovszky E, Ciofi P, Vida B, Horvath MC, Keller E, Caraty A, Bloom SR, Ghatei MA, Dhillo WS, Liposits Z, Kallo I (2010) The kisspeptin system of the human hypothalamus: sexual dimorphism and relationship with gonadotropin-releasing hormone and neurokinin B neurons. Eur J Neurosci 31(11):1984–1998. https://doi.org/10.1111/j.1460-9568.2010.07239.x

Hughes S, Jagannath A, Hankins MW, Foster RG, Peirson SN (2015) Photic regulation of clock systems. Methods Enzymol 552:125–143. https://doi.org/10.1016/bs.mie.2014.10.018

Ishizuka B, Kuribayashi Y, Murai K, Amemiya A, Itoh MT (2000) The effect of melatonin on in vitro fertilization and embryo development in mice. J Pineal Res 28(1):48–51

Iwasa T, Matsuzaki T, Murakami M, Kinouchi R, Osugi T, Gereltsetseg G, Yoshida S, Irahara M, Tsutsui K (2012) Developmental changes in the mammalian gonadotropin-inhibitory hormone (GnIH) ortholog RFamide-related peptide (RFRP) and its cognate receptor GPR147 in the rat hypothalamus. Int J Dev Neurosci 30(1):31–37. https://doi.org/10.1016/j.ijdevneu.2011.10.003

Jacobs DC, Veitch RE, Chappell PE (2016) Evaluation of immortalized AVPV- and arcuate-specific neuronal kisspeptin cell lines to elucidate potential mechanisms of estrogen responsiveness and temporal gene expression in females. Endocrinology 157(9):3410–3419

Jayasena CN, Nijher GM, Chaudhri OB, Murphy KG, Ranger A, Lim A, Patel D, Mehta A, Todd C, Ramachandran R, Salem V, Stamp GW, Donaldson M, Ghatei MA, Bloom SR, Dhillo WS (2009) Subcutaneous injection of kisspeptin-54 acutely stimulates gonadotropin secretion in women with hypothalamic amenorrhea, but chronic administration causes tachyphylaxis. J Clin Endocrinol Metab 94(11):4315–4323. https://doi.org/10.1210/jc.2009-0406

Jayasena CN, Nijher GM, Abbara A, Murphy KG, Lim A, Patel D, Mehta A, Todd C, Donaldson M, Trew GH, Ghatei MA, Bloom SR, Dhillo WS (2010) Twice-weekly administration of kisspeptin-54 for 8 weeks stimulates release of reproductive hormones in women with hypothalamic amenorrhea. Clin Pharmacol Ther 88(6):840–847. https://doi.org/10.1038/clpt.2010.204

Jensen GD, Bobbitt RA (1967) Changing parturition time in monkeys (*Macaca nemestrina*) from night to day. Lab Anim Care 17(4):379–381

Johnson MH, Lim A, Fernando D, Day ML (2002) Circadian clockwork genes are expressed in the reproductive tract and conceptus of the early pregnant mouse. Reprod Biomed Online 4 (2):140–145

Johnson MA, Tsutsui K, Fraley GS (2007) Rat RFamide-related peptide-3 stimulates GH secretion, inhibits LH secretion, and has variable effects on sex behavior in the adult male rat. Horm Behav 51(1):171–180. https://doi.org/10.1016/j.yhbeh.2006.09.009

Kaiser IH, Halberg F (1962) Circadian periodic aspects of birth. Ann N Y Acad Sci 98:1056–1068

Kalsbeek A, Buijs RM, Engelmann M, Wotjak CT, Landgraf R (1995) In vivo measurement of a diurnal variation in vasopressin release in the rat suprachiasmatic nucleus. Brain Res 682 (1–2):75–82

Karsch FJ, Bowen JM, Caraty A, Evans NP, Moenter SM (1997) Gonadotropin-releasing hormone requirements for ovulation. Biol Reprod 56(2):303–309

Kawakami M, Arita J, Yoshioka E (1980) Loss of estrogen-induced daily surges of prolactin and gonadotropins by suprachiasmatic nucleus lesions in ovariectomized rats. Endocrinology 106 (4):1087–1092. https://doi.org/10.1210/endo-106-4-1087

Kennaway DJ, Goble FC, Stamp GE (1996) Factors influencing the development of melatonin rhythmicity in humans. J Clin Endocrinol Metab 81(4):1525–1532. https://doi.org/10.1210/jcem.81.4.8636362

Kennaway DJ, Varcoe TJ, Mau VJ (2003) Rhythmic expression of clock and clock-controlled genes in the rat oviduct. Mol Hum Reprod 9(9):503–507

Kerdelhue B, Brown S, Lenoir V, Queenan JT Jr, Jones GS, Scholler R, Jones HW Jr (2002) Timing of initiation of the preovulatory luteinizing hormone surge and its relationship with the circadian cortisol rhythm in the human. Neuroendocrinology 75(3):158–163

Khan AR, Kauffman AS (2012) The role of kisspeptin and RFamide-related peptide-3 neurones in the circadian-timed preovulatory luteinising hormone surge. J Neuroendocrinol 24(1):131–143. https://doi.org/10.1111/j.1365-2826.2011.02162.x

Kivela A, Kauppila A, Leppaluoto J, Vakkuri O (1989) Serum and amniotic fluid melatonin during human labor. J Clin Endocrinol Metab 69(5):1065–1068. https://doi.org/10.1210/jcem-69-5-1065

Knutsson A (2003) Health disorders of shift workers. Occup Med (Lond) 53(2):103–108

Kotani M, Detheux M, Vandenbogaerde A, Communi D, Vanderwinden JM, Le Poul E, Brezillon S, Tyldesley R, Suarez-Huerta N, Vandeput F, Blanpain C, Schiffmann SN, Vassart G, Parmentier M (2001) The metastasis suppressor gene KiSS-1 encodes kisspeptins, the natural ligands of the orphan G protein-coupled receptor GPR54. J Biol Chem 276 (37):34631–34636. https://doi.org/10.1074/jbc.M104847200

Kouba AJ, Burkhardt BR, Alvarez IM, Goodenow MM, Buhi WC (2000) Oviductal plasminogen activator inhibitor-1 (PAI-1): mRNA, protein, and hormonal regulation during the estrous cycle and early pregnancy in the pig. Mol Reprod Dev 56(3):378–386. https://doi.org/10.1002/1098-2795(200007)56:3<378::AID-MRD8>3.0.CO;2-B

Kovanen L, Saarikoski ST, Aromaa A, Lonnqvist J, Partonen T (2010) ARNTL (BMAL1) and NPAS2 gene variants contribute to fertility and seasonality. PLoS One 5(4):e10007. https://doi.org/10.1371/journal.pone.0010007

Krajewski SJ, Burke MC, Anderson MJ, McMullen NT, Rance NE (2010) Forebrain projections of arcuate neurokinin B neurons demonstrated by anterograde tract-tracing and monosodium glutamate lesions in the rat. Neuroscience 166(2):680–697. https://doi.org/10.1016/j.neuroscience.2009.12.053

Kriegsfeld LJ, Silver R, Gore AC, Crews D (2002) Vasoactive intestinal polypeptide contacts on gonadotropin-releasing hormone neurones increase following puberty in female rats. J Neuroendocrinol 14(9):685–690

Kriegsfeld LJ, Mei DF, Bentley GE, Ubuka T, Mason AO, Inoue K, Ukena K, Tsutsui K, Silver R (2006) Identification and characterization of a gonadotropin-inhibitory system in the brains of mammals. Proc Natl Acad Sci USA 103(7):2410–2415. https://doi.org/10.1073/pnas.0511003103

Kriegsfeld LJ, Gibson EM, Williams WP 3rd, Zhao S, Mason AO, Bentley GE, Tsutsui K (2010) The roles of RFamide-related peptide-3 in mammalian reproductive function and behaviour. J Neuroendocrinol 22(7):692–700. https://doi.org/10.1111/j.1365-2826.2010.02031.x

Kriegsfeld LJ, Jennings KJ, Bentley GE, Tsutsui K (2018) Gonadotrophin-inhibitory hormone and its mammalian orthologue RFamide-related peptide-3: discovery and functional implications for reproduction and stress. J Neuroendocrinol 30(7):e12597. https://doi.org/10.1111/jne.12597

Lawson CC, Whelan EA, Lividoti Hibert EN, Spiegelman D, Schernhammer ES, Rich-Edwards JW (2011) Rotating shift work and menstrual cycle characteristics. Epidemiology 22(3):305–312. https://doi.org/10.1097/EDE.0b013e3182130016

Lawson CC, Rocheleau CM, Whelan EA, Lividoti Hibert EN, Grajewski B, Spiegelman D, Rich-Edwards JW (2012) Occupational exposures among nurses and risk of spontaneous abortion. Am J Obstet Gynecol 206(4):327.e321–328. https://doi.org/10.1016/j.ajog.2011.12.030

Le WW, Berghorn KA, Rassnick S, Hoffman GE (1999) Periventricular preoptic area neurons coactivated with luteinizing hormone (LH)-releasing hormone (LHRH) neurons at the time of the LH surge are LHRH afferents. Endocrinology 140(1):510–519. https://doi.org/10.1210/endo.140.1.6403

Lee JH, Welch DR (1997) Suppression of metastasis in human breast carcinoma MDA-MB-435 cells after transfection with the metastasis suppressor gene, KiSS-1. Cancer Res 57 (12):2384–2387

Lee JH, Miele ME, Hicks DJ, Phillips KK, Trent JM, Weissman BE, Welch DR (1996) KiSS-1, a novel human malignant melanoma metastasis-suppressor gene. J Natl Cancer Inst 88 (23):1731–1737

Lehman MN, Silver R, Gladstone WR, Kahn RM, Gibson M, Bittman EL (1987) Circadian rhythmicity restored by neural transplant. Immunocytochemical characterization of the graft and its integration with the host brain. J Neurosci 7(6):1626–1638

Levine JE (1997) New concepts of the neuroendocrine regulation of gonadotropin surges in rats. Biol Reprod 56(2):293–302

Levine JE, Ramirez VD (1982) Luteinizing hormone-releasing hormone release during the rat estrous cycle and after ovariectomy, as estimated with push-pull cannulae. Endocrinology 111 (5):1439–1448

Lin YC, Chen MH, Hsieh CJ, Chen PC (2011) Effect of rotating shift work on childbearing and birth weight: a study of women working in a semiconductor manufacturing factory. World J Pediatr 7(2):129–135. https://doi.org/10.1007/s12519-011-0265-9

Lucas RJ, Hattar S, Takao M, Berson DM, Foster RG, Yau KW (2003) Diminished pupillary light reflex at high irradiances in melanopsin-knockout mice. Science 299(5604):245–247. https://doi.org/10.1126/science.1077293

Lukas M, Bredewold R, Neumann ID, Veenema AH (2010) Maternal separation interferes with developmental changes in brain vasopressin and oxytocin receptor binding in male rats. Neuropharmacology 58(1):78–87. https://doi.org/10.1016/j.neuropharm.2009.06.020

Magiakou MA, Mastorakos G, Rabin D, Margioris AN, Dubbert B, Calogero AE, Tsigos C, Munson PJ, Chrousos GP (1996) The maternal hypothalamic-pituitary-adrenal axis in the third trimester of human pregnancy. Clin Endocrinol 44(4):419–428

Mahoney MM (2010) Shift work, jet lag, and female reproduction. Int J Endocrinol 2010:813764. https://doi.org/10.1155/2010/813764

Mahoney MM, Sisk C, Ross HE, Smale L (2004) Circadian regulation of gonadotropin-releasing hormone neurons and the preovulatory surge in luteinizing hormone in the diurnal rodent, Arvicanthis niloticus, and in a nocturnal rodent, *Rattus norvegicus*. Biol Reprod 70 (4):1049–1054. https://doi.org/10.1095/biolreprod.103.021360

Mai LM, Shieh KR, Pan JT (1994) Circadian changes of serum prolactin levels and tuberoinfundibular dopaminergic neuron activities in ovariectomized rats treated with or without estrogen: the role of the suprachiasmatic nuclei. Neuroendocrinology 60(5):520–526. https://doi.org/10.1159/000126789

Malassine A, Frendo JL, Evain-Brion D (2003) A comparison of placental development and endocrine functions between the human and mouse model. Hum Reprod Update 9(6):531–539

Matsui H, Takatsu Y, Kumano S, Matsumoto H, Ohtaki T (2004) Peripheral administration of metastin induces marked gonadotropin release and ovulation in the rat. Biochem Biophys Res Commun 320(2):383–388. https://doi.org/10.1016/j.bbrc.2004.05.185

McEwen BS, Jones KJ, Pfaff DW (1987) Hormonal control of sexual behavior in the female rat: molecular, cellular and neurochemical studies. Biol Reprod 36(1):37–45

Mendez N, Abarzua-Catalan L, Vilches N, Galdames HA, Spichiger C, Richter HG, Valenzuela GJ, Seron-Ferre M, Torres-Farfan C (2012) Timed maternal melatonin treatment reverses circadian disruption of the fetal adrenal clock imposed by exposure to constant light. PLoS One 7(8): e42713. https://doi.org/10.1371/journal.pone.0042713

Menon R, Bonney EA, Condon J, Mesiano S, Taylor RN (2016) Novel concepts on pregnancy clocks and alarms: redundancy and synergy in human parturition. Hum Reprod Update 22 (5):535–560. https://doi.org/10.1093/humupd/dmw022

Meyer-Bernstein EL, Jetton AE, Matsumoto SI, Markuns JF, Lehman MN, Bittman EL (1999) Effects of suprachiasmatic transplants on circadian rhythms of neuroendocrine function in golden hamsters. Endocrinology 140(1):207–218. https://doi.org/10.1210/endo.140.1.6428

Millar RP, Roseweir AK, Tello JA, Anderson RA, George JT, Morgan K, Pawson AJ (2010) Kisspeptin antagonists: unraveling the role of kisspeptin in reproductive physiology. Brain Res 1364:81–89. https://doi.org/10.1016/j.brainres.2010.09.044

Miller BH, Takahashi JS (2013) Central circadian control of female reproductive function. Front Endocrinol 4:195. https://doi.org/10.3389/fendo.2013.00195

Miller BH, Olson SL, Turek FW, Levine JE, Horton TH, Takahashi JS (2004) Circadian clock mutation disrupts estrous cyclicity and maintenance of pregnancy. Curr Biol 14(15):1367–1373. https://doi.org/10.1016/j.cub.2004.07.055

Miller BH, Olson SL, Levine JE, Turek FW, Horton TH, Takahashi JS (2006) Vasopressin regulation of the proestrous luteinizing hormone surge in wild-type and clock mutant mice. Biol Reprod 75(5):778–784. https://doi.org/10.1095/biolreprod.106.052845

Mittelman-Smith MA, Williams H, Krajewski-Hall SJ, Lai J, Ciofi P, McMullen NT, Rance NE (2012) Arcuate kisspeptin/neurokinin B/dynorphin (KNDy) neurons mediate the estrogen suppression of gonadotropin secretion and body weight. Endocrinology 153(6):2800–2812. https://doi.org/10.1210/en.2012-1045

Mittelman-Smith MA, Krajewski-Hall SJ, McMullen NT, Rance NE (2016) Ablation of KNDy neurons results in hypogonadotropic hypogonadism and amplifies the steroid-induced LH surge in female rats. Endocrinology 157(5):2015–2027. https://doi.org/10.1210/en.2015-1740

Moenter SM, Brand RC, Karsch FJ (1992) Dynamics of gonadotropin-releasing hormone (GnRH) secretion during the GnRH surge: insights into the mechanism of GnRH surge induction. Endocrinology 130(5):2978–2984

Moline ML, Albers HE, Todd RB, Moore-Ede MC (1981) Light-dark entrainment of proestrous LH surges and circadian locomotor activity in female hamsters. Horm Behav 15(4):451–458

Molnar CS, Kallo I, Liposits Z, Hrabovszky E (2011) Estradiol down-regulates RF-amide-related peptide (RFRP) expression in the mouse hypothalamus. Endocrinology 152(4):1684–1690. https://doi.org/10.1210/en.2010-1418

Mong JA, Pfaff DW (2003) Hormonal and genetic influences underlying arousal as it drives sex and aggression in animal and human brains. Neurobiol Aging 24(Suppl 1):S83–S88. discussion S91-2

Moore TR, Iams JD, Creasy RK, Burau KD, Davidson AL (1994) Diurnal and gestational patterns of uterine activity in normal human pregnancy. The Uterine Activity in Pregnancy Working Group. Obstet Gynecol 83(4):517–523

Moore RY, Speh JC, Leak RK (2002) Suprachiasmatic nucleus organization. Cell Tissue Res 309 (1):89–98. https://doi.org/10.1007/s00441-002-0575-2

Moore AM, Lucas KA, Goodman RL, Coolen LM, Lehman MN (2018) Three-dimensional imaging of KNDy neurons in the mammalian brain using optical tissue clearing and multiple-label immunocytochemistry. Sci Rep 8(1):2242. https://doi.org/10.1038/s41598-018-20563-2

Mounien L, Bizet P, Boutelet I, Gourcerol G, Fournier A, Vaudry H, Jegou S (2006) Pituitary adenylate cyclase-activating polypeptide directly modulates the activity of proopiomelanocortin neurons in the rat arcuate nucleus. Neuroscience 143(1):155–163. https://doi.org/10.1016/j.neuroscience.2006.07.022

Muir AI, Chamberlain L, Elshourbagy NA, Michalovich D, Moore DJ, Calamari A, Szekeres PG, Sarau HM, Chambers JK, Murdock P, Steplewski K, Shabon U, Miller JE, Middleton SE, Darker JG, Larminie CG, Wilson S, Bergsma DJ, Emson P, Faull R, Philpott KL, Harrison DC (2001) AXOR12, a novel human G protein-coupled receptor, activated by the peptide KiSS-1. J Biol Chem 276(31):28969–28975. https://doi.org/10.1074/jbc.M102743200

Nakamura Y, Tamura H, Kashida S, Takayama H, Yamagata Y, Karube A, Sugino N, Kato H (2001) Changes of serum melatonin level and its relationship to feto-placental unit during pregnancy. J Pineal Res 30(1):29–33

Nakamura TJ, Sellix MT, Kudo T, Nakao N, Yoshimura T, Ebihara S, Colwell CS, Block GD (2010) Influence of the estrous cycle on clock gene expression in reproductive tissues: effects of fluctuating ovarian steroid hormone levels. Steroids 75(3):203–212. https://doi.org/10.1016/j.steroids.2010.01.007

Navarro VM, Castellano JM, Fernandez-Fernandez R, Tovar S, Roa J, Mayen A, Barreiro ML, Casanueva FF, Aguilar E, Dieguez C, Pinilla L, Tena-Sempere M (2005a) Effects of KiSS-1 peptide, the natural ligand of GPR54, on follicle-stimulating hormone secretion in the rat. Endocrinology 146(4):1689–1697. https://doi.org/10.1210/en.2004-1353

Navarro VM, Castellano JM, Fernandez-Fernandez R, Tovar S, Roa J, Mayen A, Nogueiras R, Vazquez MJ, Barreiro ML, Magni P, Aguilar E, Dieguez C, Pinilla L, Tena-Sempere M (2005b) Characterization of the potent luteinizing hormone-releasing activity of KiSS-1 peptide, the natural ligand of GPR54. Endocrinology 146(1):156–163. https://doi.org/10.1210/en.2004-0836

Navarro VM, Gottsch ML, Chavkin C, Okamura H, Clifton DK, Steiner RA (2009) Regulation of gonadotropin-releasing hormone secretion by kisspeptin/dynorphin/neurokinin B neurons in the arcuate nucleus of the mouse. J Neurosci 29(38):11859–11866. https://doi.org/10.1523/JNEUROSCI.1569-09.2009

Ngwenya S, Lindow SW (2004) 24 hour rhythm in the timing of pre-labour spontaneous rupture of membranes at term. Eur J Obstet Gynecol Reprod Biol 112(2):151–153

Novakova M, Sladek M, Sumova A (2010) Exposure of pregnant rats to restricted feeding schedule synchronizes the SCN clocks of their fetuses under constant light but not under a light-dark regime. J Biol Rhythm 25(5):350–360. https://doi.org/10.1177/0748730410377967

Nunez AA, Stephan FK (1977) The effects of hypothalamic knife cuts on drinking rhythms and the estrus cycle of the rat. Behav Biol 20(2):224–234

Nurminen T (1998) Shift work and reproductive health. Scand J Work Environ Health 24(Suppl 3):28–34

Ohtaki T, Shintani Y, Honda S, Matsumoto H, Hori A, Kanehashi K, Terao Y, Kumano S, Takatsu Y, Masuda Y, Ishibashi Y, Watanabe T, Asada M, Yamada T, Suenaga M, Kitada C, Usuki S, Kurokawa T, Onda H, Nishimura O, Fujino M (2001) Metastasis suppressor gene KiSS-1 encodes peptide ligand of a G-protein-coupled receptor. Nature 411(6837):613–617. https://doi.org/10.1038/35079135

Ohtsuka S, Miyake A, Nishizaki T, Tasaka K, Tanizawa O (1988) Vasoactive intestinal peptide stimulates gonadotropin-releasing hormone release from rat hypothalamus in vitro. Acta Endocrinol (Copenh) 117(3):399–402

Okamura H, Yamamura T, Wakabayashi Y (2017) Mapping of KNDy neurons and immunohistochemical analysis of the interaction between KNDy and substance P neural systems in goat. J Reprod Dev 63(6):571–580. https://doi.org/10.1262/jrd.2017-103

Okatani Y, Okamoto K, Hayashi K, Wakatsuki A, Tamura S, Sagara Y (1998) Maternal-fetal transfer of melatonin in pregnant women near term. J Pineal Res 25(3):129–134

Olcese J (2012) Circadian aspects of mammalian parturition: a review. Mol Cell Endocrinol 349 (1):62–67. https://doi.org/10.1016/j.mce.2011.06.041

Olcese J, Domagalski R, Bednorz A, Weaver DR, Urbanski HF, Reuss S, Middendorff R (2003) Expression and regulation of mPer1 in immortalized GnRH neurons. Neuroreport 14 (4):613–618. https://doi.org/10.1097/01.wnr.0000063508.18654.3c

Olcese J, Lozier S, Paradise C (2013) Melatonin and the circadian timing of human parturition. Reprod Sci 20(2):168–174. https://doi.org/10.1177/1933719112442244

Palm IF, Van Der Beek EM, Wiegant VM, Buijs RM, Kalsbeek A (1999) Vasopressin induces a luteinizing hormone surge in ovariectomized, estradiol-treated rats with lesions of the suprachiasmatic nucleus. Neuroscience 93(2):659–666

Palm IF, van der Beek EM, Wiegant VM, Buijs RM, Kalsbeek A (2001) The stimulatory effect of vasopressin on the luteinizing hormone surge in ovariectomized, estradiol-treated rats is time-dependent. Brain Res 901(1–2):109–116

Pan JT, Gala RR (1985) Central nervous system regions involved in the estrogen-induced afternoon prolactin surge. II. Implantation studies. Endocrinology 117(1):388–395. https://doi.org/10.1210/endo-117-1-388

Panda S, Sato TK, Castrucci AM, Rollag MD, DeGrip WJ, Hogenesch JB, Provencio I, Kay SA (2002) Melanopsin (Opn4) requirement for normal light-induced circadian phase shifting. Science 298(5601):2213–2216. https://doi.org/10.1126/science.1076848

Pau KY, Berria M, Hess DL, Spies HG (1993) Preovulatory gonadotropin-releasing hormone surge in ovarian-intact rhesus macaques. Endocrinology 133(4):1650–1656

Piet R, Fraissenon A, Boehm U, Herbison AE (2015) Estrogen permits vasopressin signaling in preoptic kisspeptin neurons in the female mouse. J Neurosci 35(17):6881–6892. https://doi.org/10.1523/JNEUROSCI.4587-14.2015

Piet R, Dunckley H, Lee K, Herbison AE (2016) Vasoactive intestinal peptide excites GnRH neurons in male and female mice. Endocrinology 157(9):3621–3630. https://doi.org/10.1210/en.2016-1399

Pilorz V, Steinlechner S (2008) Low reproductive success in Per1 and Per2 mutant mouse females due to accelerated ageing? Reproduction 135(4):559–568. https://doi.org/10.1530/REP-07-0434

Poletini MO, McKee DT, Kennett JE, Doster J, Freeman ME (2007) Knockdown of clock genes in the suprachiasmatic nucleus blocks prolactin surges and alters FRA expression in the locus coeruleus of female rats. Am J Phys Endocrinol Metab 293(5):E1325–E1334. https://doi.org/10.1152/ajpendo.00341.2007

Poletini MO, Kennett JE, McKee DT, Freeman ME (2010) Central clock regulates the cervically stimulated prolactin surges by modulation of dopamine and vasoactive intestinal polypeptide release in ovariectomized rats. Neuroendocrinology 91(2):179–188. https://doi.org/10.1159/000254379

Poling MC, Kim J, Dhamija S, Kauffman AS (2012) Development, sex steroid regulation, and phenotypic characterization of RFamide-related peptide (Rfrp) gene expression and RFamide receptors in the mouse hypothalamus. Endocrinology 153(4):1827–1840. https://doi.org/10.1210/en.2011-2049

Rance NE (2009) Menopause and the human hypothalamus: evidence for the role of kisspeptin/neurokinin B neurons in the regulation of estrogen negative feedback. Peptides 30(1):111–122. https://doi.org/10.1016/j.peptides.2008.05.016

Ratajczak CK, Boehle KL, Muglia LJ (2009) Impaired steroidogenesis and implantation failure in Bmal1-/- mice. Endocrinology 150(4):1879–1885. https://doi.org/10.1210/en.2008-1021

Ratajczak CK, Herzog ED, Muglia LJ (2010) Clock gene expression in gravid uterus and extra-embryonic tissues during late gestation in the mouse. Reprod Fertil Dev 22(5):743–750. https://doi.org/10.1071/RD09243

Ratajczak CK, Asada M, Allen GC, McMahon DG, Muglia LM, Smith D, Bhattacharyya S, Muglia LJ (2012) Generation of myometrium-specific Bmal1 knockout mice for parturition analysis. Reprod Fertil Dev 24(5):759–767. https://doi.org/10.1071/RD11164

Reppert SM, Henshaw D, Schwartz WJ, Weaver DR (1987) The circadian-gated timing of birth in rats: disruption by maternal SCN lesions or by removal of the fetal brain. Brain Res 403(2):398–402

Resuehr D, Wildemann U, Sikes H, Olcese J (2007) E-box regulation of gonadotropin-releasing hormone (GnRH) receptor expression in immortalized gonadotrope cells. Mol Cell Endocrinol 278(1–2):36–43. https://doi.org/10.1016/j.mce.2007.08.008

Rivkees SA, Reppert SM (1991) Appearance of melatonin receptors during embryonic life in Siberian hamsters (Phodopus sungorous). Brain Res 568(1–2):345–349

Rizzo P, Raffone E, Benedetto V (2010) Effect of the treatment with myo-inositol plus folic acid plus melatonin in comparison with a treatment with myo-inositol plus folic acid on oocyte quality and pregnancy outcome in IVF cycles. A prospective, clinical trial. Eur Rev Med Pharmacol Sci 14(6):555–561

Robertson MC, Friesen HG (1981) Two forms of rat placental lactogen revealed by radioimmunoassay. Endocrinology 108(6):2388–2390. https://doi.org/10.1210/endo-108-6-2388

Ronnekleiv OK, Kelly MJ (1988) Plasma prolactin and luteinizing hormone profiles during the estrous cycle of the female rat: effects of surgically induced persistent estrus. Neuroendocrinology 47(2):133–141. https://doi.org/10.1159/000124903

Ronnekleiv OK, Fang Y, Zhang C, Nestor CC, Mao P, Kelly MJ (2014) Research resource: gene profiling of G protein-coupled receptors in the arcuate nucleus of the female. Mol Endocrinol 28 (8):1362–1380. https://doi.org/10.1210/me.2014-1103

Ruka KA, Burger LL, Moenter SM (2013) Regulation of arcuate neurons coexpressing kisspeptin, neurokinin B, and dynorphin by modulators of neurokinin 3 and kappa-opioid receptors in adult male mice. Endocrinology 154(8):2761–2771. https://doi.org/10.1210/en.2013-1268

Russo KA, La JL, Stephens SB, Poling MC, Padgaonkar NA, Jennings KJ, Piekarski DJ, Kauffman AS, Kriegsfeld LJ (2015) Circadian control of the female reproductive axis through gated responsiveness of the RFRP-3 system to VIP signaling. Endocrinology 156(7):2608–2618. https://doi.org/10.1210/en.2014-1762

Saedi S, Khoradmehr A, Mohammad Reza JS, Tamadon A (2018) The role of neuropeptides and neurotransmitters on kisspeptin/kiss1r-signaling in female reproduction. J Chem Neuroanat 92:71–82. https://doi.org/10.1016/j.jchemneu.2018.07.001

Sampaio RV, Conceicao S, Miranda MS, Sampaio Lde F, Ohashi OM (2012) MT3 melatonin binding site, MT1 and MT2 melatonin receptors are present in oocyte, but only MT1 is present in bovine blastocyst produced in vitro. Reprod Biol Endocrinol 10:103. https://doi.org/10.1186/1477-7827-10-103

Samson WK, Burton KP, Reeves JP, McCann SM (1981) Vasoactive intestinal peptide stimulates luteinizing hormone-releasing hormone release from median eminence synaptosomes. Regul Pept 2(4):253–264

Sarkar DK, Chiappa SA, Fink G, Sherwood NM (1976) Gonadotropin-releasing hormone surge in pro-oestrous rats. Nature 264(5585):461–463

Satake H, Hisada M, Kawada T, Minakata H, Ukena K, Tsutsui K (2001) Characterization of a cDNA encoding a novel avian hypothalamic neuropeptide exerting an inhibitory effect on gonadotropin release. Biochem J 354(Pt 2):379–385

Sawai N, Iijima N, Takumi K, Matsumoto K, Ozawa H (2012) Immunofluorescent histochemical and ultrastructural studies on the innervation of kisspeptin/neurokinin B neurons to tuberoinfundibular dopaminergic neurons in the arcuate nucleus of rats. Neurosci Res 74 (1):10–16. https://doi.org/10.1016/j.neures.2012.05.011

Sawai N, Iijima N, Ozawa H, Matsuzaki T (2014) Neurokinin B- and kisspeptin-positive fibers as well as tuberoinfundibular dopaminergic neurons directly innervate periventricular hypophyseal dopaminergic neurons in rats and mice. Neurosci Res 84:10–18. https://doi.org/10.1016/j.neures.2014.05.002

Schafer D, Kane G, Colledge WH, Piet R, Herbison AE (2018) Sex- and sub region-dependent modulation of arcuate kisspeptin neurons by vasopressin and vasoactive intestinal peptide. J Neuroendocrinol:e12660. https://doi.org/10.1111/jne.12660

Schenker S, Yang Y, Perez A, Acuff RV, Papas AM, Henderson G, Lee MP (1998) Antioxidant transport by the human placenta. Clin Nutr 17(4):159–167

Schoeller EL, Clark DD, Dey S, Cao NV, Semaan SJ, Chao LW, Kauffman AS, Stowers L, Mellon PL (2016) Bmal1 is required for normal reproductive behaviors in male mice. Endocrinology 157(12):4914–4929. https://doi.org/10.1210/en.2016-1620

Schwartz WJ, Coleman RJ, Reppert SM (1983) A daily vasopressin rhythm in rat cerebrospinal fluid. Brain Res 263(1):105–112

Seminara SB, Messager S, Chatzidaki EE, Thresher RR, Acierno JS Jr, Shagoury JK, Bo-Abbas Y, Kuohung W, Schwinof KM, Hendrick AG, Zahn D, Dixon J, Kaiser UB, Slaugenhaupt SA, Gusella JF, O'Rahilly S, Carlton MB, Crowley WF Jr, Aparicio SA, Colledge WH (2003) The GPR54 gene as a regulator of puberty. N Engl J Med 349(17):1614–1627. https://doi.org/10.1056/NEJMoa035322

Seron-Ferre M, Ducsay CA, Valenzuela GJ (1993) Circadian rhythms during pregnancy. Endocr Rev 14(5):594–609. https://doi.org/10.1210/edrv-14-5-594

Seron-Ferre M, Riffo R, Valenzuela GJ, Germain AM (2001) Twenty-four-hour pattern of cortisol in the human fetus at term. Am J Obstet Gynecol 184(6):1278–1283
Sharkey JT, Puttaramu R, Word RA, Olcese J (2009) Melatonin synergizes with oxytocin to enhance contractility of human myometrial smooth muscle cells. J Clin Endocrinol Metab 94 (2):421–427. https://doi.org/10.1210/jc.2008-1723
Shimizu M, Bedecarrats GY (2010) Activation of the chicken gonadotropin-inhibitory hormone receptor reduces gonadotropin releasing hormone receptor signaling. Gen Comp Endocrinol 167(2):331–337. https://doi.org/10.1016/j.ygcen.2010.03.029
Shiu RP, Kelly PA, Friesen HG (1973) Radioreceptor assay for prolactin and other lactogenic hormones. Science 180(4089):968–971
Silver R, LeSauter J, Tresco PA, Lehman MN (1996) A diffusible coupling signal from the transplanted suprachiasmatic nucleus controlling circadian locomotor rhythms. Nature 382 (6594):810–813. https://doi.org/10.1038/382810a0
Simonneaux V, Bahougne T (2015) A multi-oscillatory circadian system times female reproduction. Front Endocrinol (Lausanne) 6:157. https://doi.org/10.3389/fendo.2015.00157
Simonneaux V, Bahougne T, Angelopoulou E (2017) Daily rhythms count for female fertility. Best Pract Res Clin Endocrinol Metab 31(5):505–519. https://doi.org/10.1016/j.beem.2017.10.012
Smarr BL, Gile JJ, de la Iglesia HO (2013) Oestrogen-independent circadian clock gene expression in the anteroventral periventricular nucleus in female rats: possible role as an integrator for circadian and ovarian signals timing the luteinising hormone surge. J Neuroendocrinol 25 (12):1273–1279. https://doi.org/10.1111/jne.12104
Smarr BL, Grant AD, Perez L, Zucker I, Kriegsfeld LJ (2017) Maternal and early-life circadian disruption have long-lasting negative consequences on offspring development and adult behavior in mice. Sci Rep 7(1):3326. https://doi.org/10.1038/s41598-017-03406-4
Smith JT, Cunningham MJ, Rissman EF, Clifton DK, Steiner RA (2005a) Regulation of Kiss1 gene expression in the brain of the female mouse. Endocrinology 146(9):3686–3692. https://doi.org/10.1210/en.2005-0488
Smith JT, Dungan HM, Stoll EA, Gottsch ML, Braun RE, Eacker SM, Clifton DK, Steiner RA (2005b) Differential regulation of KiSS-1 mRNA expression by sex steroids in the brain of the male mouse. Endocrinology 146(7):2976–2984. https://doi.org/10.1210/en.2005-0323
Smith JT, Popa SM, Clifton DK, Hoffman GE, Steiner RA (2006) Kiss1 neurons in the forebrain as central processors for generating the preovulatory luteinizing hormone surge. J Neurosci 26 (25):6687–6694. https://doi.org/10.1523/JNEUROSCI.1618-06.2006
Smith JT, Coolen LM, Kriegsfeld LJ, Sari IP, Jaafarzadehshirazi MR, Maltby M, Bateman K, Goodman RL, Tilbrook AJ, Ubuka T, Bentley GE, Clarke IJ, Lehman MN (2008) Variation in kisspeptin and RFamide-related peptide (RFRP) expression and terminal connections to gonadotropin-releasing hormone neurons in the brain: a novel medium for seasonal breeding in the sheep. Endocrinology 149(11):5770–5782. https://doi.org/10.1210/en.2008-0581
Sonigo C, Bouilly J, Carre N, Tolle V, Caraty A, Tello J, Simony-Conesa FJ, Millar R, Young J, Binart N (2012) Hyperprolactinemia-induced ovarian acyclicity is reversed by kisspeptin administration. J Clin Invest 122(10):3791–3795. https://doi.org/10.1172/JCI63937
Stevens RG, Brainard GC, Blask DE, Lockley SW, Motta ME (2013) Adverse health effects of nighttime lighting: comments on American Medical Association policy statement. Am J Prev Med 45(3):343–346. https://doi.org/10.1016/j.amepre.2013.04.011
Strauss JF 3rd, Martinez F, Kiriakidou M (1996) Placental steroid hormone synthesis: unique features and unanswered questions. Biol Reprod 54(2):303–311
Summa KC, Vitaterna MH, Turek FW (2012) Environmental perturbation of the circadian clock disrupts pregnancy in the mouse. PLoS One 7(5):e37668. https://doi.org/10.1371/journal.pone.0037668
Sun Y, Shu J, Kyei K, Neal-Perry GS (2012) Intracerebroventricular infusion of vasoactive intestinal peptide rescues the luteinizing hormone surge in middle-aged female rats. Front Endocrinol (Lausanne) 3:24. https://doi.org/10.3389/fendo.2012.00024

Szawka RE, Ribeiro AB, Leite CM, Helena CV, Franci CR, Anderson GM, Hoffman GE, Anselmo-Franci JA (2010) Kisspeptin regulates prolactin release through hypothalamic dopaminergic neurons. Endocrinology 151(7):3247–3257. https://doi.org/10.1210/en.2009-1414

Takahashi JS (2015) Molecular components of the circadian clock in mammals. Diabetes Obes Metab 17(Suppl 1):6–11. https://doi.org/10.1111/dom.12514

Takayama H, Nakamura Y, Tamura H, Yamagata Y, Harada A, Nakata M, Sugino N, Kato H (2003) Pineal gland (melatonin) affects the parturition time, but not luteal function and fetal growth, in pregnant rats. Endocr J 50(1):37–43

Tamura H, Nakamura Y, Terron MP, Flores LJ, Manchester LC, Tan DX, Sugino N, Reiter RJ (2008) Melatonin and pregnancy in the human. Reprod Toxicol 25(3):291–303. https://doi.org/10.1016/j.reprotox.2008.03.005

Teclemariam-Mesbah R, Kalsbeek A, Pevet P, Buijs RM (1997) Direct vasoactive intestinal polypeptide-containing projection from the suprachiasmatic nucleus to spinal projecting hypothalamic paraventricular neurons. Brain Res 748(1–2):71–76

Thomas L, Purvis CC, Drew JE, Abramovich DR, Williams LM (2002) Melatonin receptors in human fetal brain: 2-[(125)I]iodomelatonin binding and MT1 gene expression. J Pineal Res 33 (4):218–224

Thompson EL, Patterson M, Murphy KG, Smith KL, Dhillo WS, Todd JF, Ghatei MA, Bloom SR (2004) Central and peripheral administration of kisspeptin-10 stimulates the hypothalamic-pituitary-gonadal axis. J Neuroendocrinol 16(10):850–858. https://doi.org/10.1111/j.1365-2826.2004.01240.x

Tian XZ, Wen Q, Shi JM, Liang W, Zeng SM, Tian JH, Zhou GB, Zhu SE, Liu GS (2010) Effects of melatonin on in vitro development of mouse two-cell embryos cultured in HTF medium. Endocr Res 35(1):17–23. https://doi.org/10.3109/07435800903539607

Topaloglu AK, Reimann F, Guclu M, Yalin AS, Kotan LD, Porter KM, Serin A, Mungan NO, Cook JR, Imamoglu S, Akalin NS, Yuksel B, O'Rahilly S, Semple RK (2009) TAC3 and TACR3 mutations in familial hypogonadotropic hypogonadism reveal a key role for Neurokinin B in the central control of reproduction. Nat Genet 41(3):354–358. https://doi.org/10.1038/ng.306

Torres-Farfan C, Rocco V, Monso C, Valenzuela FJ, Campino C, Germain A, Torrealba F, Valenzuela GJ, Seron-Ferre M (2006) Maternal melatonin effects on clock gene expression in a nonhuman primate fetus. Endocrinology 147(10):4618–4626. https://doi.org/10.1210/en.2006-0628

Torres-Farfan C, Valenzuela FJ, Mondaca M, Valenzuela GJ, Krause B, Herrera EA, Riquelme R, Llanos AJ, Seron-Ferre M (2008) Evidence of a role for melatonin in fetal sheep physiology: direct actions of melatonin on fetal cerebral artery, brown adipose tissue and adrenal gland. J Physiol 586(16):4017–4027. https://doi.org/10.1113/jphysiol.2008.154351

Trevisan CM, Montagna E, de Oliveira R, Christofolini DM, Barbosa CP, Crandall KA, Bianco B (2018) Kisspeptin/GPR54 system: what do we know about its role in human reproduction? Cell Physiol Biochem 49(4):1259–1276. https://doi.org/10.1159/000493406

Tsutsui K, Saigoh E, Ukena K, Teranishi H, Fujisawa Y, Kikuchi M, Ishii S, Sharp PJ (2000) A novel avian hypothalamic peptide inhibiting gonadotropin release. Biochem Biophys Res Commun 275(2):661–667. https://doi.org/10.1006/bbrc.2000.3350

Tsutsui K, Saigoh E, Yin H, Ubuka T, Chowdhury VS, Osugi T, Ukena K, Sharp PJ, Wingfield JC, Bentley GE (2009) A new key neurohormone controlling reproduction, gonadotrophin-inhibitory hormone in birds: discovery, progress and prospects. J Neuroendocrinol 21 (4):271–275. https://doi.org/10.1111/j.1365-2826.2009.01829.x

Ubuka T, Tsutsui K (2018) Comparative and evolutionary aspects of gonadotropin-inhibitory hormone and FMRFamide-like peptide systems. Front Neurosci 12:747. https://doi.org/10.3389/fnins.2018.00747

Ubuka T, Inoue K, Fukuda Y, Mizuno T, Ukena K, Kriegsfeld LJ, Tsutsui K (2012) Identification, expression, and physiological functions of Siberian hamster gonadotropin-inhibitory hormone. Endocrinology 153(1):373–385. https://doi.org/10.1210/en.2011-1110

Unfer V, Raffone E, Rizzo P, Buffo S (2011) Effect of a supplementation with myo-inositol plus melatonin on oocyte quality in women who failed to conceive in previous in vitro fertilization cycles for poor oocyte quality: a prospective, longitudinal, cohort study. Gynecol Endocrinol 27 (11):857–861. https://doi.org/10.3109/09513590.2011.564687

van der Beek EM, van Oudheusden HJ, Buijs RM, van der Donk HA, van den Hurk R, Wiegant VM (1994) Preferential induction of c-fos immunoreactivity in vasoactive intestinal polypeptide-innervated gonadotropin-releasing hormone neurons during a steroid-induced luteinizing hormone surge in the female rat. Endocrinology 134(6):2636–2644. https://doi.org/10.1210/endo.134.6.8194489

Van der Beek EM, Horvath TL, Wiegant VM, Van den Hurk R, Buijs RM (1997) Evidence for a direct neuronal pathway from the suprachiasmatic nucleus to the gonadotropin-releasing hormone system: combined tracing and light and electron microscopic immunocytochemical studies. J Comp Neurol 384(4):569–579

van der Horst GT, Muijtjens M, Kobayashi K, Takano R, Kanno S, Takao M, de Wit J, Verkerk A, Eker AP, van Leenen D, Buijs R, Bootsma D, Hoeijmakers JH, Yasui A (1999) Mammalian Cry1 and Cry2 are essential for maintenance of circadian rhythms. Nature 398(6728):627–630. https://doi.org/10.1038/19323

Vanecek J (1988) The melatonin receptors in rat ontogenesis. Neuroendocrinology 48(2):201–203. https://doi.org/10.1159/000125008

Vatish M, Steer PJ, Blanks AM, Hon M, Thornton S (2010) Diurnal variation is lost in preterm deliveries before 28 weeks of gestation. BJOG 117(6):765–767. https://doi.org/10.1111/j.1471-0528.2010.02526.x

Vida B, Deli L, Hrabovszky E, Kalamatianos T, Caraty A, Coen CW, Liposits Z, Kallo I (2010) Evidence for suprachiasmatic vasopressin neurones innervating kisspeptin neurones in the rostral periventricular area of the mouse brain: regulation by oestrogen. J Neuroendocrinol 22 (9):1032–1039. https://doi.org/10.1111/j.1365-2826.2010.02045.x

Vilches N, Spichiger C, Mendez N, Abarzua-Catalan L, Galdames HA, Hazlerigg DG, Richter HG, Torres-Farfan C (2014) Gestational chronodisruption impairs hippocampal expression of NMDA receptor subunits Grin1b/Grin3a and spatial memory in the adult offspring. PLoS One 9(3):e91313. https://doi.org/10.1371/journal.pone.0091313

Waddell BJ, Wharfe MD, Crew RC, Mark PJ (2012) A rhythmic placenta? Circadian variation, clock genes and placental function. Placenta 33(7):533–539. https://doi.org/10.1016/j.placenta.2012.03.008

Wakabayashi Y, Nakada T, Murata K, Ohkura S, Mogi K, Navarro VM, Clifton DK, Mori Y, Tsukamura H, Maeda K, Steiner RA, Okamura H (2010) Neurokinin B and dynorphin A in kisspeptin neurons of the arcuate nucleus participate in generation of periodic oscillation of neural activity driving pulsatile gonadotropin-releasing hormone secretion in the goat. J Neurosci 30(8):3124–3132. https://doi.org/10.1523/JNEUROSCI.5848-09.2010

Wakabayashi Y, Yamamura T, Sakamoto K, Mori Y, Okamura H (2013) Electrophysiological and morphological evidence for synchronized GnRH pulse generator activity among Kisspeptin/neurokinin B/dynorphin a (KNDy) neurons in goats. J Reprod Dev 59(1):40–48

Wang Y, Gu F, Deng M, Guo L, Lu C, Zhou C, Chen S, Xu Y (2016) Rotating shift work and menstrual characteristics in a cohort of Chinese nurses. BMC Womens Health 16(1):24. https://doi.org/10.1186/s12905-016-0301-y

Welsh DK, Yoo SH, Liu AC, Takahashi JS, Kay SA (2004) Bioluminescence imaging of individual fibroblasts reveals persistent, independently phased circadian rhythms of clock gene expression. Curr Biol 14(24):2289–2295. https://doi.org/10.1016/j.cub.2004.11.057

Wharfe MD, Mark PJ, Waddell BJ (2011) Circadian variation in placental and hepatic clock genes in rat pregnancy. Endocrinology 152(9):3552–3560. https://doi.org/10.1210/en.2011-0081

Wiegand SJ, Terasawa E (1982) Discrete lesions reveal functional heterogeneity of suprachiasmatic structures in regulation of gonadotropin secretion in the female rat. Neuroendocrinology 34 (6):395–404. https://doi.org/10.1159/000123335

Wiegand SJ, Terasawa E, Bridson WE, Goy RW (1980) Effects of discrete lesions of preoptic and suprachiasmatic structures in the female rat. Alterations in the feedback regulation of gonadotropin secretion. Neuroendocrinology 31(2):147–157. https://doi.org/10.1159/000123066
Williams WP 3rd, Kriegsfeld LJ (2012) Circadian control of neuroendocrine circuits regulating female reproductive function. Front Endocrinol 3:60. https://doi.org/10.3389/fendo.2012.00060
Williams LM, Martinoli MG, Titchener LT, Pelletier G (1991) The ontogeny of central melatonin binding sites in the rat. Endocrinology 128(4):2083–2090. https://doi.org/10.1210/endo-128-4-2083
Williams WP 3rd, Jarjisian SG, Mikkelsen JD, Kriegsfeld LJ (2011) Circadian control of kisspeptin and a gated GnRH response mediate the preovulatory luteinizing hormone surge. Endocrinology 152(2):595–606. https://doi.org/10.1210/en.2010-0943
Wintermantel TM, Elzer J, Herbison AE, Fritzemeier KH, Schutz G (2006) Genetic dissection of estrogen receptor signaling in vivo. Ernst Schering Found Symp Proc 1:25–44
Wu M, Dumalska I, Morozova E, van den Pol AN, Alreja M (2009) Gonadotropin inhibitory hormone inhibits basal forebrain vGluT2-gonadotropin-releasing hormone neurons via a direct postsynaptic mechanism. J Physiol 587(Pt 7):1401–1411. https://doi.org/10.1113/jphysiol.2008.166447
Xie M, Utzinger KS, Blickenstorfer K, Leeners B (2018) Diurnal and seasonal changes in semen quality of men in subfertile partnerships. Chronobiol Int 35(10):1375–1384. https://doi.org/10.1080/07420528.2018.1483942
Yamazaki S, Numano R, Abe M, Hida A, Takahashi R, Ueda M, Block GD, Sakaki Y, Menaker M, Tei H (2000) Resetting central and peripheral circadian oscillators in transgenic rats. Science 288(5466):682–685
Yano T, Iijima N, Kakihara K, Hinuma S, Tanaka M, Ibata Y (2003) Localization and neuronal response of RFamide related peptides in the rat central nervous system. Brain Res 982(2):156–167
Yeo SH, Colledge WH (2018) The role of Kiss1 neurons as integrators of endocrine, metabolic, and environmental factors in the hypothalamic-pituitary-gonadal axis. Front Endocrinol 9:188. https://doi.org/10.3389/fendo.2018.00188
Yip SH, Boehm U, Herbison AE, Campbell RE (2015) Conditional viral tract tracing delineates the projections of the distinct Kisspeptin neuron populations to gonadotropin-releasing hormone (GnRH) neurons in the mouse. Endocrinology 156(7):2582–2594. https://doi.org/10.1210/en.2015-1131
Yogev L, Terkel J (1980) Effects of photoperiod, absence of photic cues, and suprachiasmatic nucleus lesions on nocturnal prolactin surges in pregnant and pseudopregnant rats. Neuroendocrinology 31(1):26–33. https://doi.org/10.1159/000123046
Young J, Bouligand J, Francou B, Raffin-Sanson ML, Gaillez S, Jeanpierre M, Grynberg M, Kamenicky P, Chanson P, Brailly-Tabard S, Guiochon-Mantel A (2010) TAC3 and TACR3 defects cause hypothalamic congenital hypogonadotropic hypogonadism in humans. J Clin Endocrinol Metab 95(5):2287–2295. https://doi.org/10.1210/jc.2009-2600
Zahn V, Hattensperger W (1993) Circadian rhythm of pregnancy contractions. Z Geburtshilfe Perinatol 197(1):1–10
Zhao S, Kriegsfeld LJ (2009) Daily changes in GT1-7 cell sensitivity to GnRH secretagogues that trigger ovulation. Neuroendocrinology 89(4):448–457. https://doi.org/10.1159/000192370
Zhao L, Zhong M, Xue HL, Ding JS, Wang S, Xu JH, Chen L, Xu LX (2014) Effect of RFRP-3 on reproduction is sex- and developmental status-dependent in the striped hamster (*Cricetulus barabensis*). Gene 547(2):273–279. https://doi.org/10.1016/j.gene.2014.06.054

Glossary

5-Methylcytosine A genomic cytosine nucleotide that has been methylated.

Amygdala A brain region considered essential to a variety of social and emotional behaviors and which shows sex differences in anatomy and physiology.

Androgens A class of steroid hormones that includes testosterone and dihydrotestosterone. Testosterone can be aromatized into estradiol but dihydrotestosterone cannot.

Anisotropic growth Growth and differentiation from a progenitor population that is significantly different (greater) than growth/differentiation from adjacent progenitors.

Anorexigenic Inhibiting appetite and promoting energy expenditure.

Anterior lobe of the pituitary gland or adenohypophysis Contains pituitary hormone-producing cell types.

Anterioventral periventricular nucleus (AVPV) A small subnucleus in the preoptic area which is larger in females and projects to the GnRH neurons that control ovulation.

Antiphase When comparing two or more oscillations, the peak of one oscillation corresponds to the trough of another.

Arcuate nucleus An aggregation of neurons in the mediobasal hypothalamus, which is located adjacent to the third ventricle and the median eminence; orexigenic, appetite-stimulating; anorexigenic, appetite-suppressing.

Aromatase An enzyme that converts testosterone into estradiol. It is found at high concentrations in the ovary and in discrete regions of the brain.

Astrocytes A type of glia dispersed broadly throughout the brain that play important roles in modulating neuronal health and activity.

Autosynaptic feedback A chemical or electrical synapse formed by the axon of a neuron on its own dendrites or soma.

bHyp (basal Hypothalamic) cells A proliferating cell population that develops from RDVM cells, gives rise to anterior, mammillary and infundibular progenitors, and is retained through life. Potentially harbours a neural stem cell.

Bisulfite conversion of genomic DNA Treatment with sodium bisulfite to promote the deamination of cytosine residues, which is followed by desulfonation at elevated pH, thus leading to conversion into uracil. Methylated cytosines are resistant to bisulfite conversion.

S. Wray, S. Blackshaw (eds.), *Developmental Neuroendocrinology*, Masterclass in Neuroendocrinology 9, https://doi.org/10.1007/978-3-030-40002-6

Caudal brainstem Including the medulla (medulla oblongata), aka the hindbrain. Houses the area postrema (AP), nucleus of the solitary tract (NTS), dorsal motor nucleus of the vagus (DMV), medullary reticular formation (RF), and ventrolateral medulla (VLM).

Cell lineage All the cells born from one dividing cell (or from one specific kind of dividing cell) form a "cell lineage."

Cellular central oscillator Activity of the core circadian oscillator at the cellular (as opposed to organ or organismal) level.

CHIP-seq High-throughput sequencing method combining chromatin immunoprecipitation technique with deep sequencing. It allows the identification of specific DNA sequences bound to proteins of interest in vivo. Benefits include single nucleotide resolution, relatively low amount of DNA input, and high dynamic range for quantification.

Chronobiology The study of endogenous biological rhythms and the entrainment by environmental cues. Also referred to as biochronometry.

Chronotype Intrinsic, individual differences in the synchrony between light cycles and activity rhythms. Manifests as "morningness" (phase-advanced) and "eveningness" (phase-delayed sleep/wake rhythms).

Circadian An endogenous behavioural, biochemical, or physiological oscillation that approximates a 24-h period.

Circadian disruption An incongruence between endogenous circadian rhythms and external time.

Circadian entrainment This occurs when rhythmic physiological or behavioral events match their period to that of an environmental oscillation. Entraining signals can include light, food, and neurotransmitters.

Circadian phase synchrony and stabilization Circadian phase synchrony occurs when cellular clocks are synchronized and run in phase, typically in response to an external cue. Circadian phase stabilization indicates that this synchrony is robust and reproducible over many phases.

Circadian rhythms Endogenously generated behavioral, physiological, endocrine, metabolic, and/or molecular rhythms having a ~24-h period that is synchronized to environmental time and persist in the absence of environmental cues, but can be entrained by light, food and other external cues.

Clock gene A gene that is critical for generating circadian timekeeping at the cellular level.

Collateral inhibition Neural networks in which inhibitory (usually GABAergic) neurons inhibit the activity of synaptically connected neurons.

Conditional mutant A transgenic mouse mutant line engineered so that gene A (gene of interest) will become inactive only in a certain specific tissue domain where it is co-expressed with gene B ("Cre-driver gene").

Core circadian oscillator A transcriptional feedback loop in which increased expression of Per and Cry proteins inhibits activity of Bmal1 and Clock. It is active in virtually all mammalian cells.

Corticotropes Anterior lobe cells that produce adrenocorticotropin (ACTH) from pro-opiomelanocortin (POMC).

CpG island DNA sequence of more than 200 bp with >50% GC-content and a CpG dinucleotide observed/expected ratio of >60%.

CPHD Combined Pituitary Hormone Deficiency (CPHD) is multiple pituitary hormone deficiencies, usually involving GH and at least one other hormone.

Critical period A developmental window during which endogenous or exogenous stimuli impact the trajectory of growth. Examples of such stimuli are light or hormones, both of which direct changes in synaptic circuitry that are essential to normal development of the visual system and reproductive physiology and behavior, respectfully. Exposure to the stimuli must occur during the critical period for normal development to ensue.

Cyclooxygenase enzymes (COX-1 and COX-2) Enzymes which convert arachidonic acid derived from lipid membranes into short levied precursors of prostaglandins, which are synthesized by specific enzymes for each prostaglandin.

Cyclopia A birth defect characterized by the presence of one single eye, positioned in the middle of the face. Cyclopia is caused by the failure of the developing forebrain to separate into the left and right hemispheres and accompanies some kinds of *holoprosencephaly*.

De novo motif analysis A computational method to identify conserved sequence features in large amounts of CHIP-seq data.

Defeminization A biological process by which the default neural underpinnings of female physiology and behavior is removed from the male brain.

Developmental regulatory gene A gene encoding a transcription factor or a signaling protein, which is expressed during development in a specific moment in specific patterns, and is able to regulate expression of other genes and control patterning and morphogenesis of specific regions.

Diencephalon The posterior region of the forebrain. Gives rise to thalamus and hypothalamus.

Differentially methylated region (DMR) CpG island with different levels of DNA methylation on its two alleles.

DNA methylation Addition of a methyl group to DNA, which affects gene activity without alteration of the DNA sequence. DNA methylation is found primarily on the cytosine of cytosine-guanine (CpG) dinucleotides.

DNA methyltransferase These are a family of enzymes that are effectors of DNA methylation and catalyze either de novo or maintenance methylation of hemi-methylated DNA. There are 3 active DNMTs in mammalian species, DNMT1, DNMT3A, and DNMT3B. These enzymes place a methyl group onto cytosines that are proximal to guanines in the DNA, thereby changing the probability that the surrounding genome will be transcribed.

Dysmasculinization The disruption of the normal process of masculinization, which can therefore only occur in males.

Embryonic development The period of an animal's development in utero (also referred to as fetal development in higher vertebrates).

Emetic reflex The vomiting reflex, present in most mammalian species (including humans), but absent in non-emetic rodents (e.g., rats and mice).

Endocannabinoids A class of signaling molecules derived from membrane lipids and that bind to the same receptors as the active compounds in marijuana.

Endogenous Naturally occurring, e.g., hormones released from endocrine tissues.

Entrainment Synchronization of endogenously generated circadian rhythms to the external environment.

Epigenetic modification Biochemical change to the genome or chromatin template that alters the conformational structure and induces either a permissive or an inhibitory state for gene transcription. Examples include DNA methylation, histone acetylation and histone methylation.

Epigenetic regulator Chromatin remodeling enzymes and enzymes that place or remove methylation and acetylation marks on histones bound to chromatin, altering access of transcription factors to bind DNA.

Epigenetics It refers to modifications to the DNA that impact gene expression but do not change the genetic code (i.e., G, C, A & T). These can include direct methylation of the DNA and indirect via modifications of the surrounding histones that impact accessibility of transcription factors to the DNA.

Epithelial to mesenchymal-like transition Epithelial cells lose their polarity, release cell–cell adhesion, change shape and acquire migratory properties, a process that occurs in development of many organs and in cancer.

Estradiol A sex steroid of the estrogen family, synthesized in, and secreted from, the ovary.

Estrogens A class of steroid hormones that includes estradiol, estriol, and estrone. In adulthood estradiol is produced in high quantities by the ovary. During development estradiol is produced in the brain by being derived from testosterone in a process called aromatization. In rodents, estrogens are masculinizing hormones during development.

Exogenous Artificially sourced, e.g., drugs, synthetic peptides, and other chemical agents.

Exteroceptive Sensory signals arising from the environment (e.g., stimuli conveyed to the central nervous system through visual or auditory pathways) rather than from within the body.

Fast neurotransmitter Fast-acting neurotransmitters such as glutamate (excitatory) and gamma-aminobutyric acid (GABA) (inhibitory), modulate their target cells in under one millisecond by directly activating ligand-gated ion channels. Monoamines and neuropeptides, which act via G-protein coupled receptors (GPCRs), act more slowly.

Fate mapping Any technique allowing for the labeling of a *cell lineage*, i.e., the labeling of one specific kind of dividing cell and all the cells that derive from it.

Floor plate A tissue domain, that acts of a source of morphogens, occupying the ventral midline of the developing neural tube and formed by specific cells (not neuroepithelial cells) with essential functions in neural tube fate specification. The ventral midline of the most rostral part of the neural tube (corresponding to the hypothalamus) shows important differences with the floor plate elsewhere and usually receives a different name (see *Hypothalamic floor plate*, *Rostral diencephalic ventral midline*).

Follicle stimulating hormone A hormone secreted by the anterior pituitary that stimulates the growth of ovarian follicles and sperm maturation.

Folliculostellate cells Non-endocrine cells in the anterior pituitary with star-like cytoplasmic processes.

Forebrain Consists of structures derived from the telencephalon and diencephalon. It is rostral to the mesencephalon and contains the centers involved in the control of sensorimotor, autonomic and endocrine functions, in emotional behavior, as well as in cognitive functions such as learning and memory.

Fos The protein product of the immediate-early gene, c-fos. Fos protein accumulates within the nuclei of neurons and other cells due to increased intracellular calcium. Increased neuronal Fos labeling (compared to baseline levels) is interpreted as evidence that the neuron was stimulated/activated (e.g., received increased excitatory synaptic input).

Free run A self-sustained circadian rhythm that is not synchronized to environmental time.

GABA Gamma-aminobutyric acid—The primary inhibitory neurotransmitter in the adult brain and a secreted factor that regulates cell function during development.

GAD Glutamic acid decarboxylase—Enzyme that converts glutamate to GABA (two forms GAD65 and GAD67).

GATA2 Transcription factor that drives thyrotrope and gonadotrope cell fate.

Genomic imprinting Parent-of-origin-dependent, monoallelic gene expression caused by epigenetic marks (i.e., DNA methylation) set in the parental germ cells and maintained in the somatic cells of the offspring.

GHRH-neurons Arcuate nucleus neurons that stimulate linear growth; Kiss1-neurons, arcuate nucleus neurons that directs reproduction.

Gliogenesis Generation of new glial cells from progenitor cells.

Glycoprotein hormones Bioactive LH, FSH, and TSH are composed of specific beta subunits and a common alpha subunit, CGA or chorionic gonadotropin alpha. Both alpha and beta subunits are glycosylated.

GnRH Gonadotropin-releasing hormone—A hormone secreted into anterior pituitary portal system that stimulates the release of gonadotropins from the anterior pituitary into the general circulation and as such regulates reproductive function.

Gonadotropes Anterior lobe cells that produce follicle-stimulating hormone (FSH) and luteinizing hormone (LH).

Gonadotropin-inhibitory hormone An RFamide peptide that is inhibitory to the synthesis and secretion of gonadotropins from the anterior pituitary; a ligand for GPR147 also called RFamide-related peptide-3 (RFRP-3) in mammals.

Gonadotropins Hormones synthesized and secreted from the anterior pituitary that regulate sex steroid production and gametogenesis—follicle stimulating hormone and luteinizing hormone, respectively.

HESX1 Transcription factor involved in development of the eye, forebrain, and Rathke's pouch.

Histamine An amino acid that acts as a signaling molecule in inflammatory reactions and in the nervous system to promote arousal. It can be synthesized and released by neurons in a neurotransmitter fashion or from mast cells following degranulation.

Histone acetylase Acetylate-specific lysine residues in histone substrates.

Histone deacetylase Remove acetyl residues from histone substrates.

Histone modification A covalent posttranslational modification to histone proteins resulting in methylation, phosphorylation, acetylation, ubiquitylation, and sumoylation. These modifications result in chromatin structure to alter gene expression levels.

Histones Proteins that form the basis of the nucleosomes, tightly wound compact units of DNA that impact the accessibility of transcription factors to the DNA. The addition of modifications to histone "tails", determine how tightly or loosely the DNA is coiled within the nucleosomes.

Holoprosencephaly A congenital malformation of the brain and face, sometimes resulting in *cyclopia*. One major cause of holoprosencephaly is genetic mutations affecting the *Shh* signaling pathway.

Homology Structures are proposed to be homologous in different animals whenever it is thought that the same structures were present in their common ancestor, being not necessary a functional similarity.

Hypophagia Reduced food intake.

Hypophysiotropic Acting on the anterior pituitary gland (hypophysis).

Hypothalamic basal plate The *Shh* expression domain forming a lateral band in the developing forebrain.

Hypothalamic floor plate The portion of the floor plate corresponding to the presumptive hypothalamus. Unlike the floor plate in the rest of the neural tube, hypothalamic floor plate cells proliferate; they also express markers other than those typical of the floor plate elsewhere.

Hypothalamus A brain region critical to the control of the anterior pituitary and reproductive behaviors.

Hypothalamus Pituitary Gonadal (HPG) axis Neuroendocrine network that consists of functional interactions between the hypothalamus, anterior pituitary gland, and gonads. The decapeptide gonadotropin-releasing hormone (GnRH) is the hypothalamic peptide controlling reproduction that stimulates the secretion of the anterior pituitary hormones—luteinizing hormone (LH) and follicle-stimulating hormone (FSH), which ultimately controls the production of gonadal steroids and gametogenesis.

IGHD Isolated Growth Hormone Deficiency (IGHD) is a deficiency in circulating GH without deficiency of other pituitary hormones.

Imprinting Control Region (ICR) DMR that is already established in the germ cells, maintained in somatic cells of the offspring and has been proven to control the imprinting of neighboring genes within a cluster.

Interoceptive Sensory signals arising within the body (e.g., from cardiovascular, gastrointestinal, and endocrine systems) rather than from the environment.

Intrinsically photosensitive retinal ganglion cells (ipRGCs) Opn4-expressing retinal ganglion cells that send postsynaptic projections to the SCN, as well as other hypothalamic regions that regulate sleep. They are the sole source of light input to the SCN.

Ki67 Protein associated with cell proliferation.

Kisspeptin An RFamide peptide encoded by the Kiss1 gene that is an essential, stimulatory signal residing upstream of the GnRH system. A ligand for GPR54 and formerly known as metastin.

Lactotropes Anterior lobe cells that produce prolactin (PRL).

Lateral Progenitor Domain In the context of this review, the tissue domain expressing Shh as a lateral band in the presumptive hypothalamus. It is the neuroepithelial domain where the mammillary and tuberal regions are generated and receives also the name of *hypothalamic basal plate*.

LHX3 Transcription factor necessary for expansion of Rathke's pouch and functional overlap with LHX3.

LHX4 Transcription factor necessary for expansion of Rathke's pouch.

Lordosis A posture adopted by female rodents in response to mounting from a male if they are sexually receptive. The inward dorsoflexion of the spine led to the use of the term lordosis.

LSD1 Lysine-specific demethylase, officially known as Kdm1a, implicated in differentiation of somatotropes and lactotropes.

Luteinizing hormone (LH) A hormone secreted by the anterior pituitary that stimulates sex steroid production in males and females and facilitates ovarian follicle development and ovulation in females.

Mast cells Innate immune cells which are found throughout the body but include a brain resident population. They contain large secretory granules which upon degranulation release their contents into the surround. The contents include histamine as well as serotonin and other signaling molecules. They also make and release cytokines and prostaglandins.

Melanopsin A photosensitive pigment and member of the G-protein coupled receptor opsin family expressed in intrinsically photosensitive retinal ganglion cells involved in transmitting light information to entrain the circadian system.

Melanotropes Intermediate lobe cells that produce melanocyte-stimulating hormone (MSH) from POMC.

Meninges A membranous structure around the brain that forms a barrier with the periphery but is also a gateway for specific trafficking of molecules and cells.

Methylcytosine dioxygenase A family of enzymes that initiates the molecular cascade to demethylate DNA via the conversion of methylated cytosine residues into a hydromethylated state.

Microglia Specialized cells derived from macrophages and that constitute the brains innate immune system. Despite the name, microglia are not related to the other glia in the brain, astrocytes and oligodendrocytes.

Morphogenetic field Higher order module, specified by a particular combination of developmental regulatory genes, that gives rise to a particular region. It constitutes a major unit of development and evolution, and the natural comparison character for homology considerations.

Neural precursor An immature postmitotic neuron.

Neural progenitor A proliferating cell that gives rise to neurons.

Neural Shh Shh secreted by the neural tube, as opposed to prechordal Shh, which is secreted by the prechordal plate.

Neuroepithelial Shh(+) progenitors A dynamic progenitor pool that develops from bHyp cells, whose size is constantly maintained through generation and differentiation.

Neuroepithelium Cells of the neural tube that have not yet initiated neurogenesis.

Non-rapid eye movement (NREM) sleep Also known as quiescent sleep. Comprises ~80% of sleep time in young adult humans. During NREM sleep, the brain waves as measured by electroencephalographic recording are low frequency and high amplitude, and the sleeper is relatively still, but maintains muscle tone. Dreaming is infrequent.

NR5A1 Transcription factor that drives gonadotrope cell fate.

OE Main olfactory epithelium that projects to the olfactory bulb.

Orexigenic Promoting appetite and inhibiting energy expenditure.

Oscillation A recurrent event that is characterized by its period, frequency, amplitude and phase.

PAX7 Pioneering transcription factor that drives melanotrope cell fate and opens chromatin, creating access for other transcription factors to bind.

Period The length of time to complete one cycle of a rhythm.

Phase-advanced Characterized by early waking and sleeping times.

Phase-delayed Characterized by late waking and sleeping times.

Pituicytes Glial-like cells in the posterior pituitary that assists the axons producing oxytocin and vasopressin Posterior Lobe of the Pituitary Gland or Neurohypophysis: derived from neural ectoderm. Contains axon terminals for oxytocin and vasopressin and pituicytes.

PITX1 Transcription factor expressed in early pituitary development, enriched in thyrotropes and gonadotropes, with overlapping function with PITX2.

PITX2 Transcription factor important for expansion of Rathke's pouch and later, for expression of *Gata2*, *Nr5a1*, and *Pou1f1*.

Placental lactogen A placental hormone of the somatotropin family that modifies the maternal metabolic state to supply energy to the developing fetus.

POMC-neurons Arcuate nucleus neurons that suppress appetite; AgRP-neurons, arcuate nucleus neurons that stimulate appetite.

POU1F1 Pituitary-specific transcription factor necessary for promotion of the thyrotrope, somatotrope, and lactotrope cell fates.

Prechordal mesendoderm A fan-shaped structure that forms anterior to the notochord, and underlies the ventral midline of the diencephalon. Posteriorly, it is composed of axial prechordal mesoderm, and anteriorly, it is composed of a mix of axial prechordal mesoderm and endoderm.

Prechordal plate The portion of the axial mesoderm underlying the presumptive hypothalamus. Shh from the prechordal plate is required for hypothalamic specifications.

Preoptic Area (POA) A brain region critical to the control of male sexual behavior and female maternal behavior. It is also the site of the sexually dimorphic nucleus (SDN), and a variety of other sex differences in anatomy and physiology.

Primary wake-promoting neurons Neurons that rapidly (i.e., within seconds to minutes) induce wakefulness following activation.

Progenitor cell A proliferative cell that can undergo a limited number of cell divisions, and gives rise to a limited type of differentiated cells.

Progenitor domain A restricted region of the early neural tube, which generates the cells for a certain subdivision of the central nervous system.

PROP1 Pituitary-specific transcription factor that marks anterior and intermediate lobe progenitors and is required for expansion of the *Pou1f1* lineage.

Prostaglandin E2 (PGE2) A membrane derived signaling molecule usually associated with inflammation but which also has normal roles in brain development, including the promoting of synapse formation in the male POA.

Protein kinase-A (PKA) An enzyme which phosphorylates other proteins to modify their function, such as subunits of glutamate receptors, which impacts how they traffic to the membrane.

Puberty Developmental period when an individual becomes capable of sexual reproduction. It is marked by genital organ maturation and development of secondary sexual characteristics.

PVN Paraventricular nucleus of the hypothalamus—Heterogeneous grouping of cells in the hypothalamus known to function in the regulation of blood pressure, appetite, stress responses, sexual function, cardiac output, and organismal growth.

Rapid eye movement (REM) sleep Comprises ~20% of sleep time in young adult humans. During REM sleep, brain waves as measured by electroencephalographic recording are typically high frequency and low amplitude, and the sleeper exhibits flaccid paralysis, with exception of the ocular muscles, which move rapidly. Dreaming is frequent and vivid.

Rathke's pouch An invagination of oral ectoderm that forms the anterior and intermediate lobes of the pituitary gland.

Retinohypothalamic tract Axonal projections of the SCN to the hypothalamus.

RFamide peptide A peptide that contains Arg-Phe-NH2 motif at its C-terminus.

RFRP-3 An RFamide peptide synthesized and secreted from the hypothalamus that inhibits the release of gonadotropins from the anterior pituitary; mammalian ortholog of gonadotropin-inhibitory hormone (GnIH).

Rhythmic epigenetics The study of the biochemical, physiological and environmental regulation of epigenetic enzyme oscillations and the functional significance of predictable changes in epigenetic modifications across different biological periods (e.g. seasonal).

RNA-seq High-throughput sequencing method for mapping and quantifying transcriptomes. Benefits include: Simultaneous mapping of transcribed regions, dynamic range of gene expression level (>8000-fold), ability to distinguish between isoforms and allelic expression, and a relatively low cost for transcriptomes of large genomes.

Rostral diencephalic ventral midline (RDVM) cells A subset of medial neural plate cells that constitute the precursors of the hypothalamic floor plate. They are a transient embryonic population that expresses Shh and will upregulate Fgf10 to develop into bHyp cells under the influence of prechordal mesendoderm-derived signals.

scRNA-seq A method to define the expression profiles of individual cells by performing next generation sequencing at single cell levels.

Seasonal rhythms A biological rhythm that is entrained to a 12-month period.

Sensitive period A developmental window during which the brain is responsive to exogenous stimuli that can alter the trajectory of growth. Sensitive periods often overlap with critical periods.

Sensory processing The term used to describe how the brain converts sensory information (e.g., olfactory cues) into an appropriate behavioral response.

Sexual dimorphism A sex-related difference in structure, function, and/or behavior.

Sexually dimorphic nucleus of the preoptic area (SDN-POA) A collection of neurons in a subnucleus that is 3–5 times larger in males than females, it was the first major neuroanatomical sex difference reported in a mammal.

SHRP (stress-hyporesponsive period) In rodents, an early postnatal developmental period in which stress-induced activation of the hypothalamic–pituitary–adrenal (HPA) axis is suppressed.

Skeleton photoperiod An external light-dark cycle in which a long period of light (the day phase) is not present, and instead short periods of light mark the beginning and end of the day phase.

Sleep homeostasis The amount of daily sleep time required by an organism. A sleep deficit elicits a compensatory increase in the intensity and duration of sleep, while excessive sleep reduces sleep propensity. NREM and REM homeostasis are regulated separately. There is considerable age and species-dependent variation in sleep homeostasis.

Sleep pressure The consists of the sum of two components: (S), a drive to sleep that increases with time spent awake and is dissipated by time spent asleep and (C), a drive to sleep that is under direct control of the central circadian oscillator in the suprachiasmatic nucleus.

Somatotropes Anterior lobe cells that produce somatotropin or growth hormone (GH).

SOX2 Transcription factor in pituitary stem cells.

Specific *Cre* lines Mouse lines expressing the Cre recombinase in specific cell types, which are utilized to delete genes of interests in specific cell types.

Spinal nucleus of the bulbocavernosus (SNB) A collection of motor neurons in the spinal cord that controls the penis. The motor neurons are largely absent or substantially diminished in females.

Stem cell A proliferative cell that can self-renew indefinitely and give rise to many different types of differentiated cells.

Stress responses Behavioral and physiological (endocrine, autonomic) responses to stress.

Stress Referring to stimuli or conditions that threaten (or are perceived to threaten) physiological homeostasis and well-being; such stimuli and conditions are "stressful."

Stress-threat system Brain nuclei that regulate an animal's responses to stressors and external threats.

Suprachiasmatic nucleus (SCN) Small, paired hypothalamic nuclei (of about 20,000 neurons) found dorsal to the optic chiasm that act as the regulator of physiological circadian rhythms in mammals. The SCN is directly regulated by light, relayed from the retina via ipRGCs. It is the master circadian pacemaker in mammals.

Synaptic plasticity Changes in synaptic architecture or strength in response to environmental or physiologic cues.

Tanycytes Specialized cells that are a type of ependymal (ventricle lining) cell with process that extend into the basal hypothalamus; among the hypothesized functions are sensing both circulating and cerebrospinal fluid content and regulating neuroendocrine peptide release (including physically impeding axonal contacts).

Telencephalon The most anterior major division of the neural tube, and anterior region of the forebrain. Gives rise to cortex, hippocampus, amygdala, and septal nuclei.

Testicular hormones They are steroids derived from the testis, predominantly androgens which are then aromatized to estrogens in the brain or act directly on androgen receptors (AR).

Tet (Ten-eleven translocation) enzymes Oxidize the methyl group of 5-methylcytosine (5mC) to yield 5-hydroxymethylcytosine (5hmC). There are 3 Tet enzymes, Tet 1, Tet 2, and Tet 3.

Thyrotropes Anterior lobe cells that produce thyrotropin (Thyroid-Stimulating Hormone, TSH).

TPIT Transcription factor that drives corticotrope and melanotrope cell fate.

Transcription factors Proteins that bind to regulatory regions of DNA in a highly specific manner and participate in a complex with other proteins and enzymes to transcribe the DNA into RNA.

Ventricular zone The region of the developing brain lining the ventricles where neurons are born.

VMH Ventromedial nucleus of the hypothalamus—Heterogeneous grouping of cells in the hypothalamus known to function in the regulation of appetite, fear, thermoregulation, social and sexual behaviors (also referred to as VMN).

VNO Vomeronasal organ—Putative site of origin for GnRH neurons and origination of olfactory fibers to the accessory olfactory bulb.

ZEB1 Transcription factor implicated in differentiation of lactotropes from somatotropes.

ZEB2 Transcription factor involved in epithelial to mesenchymal-like transition of pituitary progenitors.

Zona incerta (ZI) A bilateral elongated nucleus in dorsal hypothalamus. It regulates a diverse range of physiological processes, including encoding sleep pressure.

Index

S. Wray, S. Blackshaw (eds.), *Developmental Neuroendocrinology*, Masterclass in Neuroendocrinology 9, https://doi.org/10.1007/978-3-030-40002-6

T

The manufacturer's authorised representative in the EU is Springer Nature Customer Service Centre GmbH, Europaplatz 3, 69115 Heidelberg, Germany. If you have any concerns regarding our products, please contact ProductSafety@springernature.com

Printed and bound by CPI Group (UK) Ltd, Croydon, CR0 4YY
22/09/2026
02220956-0013